THE NUCLEON OPTICAL MODEL

Other Books by P. E. Hodgson

The Optical Model of Elastic Scattering, 1963

Nuclear Reactions and Nuclear Structure, 1971

Nuclear Heavy-Ion Reactions, 1978

Growth Points in Nuclear Physics, (3 vols.) 1980, 1981

Nucleon Momentum and Density Distributions in Nuclei, 1988
(*with A. N. Antonov and I. Zh. Petkov*)

Pre-Equilibrium Nuclear Reactions, 1992
(*with E. Gadioli*)

Nucleon Correlations in Nuclei, 1993
(*with A. N. Antonov and I. Zh. Petkov*)

Spacetime and Electromagnetism, 1990
(*with J. R. Lucas*)

Nuclear Physics in Peace and War, 1961

Our Nuclear Future?, 1983

THE NUCLEON OPTICAL MODEL

P. E. Hodgson

University of Oxford

World Scientific

Singapore • New Jersey • London • Hong Kong

Published by

World Scientific Publishing Co. Pte. Ltd.

P O Box 128, Farrer Road, Singapore 9128

USA office: Suite 1B, 1060 Main Street, River Edge, NJ 07661

UK office: 73 Lynton Mead, Totteridge, London N20 8DH

We are grateful to the following publishers for their permission to reproduce the articles found in this volume:

Éditions du Centre National de la Recherche Scientifique/CNRS
Elsevier Science Publishers B V
Institute of Physics Publishing Ltd
Organisation for Economic Cooperation and Development/OECD (*Nuclear Energy Agency/NEA*)
Società Italiana di Fisica
Springer-Verlag

THE NUCLEON OPTICAL MODEL

ISBN 981-02-1722-6

Printed in Singapore by Utopia Press.

PREFACE

It is a remarkable feature of nuclear physics that very simple models are often able to describe at least some aspects of extremely complicated phenomena. A striking example is the success of the optical model of the elastic scattering of nucleons of moderate energy by nuclei. The force on the incident nucleon is the combined effect of all the nucleons in the nucleus, and the nucleon-nucleon interaction contains central, exchange, spin-orbit and tensor components. The target nucleus has many excited states, and these all affect the scattering. Viewed in this way, it is an extremely complicated interaction to describe.

And yet if we look at the differential cross-sections for elastic scattering, we find that they vary rather smoothly with incident energy and from one nucleus to the next. Partly this is because the energy spread of the incident beam averages over the detailed resonance structure characteristic of each target nucleus. The resulting energy-averaged cross-sections then depend more on the size of the nucleus than on its detailed structure, and they can be described by a one-body complex optical potential with parameters that vary smoothly with energy and nuclear radius.

This is essentially the optical model, and it has been developed and extended so that it can now describe a wide range of experimental data with surprising accuracy. The optical potential is the extension to positive energies of the simple shell model potential that give so many properties of nuclei in their ground states. Thus the model unifies many aspects of nuclear structure and nuclear reactions in a simple way.

The model may be extended by the coupled-channels to include a description of inelastic scattering, and this requires a model of the states excited. Furthermore, the optical model wavefunctions are needed by many theories of nucleon transfer reactions that are such powerful ways of determining the properties of excited nuclear states.

The development of the optical model over the last thirty years has made it possible to analyse a wide range of nuclear reactions, and this has increased our knowledge of both the reaction mechanisms and the structures of the participating nuclei. It may therefore be useful to provide an overall survey in the form of review articles written on various aspects of the optical model. To preserve their individuality the articles have been reproduced unchanged, although this inevitably introduces some repetition. Each article thus remains complete in itself, with full references to the original work. Taken together, they show how the model has developed over the years, and how it was used at different times.

As in all scientific research, the optical model is due to the concerted efforts of a large number of people. Collecting together and re-reading these articles has been an occasion to recall many friendships and research collaborations, and the many colleagues all over the world who have cooperated in building up our knowledge. It is a pleasure to thank them all for their help and friendship over the years.

I am particularly grateful to Franco Malaguti, Arnaldo Uguzzoni, Ettore Verondini and Enrico Sartori who have kindly permitted me to include two of our joint articles. I thank also the original publishers for their kind permission to reproduce the articles here.

P. E. Hodgson
Oxford 1994

CONTENTS

INTRODUCTION

THE NUCLEON OPTICAL MODEL

P. E. HODGSON

1. Introduction

The interaction between a nucleon and a nucleus is one of the most important in nuclear physics. It governs the elastic and inelastic scattering of nucleons by nuclei, and determines the behaviour of the nucleon channels in nuclear reactions. It is conveniently described by a potential $V(r)$ that depends on the distance r between the nucleon and the centre of the target nucleus. If $V(r)$ is known, standard quantum-mechanical formalism enables the elastic scattering cross-section to be calculated. Also obtained is the wavefunction in the region of the target nucleus, which is essential for nuclear reaction calculations.

It is therefore of great theoretical and practical importance to determine, as precisely as possible, the form of the potential $V(r)$. Ideally, from a knowledge of $V(r)$ it should be possible to calculate the elastic scattering cross-sections as a function of energy for all target nuclei for nucleons and as a further development, for composite projectiles. This can be very useful for practical applications, and also for reaction calculations where the distorted waves in many reaction channels are required.

The same potential also describes the motion of bound nucleons in the nucleus; this is the simple shell model. We can thus consider $V(r)$ as the nuclear mean field that determines the properties of scattered and bound nucleons, and thus unifies nuclear reactions and nuclear structure.

There are two ways of determining $V(r)$, the theoretical and the phenomenological. The former uses the nucleon-nucleon interaction to solve the many-body problem of the nucleon-nucleus interaction, while the latter starts from a physically-reasonable form for $V(r)$ and adjusts its parameters to optimise the fit to the experimental data.

Each method has its advantages and disadvantages. There are formidable mathematical difficulties in the theoretical method, and approximations have to be made to render the computation practicable. The phenomenological method depends on the choice of a suitable form for $V(r)$, and is subject to ambiguities in that several parameter sets often give equally good fits to the data. It is however possible to some extent to use the theoretical method to give a physically-reasonable form for the potential, and then optimise its parameters phenomenologically.

Both methods have been used extensively over the last thirty years, with considerable success, to determine the nucleon-nucleus potential. Here we are concerned mainly with the phenomenological method.

When a nucleon interacts with a nucleus many different interactions take place. At low incident energies, the cross-sections show a strong resonance structure as a function of the incident energy due to the successive excitation of states in the compound system of nucleon plus target nucleus. These are therefore highly characteristic of the target nucleus. As the incident energy increases, these resonances become broader and denser, until it is no longer possible to resolve them. The number of non-elastic channels also increases rapidly, removing flux and thus affecting the elastic scattering cross-section.

The elastic scattering cross-sections thus depend in a highly specific way on the structure of the target nucleus, and in principle this makes it very difficult to describe them theoretically with any accuracy. In practice, however, the cross-section can be described in quite a simple way that is still extremely useful.

At low energies, the cross-section can be averaged over the resonances and at somewhat higher energies the energy-averaging occurs naturally as the resonances are not resolved. The energy-averaged cross-sections are found to vary quite smoothly with energy and from one target nucleus to another. Furthermore, these energy-averaged cross-sections can be easily calculated from quite simple forms of the nucleon-nucleus potential $V(r)$. The optical model is thus essentially concerned with calculating energy-averaged cross-sections.

There are thus two conflicting requirements in our choice of potential: on the one hand we want it to be simple to define and use, and on the other we want it to fit as wide a range of data as possible. If we want to reproduce all the resonance structure at low energies we have to include the resonance states; this is practicable when the resonances are resolved but not thereafter. At higher energies where the resonances are not resolved there still remain some departures from smooth behaviour with energy and from one nucleus to the rest. Thus we have particular potentials that are able to fit selected data to high accuracy, and global potentials that fit a wider range of data to less accuracy.

The structure of the target nucleus affects the elastic scattering in two ways. At lower energies the incident nucleon can be captured by the target nucleus to form a compound system that may decay in various ways, including emission of the incident nucleon at its full incident energy; this is compound elastic scattering. At higher energies the nucleon is usually scattered directly, but the elastic scattering cross-section is affected by the reactions that can also take place, and these depend on nuclear structure.

The compound elastic cross-section falls rapidly with increasing energy as more and more competing reaction channels become energetically open. Thus for energies above about 5 to 10 MeV, depending on the nucleus there is only direct or shape elastic scattering. It is then possible to formulate the interaction quantum mechanically in a way that takes the nuclear structure into account by expanding the total wavefunction as a sum of all the wavefunctions in the reaction channels. This gives a set of coupled equations that can be solved to give not only the elastic scattering, but also the inelastic scattering that excites the target nucleus and also the nucleon transfer reactions. If all the reaction channels are included, the potential $V(r)$ is real, but there are usually so many channels open that the calculations are impracticable. It is however possible to consider explicitly only a few of the more important channels, and to take the flux removed by the other channels into account by allowing the potential to become complex. In the extreme truncation that considers only the elastic channel, we thus have a simple way of calculating the elastic scattering by solving the Schrödinger equation with a complex potential.

This is called the optical model by analogy with the treatment of the scattering of light by a refracting and absorbing medium using a coupled refractive index. The potential $V(r)$ can be expressed as a simple function of the radial distance r, with real and imaginary strengths adjusted to fit the experimental cross-sections. It is found that these strengths vary smoothly with energy and have rather little variation from one nucleus to another, so that the same potential fits quite well a wide range of data. This is the essential reason for the usefulness of the optical model.

In the following sections the development of the optical model will be described in more detail, and this forms an introduction and guide to the papers that follow. The first three papers show the historical development of the optical model, and the remainder are devoted to more recent developments.

2. The Optical Potential

The nucleon-nucleus potential is for convenience represented by simple analytical expressions with parameters that are adjusted to optimise the fit to experimental data. As the quality of the data has improved, so the expressions for the potential have become more complicated.

In the early work of Feshbach, Porter and Weisskopf (1954) the potential has the form

$$V(r) = (U + iW)f(r) \tag{2.1}$$

where U and W are the real and imaginary potential depths and the form factor $f(r)$ was a square well.

Subsequently this was chosen to have the physically more realistic Saxon-Woods form

$$f(r) = \frac{1}{1 + \exp\left(\frac{r-R}{a}\right)} \tag{2.2}$$

where R is the radius parameter and a the surface diffuseness. The radius parameter is approximately proportional to $A^{1/3}$ and so is often written $R = r_0 A^{1/3}$.

It was found that the optimum form factors of the real and imaginary potentials differ, and the spin-orbit potential can be added to enable polarisations to be calculated. The optical potential then has the form

$$V(r) = V_c(r) + Wf(r) + Wg(r) + V_{so}(r) \tag{2.3}$$

where $V_c(r)$ is the electrostatic potential that is present only for charged incident particles $f(r)$ and $g(r)$ the form factors of the real and imaginary potentials, and $V_{so}(r)$ the spin-orbit potential described in Section 3.

When the optical potential is known, the differential cross-section for elastic scattering can be calculated by solving the Schrödinger equation at the required energy. The mathematical formalism and the practical details of the calculations are given by Hodgson (1963, 1971) and references therein. These calculations can only be carried out electronically, and it is difficult but essential to ensure that the computer program is correct. Many optical model programs have now been written, and a comparison between them has established benchmark results for several specified sets of input parameters (Hodgson and Sartori, 1985). These results, given in Paper 4 enable optical model programs to be checked before use. These computer programs usually have a parameter optimisation option, which enables selected parameters to be varied to optimise the fit to the experimental data.

Phenomenological optical model analyses were initially made by optimising the four parameters U, W, r_0 and a to give the best overall fit to the data. This was often done for a series of angular distributions corresponding to nucleons of a particular energy scattered by a range of nuclei. It was found that the optimum values of r_0 and a are rather similar for all nuclei and that almost equivalent fits are obtained if they are fixed to average values, usually $r_0 = 1.25\,\text{fm}$ and $a = 0.65\,\text{fm}$ (Perey, 1963).

Analyses of proton elastic scattering at higher energies up to 180 MeV showed that better fits could be obtained if the form factors of the real and imaginary parts of the potential are allowed to differ, so that the optimum values of r_0 and **a** are not the same (Hodgson, 1961).

It is interesting to see whether these forms of the real and imaginary potentials can be derived theoretically. In the case of the real potential, a simple model is obtained by summing all the contributing nucleon-nucleon interactions,

$$U(r) = \sum_i v_i(|\mathbf{r} - \mathbf{r}_i|) \approx \int \rho(\mathbf{r}_i)\nu(|\mathbf{r} - \mathbf{r}_i|)d\mathbf{r}_i \tag{2.4}$$

where $v(\mathbf{r})$ is the nucleon-nucleon interaction and $\rho(\mathbf{r})$ the nuclear density. Since the nucleon-nucleon interaction is short-range, it may be approximated by a delta function, so that

$$U(r) = U \int \rho(\mathbf{r}_i)\delta(\mathbf{r} - \mathbf{r}_i)d\mathbf{r}_i = U\rho(r) \tag{2.5}$$

so that the potential has a form very similar to the nuclear density distribution. This argument provides an approximate justification of the form of the potential, and the values of its parameters.

Much less is known about the optimum form of the imaginary part of the potential. It is usually taken to have the Saxon-Woods or volume form (2.2) or a surface-peaked form

$$g(r) = -4a\frac{df(r)}{dr} \tag{2.6}$$

where the factor is introduced to ensure that the maximum value of $g(r)$ is unity.

Quite similar fits are often obtainable with either volume or surface-peaked absorption, although if precise data is available it is sometime found that the surface form is preferred at lower energies and the volume form at higher energies. At energies of 100 MeV or more, other form factors are often found, such as the wine-bottle shape proposed by Elton (1966).

Many analyses of the experimental elastic scattering data have been made using these potentials, and it was soon realised that the optimum fits are not unique, particularly when the data are not of high accuracy. In such cases there are several sets of parameters that give equally good fits to the data. These ambiguities, as they are called, are of two types, continuous and discrete. Continuous ambiguities occur when two parameters can be varied in a correlated way without appreciably changing the quality of the fit. Thus for example it was found that if the depth U of the real potential and its radius parameter r_n are varied so that Ur_0^2 remains constant, the fit is practically unaltered for quite substantial changes in the values of both parameters. There is a similar though less pronounced continuous ambiguity for the depth W and surface diffuseness parameter a of the imaginary potential, which can be changed without much affecting the fit provided Wa remains constant.

It can also happen that similarly good fits are obtainable with quite different sets of parameters, without any good fit for parameters of intermediate value. Thus if only the depth of the real potential is varied, good fits may be obtained for depths of 50, 100, 150 MeV, but not with intermediate values. These are the discrete ambiguities. They are particularly marked for composite particle potentials (see Section 7) but in the case of nucleons the value around 50 MeV is always chosen as it connects smoothly with the depth of the one-body potential of the simple shell model (see Section 5).

As the accuracy of the experimental data increased, the ambiguities in the optical potential become more restricted, and then attention may be concentrated on the energy variation of the potential and the small differences from one nucleus to another. Many optical model analyses have now been made, and the early results were tabulated by Perey and Perey (1974, 1976). The neutron optical potential is discussed in more detail in Paper 5.

These analyses showed that the real and imaginary potential depths vary with energy, as described in Section 5. Accurate analyses of scattering from a range of nuclei showed that the real and possibly also the imaginary parts of the potential depend on features of nuclear structure such as the nuclear asymmetry parameter $(N-Z)/A$, and this is discussed in Section 6.

There have been many calculations of the nucleon optical potential from the nucleon-nucleon interaction, and this work is briefly reviewed in Paper 6.

3. The Spin-Orbit Potential

It was soon recognised that the addition of a spin-orbit term similar to that used in the shell model to give the magic numbers also enables the polarisation of the scattered nucleons to be calculated. By analogy with the Thomas term in atoms, its form was taken to be

$$V_{so}(r) = \left(\frac{\hbar}{m_\pi c}\right)^2 V_S \frac{1}{r}\frac{df(r)}{dr}\mathbf{L}\cdot\boldsymbol{\sigma} \tag{3.1}$$

where the coefficient is the Compton wavelength of the pion, V_S is the strength of the spin-orbit potential and $f(r)$ the optical potential form factor. If this term is added to the optical potential and the wavefunction allowed to depend on the spin of the nucleon, then the spin orbit operator $\mathbf{L} \cdot \boldsymbol{\sigma}$ has two eigenvalues depending on the relative orientation of the orbital angular momentum and spin vectors. Thus, since the total angular momentum $\mathbf{J} = \mathbf{L} + \boldsymbol{\sigma}$, its absolute magnitude $J = L \pm \frac{1}{2}$. The Schrödinger equation then splits into two radial wave equations that give two sets of phase shifts that define the spin-dependent and spin-independent scattering amplitudes and hence the differential cross-section and polarisation.

The polarisation may be calculated using the same form factor as the real part of the potential for $f(r)$, and adjusting V_S to fit the measured polarisation. The fits to the data are good, and the value of V_S is quite similar to that found necessary to give the spin-orbit splitting of the bound states. More precise analyses showed that the optimum radius parameter of the spin-orbit potential is slightly smaller than that of the real part of the central potential, namely 1.1 fm instead of 1.25 fm.

4. The Bound State Potential

The optical model, as discussed so far, is concerned with the scattering of nucleons and other particles by nuclei. It describes the interaction between the nucleon and the nucleus at positive energies, and this is closely related to the corresponding interaction at negative energies, when the nucleon is bound in the nucleus. Such nucleons can only occupy discrete states characterised by a set of quantum numbers. Each nucleon in the nucleus occupies one of these states, subject to the Pauli principle that forbids more than one nucleon to occupy a state. This is the simple shell model that accounts for many features of nuclear structure including the magic numbers, the spins and parities of many nuclear ground states, and the charge and matter distributions.

The energies and quantum numbers of these single-particle states may be determined experimentally by measuring the cross-sections of one-nucleon transfer reactions that either remove a nucleon from an occupied state or put a nucleon into an unoccupied state. The energetics of the reaction give the energies of the states and the angular distributions and polarisations give their quantum numbers. For each state the depth of a potential of the form of the real part of (2.3) may be determined that gives the measured energy. It is then found that these potentials are very similar for all the states of each nucleus, so that a single potential is able to give all the bound states at or near their measured energies (Millener and Hodgson, 1973; Malaguti and Hodgson, 1973) as described in Paper 7.

The magnitude of the cross-section determines the strength or occupation number of the single particle state; these occupation numbers are required to calculate the nuclear charge and matter distributions as described in Section 9. The formalism used to analyse nucleon transfer reaction cross-sections and extract this information on single-particle states is given by Hodgson (1971, Ch.15) and in Paper 7.

This applies to the states within about 5 MeV of the Fermi surface, as these are the ones that can be studied by nucleon transfer reactions. Deeper single particle states can be studied by knock-out reactions like $(p, 2p)$ and $(e, e'p)$, and it is found that these states require a deeper potential. This is in accord with the decrease in the depth of the potential with increasing energy that is found from analysi of elastic scattering. The near-constant value in the region of the Fermi surface is a result of the surface anomaly due to the dispersion term as described in Section 8.

The imaginary potential can also be defined for negative energies by connecting it with the widths of the single-particle states. When the bound single-particle states are studied by nucleon transfer reactions it is often found that they are fragmented into several components. The energy of the single-particle state is given by the mean energy of the fragments, weighted by their spectroscopic factors. The width of the distribution is then equal to the imaginary part of the optical potential. In the case of the deeply-lying states the individual fragments are not resolved and the width is just that of the measured peak, allowing for the experimental resolution.

For an unresolved state

$$\Gamma(E) = \langle\psi_i|W(r, E)|\psi_i\rangle. \tag{4.1}$$

In the case of a fragmented state this is replaced by a sum of similar expressions for all the fragments.

5. The Energy and Radial Variations of the Optical Potential

It was soon found that the strengths of the real and imaginary parts of the optical potential vary with the incident energy. It is expected that the imaginary part increases with energy since it takes into account the flux going into the non-elastic channels, and the number of such channels increases rapidly with the incident energy.

The energy variation of the real part of the potential is more complicated. Part of it is due to the energy variation of the underlying nucleon-nucleon interactions, and part to the local form of the potential. The derivation of the optical potential from many-body theory gives a non-local potential $V(\mathbf{r}, \mathbf{r}')$, so that in place of the term $V(\mathbf{r})\psi(\mathbf{r})$ in the Schrödinger equation there is a term

$$\int V(\mathbf{r}, \mathbf{r}')\psi(\mathbf{r}')d\mathbf{r}' \tag{5.1}$$

Usually the range of the non-locality is small, so that the non-local potential tends rapidly to zero as $|\mathbf{r} - \mathbf{r}'|$ increases. It may then be expanded in a Taylor series about the point $\mathbf{r} = \mathbf{r}'$ and hence it is found that (5.1) is equivalent to a local potential with a linear energy dependence. These two effects give the overall dependence of the real optical potential on energy that may be quite well represented by the expression

$$U = 50 - 0.3E. \tag{5.2}$$

Equally good fits to the experimental data can be obtained with local and non-local potentials, so in nearly all analyses the local potentials are used for computational convenience.

Subsequent analyses over a larger energy range give small departures from linearity that may be represented by the exponential form

$$U = 46.4\exp\{-0.31(E - E_F)/46.4\} \tag{5.3}$$

where $E_F = -6$ MeV is the Fermi energy.

The study of the energy variation of the optical potential may be extended to negative energies by considering the potential that gives the bound single-particle states of the simple shell model as described in Section 4. It is found that the depths are consistent with a smooth energy variation of the potential that is continuous with the potential found at positive energies by analysis of elastic scattering as shown in Fig. 5.1. The energy variation (5.3) was found by combining bound state and scattering potentials for neutrons and ^{208}Pb from −80 to +200 MeV (Johnson, Horen and Mahaux, 1987).

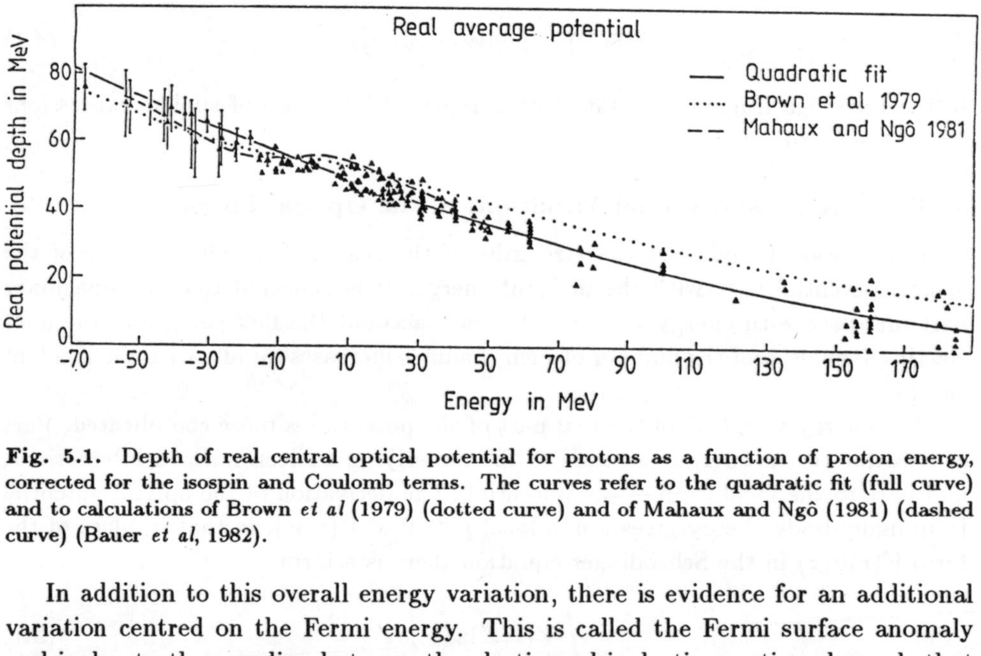

Fig. 5.1. Depth of real central optical potential for protons as a function of proton energy, corrected for the isospin and Coulomb terms. The curves refer to the quadratic fit (full curve) and to calculations of Brown *et al* (1979) (dotted curve) and of Mahaux and Ngô (1981) (dashed curve) (Bauer *et al*, 1982).

In addition to this overall energy variation, there is evidence for an additional variation centred on the Fermi energy. This is called the Fermi surface anomaly and is due to the coupling between the elastic and inelastic reaction channels that is described by the dispersion relations as discussed in Section 8.

These energy variations of the real potential can be conveniently described in terms of an effective mass m^* defined by

$$\frac{m^*}{m} = 1 + \frac{\mathrm{d}U}{\mathrm{d}E}. \tag{5.4}$$

Theoretical analyses show that the overall energy variation (5.3) and the Fermi-surface anomaly have distinct physical origins, and it is therefore useful to describe them by separate effective masses $\tilde{m}$ and $\bar{m}$, respectively defined by

$$\frac{\tilde{m}}{m} = \left(1 - \frac{m}{k}\frac{\partial U}{\partial k}\right)^{-1} \quad \text{and} \quad \frac{\bar{m}}{m} = 1 + \frac{\partial U}{\partial E}. \tag{5.5}$$

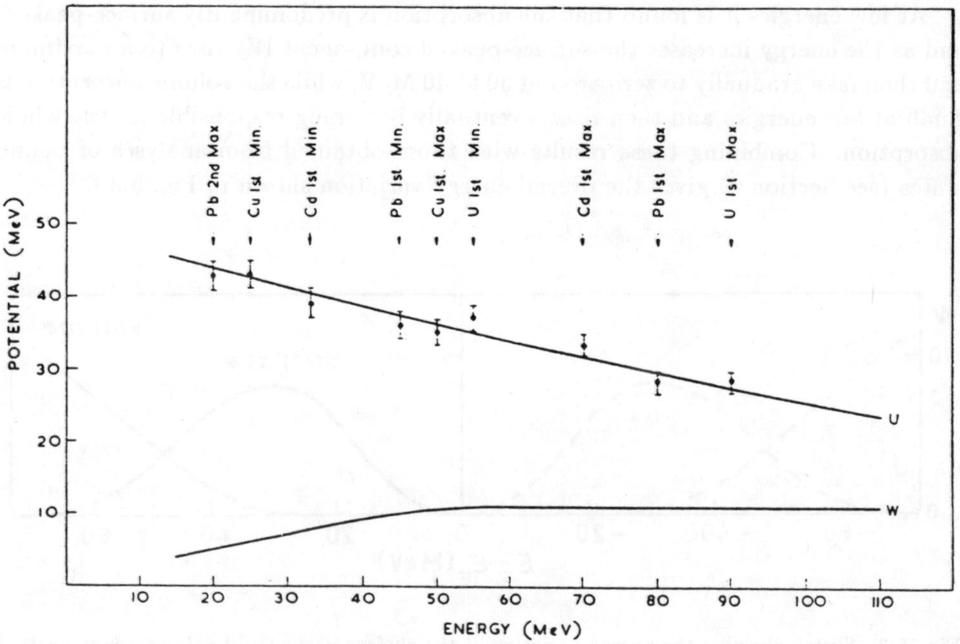

Fig. 5.2. Values of U and W derived from the analysis of total and absorption cross-sections of copper, cadmium, lead and uranium. (Bowen *et al*, 1961).

The overall energy variation (5.3) then gives

$$\frac{\tilde{m}}{m} = 0.73 + 0.0007E. \qquad (5.6)$$

The Fermi-surface anomaly may be described by an effective mass that peaks at the Fermi energy and Brown, Dehesa and Speth (1979) have proposed for the total effective mass the empirical expression

$$\frac{m^*}{m} = 0.64 + \frac{0.36}{[1 + |E - E_F|/2\hbar\omega_0]^2}, \qquad (5.7)$$

where $\hbar\omega_0 = 41A^{-\frac{1}{3}}$. The energy variation corresponding to this expression is shown in fig.5.1. Effective mass in nuclei is further discussed by Hodgson (1983).

It is not easy to obtain the energy variation of the imaginary part of the optical potential from many-body theory, so it has to be determined phenomenologically. As expected, it increases with increasing energy and finally attains an almost constant value. The results of an early determination of the energy variation of the real and imaginary potential strengths from the maxima and minima in the neutron total cross-section as a function of energy are shown in Fig. 5.2 (Bowen *et al*, 1961).

Careful analyses of more accurate data show that improved fits can be obtained by separating the imaginary potential into volume and surface-peaked components

$$W(E, r) = W_v(E)f_x(r) + W_s(E)g(r) \qquad (5.8)$$

At low energies it is found that the absorption is predominantly surface-peaked, and as the energy increases the surface-peaked component W_S rises to a maximum and then falls gradually to zero around 30 to 40 MeV, while the volume absorption is small at low energies and then rises, eventually becoming responsible for the whole absorption. Combining these results with those obtained from analyses of bound states (see Section 4) gives the overall energy variation shown in Fig. 5.3.

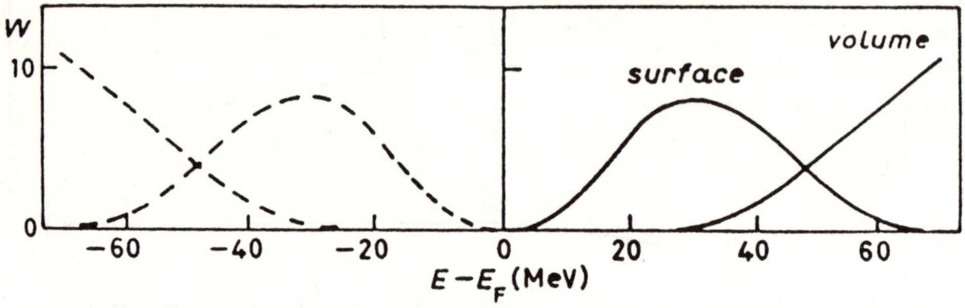

Fig. 5.3. Sketch showing the energy variation of the surface-peaked and volume components of the imaginary potential as a function of energy.

The energy variations of the real and imaginary parts of the nucleon optical potential can be connected by the dispersion relations, as described in Section 8.

The spin-orbit potential also varies with energy from negative to positive energies, but the rate of variation is very much smaller than for the real potential. Analysis of a wide range of bound state and scattering data by Cooper and Hodgson (1980) gave

$$V_{SO} = 6.5 - 0.023E \tag{5.9}$$

Some analyses have used an imaginary spin-orbit potential, but there is reason to believe that this is non-physical because it is no longer needed when the dispersion term is included, as described in Section 8.

We now consider the radial variation of the real potential. Nearly all phenomenological analyses of nucleon elastic scattering have been made using the Saxon-Woods form (2.2). Several analyses of the elastic scattering of heavier particles, in particular of alpha-particles, ahve shown that substantially better fits can be obtained with potentials of a different form, for example the Saxon-Woods squared potential, but analyses of nucleon scattering with different form factors have been rather rare.

There are, however, some indications that better fits are obtained with potentials with different form factors; initially these were not considered significant because they involved more adjustable parameters and this ineveitably improves the quality of the fit. Now, however, they fall into place in the context of the recent nuclear-matter calculations, and thus become much more credible.

It is convenient to discuss separately the evidence for non-Saxon-Woods potentials at low and at high energies.

In the first category, analyses of experimental data on the elastic scattering of 30 MeV, 40 MeV and 61.4 MeV protons by ^{208}Pb and several other nuclei by Sinha and Edwards (1970, 1971) showed that an improved fit to the differential cross-section is obtained by adding a surface-peaked potential of the Saxon-Woods derivative form to the usual Saxon-Woods potential. Initially they interpreted this additional potential as an isospin term, but this possibility was later excluded when it was found that the improvement persists in nuclei like ^{40}Ca, for which the isospin term must be zero.

An alternative explanation was provided by Ahmad and Haider (1976), who showed that the dispersion relations applied to the imaginary potential require a surface-peaked component in the real potential. This work is described in Section 8.

At higher energies, it was found that the differential cross-sections and polarizations can be fitted with Saxon-Woods potentials, provided the imaginary potential extends beyond the real part (Hodgson, 1961, 1968, 1970; Johansson, Svanberg and Hodgson, 1961). Such potentials, however, give reaction cross-sections that are too large (Satchler and Haybron, 1964). The way out of the difficulty was found by Elton (1966) who applied the phase analysis method of Venter and Frahn (1964) to obtain the scattering amplitude corresponding to the data on 180 MeV protons elastically scattered by ^{56}Fe. He found that it is possible to retain the fits to the differential cross-sections and polarizations and yet reduce the corresponding reaction cross-section to fit the experimental value by adding a repulsive core to the real part of the potential.

For over a decade this work was ignored, but recently it has received confirmation from the work of Meyer *et al* (1981) on ^{12}C(p, p) at 200 MeV, and of Bissaud (1981) on ^{40}Ca(p, p) at 155 MeV as well as the detailed and extensive analyses of the new data obtained with the Indiana cyclotron (Nadasen *et al*, 1981).

Analyses of the elastic scattering of 180 MeV and 400 MeV protons by ^{40}Ca (Schwandt, 1982) show that the Saxon-Woods potential does well at 180 MeV, but not at 400 MeV. These differences are indications that the Saxon-Woods parametrization of the optical potential at high energies is radically unsatisfactory, and this is confirmed by analyses of the spin-orbit potential. The fit to the data may be greatly improved by allowing the real potential to have a repulsive core. A suitable modified Saxon-Woods potential is obtained by adding to the usual Saxon-Woods potential another term, also of Saxon-Woods or Saxon-Woods squared form, but of smaller radius. It is difficult to distinguish between these potentials at energies less than 200 MeV, but the difference becomes increasingly important at higher energies. Data on the spin rotation function are very useful in resolving these potential ambiguities. These repulsive-core potentials, now required by the phenomenological analyses, are also given by the microscopic calculations of the potential.

6. Nuclear Structure Effects

The increasing precision of optical model analyses soon revealed small differences between the potentials for nearly nuclei, and these are attributable to differences in nuclear structure.

Some of these effects are notable in the potentials for nuclei at or near closed shells, which have few excited states at low energies and hence a small value of the absorption cross-section and imaginary potential. This is notable in the behaviour of the absorption cross-section for 4 MeV neutrons compared with the predictions of a global optical potential (Perey and Buck, 1962). The same effect is responsible for the anomalous large-angle scattering found in the elastic scattering of alpha-particles by ^{40}Ca (see Section 7).

It was also found that the real part of the optical potential depends on the nuclear asymmetry $(N-Z)$; thus an analysis by Perry (1963) of proton elastic scattering by a range of nuclei gave

$$U = 53.3 - 0.55E - 27\frac{N-Z}{A} + 0.42/A^{1/3} \tag{6.1}$$

This was initially identified as an isospin term that comes from the isospin dependence of the nucleon-nucleon interaction. Subsequently however it was realised that a similar dependence could come from the dependence of the radius of the potential on A. Most analyses assume that $R = r_0 A^{1/3}$, but if this is not quite correct the effect would be a small dependence of U on A which would affect the value of the coefficient of $(N-Z)/A$ in (5.1). The possible presence of such purely geometrical effects make it preferable to refer to the $(N-Z)/A$ dependence as an asymmetry dependence. The isospin dependence of the optical potential has been reviewed by Satchler (1969). The fine structure aspects of the neutron optical potential are discussed in Paper 5.

The optical model potential is nearly always taken to be spherically symmetrical, and this is a good approximation for many nuclei. The potential however is affected for strongly-deformed nuclei since its radial variation depends on the distance from the nuclear surface and not from the nuclear centre. Deformed nuclei usually have a strongly collective character, with many low-lying states strongly coupled to the ground state, and this affects the imaginary part of the potential. These effects are particularly marked for actinide nuclei, as described in Paper 8. Although the scattering from such nuclei can be well described by the simple optical model, improved fits can be obtained by a full coupled-channels analysis (Shamu and Young, 1993).

7. The Optical Potential for Composite Particles

The success of the nucleon optical potential encouraged its application to scattering by few-nucleon or composite particles such as deuterons and alpha-particles. Although this subject is beyond the subject of this review, these analyses deserve a brief mention, since they raise new questions concerning the optical model that may also affect the nucleon potential.

Early analyses of deuteron elastic scattering used potentials very similar to those that had proved so successful for nucleons, and it was found possible to fit very well the elastic cross-sections over the rather limited angular ranges that were then available. It was then realised that the same elementary arguments that considered the nucleon potential to be essentially the sum of the individual nucleon-nucleon interactions, indicate a real potential for deuterons with twice the depth of that for nucleons, and indeed such potentials fitted the data equally well. Further studies showed that there is a whole series of potentials of depths approximately $50N$, where $N = 1, 2, 3 \ldots$, that also fit the data. It is however usually assumed that the potential obtained when N is the number of nucleons in the projectile is the 'physical' one, and this often fits the data marginally better than the others.

The imaginary potential for composite particles is treated phenomenologically, and in the case of alpha-particle scattering it is found that better fits are obtained with a surface-peaked form at lower energies and a volume form at higher energies (Van Oers, 1971). This has important implications for the dispersion relation analysis considered in Section 8.

As the angular range and accuracy if the data increased, it was found impossible to obtain fits with the simple forms of the potential, and more elaborate forms were introduced. A particular problem arose in alpha-particle scattering, where measurements showed a marked rise in the backward direction called ALAS (anomalous large-angle scattering). This occurred particularly for closed-shell nuclei such as ^{40}Ca. Many ingenious explanations for this phenomenon were proposed, but eventually it was found possible to reproduce the data very well simply by using potentials with a small imaginary part, as indeed is appropriate for closed shell nuclei that are difficult to excite and so have fewer open reaction channels at lower energies.

The deuteron-nucleus and helion-nucleus optical potentials have been reviewed by Hodgson (1966, 1968). Heavy-ion optical potentials are discussed in Chapters 2 and 3 of Hodgson (1978).

8. The Unification of the Optical Potential by Dispersion Relations

The studies of the energy variation of the optical potential described in Section 4 show that the real and imaginary parts vary smoothly with energy from the potentials at negative energies that give the bound states to those at positive energies that describe the elastic scattering. We can thus define a complex mean field that unifies our understanding of the single-particle aspects of nuclear structure and of nuclear reactions.

It had been known for a long time that since the optical potential is an analytic function it must satisfy the dispersion relations that connect its real and imaginary parts. These dispersion relations are a consequence of causality, namely that the scattering wave is not emitted before the incident wave arrives (Mahaux *et al*, 1986). More precisely, if $U(k, E)$ is a complex analytic function,

$$U(k, E) = \frac{1}{2\pi i} \oint_c \frac{U(k, E')}{E' - E} dE' \tag{8.1}$$

where the contour c encloses the singularity at $E = E'$.

Putting $U(k,E) = U(k,E) + iW(k,E)$ and separating the real and imaginary parts gives

$$U(k,E) = \frac{\mathcal{P}}{\pi}\int_{-\infty}^{\infty} \frac{W(k,E')}{E'-E}dE' \qquad (8.2)$$

and

$$W(k,E) = \frac{-\mathcal{P}}{\pi}\int_{-\infty}^{\infty} \frac{U(k,E')}{E'-E}dE' \qquad (8.3)$$

where $\mathcal{P}$ indicates the Principal part of the integral.

These dispersion relations provide in principle a way to unify the optical potential in a new way, but there is a practical difficulty that the potentials are required over the whole energy range from minus to plus infinity.

It was realised by Ahmed and Haider (1976) that this difficulty could be overcome by separating the real potential into two parts, and making use of the radial dependence. This gives

$$U(k,E,r) = U_0(k,E,r) + \frac{\mathcal{P}}{\pi}\int_{-\infty}^{\infty} \frac{W(k,E',r)}{E'-E}dE'. \qquad (8.4)$$

The first term is the Hartree-Fock term that has an almost linear energy variation due to the use of a local potential instead of the physically more realistic non-local potential and the second is the additional energy variation due to the dispersion relation. It is known that the Hartree-Fock term has the Saxon-Woods form and that the imaginary potential $W(E,r)$ can be written as the sum of volume and surface-peaked form

$$W(E,r) = W_r(E)f(r) + W_s(E)g(r) \qquad (8.5)$$

If this is inserted in the dispersion relation (8.4) the dispersion integral containing the volume absorption can be absorbed in the Hartree-Fock term, leaving only the dispersion integral containing the surface absorption. The strength of the surface absorption rises from zero at the threshold energy, attains a maximum and then falls to zero, becoming negligible around 40 MeV. The dispersion integral can then be evaluated exactly, giving a surface-peaked addition to the real volume potential.

The method of subtracted dispersion relations uses the expression

$$U(k,E) - U(k,E_F) = \frac{E-E_F}{\pi}\int_{-\infty}^{\infty} \frac{W(E')}{(E'-E)(E'-E_F)}dE' \qquad (8.6)$$

where E_F is the Fermi energy. This integral converges sufficiently rapidly for it to be evaluated, and may therefore be applied to potentials obtained in analyses using a volume potential only. It is therefore convenient to use this method for potentials obtained by phenomenological analyses in which the form factor parameters do not vary with energy. In such analyses all the effects of the Fermi-surface anomaly appear in the potential depth. The alternative method, just described, shows the origin of the anomaly in an increased radial dependence due to the surface-peaked part of the imaginary potential.

If the imaginary potential is symmetric about the Fermi energy, (8.4) becomes

$$U(k,E) = U(k) + \frac{2}{\pi}(E - E_F)\int_{E_F}^{\infty} \frac{W(E')dE'}{(E' - E_F)^2 + (E - E_F)^2} \tag{8.7}$$

This integral has convergence properties similar to (8.6) and so provides an alternative method of analysis.

The phenomenological optical model analyses determine certain moments of the potential with greater accuracy than the individual parameters. It is therefore often useful to work with radially-integrated dispersion relations. Particularly useful is the volume integral per nucleon defined by

$$J_V = \frac{4\pi}{A}\int_0^{\infty} U(r)r^2 dr \tag{8.8}$$

and a similar relation for the imaginary part of the potential.

The corresponding dispersion relation is

$$J_v(E) = J_v^{HF}(k) + \frac{\mathcal{P}}{\pi}\int_{-\infty}^{\infty} \frac{J_W(E')}{E' - E}dE' \tag{8.9}$$

In general one can define the qth moment of the potential by

$$J_v^{(q)} = \frac{4\pi}{A}\int_0^{\infty} U(r)r^q dr \tag{8.10}$$

which satisfies the dispersion relation

$$J_v^{(q)}(E) - J_v^{(q)HF}(k) + \frac{\mathcal{P}}{\pi}\int_{-\infty}^{\infty} \frac{J_W^{(q)}(E')}{E' - E}dE' \tag{8.11}$$

The inclusion of the dispersion term removed several long-standing anomalies in optical model analyses. It explains naturally why it is sometimes easier to fit the data with the addition of a small surface-peaked real potential (Sinha and Edwards, 1970, 1971) the energy-variation of the optimum radius parameter found in some precision analyses of neutron elastic scattering (Finlay *et al*, 1985; Annand *et al*, 1985), and also the energy-averaged neutron total cross-section at low energies (Zong Di and Hodgson, 1988). Most importantly, it explained what was at one time called the Fermi surface anomaly, the departure from the smooth near-linear energy dependence of the real potential centred on the Fermi energy. This is now seen to be due to analysing the data using a volume form for the real potential. If however there is in addition a small surface-peaked component (due to the dispersion term) it has the effect of increasing the radius of the potential in the region around the Fermi surface. The analysis however uses a radius that is independent of energy and so the effect appears as the change in $U(E)$ known as the Fermi surface anomaly, more accurately called the Fermi surface effect.

The physical reason for the Fermi surface effect is the coupling between the elastic and non-elastic channels. If there were no excited states of the target nucleus and no reaction channels, the imaginary potential would fall to zero and the Fermi surface effect would vanish. If for simplicity we consider only inelastic channels, then the elastic scattering can take place via virtual nuclear excitation, and this affects the elastic scattering and hence the phenomenological optical potential. This effect is greatest when the energy is comparable with that of the excited nuclear states and since the coupling is greatest to the low-lying states the effect is greatest around the Fermi surface.

The effects of inelastic scattering can be studied by considering the elastic and inelastic scattering together using the coupled-channels formalism. If this is used to generate an elastic cross-section, which is then fitted phenomenologically with an optical potential, the results show the effects of the coupling to the inelastic channels. Several analyses made in this way have confirmed that the coupling to inelastic channels does produce changes in the real potential similar to those associated with the Fermi surface effect (Gyarmati *et al*, 1981).

A detailed analysis of the elastic scattering of 9.9 to 16.9 MeV neutrons by ^{40}Ca showed that a good fit to the differential cross-sections and polarisations can be obtained using a twelve-parameter optical potential including an imaginary spin-orbit term (Honoré *et al*, 1986). It was however also found that an equally good and perhaps better fit could be obtained with a potential that included the dispersion term but omitted the imaginary spin-orbit term. This suggests that the imaginary spin-orbit term is an unphysical artifact that is able to reproduce some of the dispersion effects (Delaroche and Tornow, 1987).

The success of the dispersion analyses in removing anomalies in previous analyses and in giving precision fits to the experimental data indicates that it has a sound physical basis and so can be used in calculations of unknown cross-sections. The dispersive optical model is described in more detail in Paper 9.

9. Nuclear Charge and Matter Distributions

The nuclear potential for bound states may also be used to calculate the nuclear charge and matter distributions from the corresponding bound state wavefunctions $\psi_i(r)$ using the expression

$$\rho(r) = \sum_i a_i |\psi_i|^2 \tag{9.1}$$

where a_i are occupation numbers and the sum runs over all occupied states. If only the proton states are included this gives the nuclear charge distribution, after allowing for the charge distribution of the proton itself. Addition of the neutron states gives the nuclear matter distribution. Many nuclear charge distributions have been determined in this way, as described in Paper 10. Some nuclear matter distributions have been calculated by Ray and Hodgson (1979).

The occupation numbers a_i in (9.1) may be taken directly from the simple shell model or, for more accurate calculations from the analyses of nucleon transfer reactions. This enables partial occupation numbers to be used. The calculation of the cross-sections of the nucleon transfer reaction uses the distorted wave formalism, which requires the optical potentials for the incoming and outgoing particles.

These distributions are particularly accurate in the nuclear surface region because the wavefunctions have the correct separation energy and hence the correct asymptotic form. They may therefore be used in calculations of the form of the optical potential using the folding model

$$V(r) = \int \rho(\mathbf{r}')v(|\mathbf{r} - \mathbf{r}'|)d\mathbf{r}' \tag{9.2}$$

where $\rho(\mathbf{r}')$ is the nuclear matter distribution and $v(r)$ the nucleon-nucleon interaction.

This expression is not often used for the nucleon analyses since the scattering is not particularly sensitive to the exact form of the potential in the region of the nuclear surface. It does have important application to the calculation of potentials for heavy-ion scattering which is very sensitive to the nuclear surface (Hodgson, 1978). These heavy-ion potentials are obtained by the single folding (9.2) or the double folding

$$V(r) = V \int\int \rho_A(\mathbf{r}_A)\rho_B(\mathbf{r}_B)v(|\mathbf{r} - \mathbf{r}_1 + \mathbf{r}_2|)d\mathbf{r}_A d\mathbf{r}_B \tag{9.3}$$

where $\rho_A(\mathbf{r}_A)$ and $\rho_B(\mathbf{r}_B)$ are the matter distributions of the interacting nuclei.

References

1. I. Ahmad and W. Haider, *J. Phys.* **G2**, L157, 1976.
2. J. R. M. Annand, R. W. Finlay and F. S. Dietrich, *Nucl. Phys.* **A443**, 249, 1985.
3. M. Bauer, E. Hernandez-Saldana, P. E. Hodgson and J. Quintanilla, J., *J. Phys.* **G3**, 525, 1982.
4. I. Bissaud, *Z. Phys.* **A300**, 247, 1981.
5. P. H. Bowen, J. P. Scanlon, G. H. Stafford, J. J. Thresher and P. E. Hodgson, *Nucl. Phys.* **22**, 640, 1961.
6. G. E. Brown, J. S. Dehesa and J. Speth, *Nucl. Phys.* **A330**, 290, 1979.
7. S. Cooper and P. E. Hodgson, *J. Phys.* **G6**, L21, 1980.
8. L. R. B. Elton, *Nucl. Phys.* **89**, 69, 1966.
9. H. Feshbach, C. E. Porter and V. F. Weisskopf, *Phys. Rev.* **96**, 448, 1954.
10. R. W. Finlay, J. R. M. Annand, J. R. Petler and F. S. Dietrich, *Phys. Lett.* **B155**, 313, 1985.
11. B. Gyarmati, R. C. Lovas, T. Vertse and P. E. Hodgson, *J. Phys.* **G7**, L209, 1981.
12. P. E. Hodgson, *Phys. Rev. Lett.* **6**, 358, 1961; *The Optical Model of Elastic Scattering* Clarendon Press, Oxford, 1963; *Adv. in Phys.* **15**, 329, 1966; *Ibid* **17**, 563, 1968; *Nuclear Reactions and Nuclear Structure*, Clarendon Press, Oxford, 1971; *Nuclear Heavy-Ion Reactions*, Clarendon Press, Oxford, 1978.
13. P. E. Hodgson, *Phys. Lett.* **6**, 75, 1968; *Nucl. Phys.* **A150**, 1, 1970; *Cont. Phys.* **24**, 491, 1983.
14. P. E. Hodgson and E. Sartori, NEA Data Bank Report–198–U. INDC (NEA) 5, 1985.
15. G. M. Honoré, W. Tornow, C. R. Howell, R. S. Pedroni, R. C. Byrd, R. L. Walter and J. P. Delaroche, *Phys. Rev.* **C33**, 1129, 1986.
16. A. Johansson, U. Svanberg and P. E. Hodgson, *Ark. Fys.* **19**, 541, 1961.

17. C. H. Johnson, D. J. Horen and C. Mahaux, *Phys. Rev.* **C36**, 2252, 1987.
18. R. E. Le Levier and D. S. Saxon, *Phys. Rev.* **87**, 40, 1952.
19. C. Mahaux and H. Ngô, *Phys. Lett.* **B100**, 285, 1981.
20. C. Mahaux, H. Ngô and G.R. Satchler, *Nucl. Phys.* **A449**, 354, 1986.
21. F. Malaguti and P. E. Hodgson, *Nucl. Phys.* **A215**, 243, 1973.
22. H. O. Meyer, P. Schwandt, G.L. Moake and P.P. Singh, *Phys. Rev.* **C23**, 616, 1981.
23. D. J. Millener and P. E. Hodgson, *Nucl. Phys.* **A209**, 59, 1973.
24. A. Nadasen, P. Schwandt, P. P. Singh, W. W. Jacobs, A. D. Bacher, P. T. Debevec, M. D. Kaitchuck and J. T. Meek, *Phys. Rev.* **C23**, 1023, 1981.
25. G. Passatore, *Nucl. Phys.* **A95**, 694, 1967; **A110**, 91, 1968.
26. F. G. Perey and B. Buck, *Nucl. Phys.* **32**, 353, 1962.
27. F. G. Perey, *Phys. Rev.* **131**, 745, 1963.
28. C. M. Perey and F. G. Perey, *Phys. Rev.* **132**, 755, 1963; *Atom. Data and Nucl. Data Tables* **13** 293, 1974; *Ibid* **17**, 1, 1976.
29. L. Ray and P. E. Hodgson, *Phys. Rev.* **C20**, 2403, 1979.
30. G. R. Satchler and R. M. Haybron, *Phys. Lett.* **11**, 313, 1964.
31. G. R. Satchler, Ch.9 in *Isospin in Nuclear Physics.* Ed. D. H. Wilkinson (North-Holland, 1969).
32. P. Schwandt, invited lecture at *ICUF Workshop*, Bloomington 1982.
33. R. E. Shamu and P. G. Young, *J. Phys.* **G19**, L169, 1993.
34. B. C. Sinha and V. R. W. Edwards, *Phys. Lett.* **B31**, 273, 1970; *Ibid* **B35**, 391, 1971.
35. W. T. H. Van Oers, *Phys. Rev.* **C3**, 1550, 1971.
36. R. H. Venter and W. E. Frahn, *Ann. Phys. (N.Y.)* **27**, 385, 1964.
37. Su Zong Di and P. E. Hodgson, *J. Phys.* **G14**, 1485, 1988.

THE OPTICAL MODEL OF ELASTIC SCATTERING

P. E. HODGSON
Nuclear Physics Laboratory, Oxford

1. INTRODUCTION

The analysis of elastic scattering interactions by means of the optical model is relevant to problems of nuclear structure for two main reasons: firstly the optical model potential is simply the shell model potential extended into the region of positive energy, and secondly a knowledge of the optical potential is needed in order to extract information on nuclear structure from nuclear reaction measurements. The first lecture will be devoted to a survey of our present knowledge of optical potentials, with particular attention to the extent to which it is possible to represent these potentials by simple formulae as a function of energy and nuclear mass and charge. The following lecture will discuss the use of these potentials in nuclear structure studies. The emphasis throughout is on the physics of the analyses, and references are given to papers where the details of the mathematical development may be found.

2. THE OPTICAL MODEL

The full analysis of even a simple elastic scattering interaction, for example that of a proton by a medium weight nucleus, is prohibitively complicated. All the nucleons of the target nucleus interact with each other and with the incident proton, and each individual nucleon-nucleon interaction contains central, exchange and tensor components. The optical model makes the drastic approximation of representing this complicated interaction by a two-body potential between the incident particle and the target nucleus. The detailed structures of the target nucleus and of the incident proton are ignored. The validity of such an approximation can only be established by a close theoretical analysis of the underlying assumptions and by a detailed comparison of its consequences with experiment, but broadly it may be expected to be better for high energies than for low energies, and for medium and heavy rather than light nuclei.

There are basically two approaches to the problem of defining this potential; individually they are inadequate but taken together they can give it with satisfactory precision, at least for incident nucleons. The first approach is from a fundamental theory of nuclear structure in terms of the interactions of the constituent nucleons, for example that of Brueckner et al. [1]. Such theories work best for infinite nuclear matter of constant density, but by making a series of approximations some information on the optical potential appropriate to finite nuclei may be obtained [1]. Nuclear scattering processes are frequently sensitive to the potential in the region of the nuclear surface, and so such calculations have their greatest uncertainty in just the region where they need to be most accurate. The theoretical approach can thus give only an overall qualitative picture of the optical potential.

The alternative approach is the phenomenological one of adjusting the potential systematically until optimum agreement with the experimental data is obtained. This approach usually gives a very good fit to the data but is unsatisfactory on its own because many quite different potentials can frequently be found that all give acceptable fits to the data.

A satisfactory optical potential can be found by combining these approaches and using fundamental theories to give the overall form of the potential, and then using the phenomenological method to determine the optimum parameters of the potential of this form. Some difficulties and ambiguities still remain, and these will be discussed later on.

Fortunately the overall features of the optical potential can be established by quite simple arguments, without recourse to any detailed theory. It is known that a purely real potential produces only scattering but no absorption of the incident particles. The absorption that is known to occur in practice may be accounted for by allowing the potential to be complex, so in its simplest form the optical potential may be written

$$V(r) = -Uf(r) - iWg(r) , \tag{1}$$

where U and W are the depths of the real and imaginary parts of the optical potential, and $f(r)$ and $g(r)$ their radial form factors. If the incident particle is charged the Coulomb electrostatic potential $V_C(r)$ must also be included. For all practical purposes this may be taken to be the potential due to an uniformly charged sphere with the radius of the nuclear density distribution.

The real part $Uf(r)$ of the optical potential corresponds to the shell model potential and so may be expected to be of similar form, provided that there is no sharp energy dependence of the potential. Thus we expect $U \approx$ 50 MeV for nucleons. Since the individual nucleon-nucleon interactions between the incident particle and the target nucleus fall off exponentially at large distances it is reasonable to assume that the composite optical potential will show the same behaviour. When the incident particle has penetrated to the interior of the target nucleus it experiences no further overall force due to the short range and saturation character of the nucleon-nucleon forces, though it possesses greater kinetic energy due to the overall nuclear attraction. This is sufficient to show that the potential $f(r)$ is uniform in the nuclear interior, falls off in the surface region and finally drops exponentially to zero. It is convenient to represent this variation analytically by the Saxon-Woods form factor

$$f(r) = \frac{1}{1+\exp[(r-R)/a]} . \tag{2}$$

In this expression R measures the overall extent of the potential and is thus expected to be similar to that of the nucleus itself. To make explicit most of the dependence of R on nuclear size it is usual to put $R = r_0 A^{\frac{1}{3}}$, where A is the nuclear mass number. The parameter a controls the rapidity with which $f(r)$ falls from its maximum value; it is therefore related to the diffuseness of the nuclear surface. From general knowledge of the properties of nuclei we expect $r_0 \approx 1.3$ fm and $a \approx 0.5$ fm.

The imaginary part $Wg(r)$ of the optical potential accounts for all the non-elastic processes, without differentiating between them. It is therefore much more sensitive than the real part of the potential to the details of the interaction such as the number and relative strengths of the possible non-elastic channels. However, the elastic scattering is somewhat insensitive to the depth of the imaginary part of the potential, so differences from reaction to reaction are unimportant for many purposes. From the known nucleon-nucleon cross sections and making allowance for the reduction due to the Pauli principle, an estimate $W \approx 10$ MeV is obtained. The form factor $g(r)$ depends on the incident energy; at low energies the interaction is likely to be concentrated in the surface region and at higher energies it spreads progressively throughout the nuclear volume. A suitable analytical form for surface absorption is $g_S(r) = -4af'(r)$ (defined so that $g_S(R) = 1$), and for volume absorption $g_V(r) = f(r)$. At intermediate energies it may be necessary to consider a mixture of surface and volume absorption, so a more general form may be written.

$$Wg(r) = sW_V f(r) - 4a(1-s)\, W_D f'(r)\ , \tag{3}$$

where s is a mixing parameter, W_V and W_D the volume and surface absorbing potential depths respectively, and the dash denotes differentiation with respect to r.

It is found experimentally that the scattered nucleons are polarised even with an unpolarised incident beam. This may be included in the optical model description of the process by the addition of a spin-orbit term in the potential

$$V_{SO}(r) = (U_S + iW_S)\, h(r)\, \boldsymbol{l}\cdot\boldsymbol{\sigma}\ , \tag{4}$$

where U_S and iW_S are the real and imaginary depths of the spin-orbit potential, and $\boldsymbol{l}$ and $\boldsymbol{\sigma}$ are the angular momentum and Pauli spin operators respectively. The value of U_S may be estimated from the corresponding term in the shell model potential; this gives $U_S \approx 20$ MeV . fm^2. There is little phenomenological evidence for the imaginary part of the spin-orbit potential so it is usual to take $W_S = 0$. The spin-orbit radial form factor $h(r)$ is assumed to be proportional to the radial derivative of the real potential by analogy with the corresponding Thomas term for atoms

$$h(r) = -\frac{1}{r} f'(r)\ . \tag{5}$$

The complete optical potential is now

$$V(r) = V_C(r) - Uf(r) - isW_V f(r) + 4a(1-s)\, iW_D f'(r) - (U_S + iW_S)\, h(r)\, \boldsymbol{l}\cdot\boldsymbol{\sigma}\ . \tag{6}$$

This formula for the optical potential is expressed in terms of the potential depths U, W_V, W_D, U_s and W_s, and the form factor parameters r_0 and a, which may be fixed to the same values or allowed to vary from one form factor to another. The values of these parameters are known fairly well from the general arguments summarised above, but more accurate values can be obtained from a phenomenological analysis of experimental data. In such analyses the parameters of the optical potential are systematically varied to optimize the fit to the experimental differential elastic scattering cross sections and polarisations as a function of scattering angle, and also to the total reaction cross section. In the case of neutron scattering the total cross section may also be used. An example of the quality of the fit to the experimental data that may be obtained in this way is shown in fig. 1.

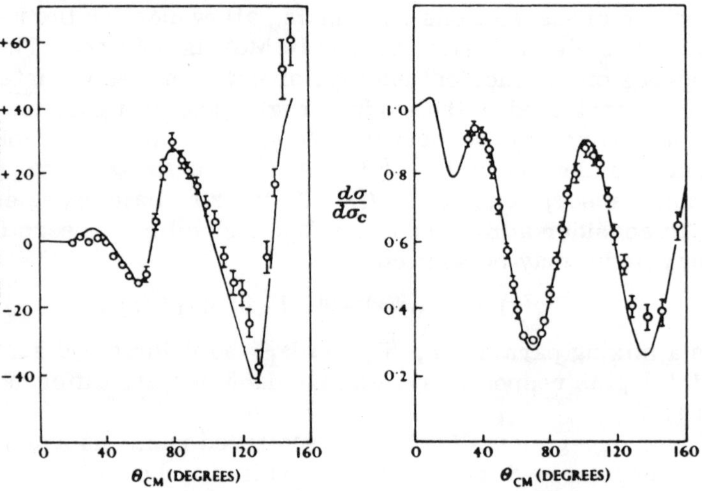

Fig. 1. Experimental data on the elastic scattering of 9.4 MeV protons by copper [2,3] compared with optical model calculations with the potential U = 37 MeV, W_{max} = 8.2 MeV, U_s = 12 MeV.fm^2, W_s = 0, r_0 = 1.3 fm and a = 0.65 fm. The form factor of the real part of the potential is Saxon-Woods and of the imaginary part its radial derivative. (By courtesy of Dr. G.W. Greenlees.)

The measurable quantities may be calculated from the optical potential by solving the Schrödinger equation

$$\nabla^2\Psi + \frac{2m}{\hbar^2}(E - V(r))\,\Psi = 0 \qquad (7)$$

subject to the condition that the asymptotic form of the total wave function Ψ is that of an incident plane wave plus a spherical scattered wave. This may be done by expanding Ψ in partial waves

$$\Psi = \sum_{l=0}^{\infty} \frac{u_l(r)}{r} P_l(\cos\theta) \qquad (8)$$

and substituting in (6) to give the radial wave equation:

$$\frac{d^2U_l}{dr^2} + \{\frac{2m}{\hbar^2}(E - V(r)) - \frac{l(l+1)}{r^2}\} U_l = 0 \ . \qquad (8)$$

These equations may be integrated numerically for each value of l contributing to the interaction and the resulting (complex) radial wave functions are matched to the known asymptotic forms beyond the nuclear field to give the scattering matrix elements S_l. The required observable quantities are then given by

$$\frac{d\sigma(\theta)}{d\Omega} = |f(\theta)|^2 \ , \qquad (9)$$

where

$$f(\theta) = f_C(\theta) + \frac{1}{2ik}\sum_{l=0}^{\infty}(2l+1)(S_l - 1)\, P_l(\cos\theta) \ , \qquad (10)$$

in which $f_C(\theta)$ is the Coulomb scattering amplitude.

The reaction cross section

$$\sigma_R = \frac{\pi}{k^2}\sum_{l=0}^{\infty}(2l+1)(1 - |S_l|^2) \qquad (11)$$

and for neutrons the total cross section

$$\sigma_T = \frac{2\pi}{k^2}\sum_{l=0}^{\infty}(2l+1)(1 - \mathrm{Re}\, S_l) \ . \qquad (12)$$

In the presence of a spin-orbit term the radial wave equation (8) is replaced by a pair of equations for $j = l \pm \frac{1}{2}$, and the matching gives corresponding matrix elements S_l^+ and S_l^-. The observable quantities, including the polarisation, may be expressed in terms of these matrix elements.

The details of this calculation have been given by Buck et al. [4] and by Hodgson [7]. The most laborious part of the calculation is the numerical integration of the radial wave equations (8), and this has to be done so many times that the use of an electronic computer is essential. Many computer programmes have been written to carry out this calculation on various types of computers. It is possible to write the programme so that starting from a given set of parameters of the optical potential the computer will automatically iterate specified parameters to optimise the fit to a given set of experimental data. Such computer programmes have been described by Maddison [8] and Perey [9].

Many sets of experimental data have been analysed in this way, and much information has been accumulated on suitable values of the parameters of the optical potential. It is found that the potentials depend on energy, while the form factors are independent of energy, apart from the systematic change of the radial variation of the imaginary part of the potential as the energy increases. In some cases it is found necessary to allow the parameters of the form factors to vary from one term to the next, in particular between the real and imaginary central potentials, but in most analyses they are kept the same. For medium and heavy nuclei the parameters of the optical potential show no systematic variation with mass number, but for light nuclei they show marked fluctuations from nucleus to nucleus since the conditions for the validity of the optical model are not adequately satisfied.

The values of the parameters of the optical potential found in these investigations show marked fluctuations about the mean values of the systematic trends. This is shown for the potential depths U, W and U_S in figs. 2-4. To some extent these fluctuations are due to the uncertainties and limitations of the experimental data, to differences in the techniques of optical model analysis and to the structural differences between the interactions

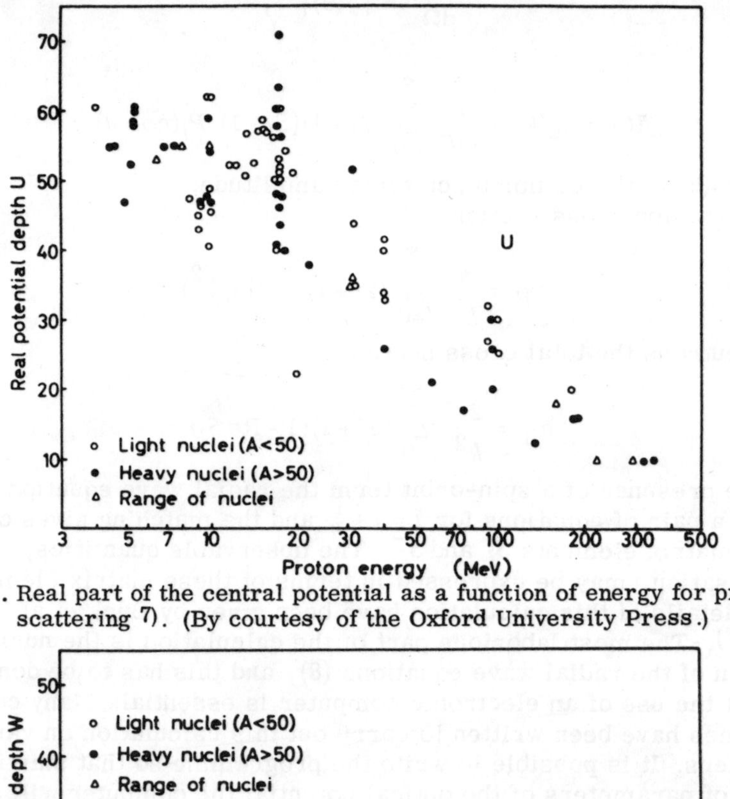

Fig. 2. Real part of the central potential as a function of energy for proton scattering [7]. (By courtesy of the Oxford University Press.)

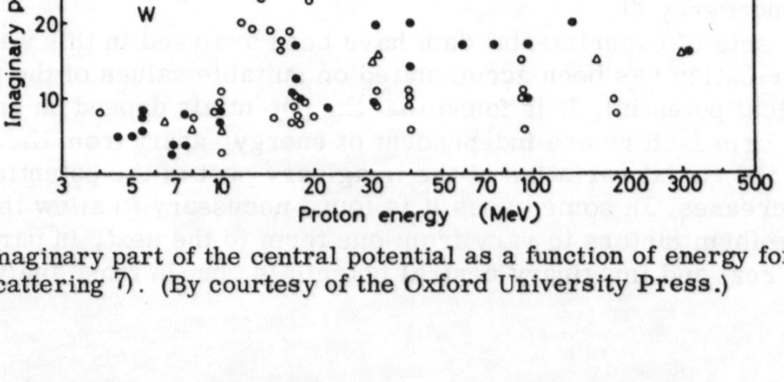

Fig. 3. Imaginary part of the central potential as a function of energy for proton scattering [7]. (By courtesy of the Oxford University Press.)

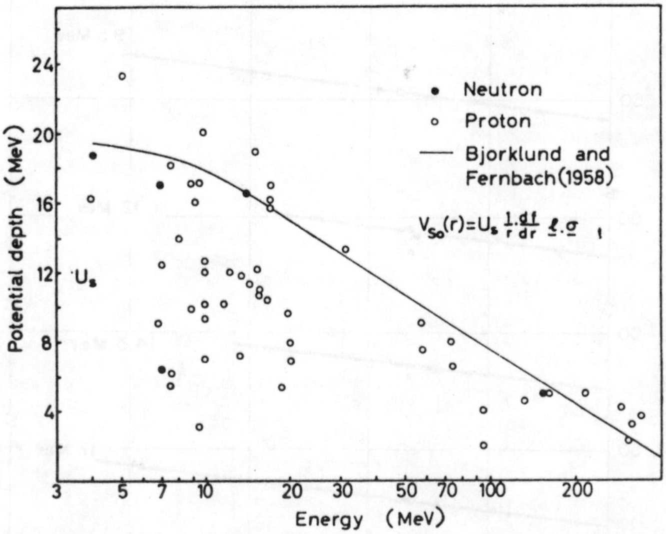

Fig. 4. Real part of the spin-orbit potential as a function of energy for neutron and proton scattering.[7]. The cluster of points around U_S = 8 - 10 MeV, E = 10 - 20 MeV is from the analysis of carbon scattering by Nodvik et al.[10]. (By courtesy of the Oxford University Press.)

studied, but it is now clear that the major source of the fluctuations is the inherent ambiguities among the optical model parameters. Even with the physically-reasonable form factors given above it is possible to find continuous and in some cases discrete sets of parameters all giving almost identical cross sections. A notable example is the Ur_0^2 ambiguity in nuclear scattering; so long as Ur_0^2 is kept constant U and r_0 can often be varied by as much as 30% without substantially affecting the fit to the experimental data. In other cases the optimum fit is in a shallow minimum in parameter space, so that when a numerical criterion of goodness-of-fit is applied to a particular set of data, slight changes in the data can give substantial variations in the parameters.

These fluctuations may be practically eliminated in most cases simply by fixing the parameters of the form factors at reasonable average values, and then repeating the analysis to determine the optimum potential depths. In most cases the fits with fixed form factors are only slightly inferior to those with variable form factors, and this is a small price to pay for the greatly improved consistency of the potentials.

The most extensive analysis with fixed form factors is Perey's work on the elastic scattering of 9-22 MeV protons by a range of medium weight nuclei. This analysis of carefully-chosen experimental data gave potentials almost entirely free from the fluctuations mentioned above, and allowed their variation with energy and from nucleus to nucleus to be studied. He confirmed earlier work showing the real potential depth U falls linearly with increasing energy and also showed the dependence of U on the nuclear symmetry parameter $(N-Z)/A$. Some of his results are shown in fig. 5 for

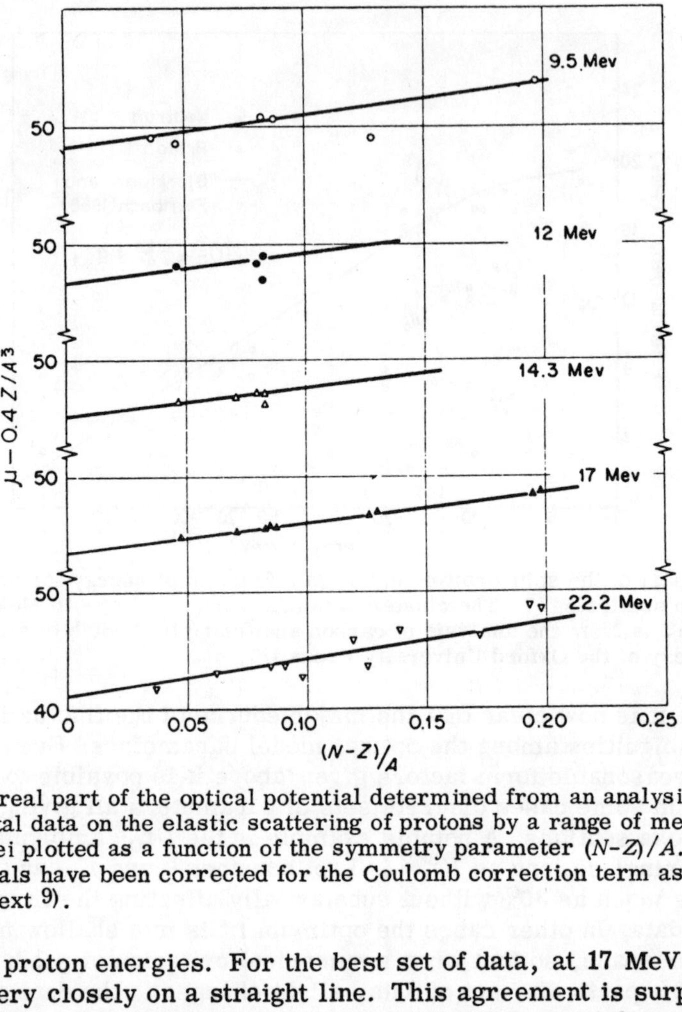

Fig. 5. The real part of the optical potential determined from an analysis of experimental data on the elastic scattering of protons by a range of medium weight nuclei plotted as a function of the symmetry parameter $(N-Z)/A$. These potentials have been corrected for the Coulomb correction term as described in the text [9].

a range of proton energies. For the best set of data, at 17 MeV, the values of U fall very closely on a straight line. This agreement is surprisingly good and shows that differences in nuclear structure not taken account of in the optical model, for example the nuclear deformations and spins, have rather a small effect on the optical potential. (This is not always the case, however, as Buck [11] found that the potentials obtained from analyses of the experimental data on the elastic scattering of 11 MeV protons by three isotopes of zinc are not in agreement with a linear function of the symmetry parameter.) An analysis of some neutron data by Hodgson [5] also showed a dependence of U on the symmetry parameter, but with the opposite sign to that found for protons.

The overall dependence of U on the symmetry parameters makes it desirable to include an additional term in the optical potential of the form

$$U_T = \pm\tfrac{1}{4}U_1\left(\frac{N-Z}{A}\right) f_1(r) \ , \qquad (13)$$

where the positive sign refers to neutrons and the negative sign to protons;

this implies that U and U_1 have opposite signs. The factor $\frac{1}{4}$ is introduced because Lane [12] has shown that a term of the form (14) arises naturally in the optical potential from the sum over the nucleon-nucleon Heisenberg operators and that for both beutrons and protons it can be written

$$U_T = U_1(\boldsymbol{T} \cdot \boldsymbol{t})\, A^{-1} f_1(r) \ , \tag{14}$$

where $\boldsymbol{T}$ and $\boldsymbol{t}$ are the isotopic spin vectors for the target nucleus and incident particle. The relation (14) then follows from (15) using the properties of these operators. There is at present no evidence for any energy variation of the isotopic spin potential.

It is also necessary to include in the proton optical potential a term $\alpha Z/A^{\frac{1}{3}}$ due to the energy dependence of the real potential U. Incident protons are retarded by the Coulomb field as they approach the nucleus, and this alters the effective potential. The constant α is equal to 1.33 $\mathrm{d}U/\mathrm{d}E$. Since $Z/A^{\frac{1}{3}}$ is closely proportional to $(N-Z)/A$ over most of the periodic table, this Coulomb correction term cannot be separated empirically from the isotopic spin term (14). It is therefore calculated and subtracted from the phenomenologically determined potentials U before the comparison with $(N-Z)/A$ is made. This is done before plotting the results shown in fig. 5.

The energy dependence of the real potential U is mainly a consequence of the neglect of the non-locality of the optical potential. The fundamental derivation of the optical potential from its constituent nucleon-nucleon interactions shows that it is essentially non-local in character. It has been shown by Perey and Buck [13] that it is possible to obtain a good fit to a wide range of neutron scattering and total cross section data for neutron energies from 1 to 25 MeV with a single non-local optical potential. Some of their results are shown in figs. 6 and 7. The essential modification introduced by the non-local potential into the optical model calculation is that the term $V\Psi$ in the Schrödinger equation is replaced by a term of the form

$$\int V(\boldsymbol{r}, \boldsymbol{r}')\, \Psi(\boldsymbol{r}')\, \mathrm{d}\boldsymbol{r}' \tag{15}$$

so that the resulting radial wave equations are integro-differential. These can be solved using an electronic computer but the numerical work is considerably longer. Since Perey and Buck also showed that to every non-local potential there corresponds an equivalent local potential it is easier in practice to work with energy-dependent local potentials than with the aesthetically more satisfying energy-independent non-local potentials. The equivalence mentioned above is only insofar as the elastic scattering is concerned; the local and non-local wavefunctions though asymptotically identical differ appreciably in the nuclear interior and this could be of importance for inelastic interactions whose matrix elements are proportional to overlap integrals of the wave functions of the interacting particles.

The best potentials available at present for various incident particles are summarised below. It should be particularly noted that these do not in general apply to light nuclei ($Z < \approx 12$), or at energies so low that individual resonances cause large fluctuations in the cross sections. Furthermore these potentials give only the shape-elastic cross section and not the experimentally-indistinguishable compound elastic cross section. This com-

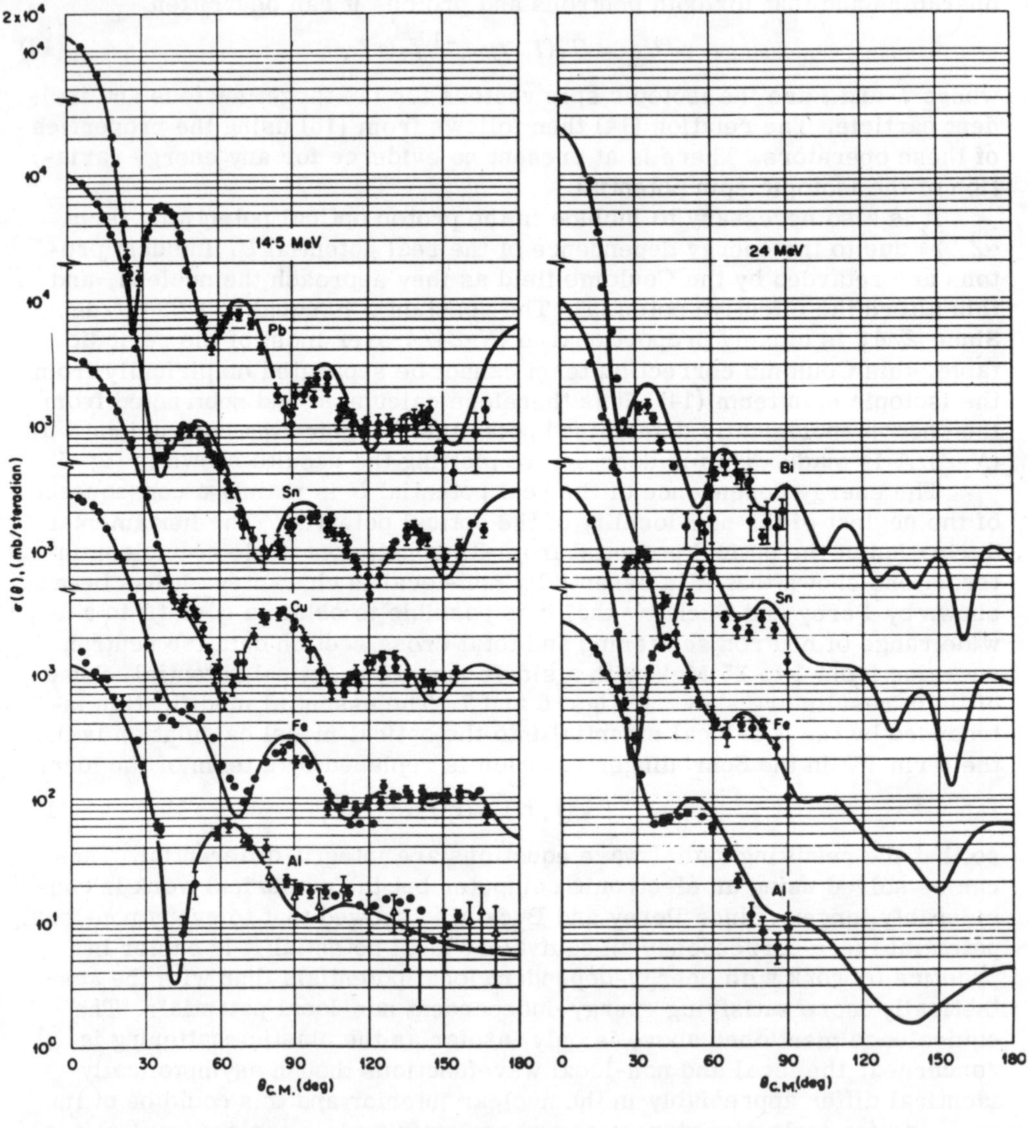

Fig. 6. Differential cross sections for the elastic scattering of 14.5 and 24 MeV neutrons by several nuclei compared with the calculations of Perey and Buck [13]) using a single non-local optical potential.

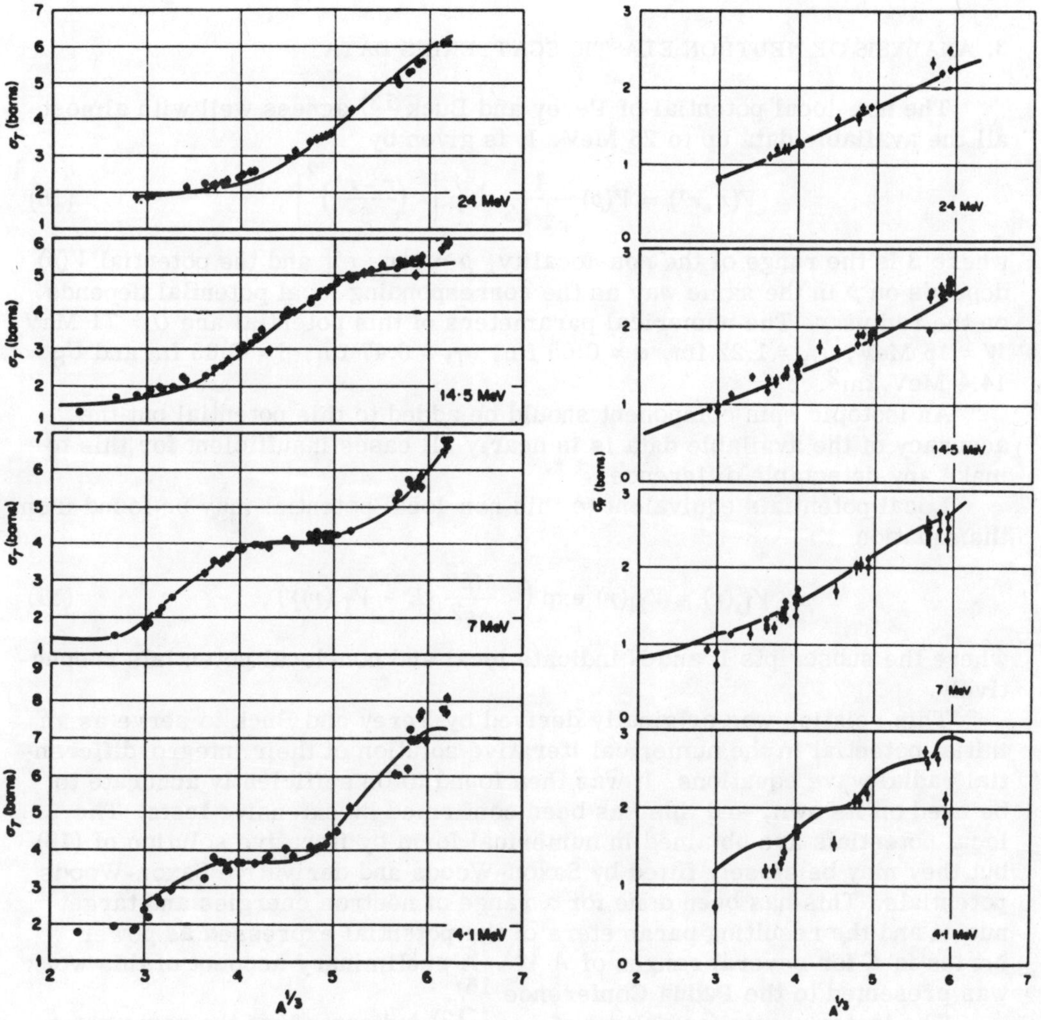

Fig. 7. **Total and reaction cross sections for neutrons of several energies interacting with a range of nuclei compared with the calculations of Perey and Buck [13] using a single non-local optical potential.**

pound elastic cross section should be evaluated and subtracted from the experimental data whenever it contributes significantly to the interaction, and only when this is done can the converted cross section be compared with that calculated from the following optical model potentials. This question of the compound elastic cross section will be considered further in the second lecture.

3. ANALYSIS OF NEUTRON ELASTIC SCATTERING DATA

The non-local potential of Perey and Buck [13] agrees well with almost all the available data up to 25 MeV. It is given by

$$V(\boldsymbol{r},\boldsymbol{r}') = V(p)\,\frac{1}{\pi^{\frac{3}{2}}\,\beta^3}\,\exp\left[-\left(\frac{\boldsymbol{r}-\boldsymbol{r}'}{\beta}\right)^2\right], \tag{16}$$

where β is the range of the non-locality, $p = \frac{1}{2}|\boldsymbol{r}+\boldsymbol{r}'|$ and the potential $V(p)$ depends on p in the same way as the corresponding local potential depends on the radius r. The numerical parameters of this potential are $U = 71$ MeV, $W = 15$ MeV, $r_0 = 1.22$ fm, $a = 0.65$ fm, $a_D = 0.47$ fm, $\beta = 0.85$ fm and $U_S = 14.4$ MeV. fm^2.

An isotopic spin component should be added to this potential but the accuracy of the available data is in nearly all cases insufficient for this to make any detectable difference.

Local potentials equivalent to this non-local potential may be found from the equation

$$V_L(r) = V_N(r)\,\exp\left(-\frac{M\beta^2}{2\hbar^2}\{E - V_L(r)\}\right), \tag{17}$$

where the subscripts L and N indicate local and non-local potentials respectively.

This relation was originally derived by Perey and Buck to serve as an initial potential in the numerical iterative solution of their integro-differential radial wave equations. It was then found to be sufficiently accurate to be used on its own, and this has been confirmed by extensive tests. The local potentials are obtained in numerical form by iterative solution of (18) but they may be closely fitted by Saxon-Woods and derivative Saxon-Woods potentials. This has been done for a range of neutron energies and target nuclei and the resulting parameters of the potential expressed as power series in E for several ranges of A [14]. A preliminary account of this work was presented to the Padua Conference [15].

The isotopic spin formalism of Lane [12] indicates that the same potential should be able to account for both neutron and proton scattering, with the appropriate change of sign of the isotopic spin term and the omission of the Coulomb correction term from the neutron potential. A study [6] of the potentials appropriate to the neutron and proton scattering in one of the few cases where good data are available for the same nucleus at nearly the same energy indicates that this is so, but more extensive studies are required to show that this is true in general. This equivalence of neutron and proton optical potentials may apply to light nuclei in cases where the optical parameters themselves are anomalous. This is being studied at the present time.

The remarks made below on the strength and form factor of the imaginary part of the proton optical potential apply also to the neutron potential, though lack of accurate neutron data has so far prevented a detailed investigation of these effects.

4. ANALYSIS OF PROTON ELASTIC SCATTERING DATA

The analysis of 9-22 MeV proton elastic scattering data by Perey referred to above gave the potential

$$V(r) = V_C(r) - (53.3 - 0.55E + 27\frac{N-Z}{A} + 0.4\frac{Z}{A^{\frac{1}{3}}})\, f(r) + (3A^{\frac{1}{3}})\, 4af'(r) - U_s h(r)\, \boldsymbol{l}\cdot\boldsymbol{\sigma} \qquad (18)$$

with $U_s = 17$ for $E \geqslant 17$ MeV and $U_s = 15$ for $E < 17$ MeV and all $r_o = 1.25$ fm, $a_U = 0.65$ fm, $a_W = 0.47$ fm, where the subscripts U and W refer to $f(r)$ and $f'(r)$ respectively.

Other work over a wider range of energies has given the following expression for the energy variation of the spin-orbit potential [16)]

$$U_s = 20 - 0.3E\,, \qquad (19)$$

which is approximately the same as the values found by Perey though the apparently anomalous energy dependence of his values of U_s remains unexplained. These uncertainties about the value of the spin-orbit potential and still more about its form factor fortunately do not affect the fit to the differential cross section, which is relatively insensitive to the spin-orbit potential. More detailed studies of the available polarisation data are needed to define this term in the potential more accurately.

The potential (19) can probably be extrapolated with some confidence to energies outside the range originally studied by Perey, though this needs to be verified in detail.

Perey also found some evidence for the dependence of the imaginary potential on the symmetry parameter according to the relations

$$\begin{aligned} W &= 9.8 + 4.6\frac{N-Z}{A} && \text{for } E = 14.3 \text{ MeV,} \\ W &= 8.2 + 4.8\frac{N-Z}{A} && \text{for } E = 17 \text{ MeV,} \qquad (20) \\ W &= 7.4 + 4.8\frac{N-Z}{A} && \text{for } E = 22 \text{ MeV.} \end{aligned}$$

The fluctuations of the phenomenological potentials about these lines is considerably greater than in the corresponding plot for the real part of the potential. It is to be expected that W is much more dependent on the target nucleus than U because it summarises all the non-elastic channels, and the number and nature of these channels is determined by the incident energy and by the structure of the target nucleus. Many more interactions may be possible for a nucleus with partially-filled shells than for one with complete proton and neutron shells, and this will be reflected in the value of W. The value of W may thus be expected to be lower in the region of closed shells, and some indication that this is so has been found from analyses of neutron strength functions [17,18]. The most sensitive way of studying these variations of W with energy is probably through the proton reaction cross sections. Perey found that in most of the cases he studied the reaction cross sections calculated from potentials fitted to the elastic scattering data only were in good accord with the experimental values. This is unlikely to remain the case when more accurate proton reaction cross sections become available, and it will then be useful to use them to determine the optimum values of W, keeping all the other parameters fixed to the above values. The results should show marked correlations with nuclear structure.

Many investigations have been made to determine the form $g(r)$ of the imaginary potential. As the elastic scattering data are insensitive to the form chosen the work has concentrated on the possibility of distinguishing phenomenologically between the extreme surface and volume forms. Several early studies fixed the radii of the form factors of the real and imaginary parts of the potential to the same value at the beginning of the analysis and thus in effect determined the result. With different radii equally good fits can be obtained with both types of form factor [5]. The results of the detailed analysis of Perey of the scattering of 17 MeV protons by a range of nuclei are shown in figs. 8 and 9. He allowed all the parameters to vary, for both surface and volume absorption. Both sets of fits are excellent, and although those for surface absorption are slightly better it is doubtful whether the optical model is sufficiently reliable for any firm conclusions to be drawn from this. There is thus hardly any phenomenological evidence in favour of one form of imaginary potential or the other, so it is preferable to choose the form in accord with theoretical analyses of the absorption process in terms of the constituent nucleon-nucleon collisions.

At high energies one would expect the interaction to be spread throughout the nuclear volume as the Pauli principle is no longer effective in inhibiting the interactions in the nuclear interior. It is found that optical model calculations with volume absorption and equal radii for the real and imaginary parts of the potential gives differential cross sections with markedly more oscillatory structure than is shown by the experimental data for protons of around 180 MeV. It has been shown [19,20] that these oscillations are removed and a differential cross section in accord with the experimental data obtained if the radius of the imaginary part of the potential exceeds that of the real part by about 1 fm (see fig. 10). This also at the same time markedly improves the fit to the polarisation, so this is probably a real effect. Further investigation is needed to interpret this result, and also to see over what range of energies such potentials have to be used.

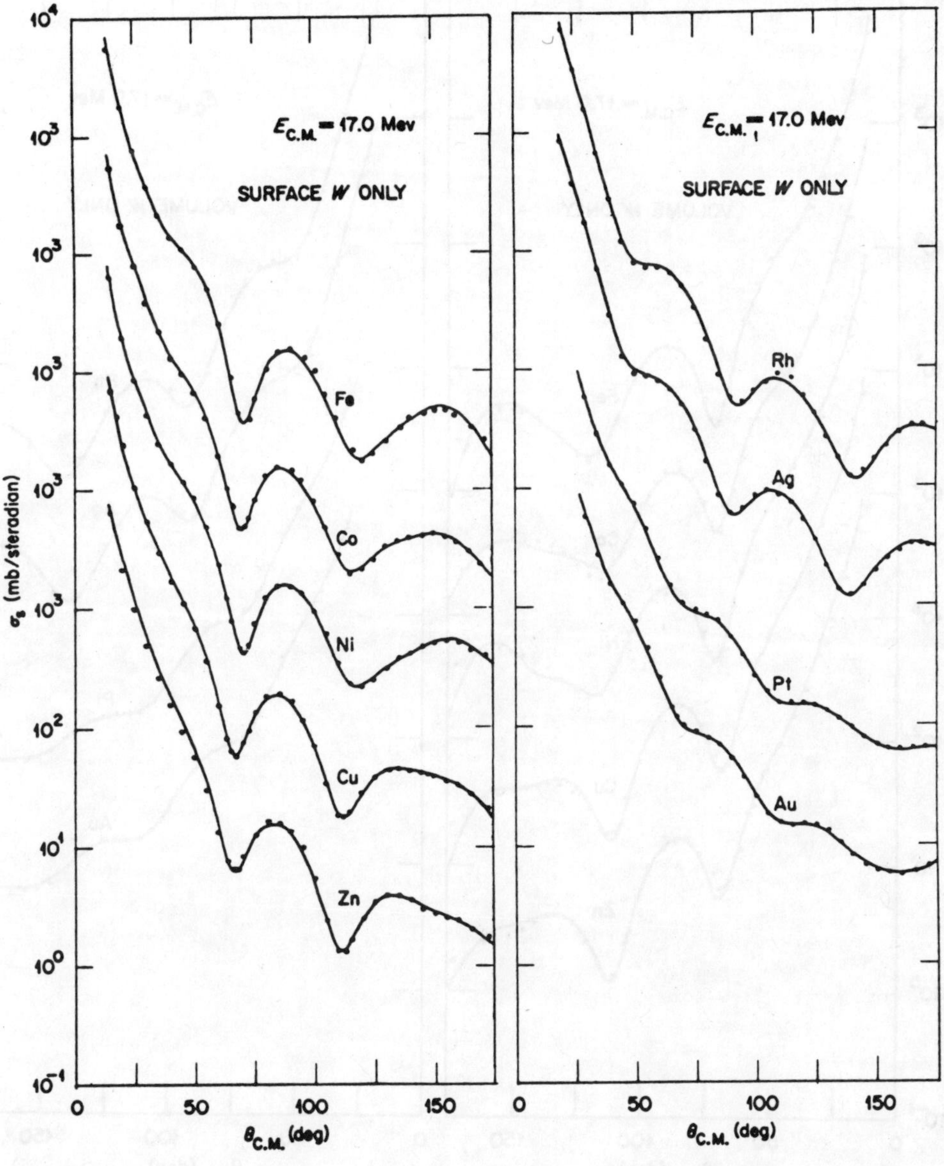

Fig. 8. Optical model fits to the experimental data on the elastic scattering of 17 MeV protons by a range of nuclei obtained by Perey [9] using a surface absorbing potential and allowing all the parameters of the potential to vary.

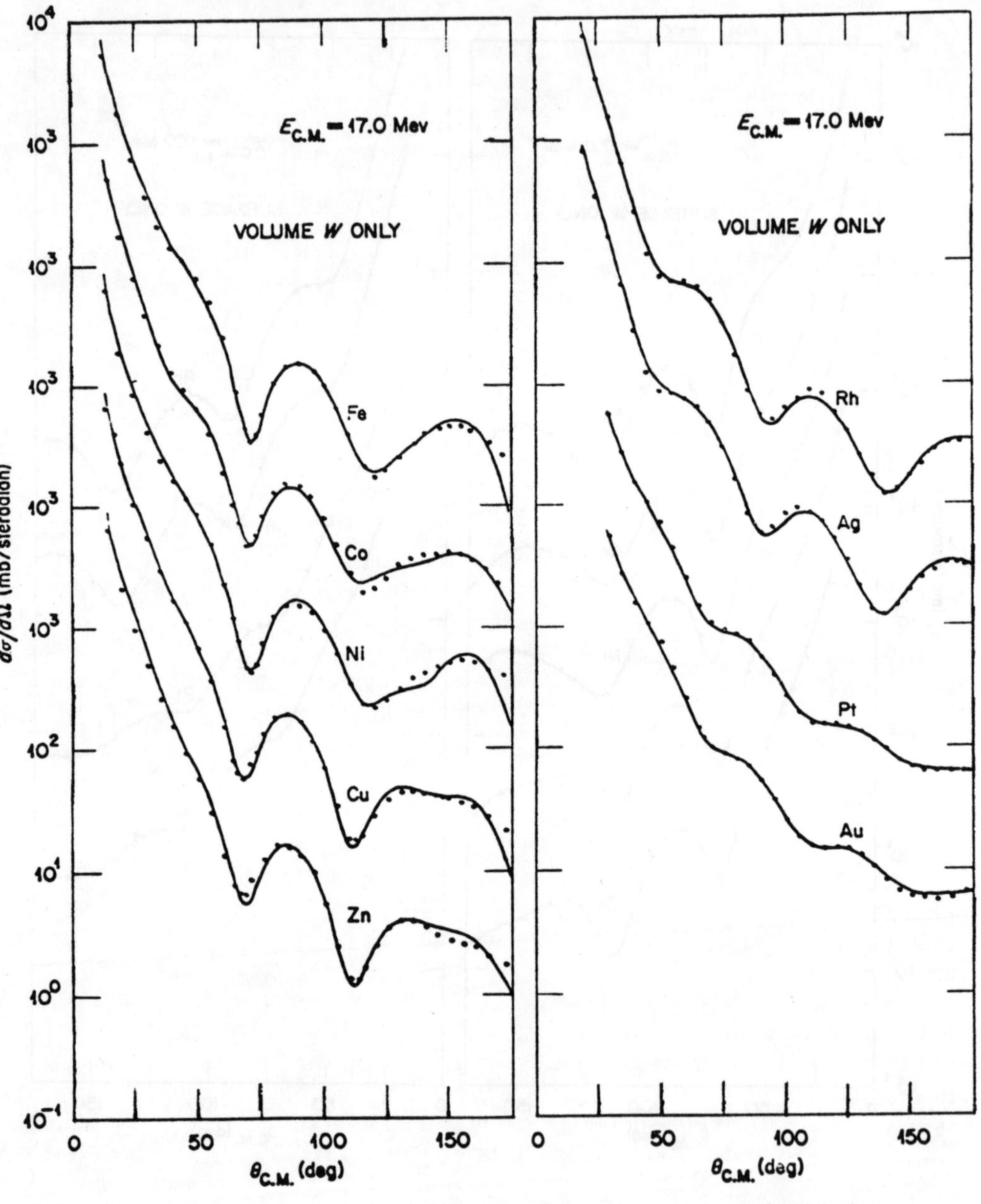

Fig. 9. Optical model fits to the experimental data on the elastic scattering of 17 MeV protons by a range of nuclei obtained by Perey [9]) using a volume absorbing potential and allowing all the parameters of the potential to vary.

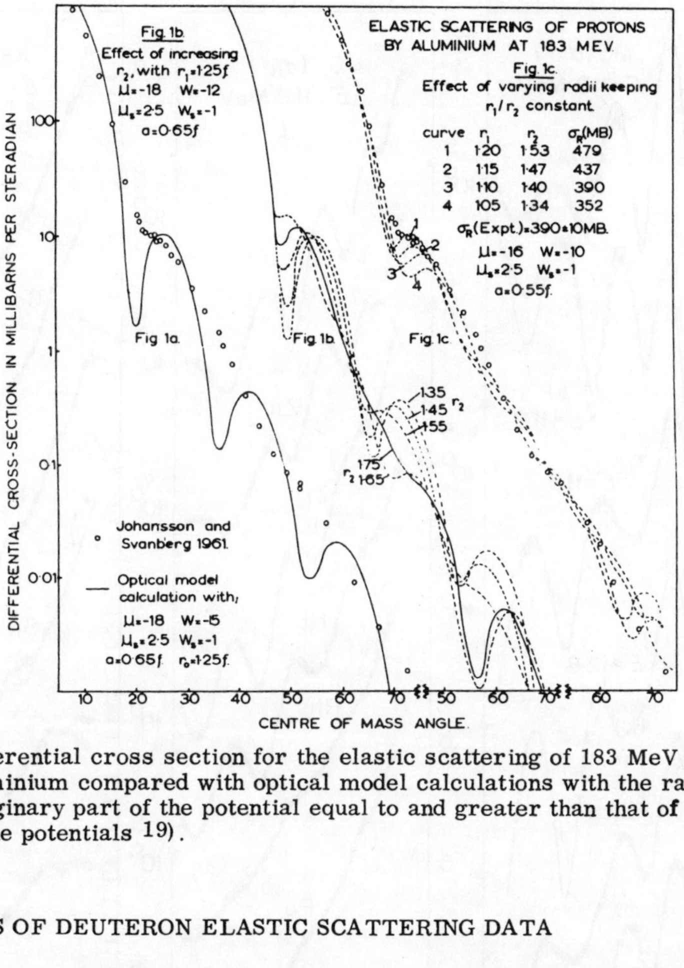

Fig. 10. Differential cross section for the elastic scattering of 183 MeV protons by aluminium compared with optical model calculations with the radius of the imaginary part of the potential equal to and greater than that of the real part of the potentials [19].

5. ANALYSIS OF DEUTERON ELASTIC SCATTERING DATA

The low binding energy of the deuteron is responsible for the special characteristics of deuteron scattering that make it more difficult to analyse than nucleon scattering. Deuterons are readily absorbed by nuclear matter and so the scattering is dominated by the potential at the nuclear surface, and is very insensitive to the potential in the nuclear interior. Analyses with the standard form factors thus show increased ambiguities among the parameters of the potential. Furthermore, since the elastic scattering depends only on the asymptotic form of the wave function at large distances there exists a discrete series of potentials of increasing depth all giving the same asymptotic wave function but having different numbers of half-waves inside the nucleus for each partial wave. In the case of nucleon scattering this ambiguity can be resolved by continuity considerations with the corresponding bound state wave functions but this cannot be done for deuteron scattering. It is therefore usual in deuteron scattering to consider only the shallowest of the series of potentials that is in accord with the observed scattering; it is found in practice that this potential usually gives

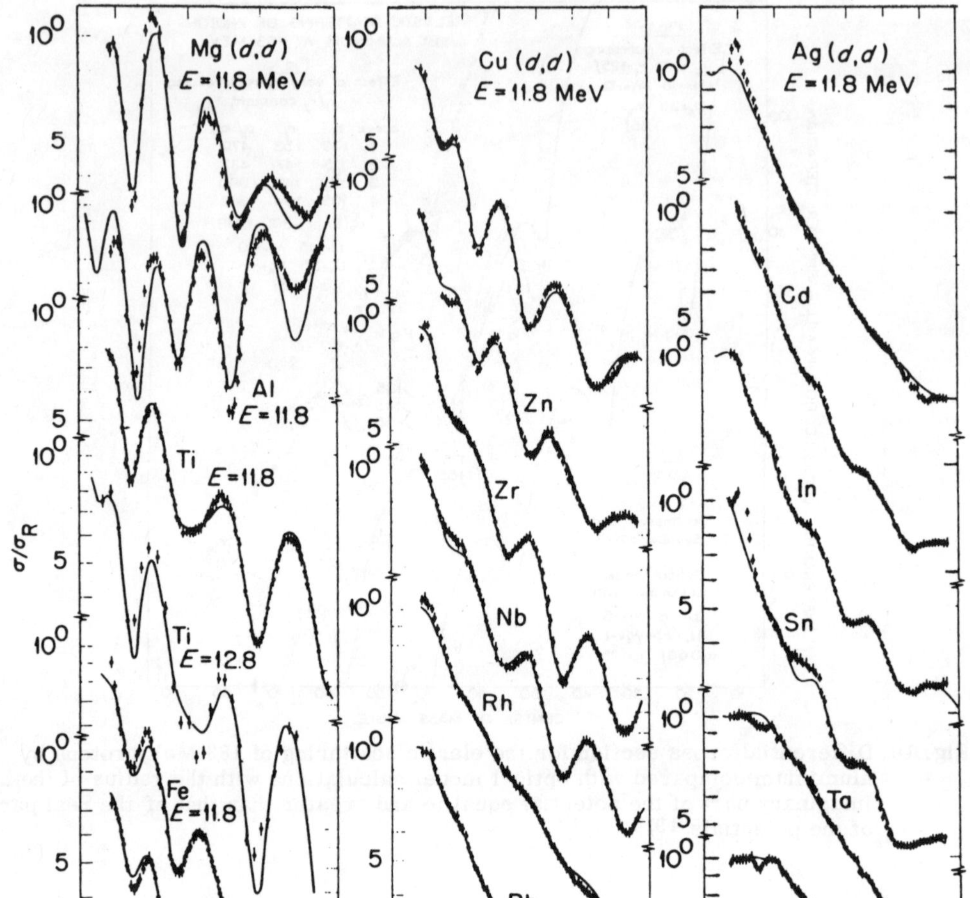

Fig. 11. Optical model fits to experimental data on the elastic scattering of 11.8 MeV deuterons by a range of nuclei [21].

slightly better overall fits than the second and subsequent potentials in the series.

An extensive series of analyses of the data on the scattering of 11 - 27 MeV deuterons by a range of medium weight and heavy nuclei by Perey and Perey [21] gives the following best set of parameters:

$$U = 42 + 3.3Z/A^{\frac{1}{3}} - 0.51E \,, \quad r_u = 1.15 \,, \quad a_u = 0.87 \,,$$
$$W_D = 11.0 + 0.24E \quad (E > 15) \,, \quad s = 0 \,, \quad r_W = 1.37 \,, \quad a_W = 0.70 \,. \tag{21}$$

For lower energies the variation of W_D with energy shows a sharp increase that remains unexplained. The results of some typical optical model calculations of deuteron scattering are shown in fig. 11.

It is notable that the radius of the real part of the optical potential is less than for nucleon scattering, contrary to the expected increase due to the larger size of the deuteron. This is offset to some extent by the higher diffuseness parameter. It is found that the quality of the fits is somewhat insensitive to the value of the radius parameter provided the remaining parameters are re-adjusted; this is similar to the Ur_o^2 ambiguity in nucleon scattering. If the value of r_o is raised to 1.3, quite good fits are still obtained, though they are definitely inferior to those with $r_o = 1.15$. As r_o is increased the dependence of U on $Z/A^{\frac{1}{2}}$ becomes less marked, finally becoming negligible around $r_o = 1.6$, where the fits are unacceptable.

The cross sections calculated from the above set of parameters differ in detail from the measured cross sections much more than in the corresponding nucleon case. This is due to the greater sensitivity of deuteron elastic scattering to differences between nuclei, and hence a careful analysis of the experimental data can give more information on nuclear structure. For example some indication of a lowered value of the absorbing potential in the region of closed shells has recently been found by Forest [22].

So far there have been no systematic studies of the spin-dependent term in the deuteron optical potential due to the paucity of relevant experimental data. Since the deuteron has spin one, tensor as well as vector polarisation has to be considered, and this complicates an already difficult analysis. Calculations with a spin-dependent term in the deuteron optical potential have shown that as for nucleon scattering the differential cross section is relatively insensitive to such a term. It is therefore legitimate to use the above spin-independent optical potential in all cases where polarisation phenomena are not being investigated. This also applies to triton and helium-3 elastic scattering.

Since the deuteron elastic scattering is dominated by the potential in the surface region it is immaterial whether volume or surface absorption is used, provided the potential has the required exponential fall-off. This choice will however affect the wave function in the nuclear interior, which may be important for inelastic interactions. The sensitivity to the nuclear surface makes it possible to detect differences in the surface diffuseness parameter from one nucleus to another. It might be expected that nuclei with closed shells have a smaller diffuseness parameter than those with unfilled shells. These comments on the sensitivity of the interaction to the surface potential apply also to the scattering of tritons, helium-3 and alpha-particles.

6. TRITON, HELIUM-3 AND ALPHA ELASTIC SCATTERING

Insufficient analyses have been made for any general conclusions on the triton optical potential to be drawn. An analysis of some data on the elastic scattering of 7 MeV tritons by carbon, oxygen, fluorine and calcium has recently been made by Pullen et al. [23] and potentials similar to the helium-3 potentials have been found.

A series of analyses [15,24] of data on the elastic scattering of 29 MeV helium-3 nuclei by a range of nuclei showed that the real and imaginary potentials are similar in magnitude and strongly dependent on A. This variation can be approximately represented by the relation

$$U = W = 220\, A^{-\frac{1}{2}} \text{ MeV} \tag{22}$$

and $r_0 = 1.6$ fm, $a = 0.65$ fm.

Some indications of the dependence of the surface diffuseness parameter on nuclear structure have been found by Hodgson [19].

There are insufficient helium-3 elastic scattering data at other energies for any conclusions about the possible energy variation of these helium-3 optical parameters to be drawn.

It is possible that the triton and helium-3 optical potentials will eventually be unified using the isotopic spin dependence in the same way as the neutron and proton potentials.

The most extensive analysis of elastic alpha-particle scattering has been made by Igo [25], who gives the potential

$$V(r) = -\,1100 \exp\left[-\left(\frac{r - 1.17A^{\frac{1}{3}}}{0.574}\right)\right] - 45.7\, i \exp\left[-\left(\frac{r - 1.40A^{\frac{1}{3}}}{0.578}\right)\right]. \tag{23}$$

As most of the data on which this potential is based refer to energies of 20 and 40 MeV, there is little information on the possible energy variation of the alpha-particle optical model potential.

7. QUASI-ELASTIC SCATTERING

The presence of an isotopic spin dependent term in the optical potential makes it possible to calculate the cross section for (p, n) reactions in which the residual nucleus is in the isotopic analogue state of the target nucleus. Since these (p, n) cross sections are approximately proportional to the square of the strength of the isotopic spin term the study of these reactions provides an accurate way of determining this strength [26,27]. The corresponding (He^3, t) cross sections can also be calculated using this formalism [26,28].

REFERENCES

1) K. A. Brueckner, A. M. Lockett and M. Rotenberg, Phys. Rev. 121 (1961) 255.
2) A. B. Robbins, K. A. Grotowski and G. W. Greenlees, Proc. Rutherford Conf. (Heywood and Co., Ltd., London, 1961) p. 419.

3) G.W. Greenlees and D N. Jarvis, Proc. Phys. Soc. 78 (1961) 1275.
4) B. Buck, R.N. Maddison and P.E. Hodgson, Phil. Mag. 5 (1960) 1181.
P.E. Hodgson, Nuclear Phys. 21 (1960) 28.
5) P.E. Hodgson, Physics Letters 3 (1963) 352.
6) P.E. Hodgson, Physics Letters 6 (1963) 75.
7) P.E. Hodgson, The optical model of elastic scattering (Oxford University Press, 1963).
8) R.N. Maddison, Proc. Phys. Soc. 79 (1962) 264.
9) F.G. Perey, Phys. Rev. 131 (1963) 745.
10) J.S. Nodvik, C.B. Duke and M.A. Melkanoff, Phys. Rev. 125 (1962) 975.
11) B. Buck, Phys. Rev. 129 (1963) 712.
12) A.M. Lane, Phys. Rev. Letters 8 (1962) 171; Nuclear Phys. 35 (1962) 676.
13) F.G. Perey and B. Buck, Nuclear Phys. 32 (1962) 353.
14) D. Wilmore and P.E. Hodgson, in preparation.
15) P.E. Hodgson, Proc. Padua Conf. (Gordon and Breach Science Publishers, New York and London, 1962).
16) A.E.S. Green, Proc. Tallahassee Conf. (The Florida State University, 1959) p. 44.
17) H. Fiedeldey and W.E. Frahn, Ann. Phys. 16 (1961) 387.
18) P.A. Moldauer, Phys. Rev. Letters 9 (1962) 17.
19) P.E. Hodgson, Phys. Rev. Letters 6 (1961) 358; Proc. Rutherford Conf. (Heywood and Co. Ltd., London, 1961) p. 407.
20) A. Johansson, U. Svanberg and P.E. Hodgson, Arkiv Fysik 19 (1961) 541.
21) C. Perey and F.G. Perey, Phys. Rev. 132 (1963) 755.
22) A. Forest, Physics Letters 7 (1963) 130.
23) D. Pullen, R. Middleton and J.R. Rook, in preparation.
24) P.E. Hodgson, Proc. Phys. Soc. 80 (1962) 1161.
25) G. Igo, Phys. Rev. 115 (1959) 1665.
26) R.M. Drisko, R.H. Bassel and G.R. Satchler, Physics Letters 2 (1962) 318.
27) P.E. Hodgson and J.R. Rook, Nuclear Phys. 37 (1962) 632.
28) J.R. Rook and P.E. Hodgson, in preparation.

THE OPTICAL MODEL ANALYSIS OF ELASTIC SCATTERING AND DIRECT NUCLEAR INTERACTIONS

P. E. HODGSON

Nuclear Physics Laboratory, Oxford, England

Introduction.

This review summarises some recent measurements of elastic and inelastic nuclear interactions and their analysis using the optical model and direct reaction theories. In the preparation of this review use has been made not only of the papers submitted to this conference but also of the recent literature and numerous private communications both before and during the conference. Particular emphasis is placed on the growing points of the subject, and fruitful lines for future experimental and theoretical work are discussed.

A. Elastic scattering.

A.1. Nucleon elastic scattering.

A.1.*a*. *Differential Cross-sections.* Substantially all neutron elastic scattering data, including differential cross-sections, polarisations, total and reaction cross-sections, and strength functions, from the resonance region to about 25 MeV are accounted for within the accuracy of existing measurements by the energy-independent non-local potential of Perey and Buck (1962). This potential is rather inconvenient in practice since it leads to a series of integro-differential equations for the radial wave functions, and Perey and Buck have shown how the equivalent local energy dependent potentials may be calculated at any energy. These potentials have been found by Wilmore and Hodgson (1964) to be in accord with a large range of experimental differential and total neutron cross-sections.

It is likely that substantial difficulties would be encountered in fitting the data with these potentials if the accuracy of the experimental cross-sections was increased to about 3 per cent and the measurements extended over the whole angular range.

Proton elastic scattering has been extensively analysed using an optical potential of the form

$$V(r) = V_c(r) + Uf(r) + iWg(r) + \frac{1}{r} V_{so}\, h(r)\, \mathbf{L}\cdot\boldsymbol{\sigma}$$

where $V_c(r)$ is the Coulomb potential, U and W the real and imaginary potential depths, and $f(r)$ and $g(r)$ their form factors, usually taken to be Saxon-Woods and either Saxon-Woods or derivative Saxon-Woods: $f(r) = [1 + \exp\{(r - R)/a\}]^{-1}$, $g(r) = f(r)$ or $4a \exp\{(r - R)/a\}$ $[1 + \exp\{(r - R)/a\}]^{-2}$ and $R = r_0 A^{1/3}$. The form factor $h(r)$ of the spin-orbit potential $V_{so} h(r)$ is usually taken to be derivative Saxon-Woods also.

Many analyses of experimental data have shown that the quantity indicating goodness-of-fit has rather a shallow minimum in the multi-dimensional space of the parameters that are varied, so that the parameters cannot all be accurately determined.

It has thus become usual to make analyses with fixed form factors, and it is found that in most cases the fit to the data is only slightly inferior to the fit obtained when all the parameters are allowed to vary. It is usual to take for the parameters of the form factors the values $r_0 = 1.25$ fm, $a = 0.65$ fm for the Saxon-Woods (volume) potential and $a = 0.47$ fm for the derivative Saxon-Woods (surface) potential. When the form factors are fixed in this way it becomes possible to look for systematic variations of the potentials with energy and with internal nuclear properties. It is found that the real potential depends on energy and symmetry parameter according to the relation (Perey, 1963)

$$U = U_0 - \beta E + U_1 \frac{N - Z}{A} + 0.4 \frac{Z}{A^{1/3}}$$

where the positive and negative signs refer to proton and neutron scattering respectively. The symmetry term $U_1 \dfrac{N - Z}{A}$ is a consequence of the isotopic spin dependence of the nucleon-nucleus potential $4\, U_1 (\mathbf{T}\cdot\mathbf{t})/A$, where $\mathbf{T}$ and $\mathbf{t}$ are the isospin vectors of the incident particle and target nucleus (Lane, 1962). The last term, present only for protons, is due to the Coulomb potential and the energy dependence of the real potential.

There are several ways of determining the symmetry potential U_1. The more accurate depend on the analysis of differential scattering cross-sections, but estimates may also be obtained from inelastic scattering and from the positions of bound states.

Since U_0 is not known accurately all methods based on elastic scattering depend on subtracting the values U_a and U_b of the real potential obtained

from two analyses at the same energy. Typically this gives a relation of the form

$$\left(U_{\mathrm{a}}-0.4\,\frac{Z_{\mathrm{a}}}{A_{\mathrm{a}}^{1/3}}\right)-\left(U_{\mathrm{b}}-0.4\,\frac{Z_{\mathrm{b}}}{A_{\mathrm{b}}^{1/3}}\right)=U_1\left(\frac{N_{\mathrm{a}}-Z_{\mathrm{a}}}{A_{\mathrm{a}}}-\frac{N_{\mathrm{b}}-Z_{\mathrm{b}}}{A_{\mathrm{b}}}\right)$$

from which U_1 may be determined. If U is determined for many nuclei, then

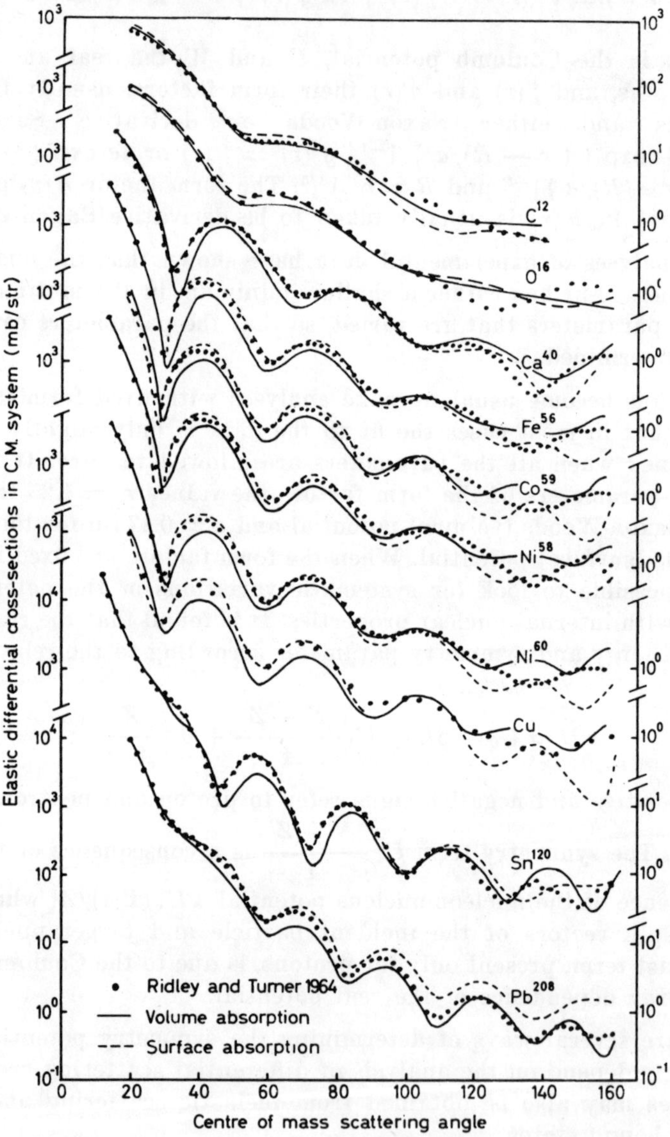

FIG. 1. — Optical model analysis of the elastic scattering of 30.3 MeV protons by a range of nuclei using fixed form factors and either volume or surface absorption (Barrett *et al.*, 1964).

$\left(U - 0.4\,\dfrac{Z}{A^{1/3}}\right)$ can be plotted against the symmetry parameter $(N - Z)/A$, and the slope of the line gives U_1. This has been done by Perey (1963) for proton scattering at several energies from 9 to 22 MeV and by Barrett *et al.* (1964) at 30 MeV. The agreement of the individual determinations with the best straight line through them confirms the overall consistency of the method. Some of the results obtained by Barrett *et al.* are shown in figs. 1 and 2.

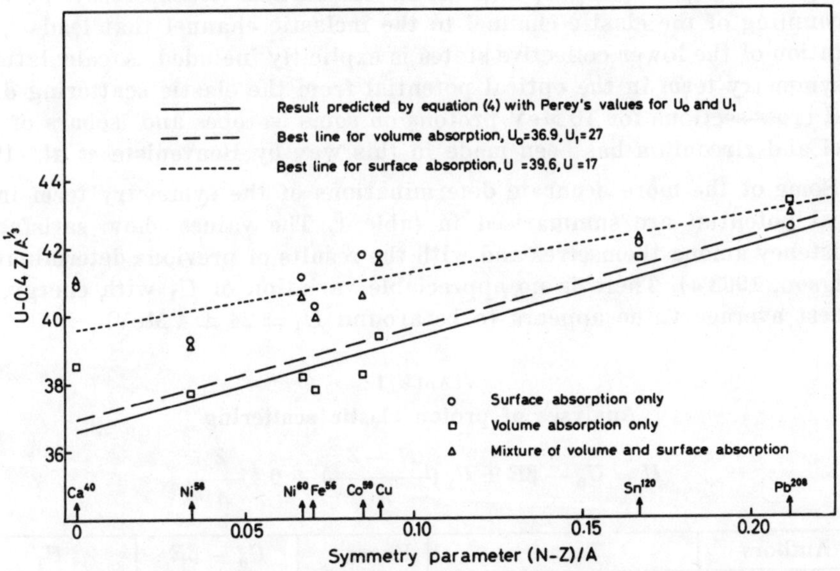

FIG. 2. — Real potential depths obtained in the optical model analysis of fig. 1 corrected for the Coulomb term and plotted as a function of the nuclear symmetry parameter $(N\text{-}Z)/A$. (Barrett *et al.*, 1964).

The main disadvantage of such difference methods is that it can only give the value of U_1 averaged over two or more nuclei and is thus insensitive to any variations of U_1 from one nucleus to another. The only way this can be completely overcome is by using proton and neutron scattering from the same nucleus, which gives:

$$\left(U_{\mathrm{p}} - 0.4\,\frac{Z}{A^{1/3}}\right) - U_{\mathrm{n}} = 2\,U_1\,\frac{N - Z}{A}\,.$$

Some determinations using this method have been made, but at present the accuracy is low as the neutron differential cross-sections are not accurately measured, though to some extent this is offset by the availability of total scattering cross-sections (Hodgson, 1963 *a*, *b*).

A method that has frequently been applied is to use the scattering from neighbouring isotopes or isobars. It is then reasonable to suppose that other features that might give an unwanted dependence on $(N - Z)/A$ are substantially less than when widely different nuclei are used. Such analyses are those of Ball *et al.* (1964) on the isotopes of zirconium, Benveniste *et al.* (1963) and Thomas *et al.* (1964) on isotopes of copper and Benveniste *et al.* (1964) on Ni^{64} and Zn^{64}.

One of the most notable differences that may exist between neighbouring nuclei is their degree of deformation, and this can affect the value obtained for the symmetry potential. It is possible to make allowance for this by making a strong coupling optical model calculation (Buck, 1963), in which the coupling of the elastic channel to the inelastic channel that leads to the excitation of the lower collective states is explicitly included. A calculation of the symmetry term in the optical potential from the elastic scattering differential cross-sections for 10 MeV protons on some isotopes and isobars of iron, nickel and zirconium has been made in this way by Benveniste *et al.* (1964).

Some of the more accurate determinations of the symmetry term in the optical potential are summarised in table 1. The values show satisfactory consistency among themselves and with the results of previous determinations (Hodgson, 1963 *a*). There is no appreciable variation of U_1 with energy, and the best average value appears to be around $U_1 = 24 \pm 4$ MeV.

TABLE 1

Analyses of proton elastic scattering

$$U = U_0 - \beta E \pm U_1 \left(\frac{N-Z}{A}\right) + 0.4 \frac{Z}{A^{1/3}}$$

Authors	Target	Energy	$U_0 - \beta E$	U_1
Benveniste* 64	$Fe^{56,58}$, $Ni^{58,64}$, Zn^{64}	10	45.1	23.1
Ball* 64	$Zr^{90,91,92,94,96}$	22.5	39.8 ± 0.8	28.8 ± 6.0
Perey 63	Range of Nuclei	9 — 22	—	27
Barrett* 64	Range of Nuclei	30	39.6	17
Thomas* 64	Cu^{63} and Cu^{65}	12.29	—	23

* Indicates multiple authorship.

The symmetry term in the optical potential may also be obtained from the cross-section for (p, n) reactions to the isobaric analogue state of the target nucleus. A recent analysis by Satchler *et al.* (1964) of some (p, n) data on medium weight nuclei yields values of $U_1 \sim 22 \pm 3$ MeV for a volume isospin potential and $U_{1\,(\mathrm{surface})} \sim 24 \pm 3$ MeV for a surface isospin potential. The overall fits to the data were markedly better for the surface potential. An analysis of (p, n) reactions at 155 MeV in some light nuclei gives $U_1 \sim 28$ MeV (Valentin, paper 4b(I)/C330).

A novel way of determining the symmetry potential for neutrons has been developed by Olness *et al.* (1964). They measured the polarisations of 1.51 MeV neutrons elastically scattered through 51.5° by a range of nuclei, and found that the results could only be accounted for by an optical potential with a symmetry term with $U_1 \sim 20$ MeV. Their analysis used a square-well potential, however, and should be repeated with a rounded well to give a more accurate result.

In most of this work on the symmetry potential it has been implicitly assumed that the form factor of the isospin term is the same as that of the real potential as a whole, namely the Saxon-Woods form. This is not necessarily the case, and Terasawa and Satchler (1963) have given reasons for expecting the isospin potential to be peaked in the region of the nuclear surface, and this has been confirmed by the analysis of data on (p, n) reactions by Satchler *et al.* referred to above. However, no substantial error is introduced through the use of the Saxon-Woods form to analyse elastic scattering data since to first order a Saxon-Woods potential with a small surface peaked symmetry term is equivalent to a Saxon-Woods potential with a radius that depends on the isospin. Terasawa and Satchler point out that the situation is quite different for the (p, n) reactions to the analogue isobaric state of the target nucleus since then the whole reaction is due only to the isospin potential. The differential cross-sections for these reactions are thus sensitive to the form of the isospin potential, having markedly more structure for the surface than for the volume potential (Drisko *et al.,* 1962). Studies of the (p, n) reactions are thus better than elastic scattering for determining the form of the isospin potential.

All these methods of determining the symmetry term in the optical potential are subject to the uncertainty that other differences between nuclei may give a spurious dependence of potential on symmetry parameter. This possibility may be investigated by studying the elastic scattering of particles with zero isotopic spin, such as alpha-particles, and seeing if any dependence on the symmetry parameter is apparent. Suitable data for this has been obtained by Darriulat *et al.* (1964), who measured the differential cross-section for the elastic scattering of 64.3 MeV alpha-particles from Ni^{58} and Fe^{58}. They chose the ratio of the differential cross-sections as a function of angle as a sensitive experimental quantity and found that this ratio was not well fitted using the optical model parameters obtained by analysing the data for either of the two nuclei separately. Further calculations were made with fixed form factors allowing the absorbing potential to differ between the two nuclei; the fit was then substantially improved but was still not satisfactory. It is not possible to draw any definite conclusions from this work about the existence of other nuclear effects that can give a dependence of the optical potential on the symmetry parameter because the shallower potentials were used. It would be instructive to repeat this analysis using deeper potentials and also to use the extended optical model in which the coupling of the elastic to the inelastic channels is explicitly taken into account.

In all these analyses it is essential to use the same forms of the optical potentials for the various nuclei being studied. If different forms are used, for example those corresponding to the overall optimum fits to the data, entirely spurious results may be obtained for the magnitude and even for the sign of the symmetry potential. This is due to the well-known ambiguities among the parameters of the optical potential. The choice of fixed form factors is thus an essential preliminary to any determination of the symmetry potential.

The analysis of Perey for proton energies from 9 to 22 MeV, and that of Barrett *et al.* (1964) at 30 MeV used form factors with the same radius for the real and imaginary parts of the potential. An analysis at 40 MeV by Fricke (1964) has however shown that it is not possible to obtain satisfactory fits with equal radius parameters at this energy, and that it is necessary to take the radius of the real potential somewhat less than that of the imaginary potential. Additional analyses at higher energies are desirable to explore the optimum optical potentials in more detail as a function of energy.

It has been recognised for some time that in principle there is no reason why the nucleon optical potential should not depend on the spin of the target nucleus. Such forces could have the forms $(\mathbf{I}\cdot\boldsymbol{\sigma})\,f(r)$ and $(\mathbf{I}\cdot\mathbf{r})\,(\boldsymbol{\sigma}\cdot\mathbf{r})\,f(r)$, as well as several velocity-dependent forms. The available data provides no evidence for the existence of such forces; for example the polarisations of protons elastically scattered from neighbouring nuclei of markedly different nuclear spins are found to be essentially the same. More direct evidence on such forces is difficult to obtain, as their effects are most prominent when the scattering takes place from aligned nuclei. Such an experiment has recently been completed using oriented Ho^{165} nuclei and measuring the change in the intensity of 350 keV transmitted polarised neutrons when the spins of the neutrons are respectively parallel and anti-parallel to those of the target nuclei (Wagner *et al.,* 1964). The measured change in cross-section was 30 ± 85 mb, which provides no evidence for any spin-spin term. If the term is of the form $-V_{ss}(\mathbf{I}\cdot\boldsymbol{\sigma})\,f(r)$ then $-0.06 < V_{ss} < 0.13$ MeV.

A.1.*b*. *Polarisations.* The accuracy of polarisation measurements has improved markedly in recent years, as the extensive series of measurements submitted to this conference testify (Craig *et al.,* paper 4b(I)/C90; Baugh *et al.,* paper 4b(I)/C114; Bercaw *et al.,* paper 4b(I)/C178; Aspelund, paper 4b(I)/C274; Pasechnik *et al.,* paper 4b(I)/C384). Unfortunately few of them have been analysed in any detail, and until this is done it is difficult to draw detailed conclusions.

The work of Perey (1963) on proton scattering in the energy range 9-22 MeV showed that in general the optical potentials fitted to the differential cross-sections give polarisations in good accord with the experimental data. This has been confirmed by analyses by Barrett *et al.* (1964) of the extensive data on 30.3 MeV proton elastic scattering obtained recently at the Rutherford Laboratory. Some of their results are shown in fig. 3; there is good overall agreement with a few systematic discrepancies. It is likely that further calcu-

lations varying the parameters of the spin-orbit potential will enable the fit to be improved in these regions, thus giving more detailed knowledge of the spin-orbit potential.

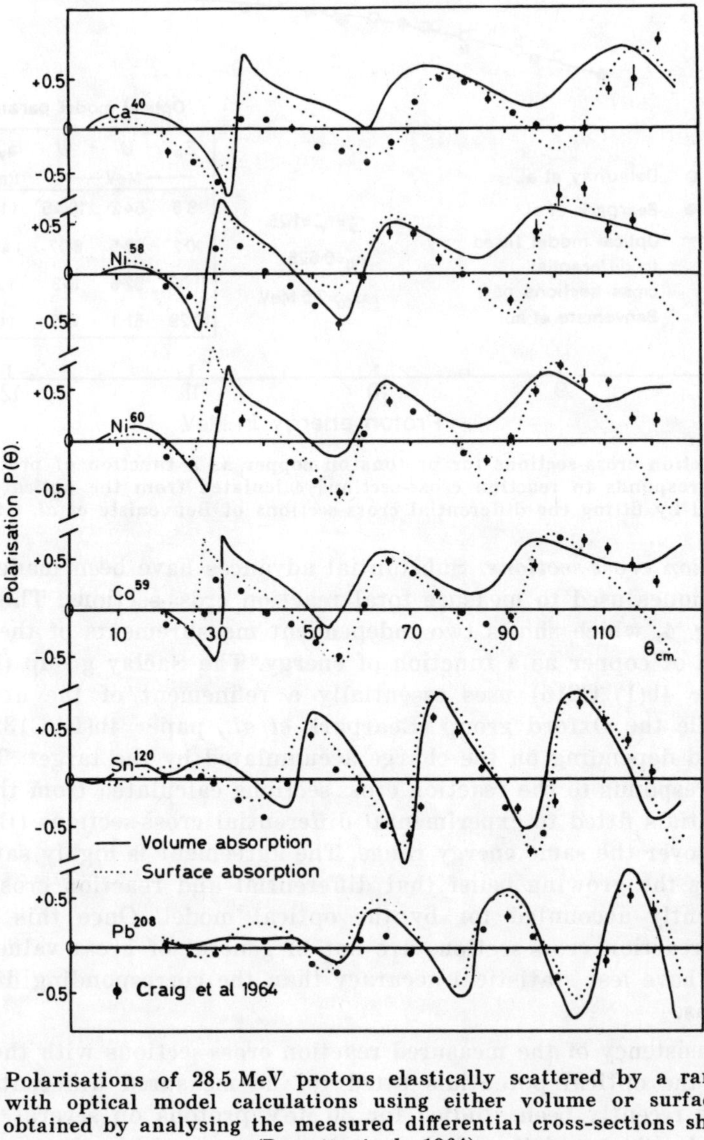

Fig. 3. — Polarisations of 28.5 MeV protons elastically scattered by a range of nuclei compared with optical model calculations using either volume or surface absorption potentials obtained by analysing the measured differential cross-sections shown in fig. 1 (Barrett *et al.*, 1964).

Some recent measurements of polarisation in neutron scattering that may be used to determine the symmetry term in the optical potential have been discussed in the previous section.

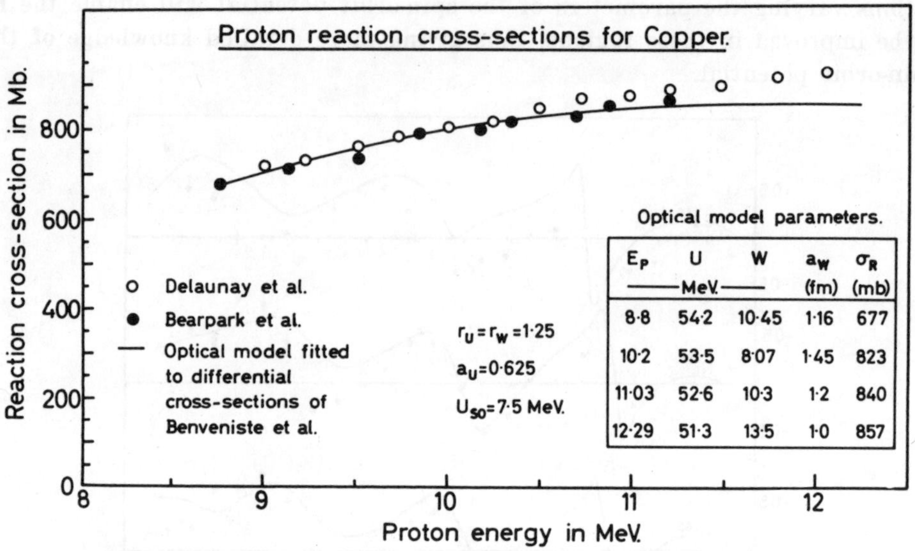

Fig. 4. — Reaction cross-sections for protons on copper as a function of proton energy. The curve corresponds to reaction cross-sections calculated from the optical potentials obtained by fitting the differential cross-sections of Benveniste *et al.* (1961).

A.1.*c*. *Reaction cross-sections.* Substantial advances have been made recently in the techniques used to measure total reaction cross-sections. This is illustrated by fig. 4, which shows two independent measurements of the reaction cross-section of copper as a function of energy. The Saclay group (Delaunay *et al.,* paper 4b(I)/C216) uses essentially a refinement of the attenuation method, while the Oxford group (Bearpark *et al.,* paper 4b(I)/C132) uses a novel method depending on the charge accumulated by the target. The curve in fig. 4 corresponds to the reaction cross-sections calculated from the optical model potentials fitted to experimental differential cross-sections (Benveniste *et al.,* 1961) over the same energy range. The agreement is highly satisfactory and confirms the growing belief that differential and reaction cross-sections are consistently accounted for by the optical model. Once this has been established reaction cross-sections are not in general of great value, as they will always have less statistical accuracy than the corresponding differential cross-sections.

The consistency of the measured reaction cross-sections with those calculated from the optical potentials fitted to the measured differential cross-sections has recently been studied for 30 MeV protons on several nuclei by Barrett *et al.* (1964) and they find overall agreement within the experimental uncertainties. Detailed analyses for 30 MeV protons on copper and gold have been made by Melkanoff *et al.* (paper 4b(I)/C147). They calculated the optimum fit to the differential cross-section as a function of reaction cross-section for various choices of radius parameters. In the case of lead the optimum fit to the differential cross-section gives a reaction cross-section consistent with the experimental value, but in the case of copper the calculated value is about 150 mb higher than the measured one. However, the reaction cross-sections of Waddell *et al.* used in this analysis are consistently about 100 mb lower than those measured by Turner *et al.*, and this may be the reason for the discrepancy.

The most useful fields for further reaction cross-section measurements would now seem to be firstly measurements for composite incident particles, which are still inadequate in number and accuracy (see section 2 c), secondly the general surveys that can bring to light anomalies and thirdly the detailed investigation of these anomalies when they occur.

The most noteworthy anomaly among the reaction cross-sections is the markedly lower value for nickel compared with neighbouring nuclei. Some of the existing data is shown in figs. 5 and 6, and it is apparent that the effect

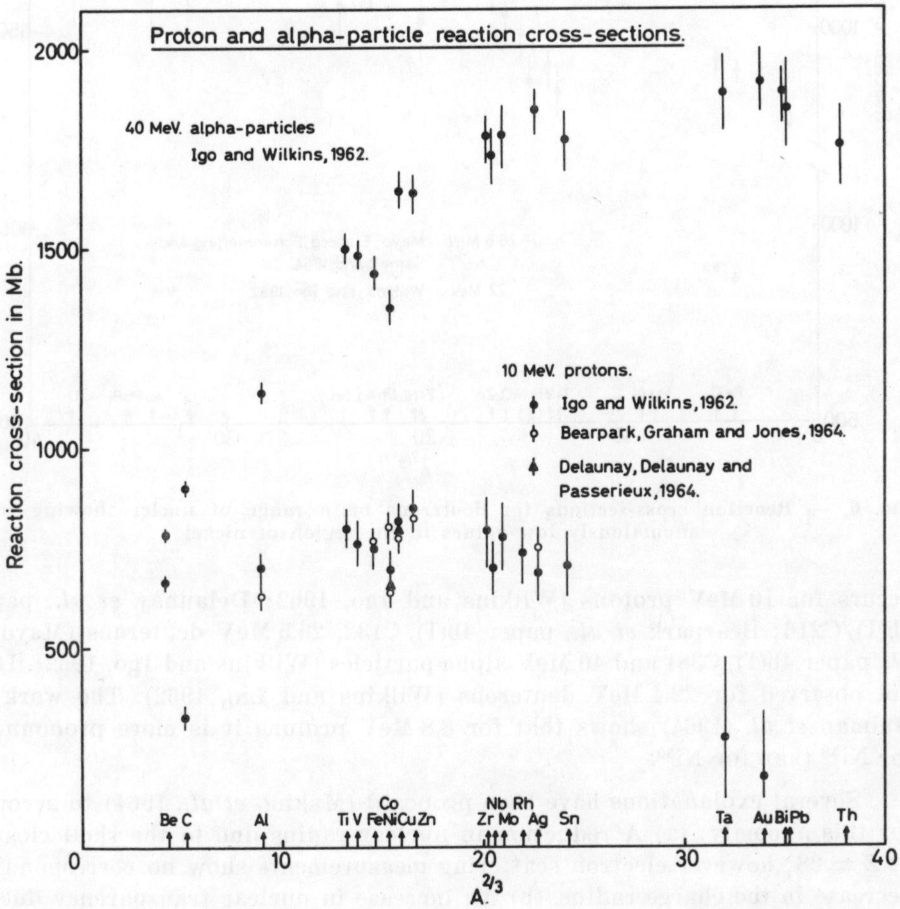

FIG. 5. — Reaction cross-sections for 40 MeV alpha-particles and 10 MeV protons on a range of nuclei showing the anomalously low values in the region of nickel.

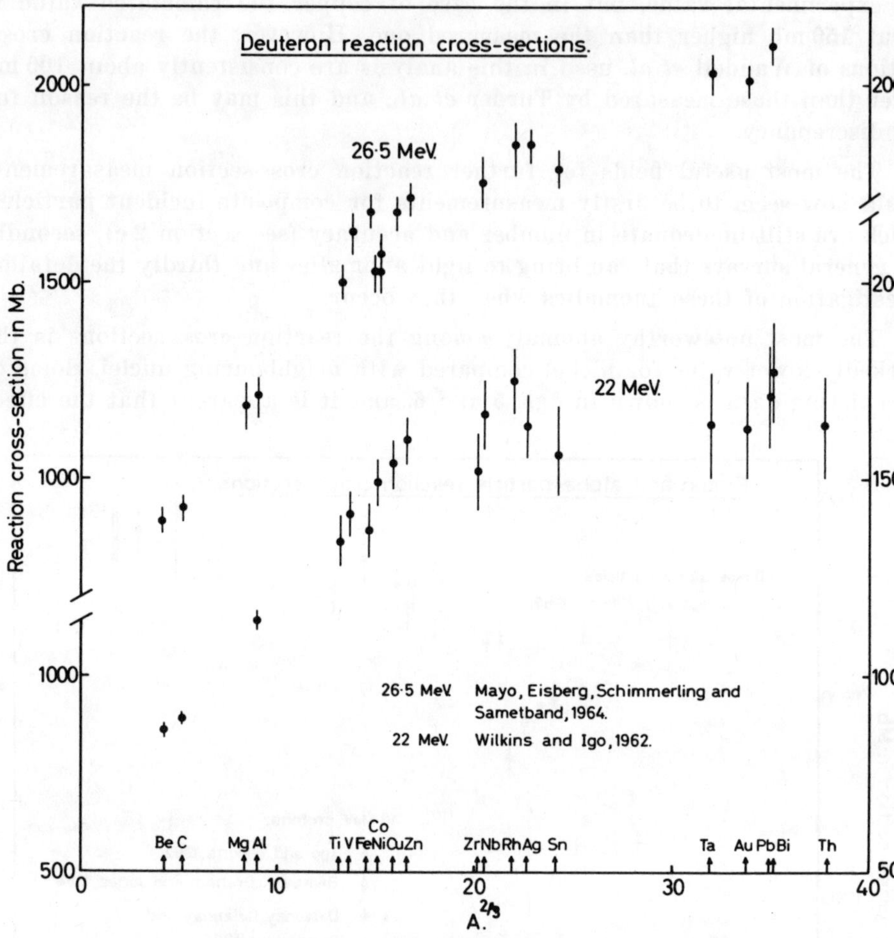

FIG. 6. — **Reaction cross-sections for deuterons on a range of nuclei showing some anomalously low values in the region of nickel.**

occurs for 10 MeV protons (Wilkins and Igo, 1962; Delaunay *et al.*, paper 4b(I)/C216; Bearpark *et al.*, paper 4b(I)/C132, 26.5 MeV deuterons (Mayo *et al.*, paper 4b(I)/C38) and 40 MeV alpha-particles (Wilkins and Igo, 1962). It is not observed for 22.4 MeV deuterons (Wilkins and Igo, 1962). The work of Bulman *et al.* (1964) shows that for 8.8 MeV protons it is more pronounced for Ni^{58} than for Ni^{60}.

Several explanations have been proposed (Makino *et al.*, 1964) to account for this anomaly: (a) A reduction in nuclear radius due to the shell closure at $Z = 28$; however electron scattering measurements show no corresponding decrease in the charge radius. (b) An increase in nuclear transparency due to a reduction in the number of accessible final states due to level density and Q-value effects. Such effects would be progressively less important at higher energies but this is not the case for the observed cross-sections. (c) The effect of the isospin term in the optical potential; however Perey (1964) has shown that this can account for a decrease of about 0.1 per cent compared with the observed decrease of about 10 per cent. (d) A variation in the absorbing potential across the region of shell closure. This can certainly cause systematic changes in the differential cross-section (see section 2 a) and may be due to changes in the nuclear deformation and surface diffuseness. Some calculations by Forest (1963) for deuteron scattering by nuclei around the $Z = 50$ shell show a decrease in the absorbing potential and a corresponding reduction in the reaction cross-section. An analysis by Wilkins (1963) of differential cross-sections for 9.85 MeV protons on nickel and 10.2 MeV protons on Cu^{63} with variable form factors gave reaction cross-sections of 667 and 853 mb respectively, in good accord with the corresponding measured values of 680 ± 30 and 840 ± 30 mb. The Gaussian form factor of the absorbing potential for nickel ($W = 16.1$ MeV, $b = 0.57$ fm) was appreciably more sharply-peaked than for copper ($W = 9.9$ MeV, $b = 0.93$ fm), in agreement with what one might expect in the region of shell closure. It is thus likely that anomalous reaction cross-sections can give useful information on the variation of the absorbing potential from nucleus to nucleus.

The potentials found by Tjin a Djie and Brockman (paper 4b(I)/C235) from the experimental differential cross-sections for the elastic scattering of deuterons by a range of nuclei give reaction cross-sections that show the observed reduction for nickel (see table 5). This suggests that the anomalous reaction cross-sections can be automatically accounted for by the optical model.

TABLE 2

Strong coupling analyses of proton scattering (Buck, 1963)

E	Nucleus	A	σ_R (mb)	β	$\sigma_{IN}^{(2+)}$
14.3	Ti	47.9	970	0.26	35
14.3	Cr	50.01	983	0.19	19
14.1	Fe	55.85	980	0.24	34
14.3	Ni	58.69	954	0.23	41
14.3	Zn	65.38	1001	0.23	36

The possibility that the anomaly is due to the variation in nuclear properties across the region of shell closure can be explored in more detail by calculations that take into account the effect of nuclear deformation by treating explicitly the coupling of the elastic and inelastic channels (Buck, 1963). Some of his results for proton scattering are shown in table 2, and it is notable that although the deformation β for nickel is not greater than that for neighbouring nuclei, the total cross-section for the excitation of the lowest $2+$ state ($\sigma_{IN}^{(2+)}$) is appreciably greater, while the reaction cross-section σ_R is appreciably lower. This suggests that the lower value observed for nickel may be explicable in this way, though the situation is likely to remain unclear until measurements on separated isotopes have been analysed.

In the case of alpha-particle scattering, measurements of differential cross-sections on separated isotopes are already available, and a strong coupling analysis by Buck (1964) gives reaction cross-sections for Ni^{58} that are somewhat lower than for neighbouring nuclei, as shown in table 3, allowance being made for the overall variation with mass number. The difference of about 40 mb is however substantially lower than the observed reduction of about 150 mb for natural nickel, so this may only be part of the explanation.

To study this anomaly in more detail, it would be useful to measure total reaction cross-sections for several isotopes of nickel and of neighbouring nuclei, together with the corresponding differential cross-sections, so that detailed optical model analyses can be made.

TABLE 3
Strong coupling analyses of alpha-particle scattering (Buck, 1964)

Data	E	Nucleus	σ_R (mb)	β	$\sigma_{IN}(2+)$
Saclay	44	Fe^{54}	1522	0.11	8.3
	44	Fe^{56}	1584	0.18	22.1
	44	Ni^{58}	1558	0.18	23.3
Argonne	43	Ni^{58}	1544	0.16	19.6
	43	Ni^{60}	1571	0.17	21.2
	43	Ni^{62}	1606	0.17	20.4
	43	Zn^{64}	1614	0.18	21.4
	43	Zn^{66}	1626	0.18	22.5
	43	Zn^{68}	1659	0.14	13.4
Seattle	41	Fe^{56}	1563	0.17	21.6
	41	Ni^{58}	1537	0.16	18.6
	41	Zn^{64}	1613	0.19	24.3

For several years there has been discussion about the form of the imaginary part of the optical potential, and it has frequently been suggested that measurements of reaction cross-sections should be useful in distinguishing between the volume and surface forms. Recent optical model analyses, however, lend considerable support to the view that if an optical potential gives a good fit to the differential cross-section then the reaction cross-section calculated from this potential, whatever the form of its imaginary part, is in agreement with the measured value. For example the analyses of proton elastic scattering by Perey (1963) gave excellent fits with both volume and with surface absorbing potentials, and the corresponding reaction cross-sections agreed within three or four per cent. Thus the reaction cross-sections cannot be used to decide the form of the imaginary potential, but once the form is fixed on other grounds they can help us to study the variations in the parameters of the imaginary potential from nucleus to nucleus, though this is essentially done by analysing the differential cross-sections.

Apart from nuclear matter calculations, the only praticable ways of finding the radial dependence of the imaginary part of the optical potential are those based on low energy neutron phenomena (Moldauer, paper 4b(I)/C68) and on direct reactions that depend on few partial waves (Lane and Stamp, 1964).

The difficulty of determining the form of the imaginary part of the optical potential from purely phenomenological analyses makes further calculations from the theory of nuclear matter particularly desirable. Most of the existing calculations were made several years ago and use semi-classical methods. Modern techniques should enable more accurate calculations to be made, and these would be particularly useful in analysing the new data in the medium energy region.

A.2. Elastic scattering of composite particles.

A.2.*a*. *Differential cross-sections.*

Due to the strong absorption of composite particles in nuclei the interaction is sensitive only to the surface region, and this greatly increases the ambiguities of the optical potentials. It is usual to retain the Saxon-Woods form factor for the real part of the potential, but in most cases it is equally easy to obtain a fit either with a volume or with a surface absorbing potential, providing they are the same in the surface region.

A more serious ambiguity occurs in the depth of the real potential. If the real potential is steadily deepened and for each depth all the other optical parameters are ajusted to optimise the fit to a given set of experimental data, it is found that the measure of goodness-of-fit passes through a whole series of minima. This is illustrated in fig. 7 for the elastic scattering of 43 MeV alpha-particles by nickel, which shows that depths of 20, 51 and 81 MeV all give acceptable fits to the data. These potentials all give essentially the same phase-shifts, and so cannot be separated by further measurements. Physically, these potentials are such that each one in the series has just one half-wave more of the radial wave function than the preceding potential. Surprisingly enough, this holds quite accurately for all the partial waves contributing to a particular interaction so that all the asymptotic wave functions, and hence the phase shifts, are essentially the same. This is illustrated by fig. 8 for the case of 11.8 MeV deuterons on copper.

In the case of nucleon scattering this type of potential anomaly is easily resolved by the requirement of continuity with the bound state potentials, which are known to be around 50 MeV. This direct method is not applicable to the potentials for composite particles, but it is important to resolve this ambiguity since most analyses of composite particle scattering are made in

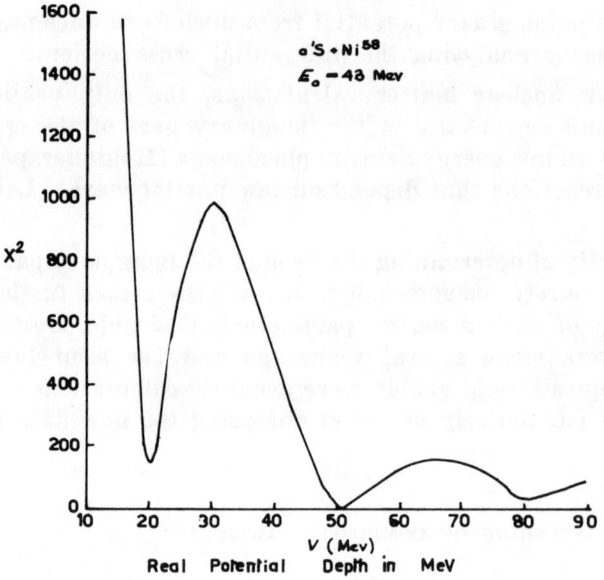

FIG. 7. — Goodness-of-fit parameter Δ for an optical model analysis of the differential cross-section for the elastic scattering of 43 MeV alpha-particles by Ni^{58} plotted against the depth of the real potential (Drisko *et al.*, 1963).

order to obtain potentials that can be used to generate distorted waves for nuclear reaction analyses. Although the asymptotic wave-functions for all the potentials of different depths are the same, so that they give the same elastic scattering, the wave-functions differ markedly in the nuclear interior, and hence give different results in nuclear reaction calculations, which depend on this region. This does in principle open up a way of resolving the ambiguity by using nuclear reaction calculations, but at present the reactions are not sufficiently understood for this to be reliable, so that another way of resolving the ambiguity has to be found.

The most promising way of tackling this problem is to calculate the composite particle potential from the known proton and neutron optical potentials; although this cannot be done very accurately, it should be sufficiently reliable to resolve this ambiguity. Such calculations have been made by Watanabe (1958) who showed that with suitable approximations the deuteron potential may be written as an overlap integral

$$V_{\mathrm{d}}(r) = \int \varphi_{\mathrm{d}}^2(\mathbf{x}) \left[V_{\mathrm{n}}\left(\mathbf{r} + \frac{1}{2}\mathbf{x}\right) + V_{\mathrm{p}}\left(\mathbf{r} - \frac{1}{2}\mathbf{x}\right) \right] \mathrm{d}\mathbf{x} \qquad (1)$$

where $\varphi_{\mathrm{d}}(\mathbf{x})$ is the deuteron wave function. Since the real part of V_{n} and V_{p}, the neutron and proton optical potentials, are constant in the nuclear interior, we can see without detailed calculation that $U_{\mathrm{d}} \simeq U_{\mathrm{n}} + U_{\mathrm{p}}$. Thus the deuteron potential that is in the region of 100 MeV is identified as the physical one. Watanabe's calculations have been repeated by Rook (1965) using accu-

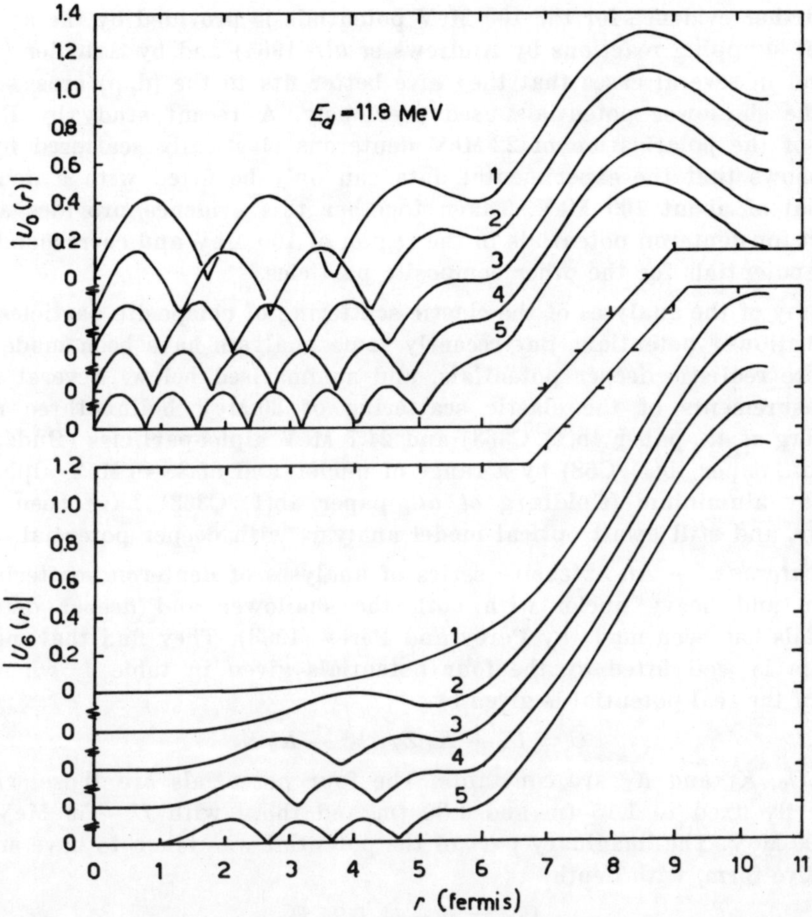

Fig. 8. — Moduli of the $l = 0$ and $l = 6$ partial waves as function of radius for the real central potentials of different depths that fit the data on the elastic scattering of 11.8 MeV protons by copper (Perey and Perey, 1963).

rate neutron and proton potentials, and he finds that the real part of the deuteron potential $U_d = 94$ MeV, and that its effective radius is only slightly larger than that for nucleons. The surface diffuseness is somewhat larger due to the spatial extent of the deuteron.

Very similar results are expected for other composite particles, and Rook has conjectured that the real potential depth should be about 50 A_p MeV, where A_p is the mass of the projectile. He studied a number of cases of deuteron, triton and alpha-particle scattering, and in all of them was able to find a potential of this type that gave an acceptable fit to the data. When applied to the reaction Ca^{40} (t, p) Ca^{42} he found that the fit obtained with triton optical potential $U = 153$ MeV is markedly better than that obtained previously with $U \sim 40$ MeV.

Further evidence for the 100 MeV potentials is provided by the analyses of (d, p) stripping reactions by Andrews *et al.* (1964) and by Satchler (1964), who find in several cases that they give better fits to the (d, p) cross-section than the shallower potentials used previously. A recent study by Raynal (1963) of the polarisation of 22 MeV deuterons elastically scattered by calcium shows that the experimental data can only be fitted with a deuteron potential of about 100 MeV. Taken together this evidence provides strong support for deuteron potentials in the region of 100 MeV and correspondingly deeper potentials for the other composite particles.

Many of the analyses of the elastic scattering of composite particles have used shallower potentials, but recently some analyses have been made with the more realistic deeper potentials, and summarised below. Several series of measurements of the elastic scattering of 30 MeV helium-three nuclei (Goldberg *et al.,* paper 4b(I)/C363) and 24.7 MeV alpha-particles (Budzanowski *et al.,* paper 4b(I)/C58) by a range of nuclei, and of 38-40 MeV alpha-particles by aluminium (Goldberg *et al.,* paper 4b(I)/C363) have been made recently, and still await optical model analysis with deeper potential.

Deuterons. — An extensive series of analyses of deuteron scattering by medium and heavy nuclei with both the shallower and deeper deuteron potentials has been made by Perey and Perey (1963). They find that most of the data is well fitted by the four potentials given in table 4, where the depth of the real potential is given by

$$U = U_0 + K_1 Z/A^{1/3} - K_2 E$$

where U_0, K_1 and K_2 are constants. The four potentials are those with r_u arbitrarily fixed to 1.15 fm and 1.30 fm and those with $U \sim 50$ MeV and $U \sim 100$ MeV. The imaginary part of the potential was taken to have surface derivative form, with depth

$$W_D = W_{OD} + 0.24\,E.$$

TABLE 4

Parameters of the optical potentials obtained in the systematic analysis of Perey and Perey (1963).

U_0	K_1	K_2	r_u	a_u	r_w	a_w
42	3.3	0.15	1.15	0.87	1.37	0.70
81	2.0	0.22	1.15	0.81	1.34	0.68
37	2.4	0.51	1.3	0.79	1.37	0.67
75	1.14	0.42	1.3	0.73	1.34	0.65

These potentials were obtained by analysing deuteron scattering by nuclei from Ca to Pb in the energy range 15-28 MeV.

It is notable that in this work the radius of the real part of the optical potentials for the better fits ($r_u = 1.15$ fm) is less than for nucleon scattering, contrary to the expected increase due to the larger size of the deuteron. This is offset to some extent by the higher diffuseness parameter. It is found that the quality of the fits is somewhat insensitive to the value of the radius parameter provided the remaining parameters are re-adjusted; this is similar to the Ur_0^n ambiguity in nucleon scattering. Quite good fits are still obtained with $r_0 = 1.3$ fm, though they are definitely inferior to those with $r_0 = 1.15$ fm. As r_0 is increased the dependence of U on $Z/A^{1/3}$ becomes less marked, finally becoming negligible around 1.6 fm where the fits are unacceptable.

The cross-sections calculated from the above set of parameters differ in detail from the measured cross-sections much more than the corresponding nucleon case. This is due to the greater sensitivity of the deuteron elastic scattering to differences between nuclei, and hence a careful analysis of the experimental data can give more information on nuclear structure.

Tjin a Djie and Brockman (paper 4b(I)/C235) have recently measured the elastic scattering of 28 MeV deuterons by a range of nuclei, and analysed their results with the deeper potentials. As they used variable form factors their optical potentials (shown in table 5) are not exactly comparable with those of Perey and Perey, but they are certainly very similar in their overall features. It is notable that the calculated reaction cross-sections for nickel are markedly below those for neighbouring nuclei, which is just the effect observed experimentally by Mayo *et al.* (paper 4b(I)/C38). As mentioned in section A.1.*c.* this is probably associated with the shell closure at $Z = 28$, and the similar effect for cerium with the shell closure at $Z = 50$.

TABLE 5

Optical model analyses of 26 MeV deuteron elastic scattering (Tjin A Djie and Brockman)

Form factors of real potentials are Saxon-Woods, and of imaginary potentials Saxon-Woods derivatives.

Nucleus	E	U	r_u	a_u	W	r_w	a_w	σ_R
C	25.9	85.3	1.11	0.71	38.1	1.37	0.671	890
Mg	25.9	81.3	1.10	0.77	45.4	1.16	0.965	1420
Fe	25.9	72.6	0.91	1.02	35.5	1.38	0.88	1840
		103.9	1.06	0.82	44.0	1.23	0.96	1850
Ni	25.8	58.3	1.04	1.01	63.2	1.41	0.64	1670
		93.6	1.12	0.85	78.7	1.38	0.62	1680
Ni^{58}	25.9	52.6	1.13	0.85	54.6	1.38	0.68	1620
Zn	25.9	71.5	0.96	0.99	56.1	1.35	0.75	1790
		108.5	1.04	0.826	59.7	1.27	0.81	1820
Zr	25.9	70.15	1.05	0.95	60.8	1.38	0.66	1890
		99.0	1.12	0.82	75.5	1.36	0.64	1890
Ag	25.3	61.3	1.13	0.94	79.0	1.33	0.66	1940
Ce	25.9	43.9	1.08	1.02	42.0	1.39	0.73	1800
		93.6	1.17	0.77	69.7	1.28	0.73	2080
Ta	25.9	41.8	1.16	1.09	66.0	1.36	0.77	2440
Au	25.3	56.6	1.06	1.05	48.9	1.35	0.77	2270

Some calculations of deuteron elastic scattering by light nuclei have been made by Satchler (1964) using a spin-dependent potential, and in many cases the fit is markedly improved.

Two distinct types of anomalies have been observed in the gross features of deuteron scattering and it is interesting to see whether they can be accounted for by the optical model. They concern the vanishing and re-appearance of maxima and minima in the differential cross-sections when these are examined as a function of energy of the incident particle and of the mass number of the scattering nucleus.

If the elastic scattering of deuterons by a range of nuclei at a particular energy is examined it is found that the angular positions of the maxima and minima in the differential cross-sections vary smoothly as a function of mass number and their number increases with increasing nuclear size. This is expected from simple diffraction theory, but when the positions of the maxima and minima are examined closely it is found that some oscillations disappear within the diffraction pattern as the energy is raised at the same time as others come in from 180°. (Freindl *et al.*, 1961; Strzalkowski, 1962; Tjin a Djie *et al.*, 1964). This behaviour is illustrated in fig. 9 for deuterons from 11 to 21.6 MeV scattered by a range of nuclei (Freindl *et al.*, 1963). In this case the mass number of the nucleus at which the oscillations vanish increases with

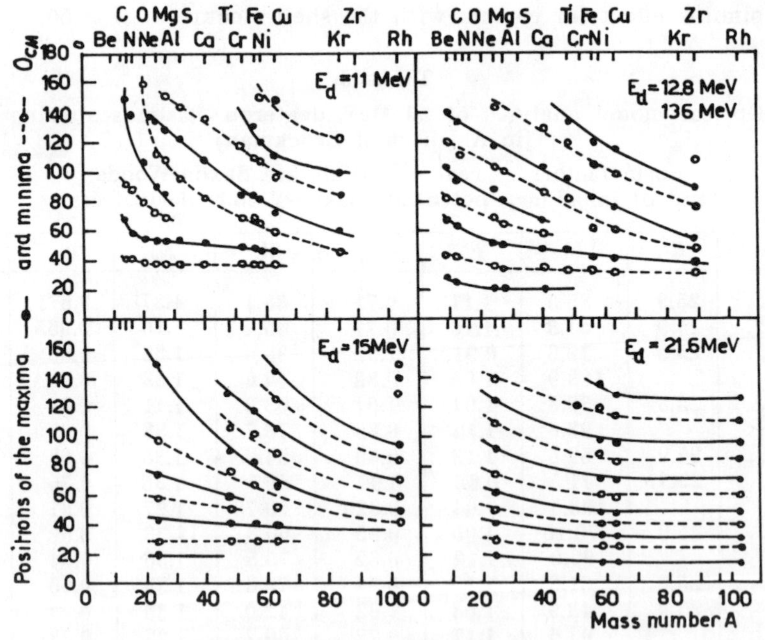

FIG. 9. — The angular positions of the maxima and minima in the differential cross-sections for the elastic scattering of deuterons by a range of nuclei at several energies (Freindl *et al.*, 1963).

increasing energy until the phenomenon no longer occurs. Similar effects are observed if the positions of maxima and minima are plotted against energy for a particular nucleus; this is equivalent to taking another cut in the angle-energy-mass number space through the surfaces giving the positions of the maxima and minima.

Lack of sufficient data makes it difficult to extend this work from the experimental side, but further studies may be made using optical potentials known to be in accord with all the available data. Such a potential is that of Perey and Buck (1962) for elastic neutron scattering. The results of some calculations of the positions of maxima and minima for neutron scattering using this potential are shown diagrammatically for titanium and silicon in

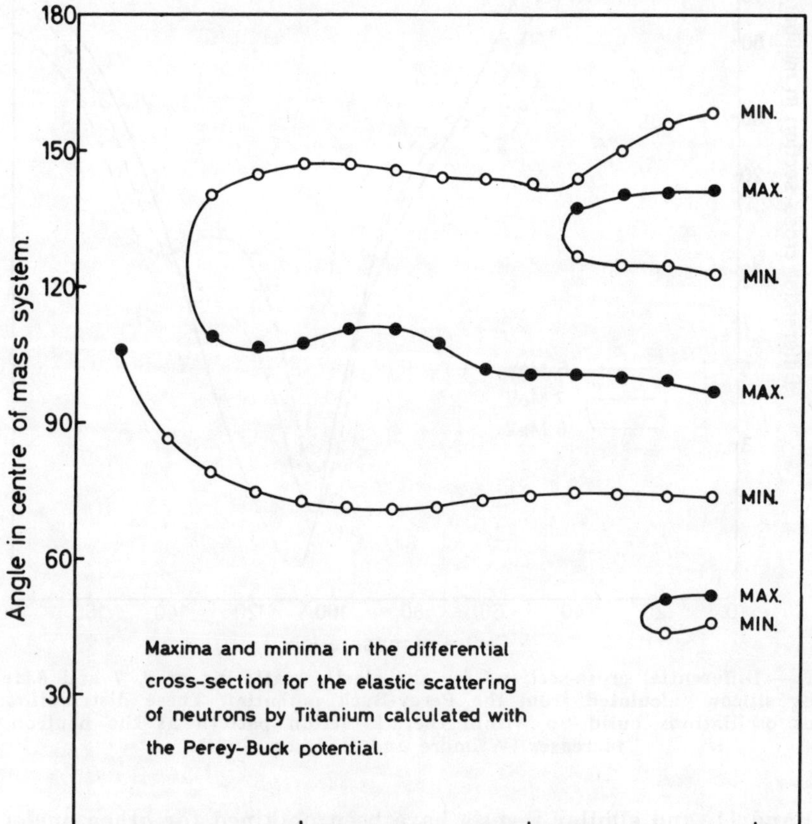

FIG. 10. — The angular positions of the maxima and minima in the differential cross-sections in the elastic scattering of neutrons by titanium as a function of neutron energy. These distributions were calculated from the Perey-Buck potential and show that the phenomenon of the vanishing maxima is given naturally by the optical model (Wilmore and Hodgson, 1964).

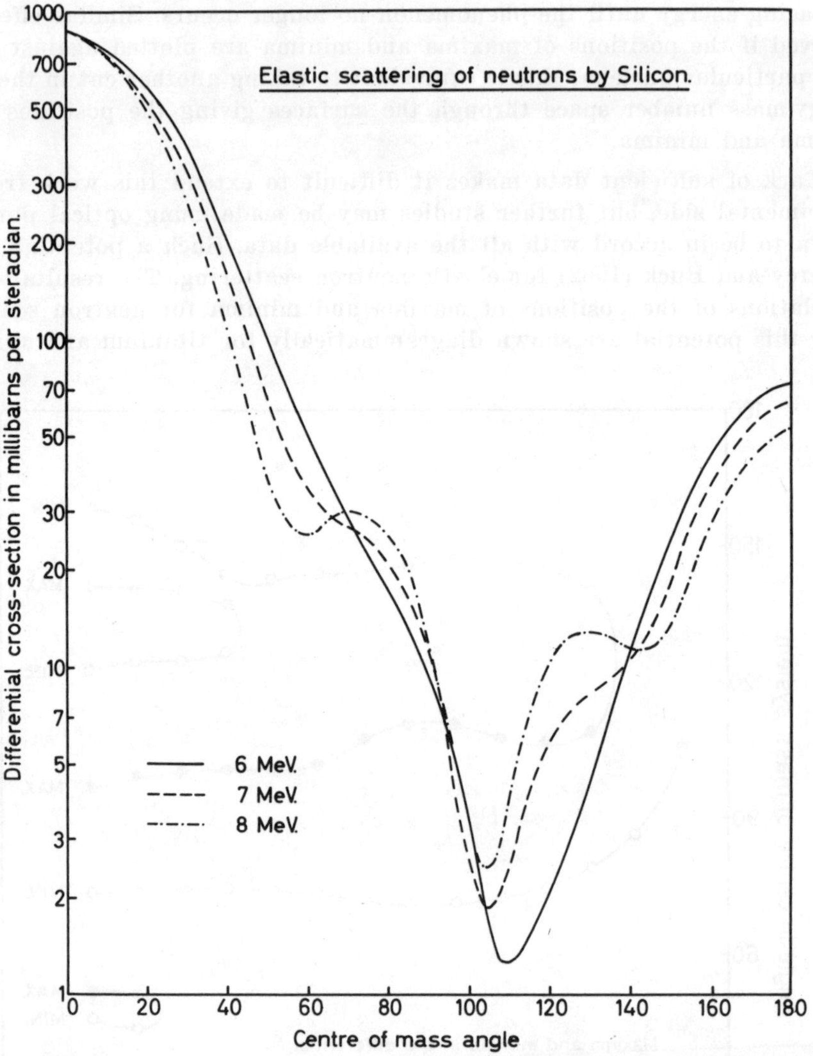

Fig. 11. — Differential cross-sections for the elastic scattering of 6, 7 and 8 MeV neutrons by silicon calculated from the Perey-Buck potential. These distributions show how the oscillations build up within the diffraction pattern as the neutron energy increases (Wilmore and Hodgson, 1964).

figs 10 and 11, and similar results have been obtained for other nuclei. Thus the optical model shows that as we go to higher energies new oscillations build up within the diffraction pattern instead of coming in from 180°. This is just the effect found experimentally for deuteron scattering. Vanishing maxima and minima are thus naturally accounted for by standard optical model calculations.

This effect has also been observed experimentally in the form of the

vanishing of the third maximum in the differential cross-section for the elastic scattering of alpha-particles of about 30 MeV by carbon, nitrogen and oxygen (Mikumo, 1961). Similar effects were observed by Budzanowski *et al.* (paper 4b(I)/C58) in their survey of the elastic scattering of alpha particles by a range of nuclei. The apparent anomaly is thus a general characteristics of elastic scattering when a large number of partial waves contributes significantly to the observed cross-section.

The second anomaly in deuteron scattering appears when the oscillations in the differential cross-section are examined for heavy nuclei when the energy of the incident deuterons is comparable with the heights of the Coulomb barriers of the target nuclei. As expected, it is found in general that the diffraction oscillations at a given energy decrease progressively in amplitude from the light to the heavy nuclei as the Coulomb field comes to dominate the nuclear field. At some energies, however, the oscillations die down gradually and vanish as the mass number increases, but then briefly reappear for even heavier nuclei before finally vanishing again. This effect was first noticed for 11.8 MeV deuterons by Igo, Lorenz and Schmidt-Rohr (1961), who found little evidence for oscillations in nuclei with $Z = 45$, 46 and 47, and for $Z > 50$, but marked oscillations in the intermediate nuclei with $Z = 48$, 49 and 50. A similar effect has been observed for 25 MeV deuterons by Tjin a Djie (1964) and for 17 MeV protons by Schrank and Pollock (1963).

This effect is basically due to shell structure of the nucleus. Around the closed shells the single-particle aspects of nuclei predominate, energy levels are widely spaced and deformations are small. These nuclei are thus harder to excite than their neighbours so that their absorbing potential is low. Since the absorbing potential has the effect of dampening the oscillations in the differential cross-section this lowering of the absorbing potential around a closed shell can lead to a reappearance of the oscillations compared with nearby nuclei. This is particularly likely to happen if the oscillations have just disappeared as the closed shell is reached. The observations of Igo *et al.* are thus explicable as due to the effect of the closed shell at $Z = 50$, those of Tjin a Djie by the sub-shell at $Z = 58$ and those of Schrank and Pollock by the double shell closure at $Z = 82$.

This explanation of the effect has been verified by optical model analyses of the differential cross-sections (Forest, 1963) which showed the expected reduction of the absorbing potential in the region of shell closure. A closer insight into this reduction is given by the coupled-channels analysis of the elastic scattering of 17 MeV protons by Ta, Pt and Bi (Perey, 1964). This showed that the observed differential cross-sections can be well-fitted by constant optical potentials across the region of shell closure, so that the reduction of the absorbing potentials found in the simple optical model analysis is due solely to the variation of the nuclear deformation parameter. This anomaly also is thus fully explicable by the optical model. It is not specific to deuterons but can occur in the elastic scattering of any particles by nuclei.

Tritons. — Very few measurements of triton elastic scattering have been made, and the only published optical model analysis (Pullen *et al.*, 1964) used the shallower potentials. However recent work by Rook (1964) has shown that the data on the elastic scattering of 6.8 MeV tritons by F^{19} and Ca^{40} can be fitted by the following potentials : F^{19} :— $U = 120$ MeV, $W = 44$ MeV, $r_0 = 1{,}33$ fm, $a = 0.7$ fm. Ca^{40} — $U = 153$ MeV, $W = 46$ MeV, $r_0 = 1.3$ fm, $a = 0.65$ fm. Both analyses used the Saxon-Woods form factor for the real and imaginary parts of the potential.

Helium-three nuclei. — Early analyses of the elastic scattering of helium-3 nuclei by a range of nuclei (Hodgson, 1963) used the shallower potentials, and it was found that the real and imaginary potentials were of about the same magnitude and decreased with increasing nuclear size. Goldberg *et al.* (paper 4b(I)/C363) have fitted similar data at 30 MeV, with

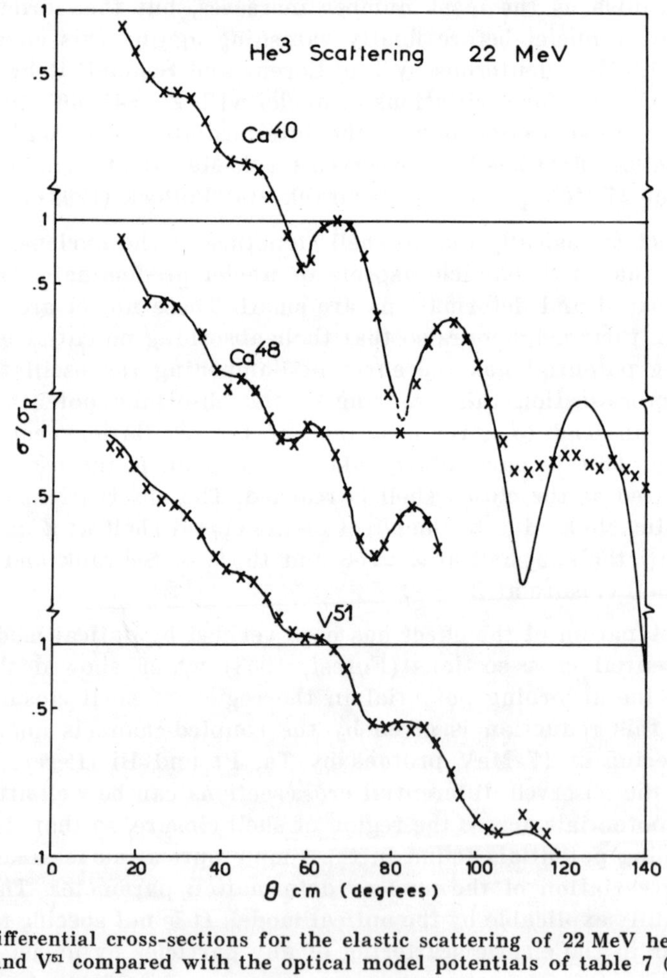

FIG. 12. — Differential cross-sections for the elastic scattering of 22 MeV helium-3 nuclei by Ca^{40}, Ca^{48} and V^{51} compared with the optical model potentials of table 7 (Bassel, 1964).

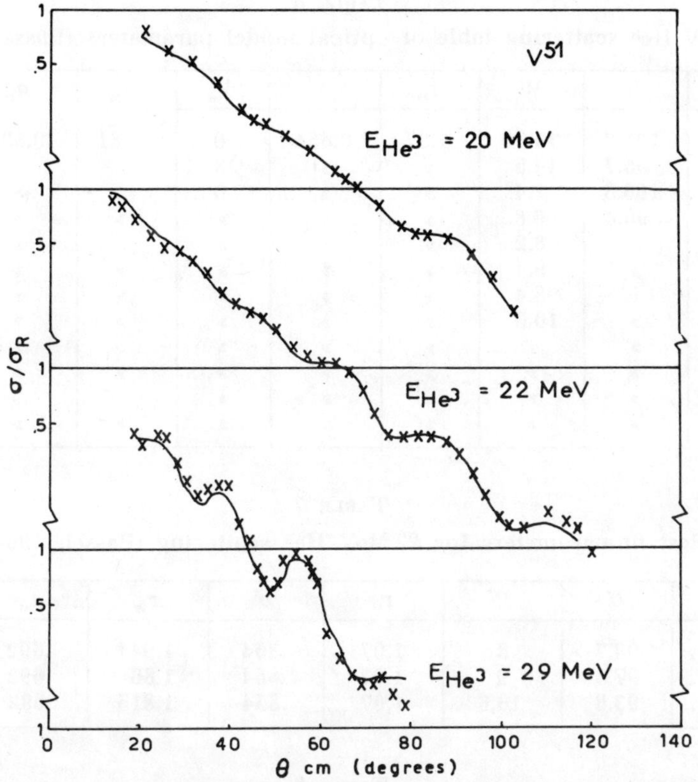

Fig. 13. — Differential cross-sections for the elastic scattering of helium-3 by V^{51} at several energies compared with the optical model potentials of table 8 (Bassel, 1964).

potentials of constant depth, but with radii increasing more slowly than $A^{1/3}$ with increasing nuclear size. These potentials are roughly equivalent due to ambiguities of the Ur_0^n type. Both analyses showed marked sensitivity of the elastic scattering to the surface diffuseness parameter; this is due to the readiness with which helium-three nuclei are absorbed by nuclear matter, so that the interaction is dominated by conditions at the nuclear surface.

An extensive analysis of helium-3 elastic scattering data has recently been carried out by Bassel (1964). Some of his potentials at 12 MeV and 22 MeV are given in tables 6 and 7 and some comparisons with the experimental data in fig. 12. The scattering from vanadium has been analysed at three energies, and the results are given in table 8 and fig. 13. Bassel's potentials have an imaginary part extending appreciably beyond the real part, and the depths of the potentials do not show a large systematic variation with nuclear size. They are thus more convenient for interpolation purposes than those reported previously. Some additional optical model analyses of helium-three, elastic scattering have recently been published by Alford *et al.* (1964), and by Klingensmith *et al.* (1964).

TABLE 6

12 MeV He^3 scattering table of optical model parameters (Bassel, 1964)

Nucleus	U	W	r_0	a	V_s	r_w	a_w	σ_r
Mg^{24}	101.2	14.76	1.07	0.854	0	1.81	0.592	994
Mg^{24}	95.7	14.5	»	»	8	»	»	988
Ca^{40}	106.3	7.4	»	»	0	»	»	790
Sc^{45}	96.4	9.8	»	»	»	»	»	810
Ti^{46}	»	8.2	»	»	»	»	»	750
Ti^{48}	»	9.4	»	»	»	»	»	792
Ti^{50}	»	8.4	»	»	»	»	»	810
V^{51}	»	10.0	»	»	»	»	»	779
Fe^{54}	»	»	»	»	»	»	»	634
Ni^{58}	»	»	»	»	»	»	»	558
Zn^{64}	»	»	»	»	»	»	»	556
Zr^{90}	»	»	»	»	»	»	»	200

TABLE 7

Best fit parameters for 22 MeV He^3 scattering (Bassel, 1964).

Nucleus	U	W	r_0	a	r_w	a_w	σ_R
Ca^{40}	98.7	8	1.07	.854	1.944	.692	1410
Ca^{48}	97.3	9	1.07	.854	1.86	.692	1507
V^{51}	93.8	10.6	1.07	.854	1.815	.692	1450

TABLE 8

Optical model parameters for He^3 scattering from V^{51} (Bassel, 1964).

E	U	W	r_0	a	r_w	a_w	σ_R
20	99.1	9.22	1.07	.854	1.815	.692	1326
22	93.8	10.55	1.07	.854	1.815	.692	1447
29	93.1	9.87	1.07	.854	1.815	.692	1561

It is notable that these absorbing potentials for helium-3 are much less than those found by Rook for tritons, whereas one might have expected them to be similar. Further analysis of the ambiguities in these potentials are needed before this difference can be shown to be real.

Alpha-particles. — Most of the optical model analyses of the elastic scattering of alpha-particles have been made with shallower potentials, though recently Rook (1964) has fitted some data for Mg and Si at 20.4 MeV with the potentials : Mg : — $U = 248$ MeV, $W = 13$ MeV, $r_0 = 1.21$ fm, $a = 0.66$ fm; Si : — $U = 181$ MeV, $W = 15$ MeV, $r_0 = 1.23$ fm, $a = 0.56$ fm. Both analyses used the Saxon-Woods form factor for the real and imaginary parts of the potentials.

The measurements of the Cracow groups (Budzanowski *et al.*, 1964) on the elastic scattering of 24.7 MeV alpha-particles by a range of nuclei are shown in figs 14 and 15. An optical model analysis of this data is in progress.

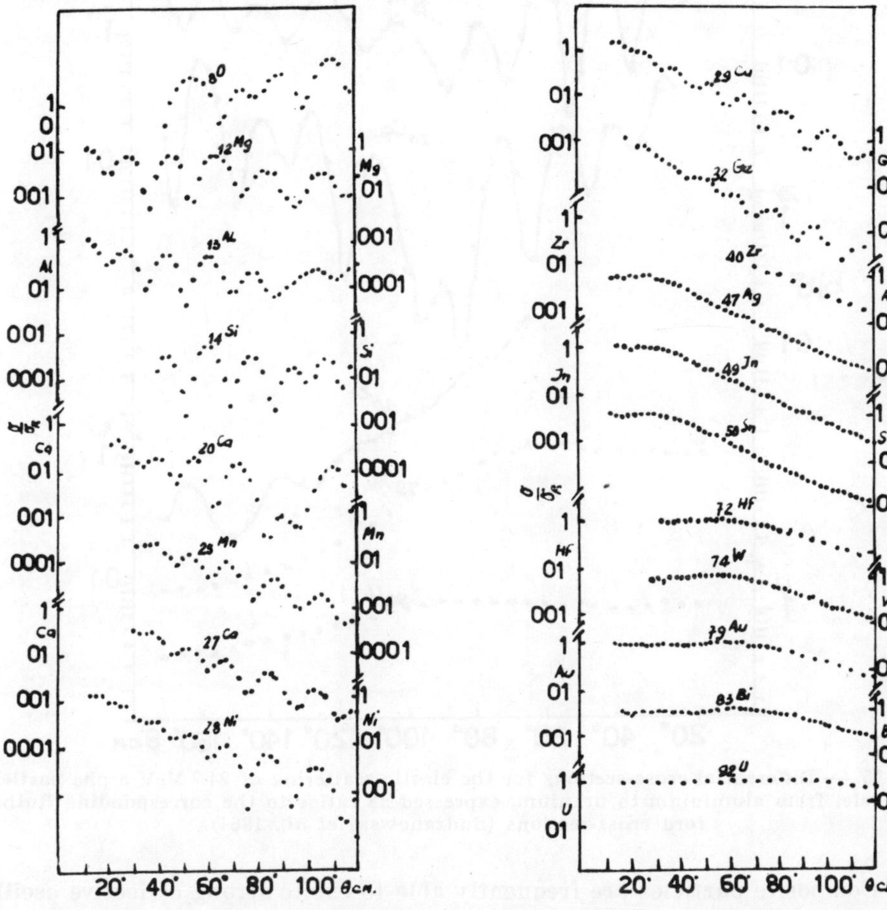

FIG. 14. — Differential cross-sections for the elastic scattering of 24.7 MeV alpha-particles by nuclei from oxygen to uranium, expressed as ratios to the corresponding Rutherford cross-sections (Budzanowski *et al.*, 1964).

Wall and Gruhn (1964) have recently found an exceptional stability of the angular positions of the maxima and minima in the differential cross-sections for the elastic scattering of alpha-particles by calcium in the energy range 31.9 to 41.5 MeV. Their results are shown in fig. 16. Such an effect is given by the optical model and may be understood as a resonance in one of the higher partial waves, with orbital angular momentum about twelve.

A detailed and extensive analysis of elastic alpha-particle scattering has been completed by Buck and Satchler but the details are not yet ready for publication.

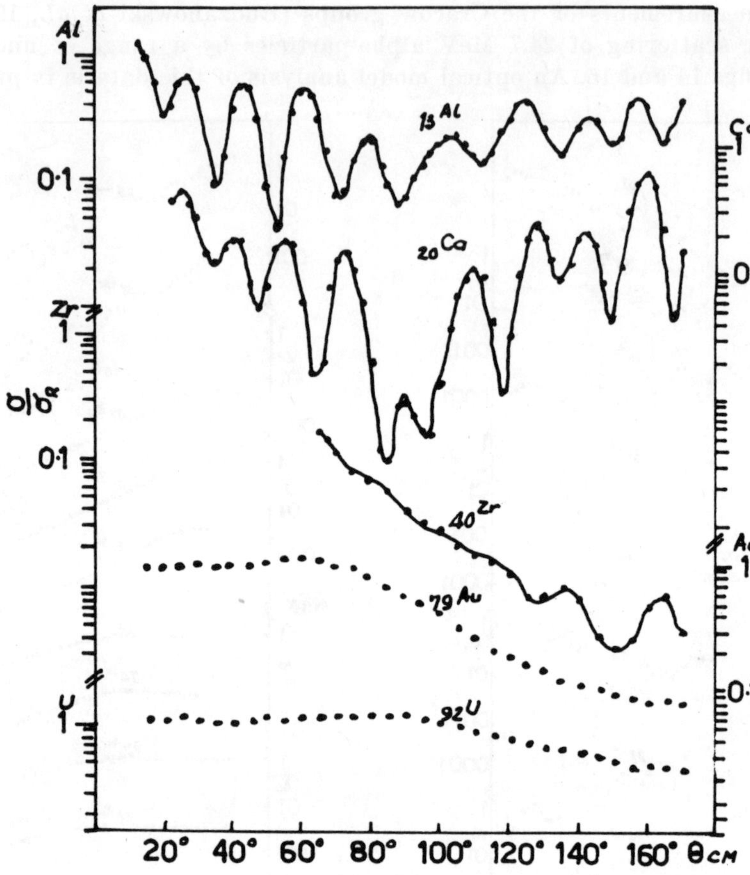

FIG. 15. — Differential cross-sections for the elastic scattering of 24.7 MeV alpha-particles by nuclei from aluminium to uranium, expressed as ratios to the corresponding Rutherford cross-sections (Budzanowski *et al.*, 1964).

Composite particles are frequently able to excite strong collective oscillations of the target nucleus, and this process couples together the elastic and inelastic scattering channels, so that it is no longer sufficiently accurate to use the simple optical model to analyse the elastic scattering data alone. This may account for some of the variations from nucleus to nucleus of the optical potentials obtained in this way. This strong coupling to inelastic scattering channels is discussed in section B.

A.2.*b*. *Polarisation.*

The vector and tensor polarisations of deuterons elastically scattered from nuclei have recently been measured by the Saclay group (Beurtey *et al.*, 1963) and analysed with optical potentials including various tensor and spin-orbit terms by Raynal (1963). His results are shown in fig. 17. He finds that

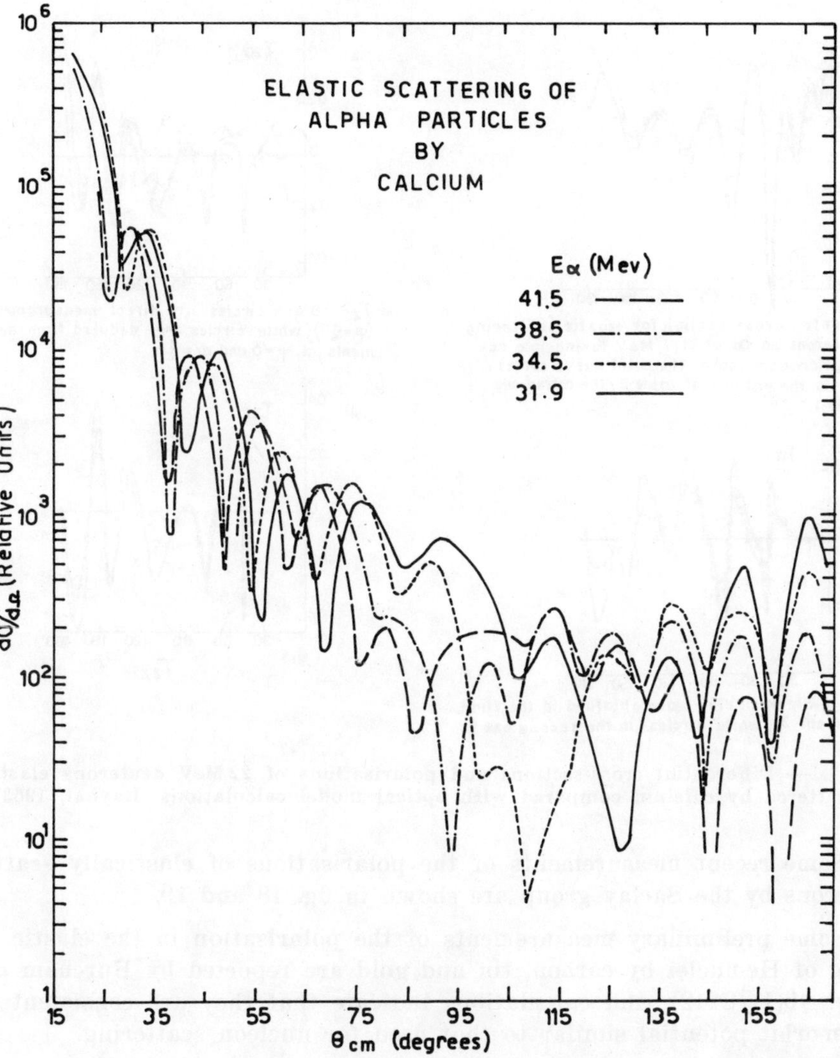

Fig. 16. — Differential cross-sections for the elastic scattering of 31.9-45.5 MeV alpha-particles by calcium, showing the stability of the positions of the maxima and minima at high angles (Wall and Gruhn, 1964).

the spin-orbit potential alone is sufficient to give a good overall agreement with the experimental data for 22 MeV deuterons on calcium, and hence there is no need to introduce the tensor terms in the potential. The best spin-orbit potentials have strengths $U_{so} \sim 13$ MeV·fm², and a derivative Saxon-Woods form factor with $r_0 \sim 0.7$ fm and $a \sim 0.4$ fm, i.e. quite sharply peaked and well within the nuclear surface. It is noteworthy that Raynal could only obtain a satisfactory fit with a real central potential of depth about 100 MeV, thus providing additional evidence for these deeper potentials.

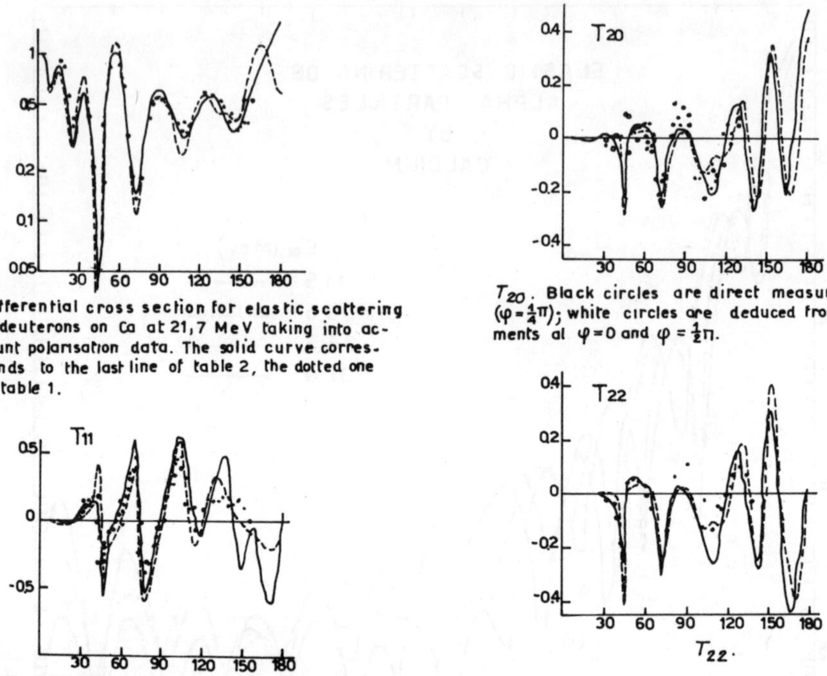

Differential cross section for elastic scattering of deuterons on Ca at 21,7 MeV taking into account polarisation data. The solid curve corresponds to the last line of table 2, the dotted one to table 1.

T_{20}. Black circles are direct measurement ($\varphi = \frac{1}{4}\pi$); white circles are deduced from measurements at $\varphi = 0$ and $\varphi = \frac{1}{2}\pi$.

Im. T_{11}. Black circles ware obtained in the first experiment 3); white circles in the second one 4)

T_{22}.

FIG. 17. — Differential cross-sections and polarisations of 22 MeV deuterons elastically scattered by calcium compared with optical model calculations (Raynal, 1963).

Some recent measurements of the polarisations of elastically scattered deuterons by the Saclay group are shown in fig. 18 and 19.

Some preliminary measurements of the polarisation in the elastic scattering of He-nuclei by carbon, tin and gold are reported by Burcham *et al.* (paper 4b(I)/C212), and calculations indicate that they are consistent with a spin-orbit potential similar to that used for nucleon scattering.

A.2.c. *Reaction cross-sections.*

There have been relatively few comparisons between measured reaction cross-sections for composite particles and those calculated from optical model potentials obtained by analysing the corresponding elastic differential cross-sections. One such comparison showed that the measured values for 22.4 MeV deuterons (Wilkins and Igo, 1962) are systematically about 10 per cent below those calculated from differential cross-sections at 21.6 MeV (Perey and Perey, 1963). A comparison for 26.5 MeV deuterons by Mayo *et al.* (paper 4b(I)/C38) showed better agreement for surface than for volume absorption potentials.

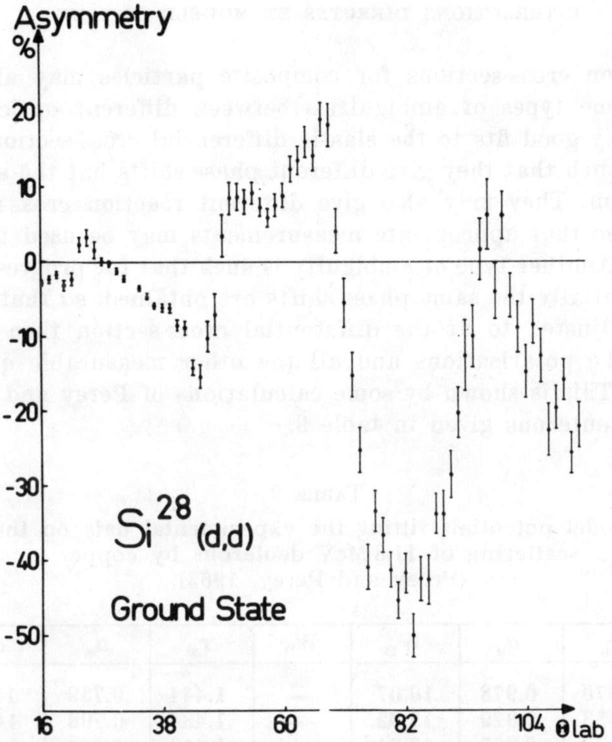

Fig. 18. — Asymmetry in the elastic scattering of 22 MeV polarised deuterons by Si^{28} (Beurtey *et al.*, 1964).

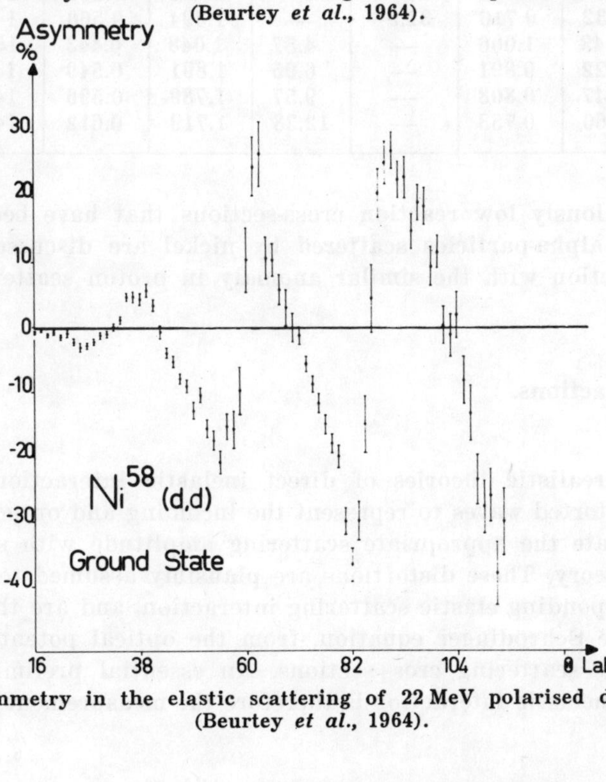

Fig. 19 — Asymmetry in the elastic scattering of 22 MeV polarised deuterons by Ni^{58} (Beurtey *et al.*, 1964).

The reaction cross-sections for composite particles may also be useful in reducing some types of ambiguities between different optical potentials that give equally good fits to the elastic differential cross-section data. Some potentials are such that they give different phase-shifts but the same differential cross-section. They may also give different reaction cross-sections (and polarisations) so that appropriate measurements may be used to distinguish between them. Another type of ambiguity is such that for progressively deeper potentials essentially the same phase-shifts are obtained, so that if the potentials are all adjusted to fit the differential cross-section then the reaction cross-section, the polarisations and all the other measurable quantities are also the same. This is shown by some calculations of Perey and Perey (1963) for 11.8 MeV deuterons given in table 9.

TABLE 9

Optical model potentials fitting the experimental data on the elastic scattering of 11.8 MeV deuterons by copper (Perey and Perey, 1963).

U	r_u	a_u	W_D	W_S	r_w	a_w	σ_R	Δ
36.8	1.070	0.978	10.07	—	1.444	0.739	1440	1.4
58.5	1.153	0.879	13.32	—	1.434	0.708	1476	0.65
90.7	1.172	0.822	18.34	—	1.410	0.661	1468	0.78
128.2	1.193	0.775	24.3	—	1.403	0.608	1460	0.84
167.8	1.232	0.716	31.9	—	1.394	0.560	1451	0.96
36.8	1.043	1.006	—	4.57	1.048	0.443	1502	0.72
61.6	1.122	0.891	—	6.95	1.891	0.549	1487	0.59
95.5	1.147	0.808	—	9.57	1.789	0.596	1477	0.67
136.9	1.160	0.753	—	12.38	1.719	0.612	1472	0.77

The anomalously low reaction cross-sections that have been found for deuterons and alpha-particles scattered by nickel are discussed in section A.1.*c*. in connection with the similar anomaly in proton scattering.

B. Direct interactions.

The more realistic theories of direct inelastic interactions frequently make use of distorted waves to represent the incoming and outgoing particles and then evaluate the appropriate scattering amplitude with some form of perturbation theory. These distortions are plausibly assumed to be the same as in the corresponding elastic scattering interaction, and are therefore generated, using the Schrödinger equation, from the optical potentials obtained from the elastic scattering cross-sections. An essential preliminary to any analysis of an inelastic interaction is therefore the measurement of the appropriate elastic cross-sections and their analysis to obtain the required optical potentials. In this way the calculation of the interaction cross-section is made with only one set of optical potentials, though there may of course be other parameters whose effect requires to be investigated. It is possible in principle to vary the optical model parameters to optimise the fit to the interaction cross-section without reference to the elastic cross-sections, but this is not usually considered to be useful due to the approximations inherent in the reaction theories and the large ambiguities among the parameters when the analysis is made in this way.

A possible exception to this is when it is desired to investigate the consistency of two sets of measurements on the same reaction, for example the differential cross-sections and the polarisations. It may then be useful to vary the optical parameters to optimise the fit to the differential cross-sections and then see if this set also gives the observed polarisations. In this case the variations in the optical potentials take account of the approximations in the reaction theory. Apart from such cases, however, the optical potentials to be used are those obtained from analyses of elastic scattering data.

In some cases it is not possible to obtain the appropriate elastic cross-sections to make this analysis, for example when the target nucleus concerned is unstable, or when it is in an excited state. The optical potentials appropriate to unstable target nuclei can sometimes be sufficiently well estimated from the studies of the variation of the potentials from one nucleus to the next, since the instability itself does not affect the optical potential. In the case of an excited nucleus the optical potential is taken to be that appropriate to the nucleus in its ground state, and the success of many calculations indicate that this is a good approximation. When reaction theory is further advanced it should be possible to explore the validity of this assumption in more detail.

The only exception to these remarks about the use of the potentials obtained from elastic scattering data occurs in the strong coupling analysis of inelastic scattering (see section B.1). This theory is essentially an extension of the same type of analysis as that used for elastic scattering, and so it is appropriate to iterate the parameters of the optical potential to give the best simultaneous fit to the elastic and inelastic cross-sections. The optical potential obtained from the analysis of the elastic scattering alone is useful as a starting potential in this iteration.

B.1. Inelastic scattering.

B.1.*a*. *Distorted wave theory.* The use of the distorted wave theory, using optical potentials obtained from the analysis of the appropriate elastic scattering cross-sections, is now well established as a method of analysing inelastic scattering. It is likely to be satisfactory in cases where the coupling between the elastic and inelastic channels is weak and when the excitation of the residual nuclear state proceeds mainly by a direct and not by a multiple process.

These calculations can be used to determine the spins of the final states by seeing which of the inelastic angular distributions calculated with possible J-values grees best with the experimental data. Such an analysis of the inelastic scattering of 55 MeV protons by Ca^{40} has recently been completed by Yagi *et al.* (1964), and fixes the spins of three excited states. The analysis also provides evidence in favour of a complex coupling interaction and determines the deformation parameters of calcium.

B.1.*b*. *Coupled channel theory*. The main deficiencies of the distorted wave theory are removed by taking into account explicitly the coupling between

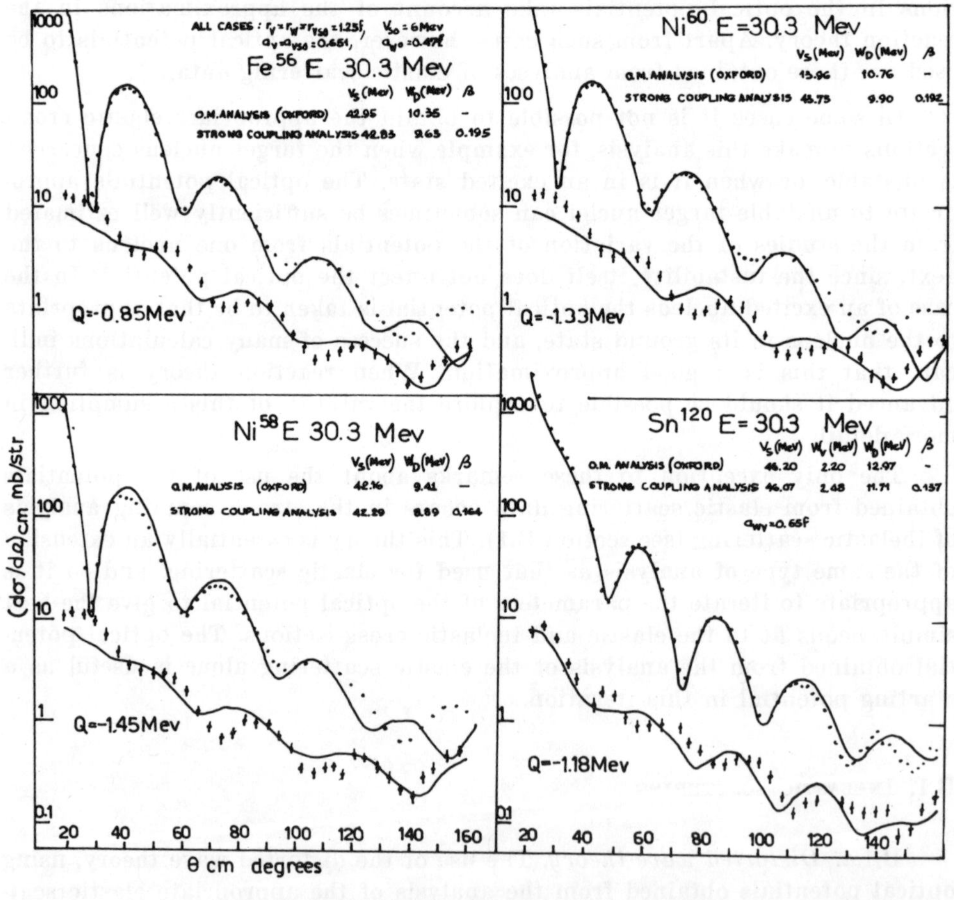

FIG. 20. — Elastic and inelastic cross-sections for 30 MeV protons scattered by several nuclei compared with a coupled channel calculation. The parameters of the optical potentials are given in table 10 (Ridley *et al.*, 1964).

the elastic and inelastic channels (Buck, 1963) using the collective (rotational or vibrational) or single-particle models of the excited states. This theory was first applied to the $0+ \rightarrow 2+$ excitation of the first excited state of an even-even nucleus, and has been used without spin-orbit forces to analyse deuteron and alpha-particle inelastic scattering, and with spin-orbit forces to analyse neutron and proton inelastic scattering. Subsequently the formalism has been extended to the $0+ \rightarrow 2+ \rightarrow 4+$ excitation of rotational nuclei (Buck *et al.*, 1963) and to the $0+ \rightarrow 2+ (0+, 2+, 4+)$ excitation of vibrational nuclei (Dickens *et al.*, 1963). Using the first of these, Buck (1962) was able to account for the apparently anomalous cross-section for the reaction $Ni^{58}(\alpha, \alpha') Ni^{58}$ in terms of a competition between direct and multiple excitation. With the vibrational model formalism it was possible to account for the inelastic scattering of 11 MeV protons by Ni^{62} and Ni^{64} with the excitation of the four lowest vibrational states. In each case the appropriate nuclear deformation parameters are determined.

A preliminary analysis of some recent data on the elastic and inelastic scattering of 30 MeV protons by several nuclei is shown in fig. 20 and the corresponding potentials and values of the deformation parameter are given in table 10. This shows that for these cases, the optical potentials are very little affected by the inclusion of the inelastic channel and also that the values of the deformation parameter β obtained from this analysis agree well with those obtained by other authors.

TABLE 10

Strong coupling analyses of 30 MeV proton elastic and inelastic scattering. (Perey form factors and $U_{so} = 12$ MeV).

Nucleus	Model	U	W_V	W_D	β (2+)	β (2+)
Fe^{56}	Optical model	42.95	—	10.35		0.233 3)
	Strong coupling (vib) ...	42.83	—	9.63	0.195	0.24 4)
	Strong coupling (rot) ...	42.86	—	9.63	0.194	0.21 5)
Ni^{58}	Optical model	42.16	—	9.41		0.19 1)
	Strong coupling (vib) ...	42.29	—	8.89	0.164	0.225 3)
Ni^{60}	Optical model	43.96		10.76	—	0.204 1)
	Strong coupling (vib) ...	43.75		9.90	0.192	±9%
Sn^{120}	Optical model	46.20	2.20	12.97		0.112 2)
	Strong coupling (vib) ...	46.06	2.48	11.71	0.137	±10%

The calculations (vib) and (rot) refer to the vibrational and rotational models respectively. The last column contains values of the deformation parameter obtained from other sources. The references are as follows : 1) 2) Coulomb excitation (STELSON and MCGOWAN, *Nucl. Phys.*, **32**, 632 (1962) and *Phys. Rev.*, **110**, 489 (1958) ; 3) Inelastic proton scattering at 11.66 MeV (BENVENISTE, MITCHELL, BUCK and FULMER, *Phys. Rev.*, **133**, B 323 (1964) ; 4) Inelastic proton scattering at 14.3 MeV and 5) at 17.3 MeV (BUCK, *Phys. Rev.*, **130**, 712).

The optical model parameters are those obtained by Barrett *et al.* (1964).

The most important recent development in the coupled channel analysis is the application of the method with odd mass number by Tamura (1964). He has formulated the theory for arbitrary spins of the target nucleus and has written the corresponding computer programme so that it can couple together up to thirty partial waves. Thus it can calculate the

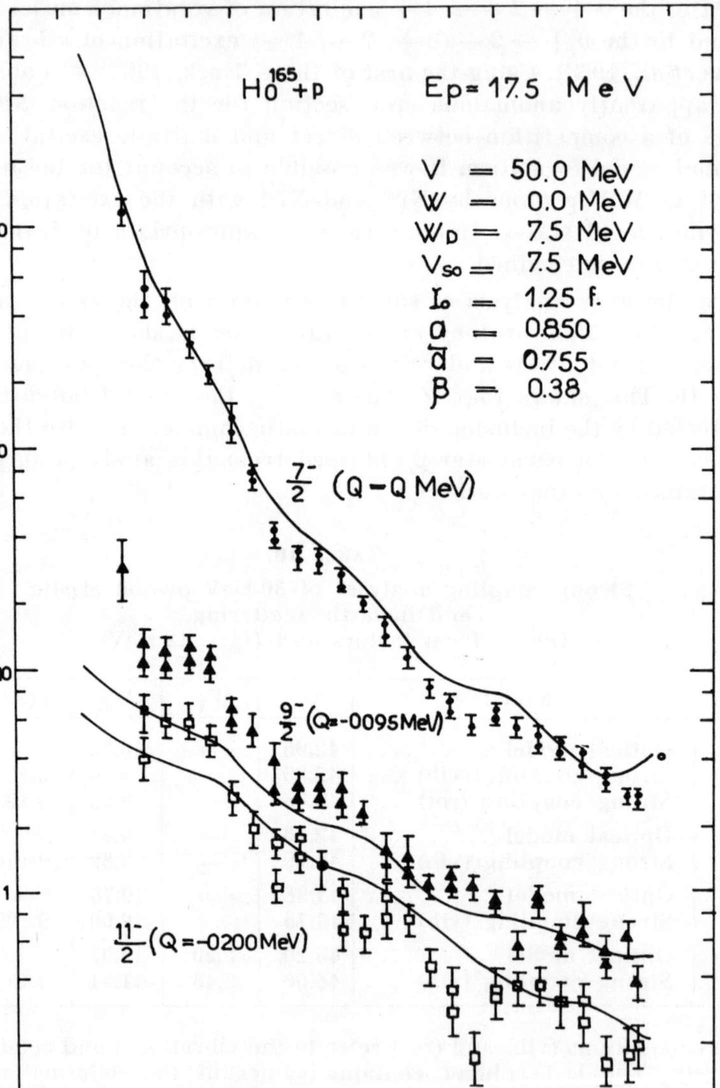

FIG. 21. — Differential cross-sections for the elastic and inelastic scattering of 17.5 MeV protons by Ho[165] compared with a coupled channel calculation (Tamura, 1964).

coupling of 0 +, 2 +, (0 +, 2 +, 4 +) and 3 — states of a vibrational even nucleus, of the 0 +, 2 +, 4 + and 6 + states in a rotational even nucleus, and of the 7/2 +, 9/2 + and 11/2 + states of an odd-*A* deformed nucleus. The ground and first two excited states of Ho^{165} form an example of the latter type of nucleus, and Tamura has calculated the elastic and inelastic scattering of protons from this nucleus. The calculation is so lengthy that it is not practicable to iterate the parameters, so a single calculation was made using average optical potentials, and the results are shown in fig. 21. This computer programme has also been applied to analyse some recent measurements of the inelastic scattering of 12 MeV protons by Te^{126} with the excitation of the 0 +, 2 +, (0 +, 2 +, 4 +) and (0 +, 2 +, 3 +, 4 +, 6 +) vibrational states (Pramila *et al.*, 1965). A further analysis of the inelastic scattering of protons by Cd^{114} (Sakai and Tamura, 1964) with the excitation of twelve states of the target nucleus gives good overall agreement with the inelastic differential cross-sections and has led to the tentative identification of several quadrupole-octupole two-phonon states.

In some cases it is possible to consider a nucleus as consisting essentially of an odd nucleon (or nucleon hole) coupled to an even-even core. The lower excited states of such nuclei are then formed by the splitting of the lowest 2 + state of the core by the coupling of the odd nucleon to the core. Examples of this are Al^{27} (Si^{28} + 1 $d_{5/2}$ proton hole) and Cu^{63} (Ni^{62} + 2 $p_{3/2}$ proton). The inelastic scattering amplitude with excitation of the 2 + state of the core is split among the components of the multiplet in the odd nucleus in the ratio of their statistical weights. Several recent investigations (Perey *et al.*, 1963; Saudinos, 1963; Niewodniczanski *et al.*, 1964) have shown that the predictions of this excited core model are in overall agreement with the observed inelastic cross-sections.

B.2. One-nucleon stripping.

B.2.*a*. *Differential cross-sections.* The measurement of the differential cross-section for the (d, p) stripping reaction is now a well-established tool in nuclear spectroscopy. It is clear that apart from very special circumstances the use of the plane wave theory to analyse experimental cross-sections is unjustifiable; if it gives the L_N-value correctly it is by a fortuitous cancellation of distorting effects, and it gives seriously inaccurate spectroscopic factors and no polarisations. Although the alternative theories involve considerable numerical work there are now computer programmes in many laboratories able to carry them out with ease, and these programmes are usually available to all who wish to use them. In these circumstances there is no longer any reason to make plane wave calculations for stripping reactions.

The distorted wave theory is a convenient and accurate way of analysing deuteron stripping and in general all one-transfer reactions. Even with ap-

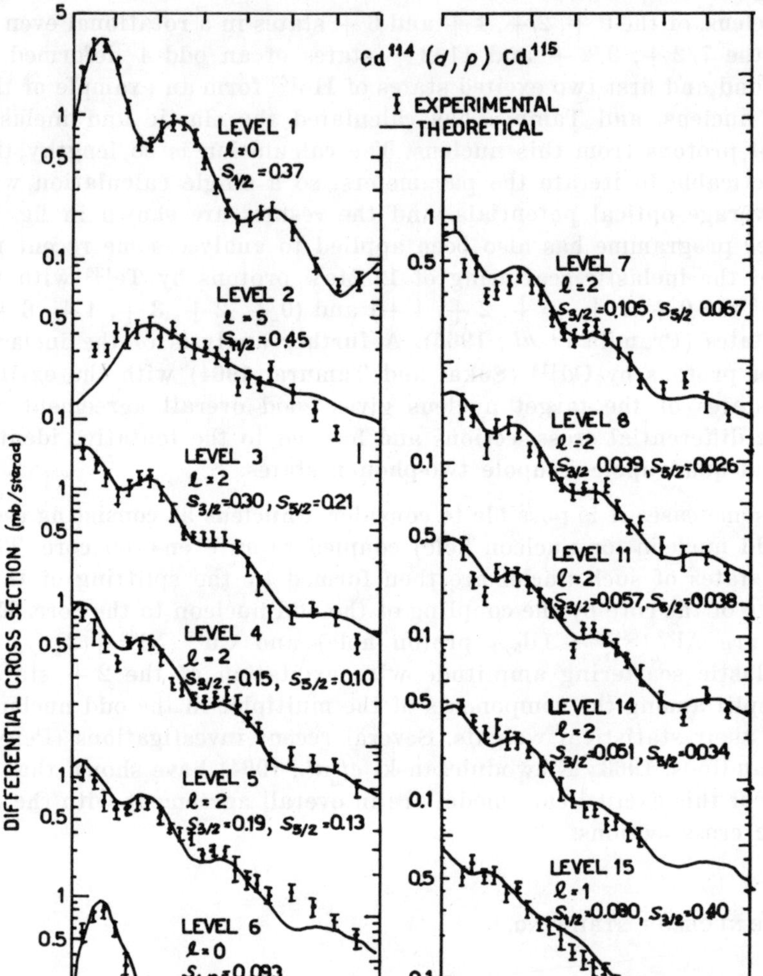

FIG. 22. — **Experimental data for the reaction Cd^{114} (d, p) Cd^{115} to sixteen states of the final nucleus compared with distorted wave calculations using optical potentials obtained from the appropriate elastic scattering data (Silva and Gordon, 1964).**

proximate optical distorting potentials it can give a unique value for the L_N-value of a transition, and with optical potentials accurately fitted to the appropriate elastic scattering data it can also give spectroscopic factors accurate to within about 20 per cent. There are the conclusions recently reached by Satchler (1964) in a comprehensive survey of the extraction of spectroscopic factors from direct interaction data.

An example of a recent (d, p) stripping calculation by Satchler (1964) is shown in fig. 22. The distorting potentials were obtained from the appropriate elastic scattering data, and no attempt was made to adjust them to fit the (d, p) cross-sections. The spectroscopic factors obtained from this analysis are in good agreement with the pairing theory.

The uncertainties in the calculation of stripping cross-sections and hence in the extraction of spectroscopic factors are as follows:

(*a*) The ambiguities in the determination of the distorting potential. Since the stripping cross-section is usually rather insensitive to the proton potential, the most important uncertainties are in the deuteron potentials. The main ambiguity in the deuteron potentials, that due to the variable number of half waves within the nucleus, has been discussed in section A.2. Some studies of the reaction Cr^{52} (d, p) Cr^{53} by Andrews *et al.* (1964) with optical potentials derived from elastic scattering data enable the sensitivity of the spectroscopic factors to the choice of the deuteron potential to be investigated. They find that the deuteron potentials with depths around 100 MeV give better overall fits to the (d, p) data than the potentials with more or fewer half-waves within the nucleus. The relative spectroscopic factors are quite insensitive to the potentials for $l = 0$, 1, and 2 transitions, while those for $l = 3$ or 4 can vary by factors of two or more.

A difficulty of obtaining suitable distorting potentials is that in some cases, particularly for light nuclei, it is not possible to fit the elastic scattering data with the optical model. For some light nuclei elastic scattering can only be fitted well with abnormal parameters that vary erratically with energy. It is not known how stable the reduced widths are in these circumstances.

(*b*) The wave function of the captured neutron is not well known. For a given form factors the neutron binding energy fixes the potential depth and hence the neutron wave function, but unfortunately this wave function and hence the stripping cross-section is still rather sensitive to the particular form factor chosen. The appropriate form factor is fairly well fixed by the shell model and by the low energy scattering data, but there seems to be no way of determining it more precisely. However the spectroscopic factors derived from distorted wave analyses are frequently compared with the predictions of particular nuclear models, and these models themselves often depend on an assumed potential. So provided a consistent potential is used throughout the whole study the final conclusion as to the validity of the nuclear model is unlikely to be sensitive to the particular form chosen for the bound state potential, provided it is within the limits set by the other bound state and scattering data. This is more likely to be the case for the overall characteristics of the bound state potential than for its finer details.

(*c*) The simplest form of distorted wave theory uses a zero-range neutron-proton interaction in order to simplify the numerical calculation. The effect of a finite range integration is to smear the wave functions in the nuclear interior so that they do not vary so sharply with radial distance. Thus to a first approximation the effect of the finite range interaction may be simulated by imposing a lower cut-off on the radial integral and omitting the contribution from the interior. In general this reduces the cross-section, provided the cut-off extends over several oscillations of the interior wave function. Many distorted wave calculations have been made with such a cut-off in the region of the nuclear surface, and in some cases the fit is improved (Satchler, 1964; Bassel *et al.,* 1963) while in others it is made worse (Andrews *et al.,* 1964). The reason for this apparent inconsistency of behaviour is presumably because the distorted wave cross-section depends on the degree of cancellation of the rapidly varying product of wave functions in the nuclear interior and this is likely to vary unpredictably from case to case. The radial cut-off is an arbitrary expedient as there is no impartial way of determining the cut-off radius, and a sharp cut-off is in any case clearly too drastic and also unphysical. Gentler and less complete cut-offs would however introduce yet more arbitrary parameters. Imposition of a cut-off markedly improves the stability of the stripping cross-section with respect to the various distorting potentials allowed by the requirement of a fit to the elastic scattering; this is to be expected since the wave functions must agree asymptotically and have their greatest differences from each other in the nuclear interior (Pullen, 1963). The cut-off can also be regarded as an approximate way of taking into account the physical unreality of allowing the wave functions of such a weakly-bound particle as a deuteron to have finite values inside the nucleus itself. Some recent finite-range calculations are discussed in the next section.

(*d*) The distorted wave theory implicitly assumes weak coupling between the incoming and outgoing channels. In some cases, particularly when collective effects are present, this may not be a good approximation.

(*e*) There are many other approximations and neglected effects in distorted wave theory, including pre- and post-excitation of the nucleons, bulk polarisation of the deuteron and of the nucleus, the D-state of the deuteron, and heavy-particle and exchange stripping, and all these contribute to the uncertainties in the derived spectroscopic factors.

It has been suggested by Goldfarb and El-Baz (paper 4b(II)/C405) that deuteron stripping at energies below the Coulomb barrier of the target nucleus may provide favourable conditions for the extraction of accurate spectroscopic factors. At these low energies the distortion of the incident deuteron wave is mainly due to the Coulomb field and its wave function is strongly attenuated in the nuclear interior. The main contribution to the radial integral in the distorted wave theory then comes from beyond the nuclear surface in the region of the classical turning point. The effects due to the structure of the deuteron and to the details of the nuclear field are not as strong as at higher energies, when the interaction takes place throughout the nuclear volume. Finite range effects may be readily allowed for, and a recent study by Kerman and Gibson (1964) shows that the inclusion of finite range for the neutron-proton potential has very little effect on the shape of the stripping cross-section but may change its magnitude by up to 16 per cent. At low energies the non-direct contributions to the reaction, as well as nuclear distortion and exchange phenomena are also small as they are inhibited by the Coulomb barrier. The stripping cross-section is low, but it is peaked at 180°, which facilitates the elimination of background. It has a simple form that can be determined from a few measurements. The greatest uncertainty in the calculations of the stripping cross-section at these energies is likely to be the forms of the neutron bound state wave function, which is decaying rapidly in the interaction region. Measurements over a range of energies are required before this effect can be evaluated, and the usefulness of low energy stripping for the determination of spectroscopic factors assessed.

One of the most important recent advances in the study of direct reactions has been the recognition of J-dependent effects in the differential cross-sections. Previously it had been assumed that the differential cross-section in for example a (d, p) reaction is sensitive to the angular momentum L_N of the transferred neutron but not to the J-value of the final state, so that even in the favourable case of even-even $0+$ target nuclei it is not possible from measurements of differential cross-sections alone to distinguish between transitions with $J = L_N + 1/2$ and $J = L_N - 1/2$. The J-value could only be determined by difficult polarisation and angular correlation measurements. Now, Lee and Schiffer (1964) have found a marked difference between the differential cross-sections for $L_N = 1$ transitions to $J = 1/2$ and $J = 3/2$ states in medium weight nuclei for deuteron energies from 8 to 12 MeV. Those to $J = 1/2$ states show a deep minimum around 135° that is not found in similar transitions to $J = 3/2$ states, as illustrated in fig. 23. This is found with remarkable constancy for a large number of such transitions in medium weight nuclei and has already been used to determine the spin of the final state in several cases where it is not known. The effect has been accounted for on a simple diffraction model by Greider (paper 4b(II)C183) that may give useful insight into the interaction in some circumstances and suggest further experiments. It is given automatically by distorted wave calculations that take account of the spin-orbit interactions between the incident deuteron and the target nucleus and between the outgoing proton and the target nucleus (Satchler, 1964, Macefield *et al.*, 1964). It has also been observed by Andrews *et al.* (1964) in their studies of the reaction Cr^{52} (d, p) Cr^{53}.

Now that a J-dependent effect has been established for $L_N = 1$ transitions, searches are in progress for similar effects for other values of L_N and in other reactions. A difference between $J = 3/2$ and $J = 5/2$ transitions for $L_N = 2$ in the reaction Se^{76} (d, p) Se^{77} had some time ago been noted by

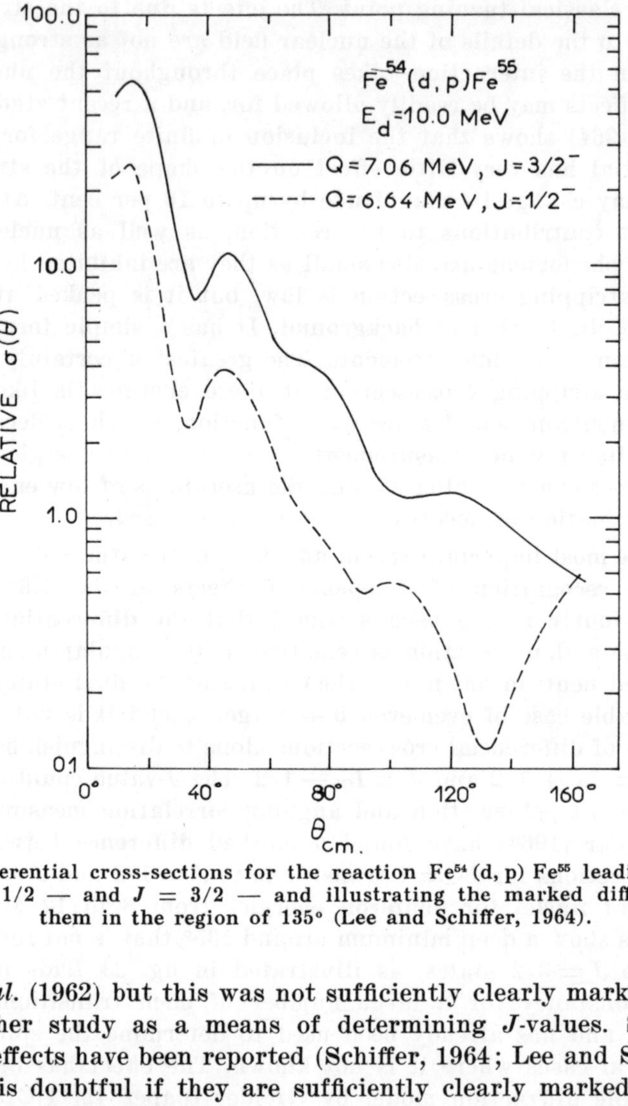

FIG. 23. — Differential cross-sections for the reaction Fe^{54} (d, p) Fe^{55} leading to states of Fe^{55} with $J = 1/2$ — and $J = 3/2$ — and illustrating the marked difference between them in the region of 135° (Lee and Schiffer, 1964).

Macefield *et al.* (1962) but this was not sufficiently clearly marked to be used without further study as a means of determining J-values. Some further J-dependent effects have been reported (Schiffer, 1964; Lee and Schiffer, 1964) but again it is doubtful if they are sufficiently clearly marked to be useful in spectroscopic investigations.

It it likely that J-dependent effects are present in other interactions, and one has already been reported for (d, t) reactions on nickel and iron by Fulmer and Daehnick (1964). As shown in fig. 24, it is very similar to that found by Lee and Schiffer for (d, p) reactions as it also consists of a marked minimum at about 125° in the differential cross-section of $L_N = 1$ reactions to final states with $J = 1/2$, with no minimum for similar reactions to $J = 3/2$ states. Forward angle differences have also been observed between the cross-sections leading to the excitation of $f_{5/2}$ and $f_{7/2}$ states in (p, d) reactions. (Sherr *et al.*, 1964; Whitten *et al.*, 1964). More J-dependent effects will probably be found for other interactions, and they could prove a most valuable tool in nuclear spectroscopy.

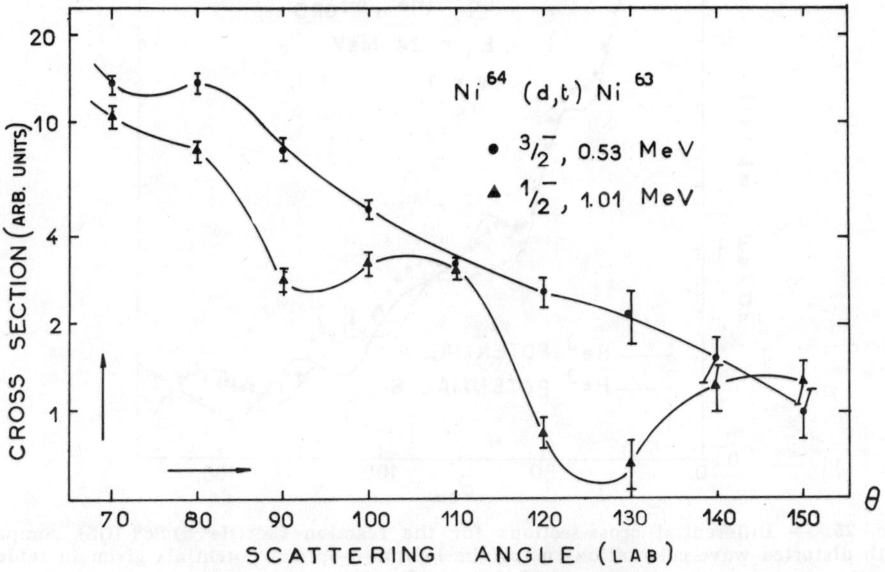

FIG. 24. — Differential cross-sections for the reaction Ni^{64} (d, t) Ni^{63} with $J = 1/2$ and $J = 3/2$, and illustrating the marked difference between them in the region of 125° (Fulmer and Daehnick, 1964).

The distorted wave theory of deuteron stripping is applicable to all single-nucleon transfer reactions, and several such analyses have been submitted to this conference.

Bassel *et al.* (paper 4b(II)/C214) have studied the reaction Ca^{40} (He^3, d) Sc^{41} at 22 MeV. The elastic scattering of helium-3 by calcium was fitted by the optical potentials given in table 11, and the deuteron potentials found from a similar analysis of deuteron elastic scattering. The corresponding zero-range distorted wave cross-sections for the (He^3, d) reaction shown in fig. 25 are in very satisfactory agreement with the experimental data. The results of similar analyses of the reactions $Ni^{62}(He^3,d)Cu^{63}$ and $Ca^{40}(d,He^3)K^{39}$ are shown in fig. 26 et 27 (Bassel, 1964).

TABLE 11

Optical model potentials for the elastic scattering of 22 MeV helium-3 by Ca^{40} (Bassel *et al.*, 1964).

Potential	U	W	r_u	a_u	r_u	a_w
A	98	7.7	1.076	0.854	1.95	0.692
B	142.4.	10.3	1.082	0.806	1.81	0.736

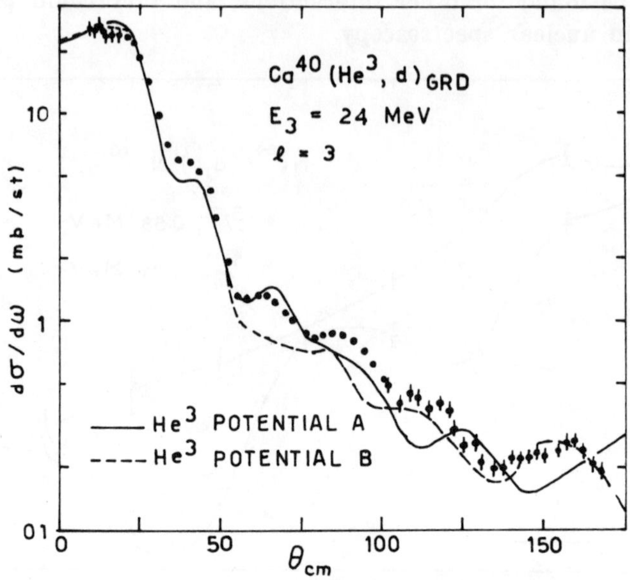

FIG. 25. — Differential cross-sections for the reaction Ca^{40} (He^3, d) Sc^{41} (GS) compared with distorted wave calculations using the helium-3 optical potentials given in table 11 (Bassel *et al.*, 1964).

A series of measurements on (n, d) reactions at 14.4 MeV has been made by Tomaš *et al.* (paper 4b(II)/C286) and analysed by the distorted wave theory to give spectroscopic factors. A finite range calculation with volume absorption and no radial cut-off was made for the reaction O^{16} (n, d) N^{15} (GS) and gives a good fit to the data; the spectroscopic factor $S = 2$ is just the shell model value. A calculation for the reaction Ti^{48} (n, d) Sc^{47} (GS) required a radial cut-off to reproduce the observed ratio of the first and second maxima.

Finite range effects in deuteron stripping. — The expression for the reaction amplitude in the distorted wave theory of deuteron stripping has the general form

$$\int \psi_p^{(-)}(\mathbf{r}_p)\, \psi_n^*(\mathbf{r}_n)\, V_{np}(\mathbf{r})\, \psi_d^{(+)}(\mathbf{r}_d)\, d\mathbf{r}\, d\mathbf{r}_n$$

where $\mathbf{r} = \mathbf{r}_p - \mathbf{r}_n$, $\mathbf{r}_d = 1/2\,(\mathbf{r}_n + \mathbf{r}_p)$. In the zero range theory this six-dimensional integral is reduced to a three-dimensional one by putting $V_{np}(\mathbf{r}) = V_{np}\,\delta(\mathbf{r})$. This greatly simplifies the numerical work but introduces an error whose magnitude should be evaluated in accurate work. The most direct way of doing this is simply to evaluate the six-dimensional integral, and this has now been done by Austern *et al.* (1964). Preliminary calculations show that the effect of taking into account the finite range of the interactions is principally to suppress the contributions to the reaction amplitude from the nuclear interior. This has little effect on the angular distribution, but can alter the absolute magnitude by 10 to 30 per cent.

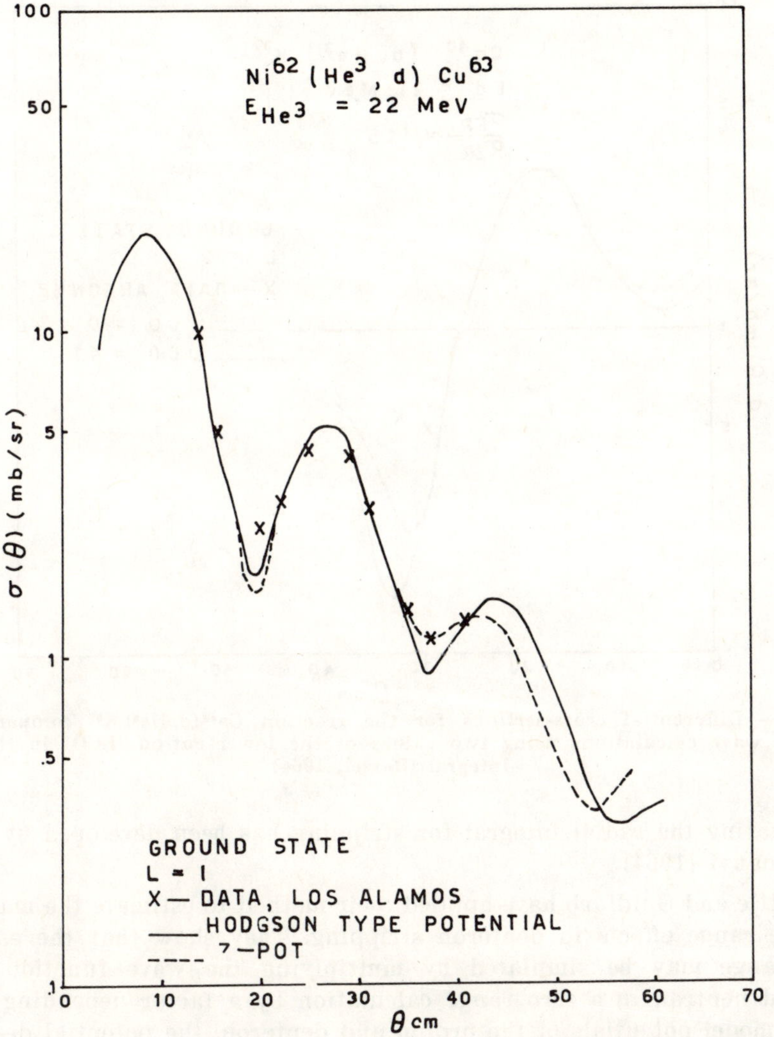

FIG. 26. — Differential cross-sections for the reaction $Ni^{62}(He^3, d)\,Cu^{63}$ compared with distorted wave calculations (Bassel, 1964).

Recently another approach has been developed independently by Buttle and Goldfarb (1964; paper 4b(II)/C404). They expand the wave functions about the radius vector $\mathbf{r}_n$ as Taylor series that may be represented symbolically:

$$\psi_p^{(-)*}(\mathbf{r}_n + \mathbf{r}) = \exp(\mathbf{r}\cdot\nabla_p)\,\psi_p^{(-)*}(\mathbf{r}_n)$$

$$\psi_d^{(+)}\left(\mathbf{r}_n + \frac{1}{2}\mathbf{r}\right) = \exp\left(\frac{1}{2}\mathbf{r}\cdot\nabla_d\right)\psi_d^{(+)}(\mathbf{r}_n) \ .$$

The integral then reduces analytically to a three-dimensional one if the higher terms in the above expansions are neglected. A very similar method

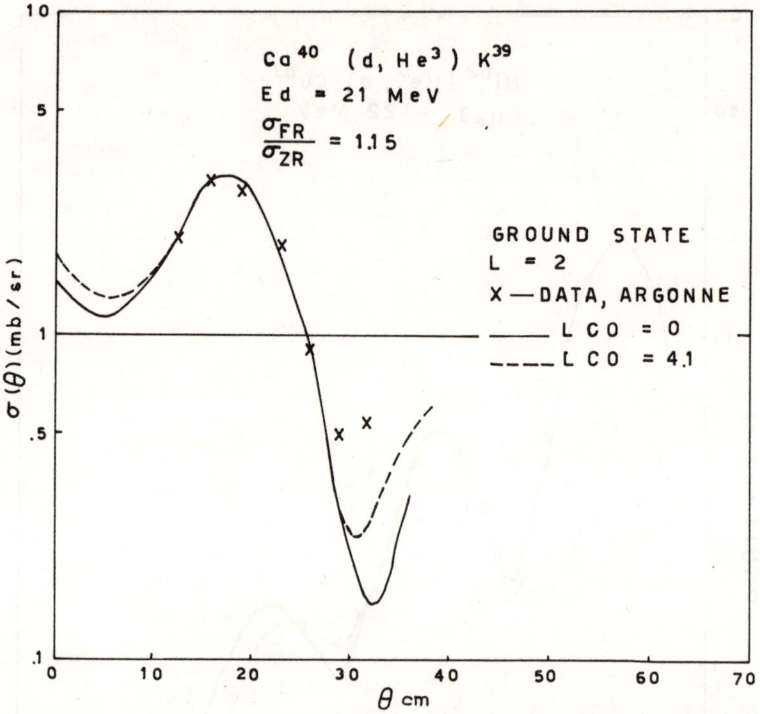

FIG. 27. — Differential cross-sections for the reaction Ca^{40} (d, He^3) K^{39} compared with distorted wave calculations using two values of the lower cut-off (LCO) in the radial integral (Bassel, 1964).

of evaluating the radial integral for stripping has been developed by Bencze and Zimanyi (1964).

Buttle and Goldfarb have applied their method to estimate the magnitude of finite range effects in deuteron stripping. They show that the effects of finite range may be simulated by multiplying the wave function of the captured neutron in a zero range calculation by a factor depending on the optical model potentials of the proton and deuteron, the potential describing the neutron binding and the binding energy of the deuteron. Numerical calculations for the reactions Mg^{24} (d, p) Mg^{25} and Cr^{52} (d, p) Cr^{53} show that the effect is small and is more marked for the shallower than for the deeper deuteron potentials.

A more accurate way of evaluating the radial stripping integral has been proposed by Perey and Saxon (1964). This is essentially a method of evaluating integrals of the form

$$I(\mathbf{r}_1, \mathbf{r}_2, \ldots \mathbf{r}_n) = \Gamma \int K(\mathbf{r}_1, \mathbf{r}_2 \ldots \mathbf{r}_n, \mathbf{r}') J(\mathbf{r}') \, d\mathbf{r}'$$

where K is a function of $(\mathbf{r}' - \mathbf{r}_n)$ that decreases rapidly to zero when $|\mathbf{r}' - \mathbf{r}_n|$ is greater than some range β. It may be used to evaluate integrals appearing in the finite range distorted wave theory of direct reactions and also in the non-local optical model. Detailed comparisons with the corresponding exact expressions are in progress to evaluate the range of applicability of this approach. It might well prove to be useful in even more complicated interactions.

B.2.*b*. *Polarisation in deuteron stripping.* Increasingly accurate measurements of the polarisation of the protons emitted in deuteron stripping interactions are now becoming available. These polarisations may be accounted for by the distorted wave theory provided the spin-orbit terms in the deuteron-nucleus and the proton-nucleus interactions are taken into account. This may be done fairly well for the protons but not for the deuterons due to the lack of optical model analyses of the polarisation of elastically scattered deuterons. Some preliminary calculations with approximate spin-dependent optical model potentials give polarisations substantially in accord with the available data (Reber and Saladin, 1964; Macefield *et al.*, 1964; Satchler *et al.*, 1964).

Further information about (d, p) reactions can be obtained by using polarised deuterons and measuring the asymmetry of the secondary protons. Some recent results of Pasechnik *et al.* (paper 4b(II)/C355) on the reactions B^{10} (d, p) B^{11} and Be^{9} (d, p) Be^{10} at 11 MeV show that for (d, p) stripping, unlike elastic scattering, the polarisation is not equal to the asymmetry, particularly at the higher angles.

It seems likely that when the spin-dependent terms in the proton and deuteron optical potentials are adequately understood the distorted wave theory will give the observed polarisations. They will then be useful to determine the *J*-value of the final state in cases where this cannot be done from the differential cross-section.

B.3. Two-nucleon stripping.

Two-nucleon stripping and pick-up reactions are being increasingly studied as they provide a useful tool for investigating the structure of nuclei further removed from the stable region than those accessible by the corresponding single-nucleon reactions.

The theoretical analysis of these interactions is difficult, and only moderate success has been achieved so far. The plane-wave theories of Newns (1960) and Glendenning (1962) usually give a satisfactory account of the forward peak, though the absolute magnitude is not well determined. An alternative formulation has been given by Gorodetzky *et al.* (paper 4b(II)/C247) and used to analyse their data on the reaction N^{14} (He^{3}, p) O^{16}. The interactions between the outgoing and captured particles are assumed to have a Gaussian form, and the ranges are adjusted to give a fair fit to the main peak of the observed angular distribution.

A series of measurements of the Ca^{40} (He^3, p) Sc^{42} reaction to many states of the residual nucleus has been made by Zurmühle *et al.* (paper 4b(II)/C229). The plane wave theory is used to determine the angular momentum transfer, and hence to set limits on the spins of the final states.

The cross-sections of several (n, t) reactions at 14.4 MeV on light nuclei have been measured by Valković and Tomaš (paper 4b(II)/C127) and it is found that the shapes of the angular distributions are characteristic of a direct interaction process.

A distorted wave theory of two-nucleon stripping has recently been formulated by Rook and Mitra (1964) and applied to analyse (t, p) reactions. Computer programmes were written for several possible double-stripping mechanisms, including stripping to single-particle states, dineutron stripping and knock-out. These all gave rather similar angular distributions characteristic of the total orbital angular momentum transfer and the agreement with the experimental data was no better than qualitative, so that it was not possible to identify the dominant mechanism. In other cases the calculations did not fit the data at all. The lack of quantitative agreement is possibly due to the assumption of zero range for the nucleon-nucleon forces binding the triton, but additional uncertainties are introduced by the presence of other possible reaction mechanisms, such as sequential stripping, that are relatively more important than the competing reactions in deuteron stripping.

Rook and Mitra found that the agreement with the experimental data is markedly improved if a cut-off is applied to the radial integrals. As in the case of the similar effect frequently found in deuteron stripping, this is mainly because the cut-off removes, in a crude way, some of the unphysical contributions from the nuclear interior that have been introduced by the approximations in the distorted wave theory, principally by the assumption of zero-range and local forces. The angular distribution is thus insensitive to the detailed structure of the bound states. This is the reason for the success of the plane wave theories of double stripping. Further progress will however come by refining the distorted wave theory by the inclusion of finite range forces and by a progressive improvement of the other assumptions of the theory.

The coupled channels theory (section B1) has been applied to (t, p) reactions by Stamp and Rook (1963) and they find that the effect of taking into account the coupling between the proton and triton channels is also to reduce the contributions from the nuclear interior. The differential cross-section for $L = 2$ transitions has the observed peak at around 30°, while the distorted wave theory incorrectly gives a peak in the forward direction. However these calculations use a zero-range force, whereas finite range forces enhance the contribution from the surface region. A finite range calculation would thus reduce the effect of the coupling between the proton and triton channels and further calculations will be required before the extent to which the conclusions depend on the finite range approximation becomes clear.

B.4. Other direct interactions.

Measurements of an extensive range of nuclear interactions are now being made, and many of the results are analysed by some form of distorted wave theory. The following reactions are the subject of communications to this conference:

(n, p). Some calculations of the differential cross-sections for the reaction Si^{28} (n, p) Al^{28} in the energy range 5-20 MeV have been made by Agodi and Schiffrer (paper 4b(II)/C264) using the distorted wave theory and a Serber and a Wigner force for the finite range effective neutron-proton interaction. Their results show how the exchange effects depend on energy and on the high momentum components of the optical model wave functions.

A series of measurements of (n, p) reactions on B^{10}, O^{16}, Co^{59}, Au and Pb at 14.4 MeV have been made by Paić *et al.* (paper 4b(II)/C126). These results have been analysed by the Butler theory and provide evidence for direct processes.

(n, α). A plane wave theory of (n, α) reactions taking into account the pick-up and the knock-out mechanisms has been formulated by Massot *et al.* (paper 4b(II)/C115). It is applied to the reactions O^{16} (n, α) C^{13} and Na^{23} (n, α) F^{20} at 14 MeV, and is in satisfactory accord with the data.

Measurements of (n, α) reaction cross-sections have also been made by Lalović *et al.* (paper 4a(II)/C63) and by Rubbino and Zubke (paper 4a(II)/C326). These results provide strong evidence for direct interactions and a detailed analysis should throw more light on the problem of alpha-particles on the nuclear surfaces.

(d, α). Some (d, α) reactions have been studied by Stanojević (paper 4a(II)/C307) and the results compared with the predictions of various direct interaction theories.

Cluster substitution reactions. A marked difference in cross-section between favoured and unfavoured cluster substitution reactions has been formed by Holmgren and Wolicki (1961), for example in the reactions C^{12} (t, α) B^{11} and C^{13} (He^{3}, α) C^{12}. Beregi *et al.* (paper 4b(II)/C86) show by the use of the Austern, Butler and McManus theory and shell model amplitudes of cluster reduced widths that this is due to constructive or destructive interference between the terms in the expression for the cross-section of the cluster substitution reaction. Further studies of such reactions, particularly in heavier nuclei, would give additional information on the possible presence of alpha-particle clusters on the nuclear surface (Wilkinson, 1962).

Conclusion.

This conference has been held at an opportune moment in the study of elastic scattering and direct interaction processes. Many detailed studies of nucleon elastic scattering have now been made, and the overall characteristics of the optical potentials established. The ambiguities in the optical model potentials for the elastic scattering of deuterons, helium-three nuclei and alpha-particles are now essentially understood, and similar analyses will be applicable to the new data on triton scattering. Extensive nucleon polarisation data is now available and the difficult problems of obtaining the corresponding data for deuterons and helium-three nuclei have recently been overcome. Reaction cross-sections can now be reliably measured and are generally found to be in accord with the optical model.

The optical potentials found from the analysis of elastic scattering data are being used to generate the distorted waves needed to analyse data on direct interactions. Together with the theories of the processes, particularly the coupled equation theory and its weak interaction limit the distorted wave Born approximation, they constitute a powerful tool that is being extensively used to extract spectroscopic factors and determine the spins and parities of the levels of a wide range of stable and unstable nuclei.

Acknowledgment.

I am grateful to all those colleagues who made this review possible by kindly sending me their measurements and calculations in advance of publication.

REFERENCES

A. Agodi and G. Schiffrer, Paris Conference, paper 4b(II)/C 264 (1964).
W. P. Alford, L. M. Blau and D. Cline, *ANL*-6848, 8 (1964).
P. T. Andrews, R. W. Clifft, L. L. Green and J. F. Sharpey-Shafer, *Nucl. Phys.*, **56**, 465 (1964).
O. Aspelund. Paris Conference, paper 4b(II)/C 274 (1964).
N. Austern, R. M. Drisko, E. C. Halbert and G. R. Satchler, *Phys. Rev.*, **133**, B 3 (1964).
J. B. Ball, C. B. Fulmer and R. H. Bassel, *Phys. Rev.*, **135**, B 706 (1964).
R. H. Bassel, private communication (1964).
R. H. Bassel, G. R. Satchler and R. M. Drisko, Paris Conference, paper 4b(II)/C 214 (1964).
R. C. Barrett, A. D. Hill and P. E. Hodgson, *Nucl. Phys.*, **62**, 133 (1964).
D. J. Baugh, G. W. Greenlees, J. S. Lilley and S. Roman, Paris Conference, paper 4b(I)/C 114 (1964).
K. Bearpark, W. R. Graham and G. Jones, Paris Conference, paper 4b(I)/C 132 (1964).

G. Bencze and J. Zimanyi, *Phys. Letters,* **9**, 246 (1964).
J. Benveniste, R. Booth and A. Mitchell, *Phys. Rev.,* **123**, 1818 (1961).
J. Benveniste, A. C. Mitchell, B. Buck and C. B. Fulmer, *Phys. Rev.,* **133**, B 323 (1964).
R. W. Bercaw, E. T. Boschitz and J. S. Vincent, Paris Conference, paper 4b(I)/C 178 (1964).
P. Beregi, V. G. Neudatchin and Yu. F. Smirnov, Paris Conference, paper 4b(II)/C 86 (1964).
R. Beurtey, R. Chaminade, A. Falcoz, M. M. Faure, R. Maillard, D. McDaniels, T. Mikumo, A. Papineau, L. Schecter and J. Thirion, *Compt. Rend.,* **256**, 922 (1963).
R. Beurtey, R. Maillard, A. Papineau and J. Thirion, *Compt. Rend.,* **257**, 1477 (1963).
R. Beurtey *et al.*, communication privée.
B. Buck, *Phys. Rev.,* **127**, 940 (1962) ; *Phys. Rev.,* **130**, 712 (1963) ; private communication (1964).
B. Buck, A. P. Stamp and P. E. Hodgson, *Phil. Mag.,* **8**, 1805 (1963).
A. Budzanowski, K. Grotowski, S. Micek, H. Niewodńiczanski, J. Sliz, A. Strzałkowski and H. Wojciechowski, Paris Conference, paper 4b(I)/C 58 (1964) and *Phys. Letters,* **11**, 74 (1964).
A. Budzanowski, K. Grotowski, L. Jarczyk, H. Niewodniczański, A. Strałkowski and Z. Wroblowa, Paris Conference, paper 4b(I)/C 59 (1964).
P. J. Bulman, G. W. Greenlees and M. J. Sametband, private communication (1964).
W. E. Burcham, J. B.A. England, J. E. Evans, A. Garcia, R. G. Harris and C. Wilne, Paris Conference, paper 4b(I)/C 212 (1964).
P. J. A. Buttle and L. J. B. Goldfarb, *Proc. Phys. Soc. (London),* **83**, 701 (1964). Paris Conference, paper 4b(II)/C 404 (1964).
R. M. Craig, J. C. Dore, G. W. Greenlees, J. S. Lilley, J. Lowe and P. C. Rowe, *Nucl. Phys.,* **58**, 515 (1964).
P. Darriulat, G. Igo, H. Pugh, J. M. Meriwether and S. Yamabe, *Phys. Rev.,* **134**, B 42 (1964).
J. Delaunay, B. Delaunay and J. P. Passerieux, Paris Conference, paper 4b(I)/C 216 (1964).
J. K. Dickens, F. G. Perey, R. J. Silva and T. Tamura, *Phys. Letters,* **6**, 53 (1963).
R. M. Drisko, R. H. Bassel and G. R. Satchler, *Phys. Letters,* **2**, 318 (1962).
R. M. Drisko, G. R. Satchler and R. H. Bassel, *Phys. Letters,* **5**, 347 (1963).
R. M. Eisberg and C. N. Waddell, private communication to Mayo *et al.* (1964).
A. E. Forest, *Phys. Letters,* **7**, 130 (1963).
L. Freindl, H. Niewodniczanski, J. Nurzynski, M. Slapa and A. Strzałkowski, Proc. Rutherford Jubilee Conf. Manchester, 529 (1961) ; *Acta Phys. Polon.,* **23**, 619 (1963).
M. P. Fricke, Thesis (Minnesota) (1964).
R. H. Fulmer and W. W. Daehnick, *Phys. Rev. Letters,* **12**, 455 (1964).
N. K. Glendenning, *Nucl. Phys.,* **29**, 109 (1962).
V. Z. Goldberg, V. P. Rudakov and I. N. Serĭkov, Paris Conference, paper 4b(I)/C 363 (1964).
L. J. B. Goldfarb, Paris Conference, paper 4b(II)/C 406 (1964).
L. J. B. Goldfarb and E. El-Baz, Paris Conference, paper 4b(II)/C405 (1964).
S. Gorodetzky, G. Bassompierre and A. Gallmann, Paris Conference, paper 4b(II)/C 247 (1964).
K. R. Greider, Paris Conference, paper 4b(II)/C 183 (1964).
P. E. Hodgson, Proc. Rutherford Jubilee Conf. Manchester, 407 (1961); *Phys. Letters,* **3**, 352 (1963) ; *Ibid.,* **6**, 75 (1963).
H. D. Holmgren and E. A. Wolicki, Proc. Rutherford Jubilee Int. Conf. Manchester, 541 (1961).

G. Igo, W. Lorenz and U. Schmidt-Rohr, *Phys. Rev.*, **124**, 832 (1961).
A. K. Kerman and F. P. Gibson, *ANL*-6848, 43 (1964).
R. W. Klingensmith, H. J. Hausman and W. D. Ploughe, Ohio State University Report (1964); *Phys. Rev.*, **134**, B 1220 (1964).
B. Lalović, V. Ajdačić, P. Strohal, P. Kulišić, D. Kovač and N. Cindro, Paris Conference, paper 4a(II)/C 63 (1964).
A. M. Lane, *Phys. Rev. Letters*, **8**, 171 (1962) ; *Nucl. Phys.*, **35**, 676 (1962).
A. M. Lane and A. P. Stamp, *Nucl. Phys.*, **57**, 444 (1964).
L. L. Lee and J. P. Schiffer, *Phys. Rev. Letters*, **12**, 108 (1964); *Phys. Rev.*, **136**, B 405 (1964).
B. E. F. Macefield, R. Middleton and D. J. Pullen, *Nucl. Phys.*, **44**, 309 (1962).
B. E. F. Macefield, J. R. Rook and P. E. Hodgson, private communication (1964).
M. Q. Makino, C. N. Waddell, R. M. Eisberg and J. Hestenes, *Phys. Letters*, **9**, 178 (1964).
J. N. Massot, E. El-Baz and J. Lafoucrière, Paris Conference, paper 4b(II)/C 115 (1964).
S. Mayo, R. M. Eisberg, W. Schimmerling and M. J. Sametband, Paris Conference, paper 4b(I)/C38 (1964).
M. A. Melkanoff, D. S. Saxon and T. Sawada, Paris Conference, paper 4b(I)/C147 (1964).
T. Mikumo, *J. Phys. Soc. Japan*, **16**, 1066 (1961).
P. A. Moldauer, Paris Conference, paper 4b(I)/C 68 (1964).
H. C. Newns, *Proc. Phys. Soc. (London)*, **76**, 489 (1960).
H. Niewodniczanski, J. Nurzynski, A. Strzałkowski, J. Wilczynski, J. R. Rook and P. E. Hodgson, *Nucl. Phys.*, **55**, 386 (1964).
R. J. Olness, K. K. Seth and H. W. Lewis, *Nucl. Phys.*, **52**, 529 (1964).
G. Paić, I. Šlaus, V. Valković and M. Cerineo, Paris Conference, paper 4b(II)/C126 (1964).
M. V. Pasechnik, N. N. Putcherov, L. S. Saltykov and D. I. Tambovtsev, Paris Conference, paper 4b(II)/C 355 (1964).
M. V. Pasechnik, N. N. Putcherov and V. I. Tchirko, paper 4b(I)/C384 (1964).
F. Perey, R. J. Silva and G. R. Satchler, *Phys. Letters*, **4**, 25 (1963).
F. Perey and D. S. Saxon, *Phys. Letters*, **10**, 107 (1964).
F. G. Perey and B. Buck, *Nucl. Phys.*, **32**, 353 (1962).
F. G. Perey, *ANL*-6848, 114 (1964).
C. Perey and F. Perey, *Phys. Rev.*, **132**, 755 (1963).
F. G. Perey, *Phys. Rev.*, **131**, 745 (1963) ; private communication to Mayo *et al.* (1964).
G. C. Pramila, R. Middleton, T. Tamura and G. R. Satchler, *Nucl. Phys.*, **61**, 448 (1965).
D. J. Pullen, Thesis Oxford (1963).
D. J. Pullen, J. R. Rook and R. Middleton, *Nucl. Phys.*, **51**, 88 (1964).
J. Raynal, *Phys. Letters*, **3**, 331 (1963) ; *Ibid.*, **7**, 281.
L. H. Reber and J. X. Saladin, *Phys. Rev.*, **133**. B 1155 (1964).
B. Ridley and J. F. Turner, *Nucl. Phys.*, **58**, 497 (1964).
J. R. Rook, *Nucl. Phys.*, **61**, 219 (1965).
J. R. Rook and D. Mitra, *Nucl. Phys.*, **51**, 96 (1964).
A. Rubbino and D. Zubke, Paris Conference, paper 4a(II)/C 326 (1964).
M. Sakai and T. Tamura, *Phys. Letters*, **10**, 323 (1964).
G. R. Satchler, Chicago Conference. *ANL*-6878, 23 (1964).
G. R. Satchler, C. M. Perey and F. G. Perey, private communication (1964).
J. Saudinos, Nuclear Spectroscopy, The Proceedings of the Nuffic Summer School (1963).
J. P. Schiffer, Chicago Conference. *ANL*-6878, 279 (1964).
G. Schrank and R. E. Pollock, *Phys. Rev.*, **132**, 2200 (1963).

R. Sherr, E. Rost and M. E. Rickey, *Phys. Rev. Letters*, **12**, 420 (1964).
R. J. Silva and G. E. Gordon, *Phys. Rev.*, **136**, B 618 (1964).
D. M. Stanojević, Paris Conference, paper 4a(II)/C307 (1964).
A. P. Stamp and J. R. Rook, private communication (1964).
A. Strzałkowski, *Phys. Letters*, **2**, 121 (1962).
T. Tamura, *ANL*-6848, 24 (1964) ; *Phys. Letters*, **9**, 334 (1964).
T. Terasawa and G. R. Satchler, *Phys. Letters*, **7**, 265 (1963).
R. N. Thomas, E. J. Burge and P. E. Hodgson, *Nucl. Phys.*, **52**, 93 (1964).
H. R. E. Tjin a Djie and K. W. Brockman, Paris Conference, paper 4b(I)/C 235 (1964) ; *Nucl. Phys.*, in press (1964).
H. R.E. Tjin a Djie, F. Udo and L.A.C.L. Koerts, *Nucl. Phys.*, **53**, 625, (1964).
P. Tomaš, G. Paić, V. Valković, M. Cerineo, I. Šlaus and D. Rendić, Paris Conference, paper 4b(II)/C 286 (1964).
J. F. Turner, B. W. Ridley, P. E. Cavanagh, G. A. Gard and A. G. Hardacre, *Nucl. Phys.*, **58**, 509 (1964).
L. Valentin, Paris Conference, paper 4b(I)/C 330 (1964).
V. Valković and P. Tomaš, Paris Conference, paper 4b(II)/C127 (1964).
R. Wagner, P. D. Miller, T. Tamura and H. Marshak, *Phys. Letters*, **10**, 316 (1964).
N. S. Wall and C. R. Gruhn, to be published (1964).
S. Watanabe, *Nucl. Phys.*, **8**, 484 (1958).
C. Whitten, E. Kashy and J. P. Schiffer, in preparation (1964).
B. Wilkins and G. Igo, *Phys. Letters*, **3**, 48 (1962); *Phys. Rev.*, **129**, 2198 (1963)
B. D. Wilkins, Thesis, California, *UCRL*-10783 (1963).
D. H. Wilkinson, Proc. Rutherford Jubilee Int. Conference Manchester, 339 (1961).
D. Wilmore and P. E. Hodgson, *Nucl. Phys.*, **55**, 673 (1964).
K. Yagi, H. Ejiri, M. Furukawa, Y. Ishizaki, M. Koike, K. Matsuda, Y. Nakejima, I. Nonaka, Y. Saji, E. Tanaka and G. R. Satchler, *Phys. Letters*, **10**, 186 (1964).
R. W. Zurmühle, C. M. Fou and W. Swenson, Paris Conference, paper 4b(II)/C229 (1964).

The nuclear optical model

P E HODGSON

Nuclear Physics Laboratory, University of Oxford, Oxford

Contents

Abstract. The main features of the nuclear optical model are reviewed, with particular attention to current growth points. The determination of the form and strength of the nucleon–nucleus interaction by calculation from the nucleon–nucleon interaction is described and the results compared with those obtained phenomenologically. The potential is applied to analyse single-particle bound states, and it is shown how this can give its isospin and state dependence. Even when it is used simply to calculate the elastic scattering, the optical model must be established in the framework of nuclear reaction theory, and allowance made for the effects of all the reaction channels, and in particular for the compound nucleus processes and the associated fluctuations in the cross sections. The model can then be applied to analyse data on the elastic scattering of nucleons, deuterons, helions, tritons and α particles by nuclei, and this enables its parameters to be determined more precisely and its range of validity to be investigated. The nucleon data is so extensive that it is possible to study several small terms in the potential, in particular the isospin term and the nuclear spin term, in addition to the familiar spin–orbit term. The scattering of the composite particles shows several apparently anomalous features that on investigation yield additional information on nuclear structure. The optical model may be extended to give the inelastic as well as the elastic scattering, and this makes it possible to analyse many apparently complicated phenomena. The application of the results of optical model analyses to the calculation of compound nucleus cross sections is also discussed, together with the use of optical model wavefunctions in calculations of direct reaction cross sections.

This review was completed in August 1971.

1. Historical introduction and summary

It is about twenty years since it was first suggested that the interaction of nucleons with nuclei could be represented by a complex potential, and since then it has been developed into a model of considerable power and generality. A wide range of scattering data can be accounted for by the model to a high degree of precision, and the corresponding wavefunctions are extensively used to extract information on nuclear structure from measurements of nuclear interaction cross sections.

In this review, the historical development of the optical model is briefly summarized, and the special features of its application to a variety of scattering problems are described. The main emphasis is on the physical ideas rather than on the mathematical techniques necessary for realistic calculations, but references to specialized treatments are given whenever appropriate.

The essential idea of the optical model is that a nucleon incident on a nucleus may be elastically scattered or it may cause a variety of different reactions. If the incident particle is represented by a wave, then in classical language it may be scattered or it may be absorbed. In optics this is analogous to the refraction and absorption of a light wave by a medium of complex refractive index, and just as the imaginary part of the refractive index takes account of the absorption of the light wave so in the nuclear case the imaginary part of the complex potential describing the interaction takes account of all the nonelastic reactions.

This simple idea was applied in a semiclassical way by Fernbach *et al* (1949) to the interpretation of the data on the scattering and absorption of 90 MeV neutrons by a range of nuclei. It was found that the total elastic and the total reaction cross sections could be well accounted for, and the real and imaginary parts of the potential were in qualitative accord with crude semiclassical estimates based on the nucleon–nucleon cross sections. A full quantum mechanical calculation of the scattering of 17 MeV protons by aluminium using a complex potential to represent the proton–nucleus interaction was made by Le Levier and Saxon (1952), and gave a good account of the rather imprecise data available at that time.

It is an essential feature of the model that nuclei are regarded as blobs of nuclear matter whose properties are determined by their size (or equivalently by the number of nucleons comprising them, since nuclear matter is almost incompressible), apart from certain small structure effects that will be mentioned in more detail later. This implies that the same potential can account for the interaction of particles of different energies with different nuclei. To the extent that this can be done, the optical model is at least a useful way of summarizing a vast body of experimental data.

If this is to be so, then the experimental data should show a regular variation with energy and atomic number, and at first sight this seems very unlikely because nuclei have very different structures, as shown by their energy level diagrams. Indeed, if measurements are made at high resolution the cross sections often show an extremely complicated and characteristic structure. If, however, the energy-averaged cross sections are examined, and these may be obtained at once from low-resolution measurements, a systematic behaviour becomes apparent. This was first shown by the neutron cross sections of Barschall (1952) (see figure 1). The energy-averaging of cross sections is thus an essential feature of the model at low energies when compound nucleus reactions are appreciable.

It was then found by Feshbach *et al* (1954) that the overall variation of these neutron cross sections with atomic weight could be represented very well by a complex neutron–nucleus potential of the form

$$\begin{aligned} V &= V_0(1+\mathrm{i}\zeta) \qquad \text{for } r<R \\ &= 0 \qquad\qquad\quad\; \text{for } r>R \end{aligned} \tag{1.1}$$

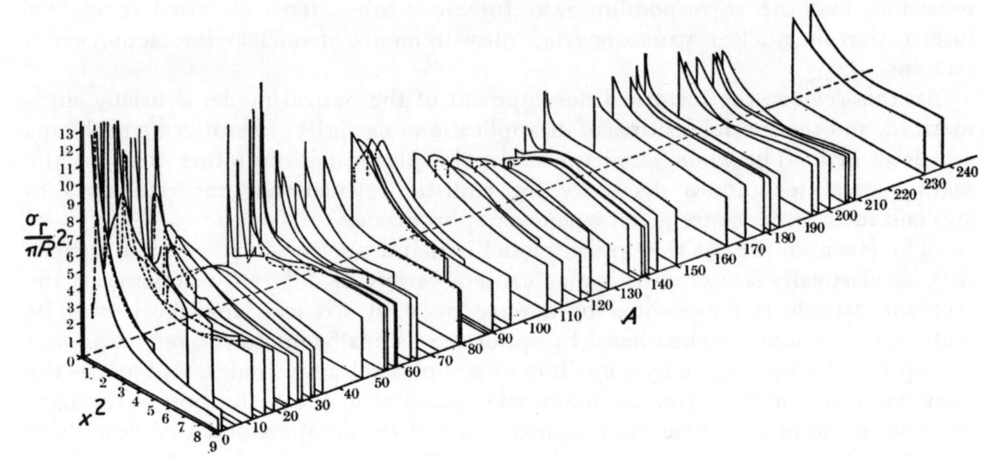

Figure 1. Observed neutron total cross sections as a function of energy and mass number, radius $R = 1{\cdot}45A^{1/3}$. From Feshbach *et al* (1954).

where $R = 1{\cdot}45 \times A^{1/3}$ fm, $V_0 = 42$ MeV, $\zeta = 0{\cdot}03$ and A is the atomic mass. The results of their calculations are shown in figure 2.

This spectacular success established the usefulness of the optical model, and thereafter it was soon applied to a wide range of nuclear interactions. As the precision of the data increased the model was refined so that it can now account not only for differential and reaction cross sections but also for polarizations to a high degree of accuracy. Many measurements have also been made of the elastic scattering of composite particles by nuclei and, as the cross sections show the same

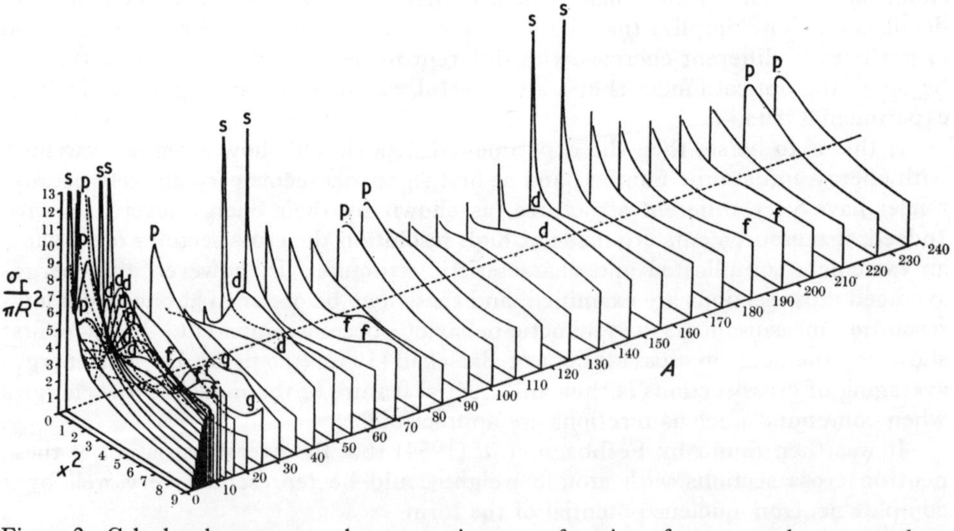

Figure 2. Calculated neutron total cross sections as a function of energy and mass number, for a potential depth $V_0 = 42$ MeV, radius $R = 1{\cdot}45A^{1/3}$, $\zeta = 0{\cdot}03$. The energy ϵ is expressed in terms of $x^2 = A^{2/3}\,A/10(A+1)\,\epsilon$, where ϵ is in MeV. From Feshbach *et al* (1954).

overall features and regularities as the corresponding nucleon cross sections, the optical model was applied to them as well, with similar success.

Apart from the rather crude basis provided by the semiclassical optical model all this work is essentially phenomenological, in that the values of the parameters of the optical potentials are found by optimizing the fit to the experimental data. But while the model can be used in this way it is desirable to put it on a more fundamental basis, and to derive it as far as possible from the known properties of the nucleus and of the nucleon–nucleon interaction. This is desirable not only to unify our theory of nuclear reactions but also to resolve certain difficulties and ambiguities in purely phenomenological analyses of data of high precision. The problem of calculating the optical potential is evidently an exceedingly complex one, since all the interactions of the incident particle with the nucleons of the target must be included. Brueckner's (1954) theory of nuclear matter was an important step forward, and Feshbach's formalism provided a useful framework for a theoretical definition of the potential. Their work established certain important features of the optical potential, in particular its nonlocality, but was unable to define its form and strength with sufficient precision. A simpler approach by Greenlees *et al* (1969) using realistic forms for the nuclear matter distribution and the nucleon–nucleon interaction gave good quantitative results, and subsequent work by Kidwai and Rook (1971) has justified the assumptions of Greenlees *et al* (1969) by a nuclear matter calculation similar to that of Brueckner. This work is summarized in § 2.

The optical model can be considered as an extension to positive energies of the nuclear shell model, so that there is a physical continuity between the shell model potential for nucleons of negative energies and the real part of the optical potential at positive energies. The single-particle states of nuclei can be used to determine the shell model potential, and this brings out certain features that are not readily obtainable from the analysis of scattering data. This work is described in § 3.

In order to use the optical model to analyse experimental data it is necessary to see how the elastic scattering cross sections obtained from the model are related to all the other reactions that may occur. In particular the role of energy-averaging needs to be precisely expressed in relation to the experimental conditions and the components of the cross sections. This is considered in § 4.

With the theoretical understanding provided by the optical model formalism, it is possible to make detailed analyses of experimental data and thus to confirm its usefulness and determine its parameters to a high degree of precision. The mathematical and computational techniques necessary to do this are summarized in § 5.

A vast body of experimental data for the elastic scattering of neutrons and protons by nuclei is now available, and many phenomenological optical model analyses have been made of the differential cross sections and polarizations, together with the associated reaction and total cross sections. The analyses give information on the energy dependence of the optical potential and on the way it depends on nuclear structure. These analyses are summarized in § 6.

The application of the optical model to the elastic scattering of composite particles, in particular deuterons, helions, tritons and α particles, presents many features that are of interest both for nuclear reactions and for nuclear structure. The corresponding optical potentials are not as well defined as for nucleons, but nevertheless some success has been achieved in calculating them from the constituent nucleon–nucleus interactions. Their stronger absorption and higher mass-to-charge ratios relative to those of protons give rise to special features that are discussed for deuterons in § 7, for helions and tritons in § 8 and for α particles in § 9.

The formal theory of nuclear reactions shows that the wavefunctions in all the open reaction channels satisfy a set of coupled differential equations, one for each channel. This set is too large to solve in practice, so it is severely truncated, and the effect of the channels not considered is taken into account by allowing the interaction potential to become complex. The most extreme truncation, that which retains only the equation for the wavefunction in the shape elastic channel, gives the usual optical model. A less severe truncation retains one or more equations for wavefunctions in nonelastic channels, and this enables both the elastic and nonelastic cross sections to be calculated simultaneously, provided a model is available for the channels considered, and for the corresponding interaction potentials. This approach is a significant improvement over the perturbation theory calculation when the coupling between the elastic and the nonelastic channels chosen is strong, and as this is most often the case for inelastic scattering with the excitation of collective states, these have most frequently been treated by the coupled-channels method. Since this is a natural extension of the optical model it is briefly considered in § 10.

The energy-averaged elastic scattering cross sections all contain a compound nucleus component, which may be important or negligible depending on the conditions. The optical model enables this to be calculated by the methods described in § 11.

Optical model wavefunctions are widely used in reaction calculations, and some of the attendant uncertainties and possible ways of overcoming them are considered in § 12.

Finally in § 13 the present state of optical model calculations is reviewed, and some lines of future development are indicated.

The nuclear optical model and its relation to other theories of nuclear reactions is discussed in more detail by Hodgson (1971). The present review concentrates on the essential physical ideas of the model with attention to some current growth points and unresolved problems.

Throughout this review, all distances are in fermis (1 fm = 10^{-13} cm) and energies in MeV.

2. The nucleon–nucleus interaction

In their pioneer work on the optical model, Feshbach *et al* (1954) used the square-well potential (equation (1.1)). This is clearly unphysical, and indeed calculations soon showed that the high reflectivity of such potentials gives absorption cross sections that are far too small. A rounded potential is clearly preferable, and since the nucleon–nucleon potential falls off exponentially at large distances it is reasonable to expect the optical potential that results from a sum of such interactions to do the same. It is also desirable to retain the feature of the square-well potential of constancy in the nuclear interior since this is to be expected from the short-range and saturation character of nuclear forces which ensures that once a nucleon is within the nucleus it experiences on the average no resultant force.

These features are conveniently represented by the Saxon–Woods form factor

$$f(r) = \frac{1}{1+\exp\{(r-R)/a\}} \tag{2.1}$$

where R is the radius parameter and a the surface diffuseness parameter. Since the nuclear radii are approximately proportional to $A^{1/3}$ it is usual to take out most of this dependence by putting $R = r_0 A^{1/3}$, while recognizing that the parameter r_0 may still have a weak dependence on A. These arguments only apply to the real part of the potential, but for convenience the same radial dependence was used in earlier calculations for the imaginary part as well. Subsequent work, outlined in § 6, has indicated substantial differences between the form factors of the real and imaginary parts of the potential.

The observation that elastically scattered nucleons are usually polarized implies that the optical potential is spin dependent, and, of the variety of possibilities, the spin–orbit form, familiar from the shell model, is the most convenient. Analogy with the spin–orbit force in atoms, substantiated by nuclear matter calculations, indicates the explicit form, applicable to particles of spin $\frac{1}{2}$

$$V_S(r) \equiv V_S f_S(r) = V_S \left(\frac{\hbar}{m_\pi c}\right)^2 \frac{1}{r} \frac{df(r)}{dr} \boldsymbol{L}.\boldsymbol{\sigma} \tag{2.2}$$

and this is able to account for most of the polarization data. More complicated forms of spin dependence are used for particles of spin greater than $\frac{1}{2}$ (see § 7). These simple considerations are adequate to establish the phenomenological form of the potential, and the parameters may be determined by systematically optimizing the fit to the experimental data, using the methods of § 5.

In practice it is frequently found that many sets of parameters give equally good fits to the data, and the question then arises whether any one of these is more physical than the others and if so which is to be preferred.

These parameter ambiguities, as they are called, are of two main types, discrete and continuous. Discrete ambiguities refer to regions of parameter space that give acceptable fits separated by unacceptable regions. The most notable example of this is the potential depth ambiguity, in which it is found, particularly for composite particles, that a whole series of potentials of increasing depth (eg 50, 100, 150 … MeV,) give almost equally good fits, the other parameters remaining the same. Continuous ambiguities refer to combinations of parameters that may be simultaneously varied subject to some constraint without significantly affecting the fit. The most notable examples of this type are the Ur_u^2 and Wa_w ambiguities in nucleon scattering; provided these products are kept constant, the individual parameters may be varied by as much as 20–30% without affecting the fit. The existence of these and other more complicated parameter ambiguities means that it is not possible to establish the optical potential by phenomenological analyses alone. It is necessary to start by constraining the potential as closely as possible by physical requirements *before* parameter optimization. This is not easy, as the physical constraints depend on complex many-body calculations, and even then there is no guarantee that they will be just adequate to resolve the ambiguities without adversely affecting the fits.

Discrete ambiguities do not occur for nucleons, since the depth of about 50 MeV appropriate to the shell model must be the physical one. In the case of deuterons and, to a lesser extent, helions, tritons and α particles the discrete ambiguities may be resolved by calculating the potential from the constituent nucleon–nucleus interactions. This is discussed more fully in the appropriate sections. Continuous ambiguities are more difficult to resolve, and it is usual to sidestep most of the trouble they cause by using standard values for the form factor parameters. This at least makes it possible to compare the potentials obtained from different analyses. Further progress thus requires a detailed theory of the optical potential in terms of the constituent nucleon–nucleus interactions. This is clearly a very difficult many-body problem, but substantial progress has been made by Brueckner (1954) and his colleagues and by Bethe (1956). The formal theory of nuclear reactions developed by Feshbach (1958, 1962) enables the contributions to the optical potential to be identified, but it is not very amenable to practical evaluation.

It would require a separate review to consider these theories in detail, but it is useful to indicate very briefly the essentials of Feshbach's formalism, as this clarifies several key features of the optical potentials. It enables the optical potential to be expressed in terms of the unperturbed hamiltonian H_0 and the potential $V(\boldsymbol{r}_0, \boldsymbol{\xi})$ between the incident particle and the target nucleus, where $\boldsymbol{r}_0$ is the position of the incident particle and $\boldsymbol{r}_1, \ldots, \boldsymbol{r}_A$ ($\equiv \boldsymbol{\xi}$) those of the particles in the target nucleus. The optical potential describes only the elastic scattering, but since the nonelastic processes remove flux from the elastic channel they must be taken into account as well. In addition there are higher-order processes, such as the excitation of virtual states and their subsequent de-excitation back to the elastic channel, that also contribute to the elastic scattering amplitude, so that the excited states $\phi_n(\boldsymbol{\xi})$ of the target must also be taken into account. The other channels involving transfer reactions can also contribute, but they are omitted here for simplicity, so that we consider only inelastic scattering.

The total hamiltonian is given by

$$H = H_0 + V(\boldsymbol{r}_0, \boldsymbol{\xi}) = H_0(\boldsymbol{\xi}) + T(\boldsymbol{r}_0) + V(\boldsymbol{r}_0, \boldsymbol{\xi}) \tag{2.3}$$

where $H_0(\boldsymbol{\xi})$ is the target hamiltonian and $T(\boldsymbol{r}_0)$ the kinetic energy of the incident particle. The total wavefunction Ψ satisfies the equation

$$(H - E)\Psi = 0 \tag{2.4}$$

and the unperturbed wavefunction Φ satisfies the equation

$$(H_0 - E)\Phi = 0. \tag{2.5}$$

Neglecting the requirements of antisymmetrization, Ψ may be expanded as a series of products of the wavefunction $\psi_\alpha(r_0)$ of the incident particle and the corresponding states of the target nucleus

$$\Psi(\boldsymbol{r}_0, \boldsymbol{\xi}) = \sum_\alpha \psi_\alpha(r_0)\,\phi_\alpha(\xi) \tag{2.6}$$

when the $\psi_\alpha(r_0)$ satisfy the equation

$$(T + \mathscr{V}(r_0) - \epsilon_\alpha)\,\psi_\alpha(r_0) = 0 \tag{2.7}$$

where $\mathscr{V}(r_0)$ is the optical potential and $\epsilon_\alpha = E - E_\alpha$, E_α being the excitation energy of the target nucleus.

We now define the operator P that projects from the total wavefunction only the ground state part of the target wavefunction

$$P\Psi = \Phi_0 = \phi_0\psi_0(r_0) \tag{2.8}$$

and also the operator

$$Q = 1 - P. \tag{2.9}$$

These operators satisfy the relations

$$P^2 = P, \quad Q^2 = Q, \quad PQ = 0. \tag{2.10}$$

Insertion of the unit operator in (2.4) gives

$$(H-E)(P+Q)\Psi = 0. \tag{2.11}$$

Multiplying by P and Q in turn and using (2.10) gives coupled equations for $P\Psi$ and $Q\Psi$

$$(E-PHP)P\Psi = PHQQ\Psi \tag{2.12}$$

$$(E-QHQ)Q\Psi = QHPP\Psi. \tag{2.13}$$

The latter equation may be solved formally to give

$$Q\Psi = \frac{1}{E-QHQ+\mathrm{i}\epsilon} QHPP\Psi \tag{2.14}$$

where the term $+\mathrm{i}\epsilon$ is a reminder that only outgoing solutions are required. In the subsequent formalism it will be omitted for convenience. Substitution of (2.14) in (2.12) gives a formal equation for the wavefunction in the incident channel,

$$\left(E-PHP-PHQ\frac{1}{E-QHQ}QHP\right)P\Psi = 0. \tag{2.15}$$

Taking the ground state of the target nucleus as the zero of energy

$$H(\xi)\phi_0 = 0 \tag{2.16}$$

and using the explicit expression for P

$$P = |\phi_0)(\phi_0| \tag{2.17}$$

where the curved brackets indicate dependence on the ξ coordinates only, gives

$$\begin{aligned} PHP &= |\phi_0)(\phi_0|H(\xi)+T+V|\phi_0)(\phi_0| \\ &= |\phi_0)\{T+(\phi_0|V|\phi_0)\}(\phi_0| \end{aligned} \tag{2.18}$$

and

$$\begin{aligned} PHQ &= |\phi_0)(\phi_0|H(\xi)+T+V|Q \\ &= |\phi_0)(\phi_0|VQ. \end{aligned} \tag{2.19}$$

Using these expressions, (2.15) becomes

$$\left\{E-T-(\phi_0|V|\phi_0)-\left(\phi_0\left|VQ\frac{1}{E-QHQ}QV\right|\phi_0\right)\right\}\psi_0(r_0) = 0. \tag{2.20}$$

Comparison with (2.7) gives an exact formal expression for the optical potential

$$\mathscr{V}(r_0) = (\phi_0|V|\phi_0)+\left(\phi_0\left|VQ\frac{1}{E-QHQ}QV\right|\phi_0\right). \tag{2.21}$$

The first term represents direct transitions from the incident to the outgoing channel while the second takes account of virtual nuclear excitations followed by de-excitations back into the elastic channel. The second term has singularities at the eigenvalues of QHQ, and to obtain a useful optical potential these must be smoothed by averaging over sufficiently large energy intervals.

The first term is purely real, while the second is complex, since it takes account of the nonelastic processes. Since the potential V can be expressed as a sum of the energy-dependent constituent nucleon–nucleon interactions then the optical potential also depends on energy. Further analysis of the second term shows it to be nonlocal, so that the effective potential at any point depends on the wavefunction throughout a surrounding region of greater or lesser extent depending on the range of the nonlocality. Mathematically, the term $V(\boldsymbol{r})\psi(\boldsymbol{r})$ in the Schrödinger equation is replaced by $\int V(\boldsymbol{r},\boldsymbol{r}')\psi(\boldsymbol{r}')\,\mathrm{d}\boldsymbol{r}'$, where $V(\boldsymbol{r},\boldsymbol{r}')$ is the nonlocal potential.

Nonlocal potentials have several features that are important for optical model analyses. Firstly, to every nonlocal potential there corresponds an equivalent local potential, in the sense that they give the same asymptotic wavefunctions and hence the same phase shifts. It is thus not possible to distinguish between a local and a nonlocal potential by a phenomenological analysis. In practice optical potentials are usually taken to be purely local for practical convenience, as the use of a nonlocal potential requires the solution of an integrodifferential equation. Secondly, the local potential equivalent to an energy independent nonlocal potential depends on the energy, and the degree of energy dependence is a function of the range of the nonlocality, so that in this sense nonlocality and energy dependence are equivalent. Thus part of the energy dependence of a local potential is intrinsic, and part is due to its nonlocality. Finally, the wavefunctions corresponding to equivalent local and nonlocal potentials differ inside the nucleus, and this has implications for reaction phenomena that depend on the wavefunctions in this region. This is further discussed in § 12.

In order to proceed further in the calculation of the optical potential, it is necessary to evaluate the expression (2.21) using realistic wavefunctions and interactions. This is a very difficult and complicated problem, particularly as regards the the second term, but considerable success has been achieved by calculating the real part of the potential from the first term only and representing the imaginary part by a phenomenological potential. The remainder of this section will be devoted to a brief summary of this work, due mainly to Greenlees and his colleagues.

If $v(|\boldsymbol{r}_i-\boldsymbol{r}_j|)$ is the nuclear–nucleon interaction and $\rho_{\mathrm{m}}(\boldsymbol{r})$ the nuclear matter distribution, then the first term of (2.21) becomes

$$V(\boldsymbol{r}) = \int\rho(\boldsymbol{r}')\,v(|\boldsymbol{r}-\boldsymbol{r}'|)\,\mathrm{d}\boldsymbol{r}' \qquad (2.22)$$

which is just the expression that would be written down classically. This accounts for the bulk of the real part of the optical potential, but it does omit terms corresponding to the real part of the second term in (2.21), in particular exchange terms due to the requirement of antisymmetrization. To evaluate (2.22), Greenlees *et al* (1969) used a Saxon–Woods form factor for the nuclear density and a phenomenological nucleon–nucleon interaction of the form

$$\begin{aligned} v(|\boldsymbol{r}_i-\boldsymbol{r}_j|)\equiv v(\boldsymbol{r}) = {} & v_{\mathrm{D}}(\boldsymbol{r})+v_{\tau}(\boldsymbol{r})\,\boldsymbol{\tau}_i\cdot\boldsymbol{\tau}_j+v_{\sigma}(\boldsymbol{r})\,\boldsymbol{\sigma}_i\cdot\boldsymbol{\sigma}_j \\ & +v_{\sigma\tau}(\boldsymbol{r})\,(\boldsymbol{\sigma}_i\cdot\boldsymbol{\sigma}_j)\,(\boldsymbol{\tau}_i\cdot\boldsymbol{\tau}_j)+\{v_{\mathrm{t}}(\boldsymbol{r})+v_{\mathrm{t}\tau}(\boldsymbol{r})\,\boldsymbol{\tau}_i\cdot\boldsymbol{\tau}_j\}\,S_{ij} \\ & +v_{\mathrm{S}}(\boldsymbol{r})\,\hbar^{-1}\{(\boldsymbol{r}_i-\boldsymbol{r}_j)\times(\boldsymbol{P}_i-\boldsymbol{P}_j)\cdot(\boldsymbol{\sigma}_i+\boldsymbol{\sigma}_j)\} \end{aligned} \qquad (2.23)$$

where S_{ij} is the tensor force operator and $\boldsymbol{\tau}_i$ the isospin operator of the ith nucleon. The real part of the optical potential may then be expressed as a sum of three terms,

$$V(\boldsymbol{r}) = V_{\mathrm{R}}(\boldsymbol{r})+V_{\mathrm{I}}(\boldsymbol{r})+V_{\mathrm{S}}(\boldsymbol{r}) \qquad (2.24)$$

where $V_{\mathrm{R}}(\boldsymbol{r})$ is independent of spin and isospin, $V_{\mathrm{I}}(\boldsymbol{r})$ is the isospin term and $V_{\mathrm{S}}(\boldsymbol{r})$

the spin–orbit term. Defining the matter distribution $\rho_m(\boldsymbol{r})$ as the sum of the neutron and proton density distributions

$$\rho_m(\boldsymbol{r}) = \rho_n(\boldsymbol{r}) + \rho_p(\boldsymbol{r}) \tag{2.25}$$

we obtain

$$V_R(r) = \int \rho_m(\boldsymbol{r}')\, v_D(|\boldsymbol{r}-\boldsymbol{r}'|)\, d\boldsymbol{r}' \tag{2.26}$$

$$V_I(r) = \tau_z \int (\rho_p(\boldsymbol{r}') - \rho_n(\boldsymbol{r}'))\, v_\tau(|\boldsymbol{r}-\boldsymbol{r}'|)\, d\boldsymbol{r}' \tag{2.27}$$

and a rather more complicated expression for $V_S(r)$.

These expressions give the optical potential in terms of interactions and density distributions that are already known, at least approximately, and thus remove many of the uncertainties of a purely phenomenological analysis. In practice it is usual to treat the radius and diffuseness of the nuclear matter distribution as adjustable parameters in order that they may be determined by comparison with experimental data.

In their calculations, Greenlees *et al* (1968, 1970a,b) used a Yukawa form for $v_D(r)$ and adjusted its strength and range to the two-body data. The imaginary part of the potential cannot easily be calculated, so they used a purely phenomenological expression. They were able to fit their calculations to experimental data on

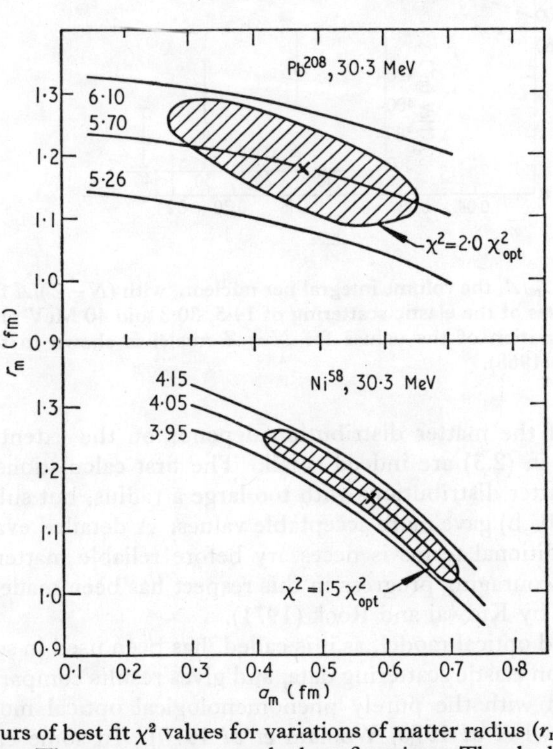

Figure 3. Contours of best fit χ^2 values for variations of matter radius (r_m) and diffuseness (a_m) parameters. The crosses represent the best-fit points. The shaded area encompasses a region within which the fits are visibly indistinguishable. Lines of constant root mean square matter radius $\langle r^2\rangle_m{}^{1/2}$, denoted by values in fermis, are seen to run nearly parallel to the sides of the shaded areas. From Greenlees *et al* (1968).

elastic scattering of 30 MeV protons by several medium and heavy nuclei, using matter distributions in accord with those found in other investigations. An important result of this analysis is that the elastic scattering data fixes the root mean square radius of the real part of the optical potential and its volume integral per nucleon much more accurately than the three parameters of the corresponding form factor. The volume integral per nucleon is found to be independent of energy and the nuclear symmetry parameter $(N-Z)/A$. These results are illustrated in figures 3 and 4.

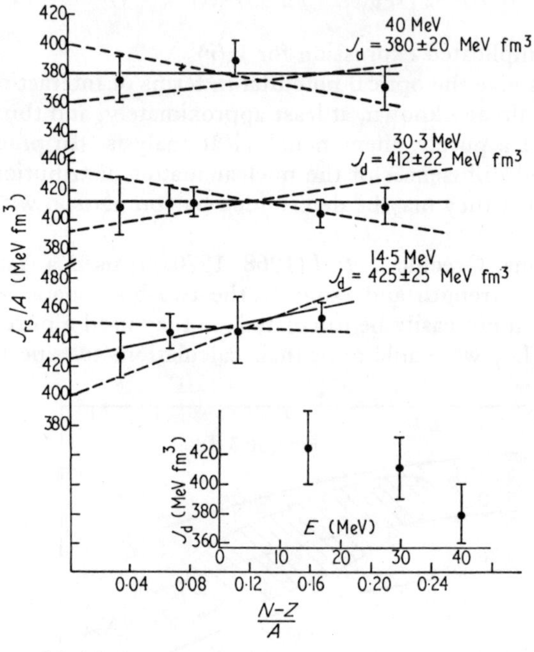

Figure 4. Variation of J_{rs}/A, the volume integral per nucleon, with $(N-Z)/A$ for the potentials found by analyses of the elastic scattering of 14·5, 30·3 and 40 MeV protons by nuclei. The energy variation of the values for $N = Z$ nuclei is shown in the inset. From Greenlees *et al* (1968).

The accuracy of the matter distribution depends on the extent to which the terms not included in (2.3) are indeed small. The first calculations of Greenlees *et al* (1968) gave matter distributions with too large a radius, but subsequent work (Greenlees *et al* 1970a,b) gave more acceptable values. A detailed evaluation of the more important additional terms is necessary before reliable matter distributions can be obtained. Encouraging progress in this respect has been made by Owen and Satchler (1970) and by Kidwai and Rook (1971).

The reformulated optical model, as it is called, has been used in several analyses of neutron and proton elastic scattering data, and gives results comparable in quality with those obtained with the purely phenomenological optical model (Pyle and Greenlees 1969, Boyd *et al* 1970, Woolam *et al* 1970). In some cases, however, particularly at back angles, the fits are not yet adequate, and it is clear that more work needs to be done on the higher-order correction terms to (2.22). Recent studies of the interactions of light nuclei using the resonating group method suggest how this might be done. Thus Tang (1969) has shown that when antisymmetrization is taken into account the wavefunction describing the interaction between two light nuclei satisfies the equation

$$\nabla^2 \psi(\boldsymbol{r}) + \frac{2m}{\hbar^2}(E - V_{\rm R}(r) - V_{\rm C}(r))\,\psi(\boldsymbol{r}) = \int K(\boldsymbol{r}, \boldsymbol{r}')\,\psi(\boldsymbol{r}')\,{\rm d}\boldsymbol{r}' \tag{2.28}$$

where $V_{\rm R}(r)$ is the direct potential obtained by folding the matter distribution and the nucleon–nucleon interaction, and $K(\boldsymbol{r}, \boldsymbol{r}')$ is a nonlocal kernel of the form

$$K(\boldsymbol{r}, \boldsymbol{r}') = g_1(r, r')\exp\{-\beta_1(\boldsymbol{r}-\boldsymbol{r}')^2\} + g_2(r, r')\exp\{-\beta_2(\boldsymbol{r}+\boldsymbol{r}')^2\} \tag{2.29}$$

where β_1 and β_2 are related to the ranges of the nonlocalities. The first of these terms is a nonlocality of the type already used by Perey and Buck (1962) in their analysis of neutron scattering, while the second gives the exchange contributions and is particularly important at back angles. It is likely that similar terms are also present in the full expression for the optical potentials for heavier nuclei. Thus inclusion of the latter term in the optical potential could improve the fits in the backward direction for the differential cross section for 30 MeV protons on ^{58}Ni, and possibly also the back angle data for α particle scattering, though in this case other effects may also be present (see § 9) (Greenlees and Tang 1971).

In all these calculations of the optical potential, the real part obtained from a folding procedure is supplemented by a phenomenological imaginary potential. Many attempts have, however, been made to calculate the imaginary potential; indeed the earliest semiclassical model already gave a simple expression in terms of the velocity of the particle and its absorption coefficient K in nuclear matter,

$$W = v\hbar K/2 \tag{2.30}$$

where

$$K = \frac{3A\sigma}{4\pi R^2} \tag{2.31}$$

R being the nuclear radius and σ the mean cross section for nuclear collisions

$$\sigma = \{Z\sigma_{\rm np}\,\alpha_{\rm np} + (A-Z)\,\sigma_{\rm nn}\,\alpha_{\rm nn}\}/A. \tag{2.32}$$

The coefficients $\alpha_{\rm np}$ and $\alpha_{\rm nn}$ allow for the reduction in cross section due to the Pauli principle, and have been evaluated semiclassically by Goldberger (1948).

These calculations of the imaginary potential have been improved by introducing more exact expressions for the nucleon–nucleon interaction and more sophisticated ways of allowing for the Pauli principle (Greenlees *et al* 1968, Afnan and Tang 1970). This confirms two important features of the phenomenological potential, namely that it increases with increasing energy and that its radial extent is substantially greater than that of the real part of the potential. The final form of the potential depends in rather a complicated way on the energy variations of the contributing interactions and on the inhibiting effect of the Pauli principle. There are still discrepancies between the calculated and phenomenological potentials, which may be due to nonlocal effects arising from antisymmetrization. Further work is thus necessary to define the form of the imaginary part of the potential and its variation with energy. The remaining uncertainty in the imaginary potential fortunately does not affect the work on the real part of the potential, since the fit to experimental data is quite insensitive to the form chosen, provided its parameters are optimized.

3. Nuclear single-particle states

Any potential of sufficient depth has a series of bound states, and their ordering and specification by the total, orbital and spin quantum numbers is familiar from the shell model. The energies of these states may easily be found experimentally using one-nucleon transfer reactions for nuclei near closed shells that often have low-lying single-particle states of high purity. Even for nuclei away from closed shells where the single-particle states are split into many fragments by the residual interactions it is still possible to define an equivalent single-particle energy

$$\bar{E} = \sum_i g_i E_i \tag{3.1}$$

where the g_i are the weights (spectroscopic factors) of the fragments of energies E_i. The cross section for a one-nucleon transfer reaction to such a state is given by

$$\frac{d\sigma}{d\Omega} = \mathscr{S} F(E, \theta) \tag{3.2}$$

where $\mathscr{S}$ is the spectroscopic factor that depends only on the structures of the initial and residual nuclei and all the energy and angular dependence is contained in the reaction factor $F(E, \theta)$, which may be calculated by the distorted wave theory. These calculations are now so well understood and calibrated that providing the direct nucleon transfer reaction is the dominant mode and the final state is largely single particle, the absolute cross section, and hence the spectroscopic factor, can be determined to within about 20%, thus providing a powerful method of obtaining the energies and strengths of single-particle states. Neutron single-particle states are most conveniently studied by the (d, p) and (p, d) reactions for the unfilled and filled levels respectively and the proton states by the (h, d) and (d, h) reactions, as in the (d, n) and (n, d) reactions it is difficult to determine the neutron energies with sufficient precision.

Single particle states may also be studied by nucleon knockout reactions, particularly (p, 2p), and although the corresponding energy measurements are usually less accurate than for the nucleon transfer reactions they do enable rather deeply lying states to be investigated. Once the energies of the single-particle states are known the depth of the corresponding potential may easily be determined, providing a suitable form factor is assumed. It is usual to take a potential of the form

$$V(r) = Vf(r) + V_S f_S(r) \tag{3.3}$$

with the form factors given by (2.1) and (2.2). For charged particles the electrostatic potential due to a uniformly charged sphere is added to (3.3). It is convenient to begin by describing the method of solution in the absence of a spin–orbit term. If this potential is inserted in the Schrödinger equation and the wavefunction expanded as a product of angular and radial functions

$$\Psi(r) = \sum_L \frac{u_L(r)}{r} \mathrm{i}^L P_L(\cos\theta) \tag{3.4}$$

the usual multiplication from the left by $P_{L'}(\cos\theta)$ and integration over the angular coordinates gives the radial wave equation

$$\frac{d^2 u_{NL}(r)}{dr^2} + \left(\frac{2m}{\hbar^2}(E - V(r)) - \frac{L(L+1)}{r^2}\right) u_{NL}(r) = 0. \tag{3.5}$$

Since the nuclear potential falls rapidly to zero the asymptotic solution to this wave equation may be obtained by solving analytically the equation without the term $V(r)$. This external solution may then be matched to the solution in the internal region, and the required bound state is defined by the condition that the interior and exterior solutions match smoothly. Since the energy is not known, an iterative method has to be used. For each value of the orbital angular quantum number L there are several internal solutions, with different numbers of nodes, that can be matched to the exterior solution. These correspond to states of different principal quantum number N, which is accordingly required to specify the bound states. The whole calculation can be carried out electronically and the convergence is rapid (Hodgson 1963).

In practice the potential contains a spin–orbit term $V_S f_S(r)$ and this gives two radial wave equations with total angular momenta $J = L \pm \frac{1}{2}$, corresponding to the two spin orientations. Each of these equations may be solved in the way already described. The energies of the two components of a $J = L \pm \frac{1}{2}$ pair, for example of the $1p_{1/2}$ and $1p_{3/2}$ states, then suffice to fix both V and V_S. The potential is thus determined for particular values of N and L, unlike the optical potential for free particles which is assumed to be the same for all N and L. The study of bound single-particle states can thus give information on the state dependence of the potential.

Many studies of single-particle states in nuclei have now been made, particularly in nuclei with closed shells plus or minus one nucleon, for then the single-particle states are particularly pure. In the case of ^{208}Pb, for example, there are four nuclei that may be found by adding or subtracting a nucleon, and in these at least 22 strongly marked single-particle states may be identified. If the parameters of the potential (3.3) are then adjusted to optimize the overall fit to these states, the usefulness of the model can be evaluated. This has been done by Rost (1968) and some of his results are shown in figure 5. The overall fit to the data is as good as can be expected with such a simple model. Similar calculations have been made by Batty and Greenlees (1969).

These single-particle potentials can be used to calculate the wavefunctions, and hence the matter and charge distributions for the particles in the observed levels, and if it is assumed that the same potential applies to the lower-lying particles as well the distributions for the whole nucleus can be obtained. The resulting charge distributions have been found to account very well for the elastic scattering of 175 and 250 MeV electrons by ^{208}Pb, and this constitutes a useful check on the model (Elton 1968).

If these single-particle states are studied throughout the periodic table, the variation of their binding energies with neutron and proton numbers may be studied. An overall picture may be obtained by calculating the energies of all the single-particle states in the potential (3.3), with all the parameters fixed, except the radius which is allowed to depend on A. The result of such a calculation is shown in figure 6. The grouping of the studies into bands characteristic of the shell model is immediately evident.

Detailed studies of bound single-particle states are made in two ways; firstly by examining as many states as possible in the same nucleus as in the work on ^{208}Pb already mentioned, and secondly by following particular states from one nucleus to the next across the periodic table.

Several analyses of the former type have been made for light and medium weight nuclei, as well as for ^{208}Pb. Thus Brown *et al* (1963) studied the single-particle states in Ni and Pb and found that the calculated levels tended to be rather more widely spaced than the observed levels. This can be accounted for if the potential depends on either the state or the energy of the particle concerned. They chose the latter alternative, and found that if this energy dependence is expressed in terms of effective mass, then m^*/m is greater than unity. This is in contrast to the situation for positive energies, where the energy dependence corresponds to an effective mass around $0{\cdot}7m$.

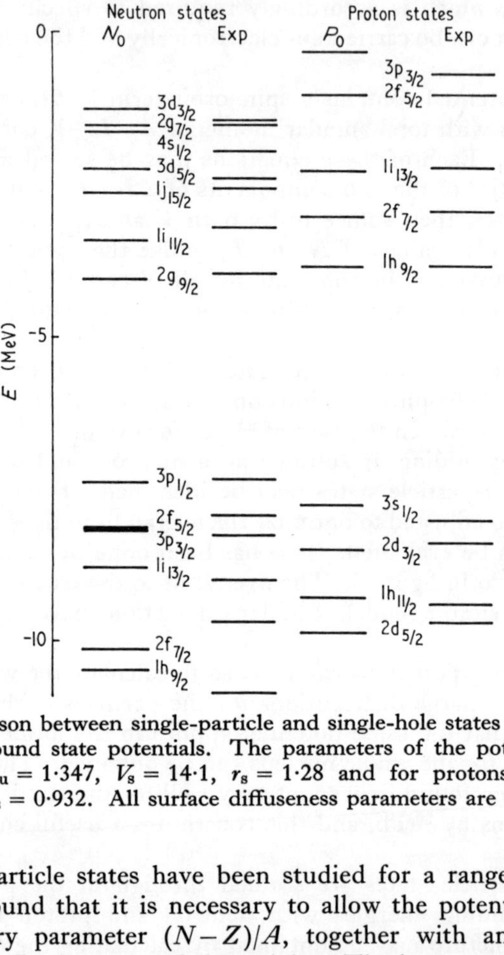

Figure 5. Comparison between single-particle and single-hole states for ^{208}Pb and the eigenvalues of bound state potentials. The parameters of the potentials are for neutrons $V = 40{\cdot}6$, $r_u = 1{\cdot}347$, $V_s = 14{\cdot}1$, $r_s = 1{\cdot}28$ and for protons $V = 58{\cdot}7$, $r_u = 1{\cdot}275$, $V_s = 11{\cdot}5$, $r_s = 0{\cdot}932$. All surface diffuseness parameters are fixed at $0{\cdot}7$. From Rost (1968).

The single-particle states have been studied for a range of nuclei by Cohen (1965) and he found that it is necessary to allow the potential to depend on the nuclear symmetry parameter $(N-Z)/A$, together with an energy dependence equivalent to an effective mass of about $1{\cdot}3m$. The large number of measurements of nucleon transfer reactions made during the last few years enables this analysis to be extended in a more precise way to a larger range of nuclei. For such work it is necessary to take into account the implications of the isospin dependence of the potential. A proton stripping reaction can excite both $T_>$ and $T_<$ states, while the condition that $T \geqslant T_Z$ implies that the neutron stripping reaction can excite only $T_>$ states, and conversely for pickup reactions. If the nucleon optical potential contains an isospin term of the form

$$V_I(r) = 4/f(r)\frac{V_1}{A}t\,.\,T\frac{4V_1}{A}t\,.\,Tf(r) \tag{3.6}$$

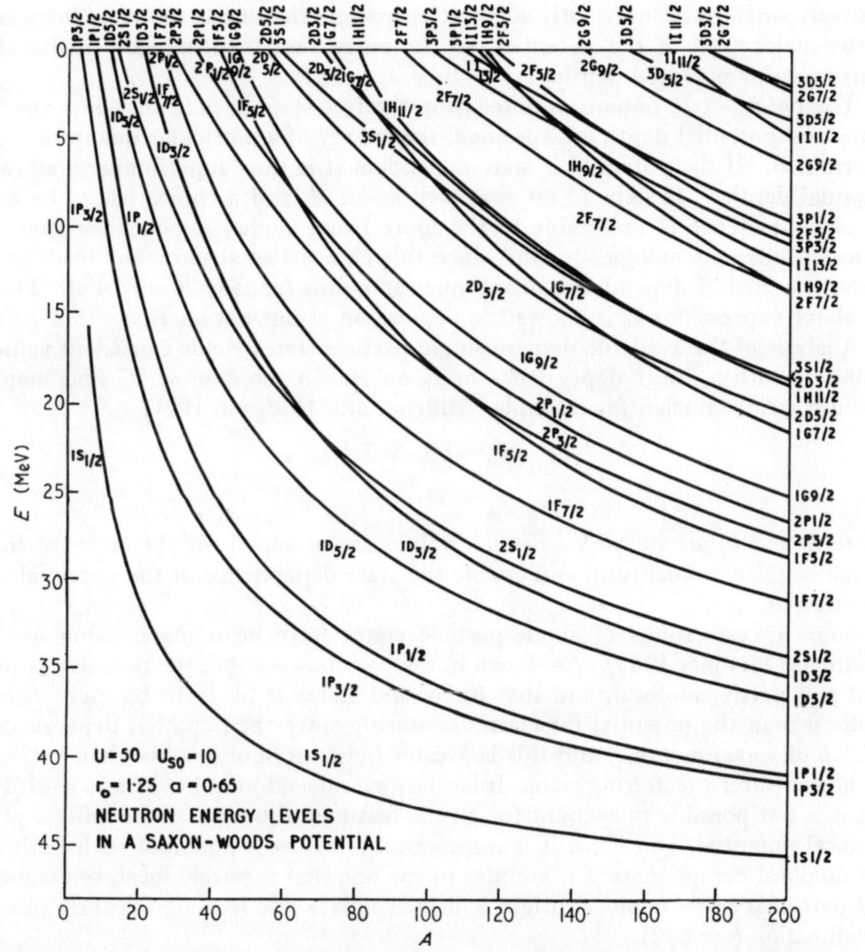

Figure 6. Single-particle states (neutron energy levels) in a Saxon–Woods potential.

where t and T are the isospins of the incident nucleon and target nucleus respectively, the resulting expressions for the potentials are, for stripping reactions,

$$\left.\begin{aligned} V_{\mathrm{N}}^{>} &= V_0 - V_1(N-Z)/A \\ V_{\mathrm{P}}^{<} &= V_0 + V_1(N-Z+2)/A + \beta Z/A^{1/3} \\ V_{\mathrm{P}}^{>} &= V_0 - V_1(N-Z)/A \end{aligned}\right\} \tag{3.7}$$

and for pickup reactions

$$\left.\begin{aligned} V_{\mathrm{N}}^{<} &= V_0 - V_1(N-Z-1)/A \\ V_{\mathrm{P}}^{>} &= V_0 + V_1(N-Z+3)/A + \beta(Z-1)/A^{1/3} \\ V_{\mathrm{N}}^{>} &= V_0 + V_1(N-Z+3)/A \end{aligned}\right\} \tag{3.8}$$

where N, Z and A refer to the target nucleus. The extra term $0{\cdot}4Z/A^{1/3}$ has been added to the proton potentials for the reasons described in §6. As this term is relatively small, it is sufficiently accurate to assume that it has the same form factor as the major part of the potential, so the correction then appears in the above expression for potential depths.

The bound-state potentials differ from the free state ones in that once the form factor and potential depth are specified, the energy of a particular quantum state is determined. If the potential is state dependent it is thus superfluous to allow the potential depth V to depend on E as well as on A, and a choice has to be made. An A dependence is preferable to the more usual energy dependence since it is indicated phenomenologically, and since this choice also absorbs any inadequacies in the assumed A dependence of the nuclear radius (usually $R = r_0 A^{1/3}$). Thus in the above expressions V_0 is allowed to depend on A but not on E.

Analysis of the available data on single-particle states yields consistent values of V_0 and V_1 with a linear dependence of V_0 on A. In the case of 2S hole states in medium weight nuclei, for example (Millener and Hodgson 1971),

$$\begin{aligned} V_0 &= 59 - 0{\cdot}2A \\ V_1 &= 37 \end{aligned} \tag{3.9}$$

where V_0 and V_1 are in MeV. Similar values are obtained for the states of higher orbital angular momentum, and enable the state dependence of the potential to be investigated.

Some investigations of single-particle states have been made using nonlocal potentials (Meldner 1969). As shown in the previous section, the potential is partly local and partly nonlocal, and that for bound states is likely to be more strongly nonlocal than the potential for elastic scattering since the potential depends on an overlap of wavefunctions, and this is greater between bound states than between a bound state and a scattering state. It has however been found by Grimm *et al* (1971) that it is not possible to account for all the features of bound states using a purely nonlocal potential, and since it is impracticable to use a potential with both local and nonlocal components it is simpler to use one that is purely local, remembering that part of it has a nonlocal origin and hence gives rise to a significantly different wavefunction (see § 12).

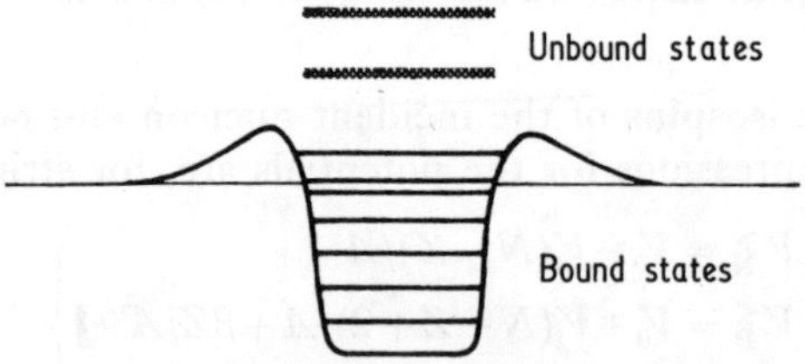

Figure 7. Bound and unbound states of a potential well.

The nucleon–nucleus potential has unbound states as well as bound states (see figure 7); these are sometimes referred to as bound states embedded in the continuum. The energies of these states may be calculated in a similar way to the corresponding bound states, except that now the interior wavefunctions are matched to the scattering wavefunctions that are solutions of the appropriate Schrödinger radial wave equation. The resonances occur at energies for which the phase shift η passes through $\pi/2$ and the rate $d\eta/dE$ at which it does so is proportional to its width and inversely proportional to its lifetime. The resonances are sharp and long-lived at low energies and become broader and shorter-lived as the energy increases. If reaction channels are open the resonances are correspondingly broadened, and this may be taken into account by allowing the potential to become complex.

These unbound states may be investigated experimentally in several ways, in particular by elastic scattering and nucleon-transfer reactions. For light nuclei at low energies the reaction channels are closed and the unbound states are sharp enough to be clearly visible on the elastic scattering excitation function, particularly at backward angles. Apart from these resonances, the cross section is frequently purely Rutherford, so the cross section may easily be analysed as a superposition of a Coulomb and a Breit–Wigner amplitude, and hence the energy, spin and width of the unbound state determined. At higher energies the nuclear scattering becomes significant, and the corresponding scattering amplitude may easily be calculated from an optical model potential adjusted to the data in the regions away from resonances. Comparison between the measured and calculated width gives immediately the single-particle strength of the state. Thus, for example, a $d_{5/2}$ resonance occurs at 1·75 MeV in the elastic scattering of protons by carbon. If the depth of a potential with standard form factors is adjusted to give a $d_{5/2}$ resonance at this energy the corresponding width is 56 keV. Since the measured width is 61 keV the state is essentially a pure single-particle one (Schiffer 1963).

Unbound single-particle states can also be studied by nucleon transfer reactions. The lifetime of the state is usually much greater than the reaction time so the reaction is unaffected by its subsequent decay. The angular distributions are characteristic of the orbital angular momentum transfer, and the absolute cross sections give the spectroscopic factor and parity of the single-particle state, in just the same way as in nucleon transfer reactions to bound states.

The calculation of the cross section of direct reactions to unbound states may be made with the same distorted wave formalism as for bound states except that the wavefunction of the captured particle has the oscillatory tail at large distances characteristic of a scattering state. Inside the nucleus the wavefunction at the resonance energy is very similar to that of a bound state. These wavefunctions may be calculated by solving the radial wave equation at the resonance energy or by using the Gamow pole functions (Bang and Zimanyi 1969). The oscillatory tail of these unbound state wavefunctions gives rise to convergence difficulties in the overlap integral for the reaction matrix elements. There are large, alternatively positive and negative, contributions from outside the reaction region and, though these eventually cancel, small numerical errors can have a large effect on the final result. In practice the integration is carried out numerically as far as an external cut-off radius, chosen so that the contribution from the region still farther out is negligible. For direct reactions to bound states it is sufficient to integrate to a few fermis beyond the nuclear surface, but for unbound states adequate convergence is often not attained by integrating out to 100 fm or more (Fortune *et al* 1969, Youngblood *et al* 1970). Several attempts have been made to improve the convergence of the integral by the application of a convergence factor (Huby and Mines 1965) and by altering the path of integration (Fortune and Vincent 1969). More recently, Huby (1970) has proposed the use of 'pseudobound' wavefunctions, obtained as unbound eigenvalues of the appropriate Saxon–Woods potential, the small oscillations at large distances being cut off exponentially near the first node outside the nucleus. Huby justifies this prescription by Weinberg's quasiparticle formalism, and the results are found to be insensitive to the precise details of the cut-off (Cole *et al* 1970). The spectroscopic factors found in this way are consistent with those obtained for the widths of the corresponding resonances excited by elastic scattering. It is thus now possible to analyse transitions to unbound states just as easily as those to bound states.

The optical potentials describing the interaction of composite particles with nuclei also have unbound states, and although they are too broad to be observable in most circumstances, it is possible that they could be found at low energies in light nuclei. A possible indication of the presence of such states is found in the α particle transfer reactions in medium weight nuclei studied by Faraggi *et al* (1971). Figure 8

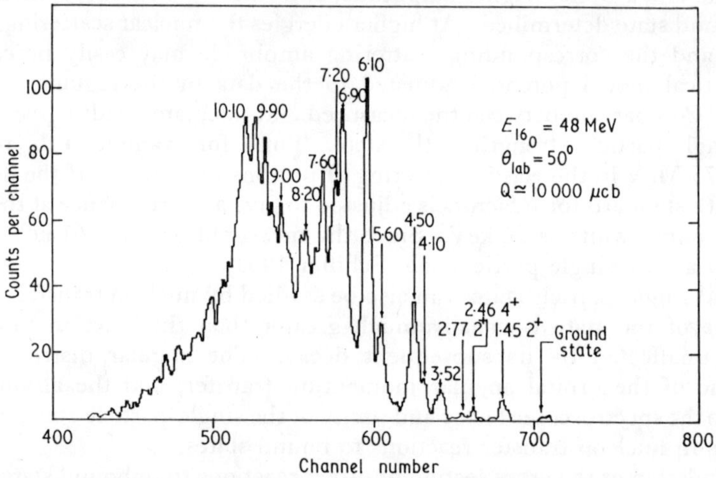

Figure 8. Energy spectrum of the outgoing particles in the reaction $^{54}Fe(^{16}O, ^{12}C)^{58}Ni$ showing the peaks ascribed to α particle resonant states. From Faraggi *et al* (1971).

shows the energy spectrum of the outgoing particles in the $^{54}Fe(^{16}O, ^{12}C)^{58}Ni$ reaction, and there are several prominent peaks at energies where the level density in the residual nucleus is so high that a smooth Maxwellian energy distribution is expected. Similar results were obtained for several medium weight nuclei. Lane (1971 private communication) has suggested that these observations could be explained if the α particle was first captured into a resonant or unbound state in an α particle optical potential and preliminary calculations (Dudek and Hodgson 1972) show that the observed density of such states is in accord with this model. The states are grouped into bands, each corresponding to a particular number of oscillator quanta, and alternate bands are composed of states with all odd J and all even J. It will be a critical test of this model to confirm this prediction, and this can be done in principle by comparing the experimental angular distributions with distorted wave calculations in the usual way.

4. Nuclear reactions

In order to define the conditions under which the optical model may be used to calculate nuclear interaction cross sections it is necessary to consider all the types of reaction that take place. This will be done for the special case of protons but the discussion is applicable to other particles as well.

A proton incident on a nucleus can undergo either a direct scattering or reaction in a time of the order of the transit time, or it can be captured to form a compound nucleus. The direct processes are (shape) elastic scattering, inelastic scattering and the whole range of charge-exchange, knockout and nucleon-transfer reactions. If the proton is captured to form the compound nucleus a complicated excitation

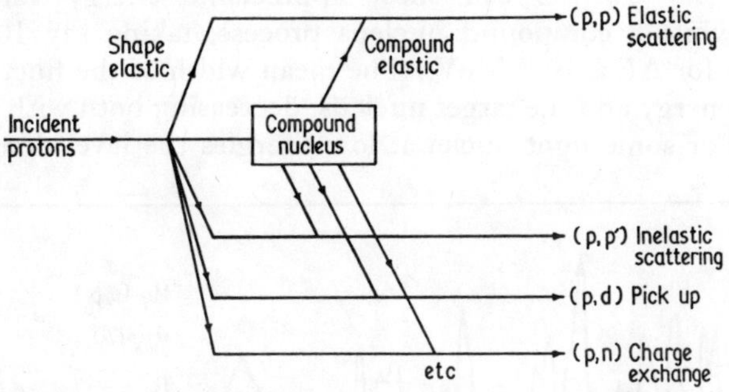

Figure 9. Direct and compound nucleus processes in nuclear reactions.

process takes place and, after a long time on the nuclear scale, it de-excites by emission of particles and eventually β and γ rays. The particles can be emitted into all the open reaction channels, including the elastic channel, and thus all cross sections have both a direct and a compound nucleus component. This is illustrated schematically in figure 9.

The cross sections of the direct process vary smoothly with energy, while those of the compound nucleus processes show rapid fluctuations over energy intervals that may be as small as a few electron volts or less. This difference arises from the

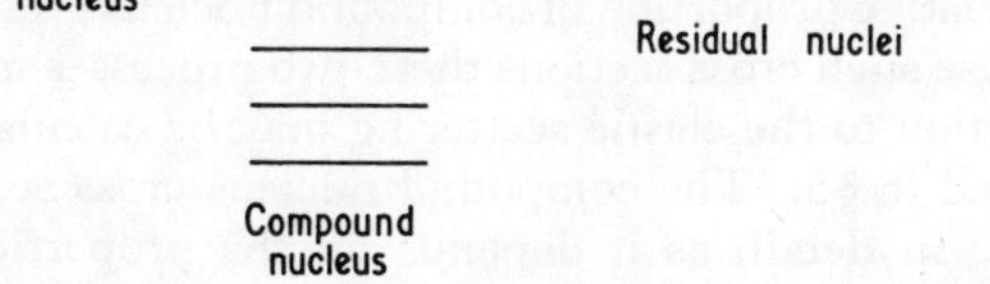

Figure 10. Energy diagram illustrating the excitation of intermediate compound nucleus states in a nuclear reaction.

characteristics of the two processes. Direct reactions proceed directly from the initial to the final state and so the cross section depends on a matrix element expressible in terms of the initial and final wavefunctions and the interaction hamiltonian; since these vary smoothly with energy so does the cross section itself. Compound nucleus reactions, on the other hand, proceed through a very large number of complicated intermediate states. The density of levels in the compound nucleus is very high, and so, as indicated schematically in figure 10, a very small change in the incident energy suffices to alter completely the intermediate states, and hence the cross section.

This difference between the energy dependence of direct and compound nucleus reactions may also be seen as a consequence of the time they take, making use of the uncertainty relation between energy and time, $\Delta E \Delta t \geqslant h$. Thus a direct process, taking say 10^{-22} s, can show appreciable energy variation only for $\Delta E \simeq 10$ MeV, while a compound nucleus process, taking say 10^{-16} s, can show energy variation for $\Delta E \approx 10^{-5}$ MeV. The mean width of the fluctuations depends on the incident energy and the target nucleus, decreasing both with energy and with nuclear mass. For some light nuclei at low energies the level density is such that

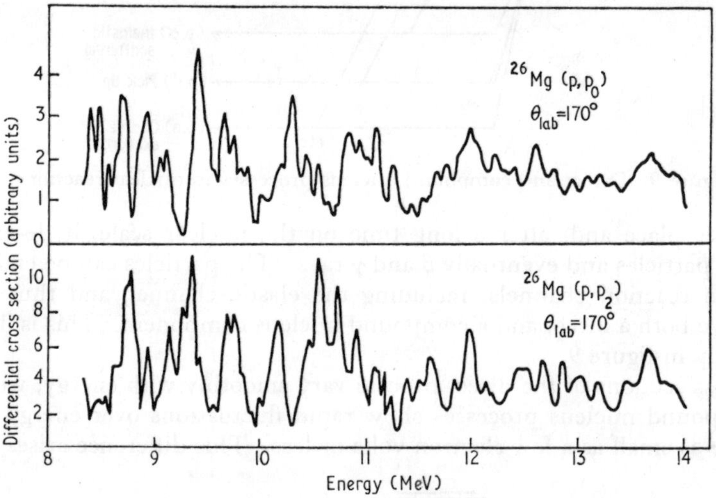

Figure 11. Excitation functions for the reactions ^{26}Mg(p, p$_0$) and ^{26}Mg(p, p$_2$) showing fluctuations. The amplitude of the fluctuations decreases with increasing energy due to the rising contribution of direct processes to the reactions. From Häusser *et al* (1964).

accurate and painstaking measurements are able to show the rapid fluctuations of the cross section with incident energy. Two examples of such cross sections are shown in figure 11 which also illustrates the gradual decline in the amplitude of the fluctuations as the relative proportion of compound nucleus cross section diminishes.

In order to analyse such cross sections these two processes must be disentangled. The direct contribution to the elastic scattering may be calculated from the optical potential as described in §5. The compound nucleus cross section is far too complicated to calculate in detail, as it depends on the properties of all the myriad compound states. It is however possible to analyse the overall statistical features of the fluctuating cross sections and hence to obtain the ratio of direct to compound contributions. A simple analysis is possible if the cross sections are averaged over energy intervals greater than the mean width of the fluctuations. The energy-averaged compound nucleus cross sections can be calculated as described in §11, and the analysis given later in this section shows that the observed cross section is simply the sum of the direct cross section and the energy-averaged compound nucleus cross section. Thus the compound nucleus component may be subtracted from the measured cross section to give the direct (shape elastic) cross section, and this may be analysed as described in §5.

At higher energies the compound nucleus contribution to a particular channel is frequently negligible, as the number of channels open to the decay of the compound nucleus increases extremely rapidly with energy. In other cases the compound nucleus contribution is appreciable and must be subtracted before an optical model analysis can be made. It is thus essential to understand the relation of the direct and compound processes before commencing an optical model analysis. This may be done by using the theory of nuclear reactions to relate the real (fluctuating) cross sections to the energy-averaged cross sections. For simplicity it is convenient to consider only the S wave ($L = 0$) neutron channel; this gives the essential physics and the results may easily be extended to the other channels.

The elastic scattering is completely described by the (energy dependent) matrix element S, and the elastic, reaction and total cross sections are then given by

$$\left.\begin{aligned}\sigma_{\mathrm{E}} &= (\pi/k^2)|1-S|^2\\ \sigma_{\mathrm{R}} &= (\pi/k^2)(1-|S|^2)\\ \sigma_{\mathrm{T}} &= (2\pi/k^2)(1-\mathrm{Re}\,S).\end{aligned}\right\} \tag{4.1}$$

The matrix element S and the corresponding cross sections fluctuate rapidly with energy. The energy-averaged quantities may similarly be described by an average matrix element $\langle S\rangle$, where the angular brackets represent averaging over an energy interval larger than the mean width of the underlying resonances. Thus

$$\left.\begin{aligned}\hat{\sigma}_{\mathrm{E}} &= (\pi/k^2)|1-\langle S\rangle|^2\\ \hat{\sigma}_{\mathrm{R}} &= (\pi/k^2)(1-|\langle S\rangle|^2)\\ \hat{\sigma}_{\mathrm{T}} &= (2\pi/k^2)(1-\mathrm{Re}\,\langle S\rangle).\end{aligned}\right\} \tag{4.2}$$

These quantities may be related by expressing the matrix element S as the sum of an average and a fluctuating component

$$S = \langle S\rangle + \hat{S}. \tag{4.3}$$

Then substituting in (4.1) and carrying out the energy-averaging

$$\left.\begin{aligned}\sigma_{\mathrm{E}} &= \hat{\sigma}_{\mathrm{E}} + \sigma_{\mathrm{F}}\\ \sigma_{\mathrm{R}} &= \hat{\sigma}_{\mathrm{R}} - \sigma_{\mathrm{F}}\\ \sigma_{\mathrm{T}} &= \hat{\sigma}_{\mathrm{T}}\end{aligned}\right\} \tag{4.4}$$

where the fluctuating cross section σ_{F} is defined by

$$\sigma_{\mathrm{F}} = (\pi/k^2)(\langle|S|^2\rangle - |\langle S\rangle|^2). \tag{4.5}$$

To investigate this fluctuating component it is necessary to make use of the expression for the S matrix element in terms of a phase shift that varies smoothly with energy and a sum over the underlying compound nucleus resonances

$$S = \mathrm{e}^{2\mathrm{i}\delta}\left(1 - \sum_s \frac{\mathrm{i}\Gamma_{\mathrm{n}}^s}{E - E_s + \mathrm{i}\Gamma^s/2}\right) \tag{4.6}$$

and hence

$$\langle S\rangle = \mathrm{e}^{2\mathrm{i}\delta}\left(1 - \frac{\pi\overline{\Gamma}_{\mathrm{n}}}{D}\right) \tag{4.7}$$

where $\bar{\Gamma}_n$ is the mean neutron width and D the mean level separation. The compound nucleus contribution to the cross section can be calculated from the Breit–Wigner resonance formalism, giving

$$\langle\sigma_{CN}\rangle = \left(\frac{\pi}{K^2}\right)\left(\frac{2\pi\bar{\Gamma}_n}{D}\right). \tag{4.8}$$

Then putting (4.7) in (4.2), neglecting terms in $(\bar{\Gamma}/D)^2$ and using (4.5) gives

$$\langle\sigma_R\rangle = \langle\sigma_{CN}\rangle - \sigma_F. \tag{4.9}$$

Since

$$\langle\sigma_{CN}\rangle = \langle\sigma_R\rangle + \langle\sigma_{CE}\rangle \tag{4.10}$$

the fluctuating component is identified as the compound elastic contribution to the cross section and this applies to the differential as well as to integrated cross sections. It is thus possible to analyse a cross section with direct and compound components by evaluating them independently and then simply adding the results. Some examples of such analyses are given in § 11.

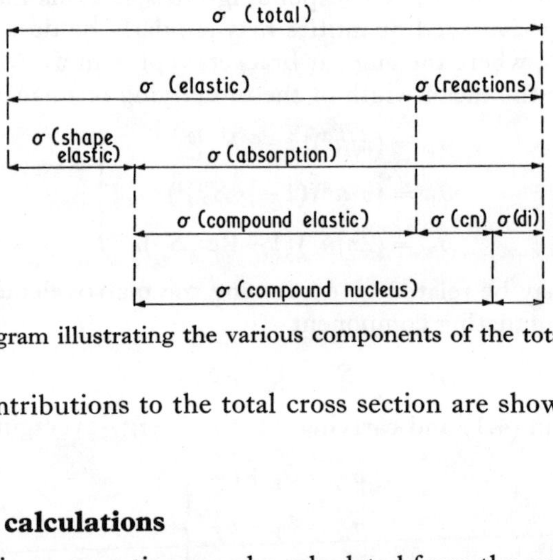

Figure 12. Diagram illustrating the various components of the total cross section.

The various contributions to the total cross section are shown schematically in figure 12.

5. Optical model calculations

The shape elastic cross section may be calculated from the optical potential $V(r)$ by solving the Schrödinger wave equation

$$\nabla^2\Psi + (2m/\hbar^2)(E - V(r))\Psi = 0 \tag{5.1}$$

subject to the boundary conditions of incident plane wave and an outgoing scattered wave

$$\Psi \sim e^{ikz} + (1/r)\,e^{ikr} f(\theta). \tag{5.2}$$

It is usual to assume the optical potential is spherically symmetrical so that the wavefunction can be expanded in Legendre polynomials

$$\Psi = \sum_L \frac{u_L(r)}{r} i^L P_L(\cos\theta) \tag{5.3}$$

where L is the orbital angular momentum quantum number. Substituting this in

(5.1) and carrying out the angular integration gives the radial wave equation for the Lth partial wave

$$\frac{d^2 u_L(r)}{dr^2} + \left(\frac{2m}{\hbar^2}(E - V(r)) + \frac{L(L+1)}{r^2}\right) u_L(r) = 0. \tag{5.4}$$

This equation may be solved for any given $V(r)$ by integrating step by step outwards from the origin using the appropriate numerical technique. This gives the wavefunction outside the nuclear field, and comparison with the corresponding wavefunction in the absence of the nuclear field immediately yields the phase shift δ_L and hence the matrix element

$$S_L = e^{2i\delta_L}. \tag{5.5}$$

This can be done for all partial waves and hence the differential cross section

$$\frac{d\sigma}{d\Omega} = |f(\theta)|^2 \tag{5.6}$$

where

$$f(\theta) = (1/2ik) \sum_{L=0}^{\infty} (2L+1)(S_L - 1) P_L(\cos\theta). \tag{5.7}$$

This outline of the calculation refers to neutral spinless particles, and may easily be generalized to charged particles with spin (Hodgson 1963). In the case of charged incident particles the Coulomb (electrostatic) field of the nucleus is added to the optical potential and allowance is made for the particular analytic properties of the Coulomb wavefunctions in the asymptotic region. If the particle has spin the total wavefunction is expanded as a product of spatial and spin wavefunctions, and the operator $\boldsymbol{L}.\boldsymbol{\sigma}$ in the spin dependent part of potential (2.2) splits each radial wave equation into $(2S+1)$ components corresponding to the possible values of $\boldsymbol{J} = \boldsymbol{L} + \boldsymbol{S}$. These may be solved in a similar way and the resulting S matrix elements combined to give not only the differential and total cross sections but also the polarization of the scattered particles as well.

The formalism may also be applied to calculate the change in polarization when the incident particles are polarized. This is frequently used in connection with the scattering of deuterons, due to the complexity of the corresponding polarization phenomena.

When realistic potentials are used, the calculations must be carried out numerically and may easily be programmed for an electronic computer. These computer programs enable the differential cross sections, polarizations and other measurable quantities to be calculated for any specified potential. Usually, however, one needs to find the potential that gives the best fit to a particular set of data. The calculation cannot be reversed, essentially because of the loss of information that occurs when the complex scattering amplitude is squared to give the cross section, so the only way to proceed is to adjust the parameters of the potential to optimize the fit to the data. Many automatic techniques have been developed to do this and they have been incorporated in the optical model programs so that the whole calculation may be done at once. It frequently happens that there are several potentials that give equally good fits to the data. Examples of these ambiguities, together with ways of overcoming them, are given in the following sections.

6. The elastic scattering of nucleons

In this and the three subsequent sections the optical model is applied to the analysis of elastic scattering cross sections and polarizations in conditions where the compound elastic contribution either is negligible or has been evaluated and subtracted by the methods of § 11. The scattering may then be calculated directly from the optical potential by the methods of § 5. In practice it is required to obtain the potential from the experimental data, and this may be done by systematically varying the parameters of the optical potential to optimize the overall fit to the data, using appropriate computer programs.

Many different expressions have been used for the nucleon optical potential. All have a real central term of Saxon–Woods or similar form, and an imaginary volume or surface-peaked term. To account for polarization data a spin–orbit term of the form (2.2) familiar from shell model studies is added. The precision of optical model analyses makes it possible to study the presence of small terms in the potential, and among these the isospin term has received much attention. Some work has also been done to investigate the possibility of terms depending on the spin and magnetic moment of the target nucleus. Apart from these small terms, the standard optical model potential may be written in the form

$$V(r) = V_C(r) + Vf_1(r) + i(W_v f_2(r) + W_s g(r)) + \left(\frac{\hbar}{m_\pi c}\right)^2 \frac{V_S}{r}\frac{df_3(r)}{dr}\boldsymbol{L}.\boldsymbol{\sigma} \qquad (6.1)$$

where $V_C(r)$ is the electrostatic potential (for protons only) which may be taken without appreciable error as that due to a uniformly charged sphere with the same radius as the real part of the potential, the $f_i(r)$ are Saxon–Woods form factors that may have different radius and diffuseness parameters, $g(r)$ is a surface-peaked form factor usually taken to be the radial derivative of a Saxon–Woods form factor, and W_v and W_s are the volume and surface absorption potentials.

The earliest analyses used square well potentials but these were replaced by the physically more realistic rounded wells as soon as electronic computers made it practicable to solve the corresponding radial wave equations numerically. Accurate analyses of differential cross sections and polarizations for the elastic scattering of 183 MeV protons by a range of nuclei soon showed that it is necessary to allow the radius parameter of the imaginary part of the potential to exceed that of the real part by about 20 % (Hodgson 1961, Johansson *et al* 1961), and that the fit is also improved by allowing the corresponding diffuseness parameters to differ as well (Satchler and Haybron 1964). Many attempts were also made to distinguish between surface and volume absorption potentials but it was found that, providing the radius and diffuseness parameters of the imaginary part are allowed to differ from those of the real part and the strength and form factors are optimized to fit the experimental data, both types of absorbing potential give practically equally good fits. Thus for practical purposes it does not matter which form is used, and to determine the form a more fundamental approach is required.

Many analyses of polarization data have shown that the optimum radius of the spin–orbit form factor is about 10 % less than that of the central term. This is a consequence of the shorter range of the nucleon–nucleon spin–orbit interaction compared with that of the central interaction.

Optical model analyses of elastic scattering and polarization data cannot give unique values of all the parameters of the potential; rather it is certain combinations that correspond to a particular set of data. Thus, for example, the fit to the data is insensitive to variations of U and r_u (the radius parameter of the real part of the potential) that keep $Ur_u{}^2$ constant, and similarly for Wa_w. Since the calculations of the potential from nuclear matter considerations are insufficiently precise to resolve these ambiguities, it is usual to fix the parameters of the form factors to average values and then to adjust the potential depths, U, W and U_S to fit the data. It is then possible to compare on the same basis the variation of these potentials with energy and from nucleus to nucleus.

Many such analyses of nucleon scattering have now been made and it is found that the potentials are quite similar for all nuclei and vary rather slowly with the incident energy. The optical model is thus a successful way of describing elastic scattering in a wide range of conditions, and this provides confirmation of the overall correctness of the derivations of the potential from more fundamental considerations.

The most extensive analysis of neutron elastic scattering was made by Perey and Buck (1962) who fitted the differential cross sections, polarizations, total and reaction cross sections and strength functions for a wide range of nuclei from 4 to 25 MeV with a nonlocal potential of the form

$$V(\boldsymbol{r},\boldsymbol{r}') = V\{\tfrac{1}{2}(\boldsymbol{r}+\boldsymbol{r}')\}\frac{1}{(\pi\beta^2)^{3/2}}\exp\left\{-\left(\frac{\boldsymbol{r}-\boldsymbol{r}'}{\beta}\right)^2\right\} \tag{6.2}$$

where β is the range of the nonlocality. The equivalent local potential has been calculated by Wilmore and Hodgson (1964) and is consistent with most existing data on neutron scattering, except from very deformed nuclei.

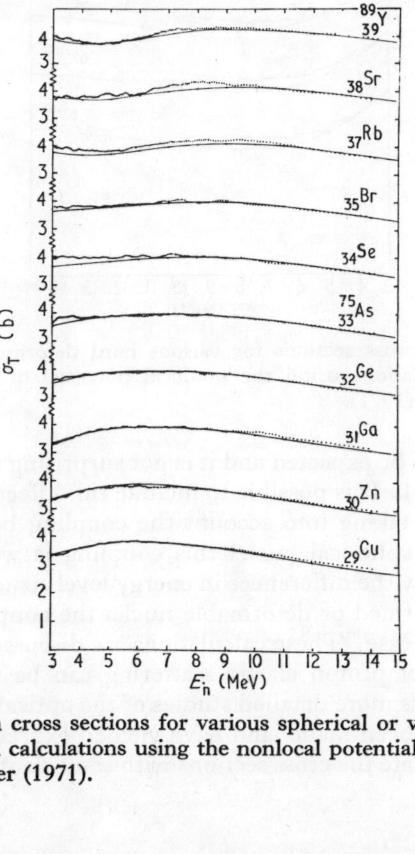

Figure 13. Total neutron cross sections for various spherical or vibrational nuclei compared with optical model calculations using the nonlocal potential of Perey and Buck. From Glasgow and Foster (1971).

A remarkable confirmation of the Perey–Buck potential is provided by the very extensive and accurate measurements of neutron and total cross sections made by Glasgow and Foster (1971). The cross sections calculated from the potential of Perey and Buck differ by less than 3% from the experimental values for 46 spherical or soft nuclei for neutrons from 3 to 15 MeV. The fit is considerably poorer, but still within about 20% for 19 hard deformed nuclei. Some of these results are shown in figures 13 and 14. This influence of the nuclear deformation

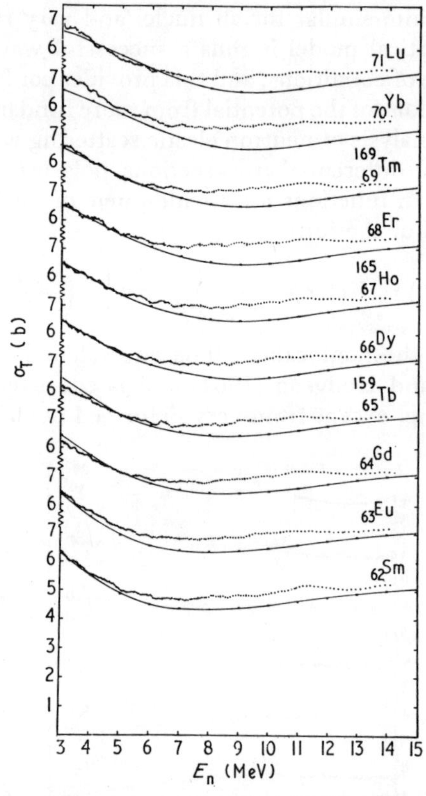

Figure 14. Total neutron cross sections for various hard deformed nuclei compared with optical model calculations using the nonlocal potential of Perey and Buck. From Glasgow and Foster (1971).

on the cross sections is to be expected and it is not surprising that it is not given by a spherical optical potential. It is possible to include this effect within the framework of the optical model by taking into account the coupling between the elastic and inelastic channels. For spherical nuclei this coupling is weak and so the elastic scattering is unaffected by the differences in energy level structure from one nucleus to the next, but for deformed or deformable nuclei the coupling may be so strong that this is no longer the case. These calculations are discussed in § 10.

The measurements of proton elastic scattering can be made to much higher accuracy, and this permits more detailed studies of the optical potential. Very many investigations have now been made, and have yielded expressions for the potential that can be used to calculate the cross sections with some confidence for many nuclei over a wide energy range. Examples of the precision of optical model fits to differential cross sections and polarizations are shown in figures 15 and 16.

The real central potential has received the most detailed study, and it has been found to decrease linearly as the energy increases, with $dU/dE \simeq -0{\cdot}2$. For proton scattering this gives an extra term in the potential (Lane 1957) since the Coulomb potential decreases the energy of the proton in the nuclear interior. The magnitude of this term may be found by averaging the Coulomb field over the nuclear interior and multiplying by dU/dE; this gives $0{\cdot}4Z/A^{1/3}$ for $r_0 = 1{\cdot}25$ fm.

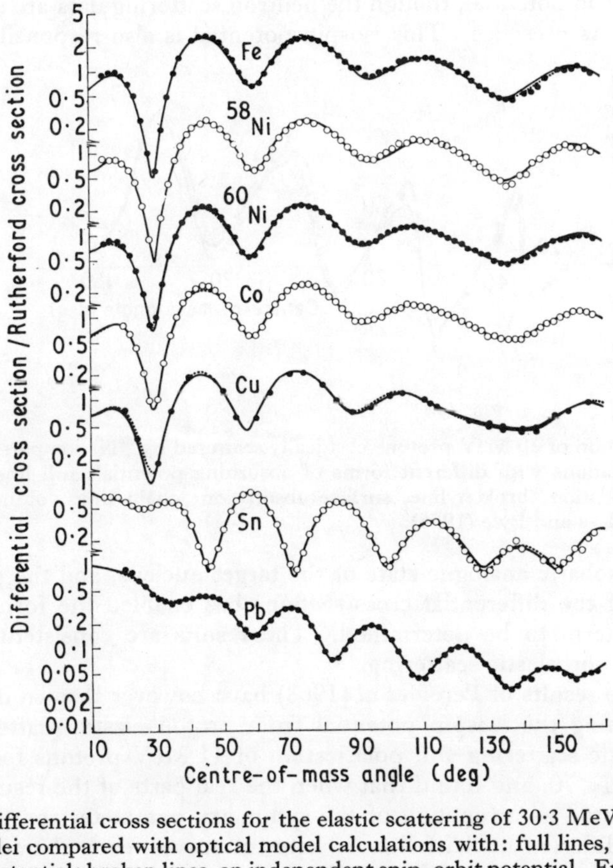

Figure 15. Differential cross sections for the elastic scattering of 30·3 MeV protons by a range of nuclei compared with optical model calculations with: full lines, a constrained spin–orbit potential; broken lines, an independent spin–orbit potential. From Satchler (1967).

An extensive analysis of the differential cross sections for the elastic scattering of 9–22 MeV protons by a range of nuclei (Perey 1963) showed that if the form factors are fixed to the average values $r_u = 1{\cdot}25$ fm and $a_u = 0{\cdot}65$ fm the depth of the real part of the potential is given in MeV by

$$V = 53{\cdot}3 - 0{\cdot}55E - 27(N-Z)/A + 0{\cdot}4Z/A^{1/3}. \tag{6.3}$$

This shows the linear dependence on the energy and on the symmetry parameter $\alpha = (N-Z)/A$. This coefficient of the energy dependence is rather high, and values between 0·2 and 0·3 are found in more recent analyses.

The dependence of the optical potential on the symmetry parameter has also been found in studies of nuclear binding energies and Lane (1962) suggested that these results can be correlated by postulating a symmetry or isospin term in the nucleon–nucleus potential of the form

$$V_{\mathrm{I}}(r) = \frac{4V_1}{A}\,\boldsymbol{t}\,.\,\boldsymbol{T}f(r) \tag{6.4}$$

where $\boldsymbol{t}$ and $\boldsymbol{T}$ are the isospin vectors of the incident particle and target nucleus respectively. This gives a term αV_1 in the proton potential and also a similar term $-\alpha V_1$ in the neutron potential, though the neutron scattering data are insufficiently accurate to show its presence. This isospin potential is also responsible for (p, n)

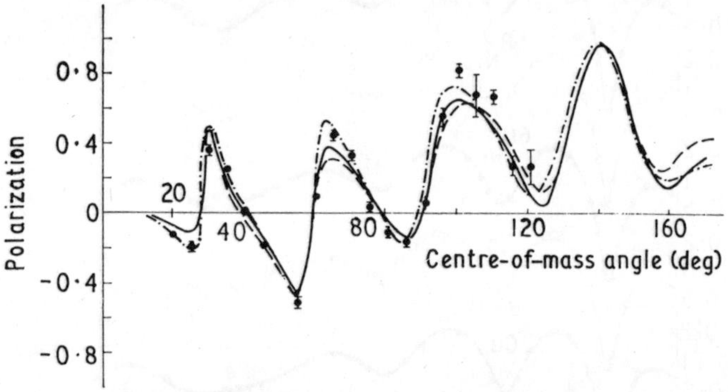

Figure 16. Polarization of 30 MeV protons elastically scattered by ^{60}Ni compared with optical model calculations with different forms of absorbing potential: full line, surface and volume absorption; broken line, surface absorption; chain line, volume absorption. From Greenlees and Pyle (1966).

reactions to the isobaric analogue state of the target nucleus, and the phenomenological analysis of the differential cross sections has enabled the form factor and strength of this term to be determined. The results are consistent with those obtained from proton elastic scattering.

Some accurate results of Perey *et al* (1968) have however thrown doubt on the method of obtaining the isospin potential from proton elastic scattering. They analysed the elastic scattering and polarization of 11 MeV protons for a range of nuclei with $50 \leqslant A \leqslant 70$, and found that when the real parts of the resulting optical potentials are plotted as a function of the symmetry parameter the points corresponding to the different nuclei fall on a series of curves, each characterized by a single value of the isospin T (see figure 17). A clue to the explanation of this result is provided by the observation of Greenlees *et al* (1968) that the integral of the real part optical potential per nucleon is constant to within a few per cent. Using a Saxon–Woods form factor this gives for the real potential depth

$$V = V_0 \Big/ \left(1 + \frac{\pi^2 a^2}{R^2}\right). \tag{6.5}$$

This may be converted to a relation giving V as a function of α and T by using the expressions $2T = \alpha A$ and $R = r_0 A^{1/3}$, so that

$$V = V_0 \Big/ \left\{1 + \frac{\pi^2 a^2}{r_0^{\,2}}\left(\frac{\alpha}{2T}\right)^{2/3}\right\}. \tag{6.6}$$

This relation is in qualitative accord with the results of figure 17 and it is remarkable that the expression is purely geometrical, with no reference to isospin. However, an isospin term must be present to account for (p, n) reactions to the isobaric analogue state, so this must be added to (6.5), together with the term $0{\cdot}4Z/A^{1/3}$ already mentioned. Thus

$$V = V_0\Big/\left(1+\frac{\pi^2 a^2}{R^2}\right)+U_1\alpha+0{\cdot}4Z/A^{1/3}. \tag{6.7}$$

This expression with the single adjustable parameter V_0 is in fair accord with the data of Perey *et al* (1968), as shown in figure 17, and a further phenomenological

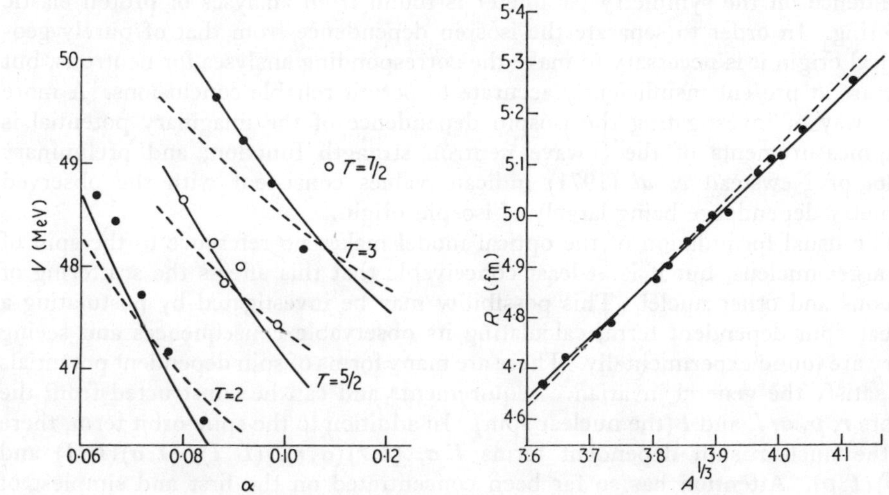

Figure 17. (*a*) Proton optical potentials, as a function of the nuclear symmetry parameter $\alpha = (N-Z)/A$, calculated from the expression for the real potential depth

$$V = \frac{106{\cdot}75A}{R^3(1+\pi^2 a^2/R^2)}+24\alpha+\frac{0{\cdot}517T(1-\alpha)}{\alpha R}$$

compared with the experimental data of Perey and Perey (1968). (*b*) Comparison of the relationship between R and A with the data of Perey and Perey. In each case the broken line refers to a value $R = 1{\cdot}285A^{1/3}$ and the full line to $R = 1{\cdot}204A^{1/3}+0{\cdot}305$. From Hodgson (1970).

adjustment gives quantitatively acceptable curves. Equation (6.7) enables the real part of the optical potential for protons to be calculated quite accurately for 11 MeV protons on nuclei with $50 \leqslant A \leqslant 70$, and a similar though more complicated expression may be obtained for other nuclei.

Examination of (6.7) shows that the original method of determining the strength of the isospin potential from proton elastic scattering is valid only for isobaric sequences, for which the A dependence of the first term is negligible. In other cases the method is not applicable unless the A dependence is accurately known in some other way and taken into account in the calculations.

In all this work on the isospin term it is assumed that its form factor is the same as that of the remainder of the real potential, but this is not unduly restrictive since the addition of a small surface-peaked potential to a volume potential of the same radius has the effect of increasing the radius, and thus is equivalent to increasing the depth. It is however possible (Sinha and Edwards 1970) that the isospin potential has a smaller radius than the rest of the real potential, in which case the above result will not hold. It is in any case likely from nuclear matter calculation that the isospin term does not have the same form for all nuclei, but tends to be surface-peaked for light nuclei and of the volume form for heavy nuclei. It is thus desirable to use a calculated form factor for the isospin potential and to determine its strength by comparison with the (p, n) data.

There has been some work on the imaginary part of the isospin potential, but lack of knowledge of its form factor prevents any definite conclusions being drawn (Satchler 1967, Hodgson 1967). If it is assumed to be surface-peaked then a marked dependence on the symmetry parameter is found from analyses of proton elastic scattering. In order to separate the isospin dependence from that of purely geometrical origin it is necessary to make the corresponding analyses for neutrons, but these are at present insufficiently accurate to permit reliable conclusions. A more direct way of investigating the isospin dependence of the imaginary potential is from measurements of the S wave neutron strength function, and preliminary results of Newstead *et al* (1971) indicate values consistent with the observed symmetry dependence being largely of isospin origin.

The usual formulation of the optical model makes no reference to the spin of the target nucleus, but it is at least conceivable that this affects the scattering of nucleons and other nuclei. This possibility may be investigated by postulating a nuclear spin dependent term, calculating its observable consequences and seeing if they are found experimentally. There are many forms of spin dependent potentials that satisfy the general invariance requirements and can be constructed from the vectors $\boldsymbol{r}$, $\boldsymbol{p}$, $\boldsymbol{\sigma}$, $\boldsymbol{L}$ and $\boldsymbol{I}$ (the nuclear spin). In addition to the spin–orbit term, there are the nuclear spin dependent terms $\boldsymbol{I}.\boldsymbol{\sigma}$, $(\boldsymbol{I}.\boldsymbol{r})(\boldsymbol{\sigma}.\boldsymbol{r})$, $(\boldsymbol{L}.\boldsymbol{I})$, $(\boldsymbol{I}.\boldsymbol{\sigma})(\boldsymbol{L}.\boldsymbol{I})$ and $(\boldsymbol{\sigma}.\boldsymbol{p})(\boldsymbol{I}.\boldsymbol{p})$. Attention has so far been concentrated on the first and simplest of these, which is written in the form

$$V_{SS}(r) = V_{SS}\, f_{SS}(r) \frac{\boldsymbol{I}.\boldsymbol{\sigma}}{I} \tag{6.8}$$

usually with a Saxon–Woods form factor $f_{SS}(r)$. If the form factor parameters are chosen to be the same as the real central potential the spin–spin term may conveniently be included in the expression for the real potential depth, and as there are at present insufficient data to determine the form factor this is usually done.

It has not proved easy to find any evidence for the presence of this term. The polarization is directly dependent on the spin–orbit term and so analyses of polarization data, together with differential cross sections, enables its characteristics to be determined. The polarization might also be expected to be sensitive to the spin–spin term (6.8) but the measured polarizations of protons elastically scattered from nuclei with very different spin but otherwise similar characteristics show almost identical polarizations, while calculations indicate that the spin–spin term has no first-order effect on the polarization. Thus, at best, the differential cross sections and polarizations show only a second-order dependence on the spin–spin term, and so cannot be used to obtain a reliable indication of its presence.

The scattering phenomena directly sensitive to the target spin require polarized or at least aligned target nuclei, and such experiments are technically very difficult. Nevertheless several attempts to perform them have been made. The techniques of nuclear alignment require a fairly thick sample, complicated surrounding apparatus and a high nuclear spin, so most investigations have used neutrons of about 1 MeV scattered by ^{59}Co and ^{165}Ho. No evidence for the presence of a spin–spin term has yet been obtained, though it is possible to set an upper limit to the value of the spin–spin potential V_{SS}. Thus, for example, the work of Nagamine *et al* (1970) on the total cross section of 1 MeV polarized neutrons on polarized ^{59}Co showed that $|V_{SS}| < 1{\cdot}0$ MeV.

Some measurements have been made by Batty and Tschalär (1970) of the depolarization parameter D, which is a measure of the change in polarization occurring in elastic scattering. Their results for 50 MeV protons elastically scattered by a range of nuclei gave no indication of the presence of a spin–spin term in the optical potential.

The introduction of the spin–spin term into the Schrödinger equation splits the single-particle states of orbital angular momentum L into components with total angular momenta $J = |L - S|$ to $(L+S)$, for $L \geqslant S$, where $\boldsymbol{S} = \boldsymbol{I} + \frac{1}{2}$ is the channel spin. The corresponding phase shifts thus pass through resonances at the appropriate energies, but in the normal scattering situation, when many reaction channels are open, the absorbing potential so broadens these states that they cannot be identified. At low energies in light nuclei, however, most if not all these reaction channels are closed, so the resonances are quite sharp, and this may provide one of the most favourable circumstances for the investigation of the spin–spin term.

A suitable casc is the elastic scattering of protons and neutrons of a few MeV by ^{12}C. The cross sections are well known experimentally and so few phase shifts contribute that they can be obtained from the experimental data as a function of incident energy. Some of the resonances are due to single-particle states in the

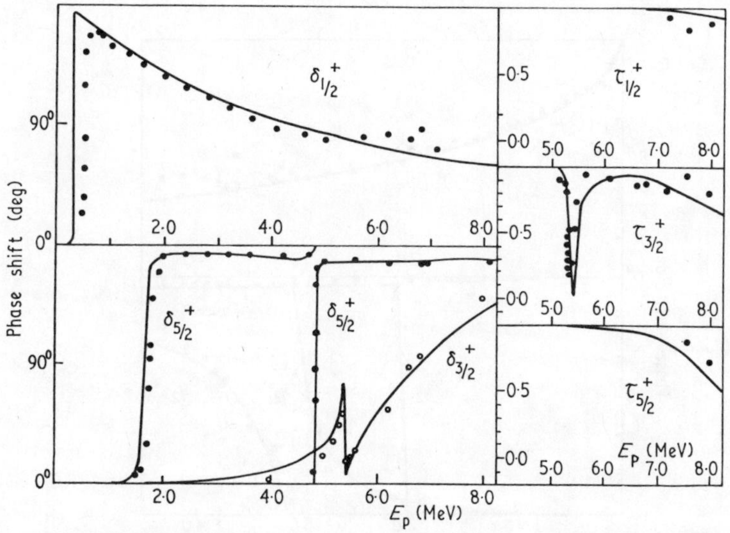

Figure 18. Comparison between the experimental phase shifts for the elastic scattering of protons by ^{12}C and coupled calculations using a rotational model with $\beta = -0{\cdot}5$ and a spin–spin term with $V_{ss} = 0{\cdot}37$ MeV. The quantity δ_{J^π} is the real part of the phase shift for the J^π partial wave, and $\tau_{J^\pi} = \exp(-2\eta)$ where η is the imaginary part of the phase shift for the J^π partial wave. From Mikoshiba *et al* (1971).

continuum, while others result from coupling to collective states of the target nucleus. Such a situation may be treated theoretically by the coupled channels theory discussed in § 10, and a series of studies of increasing refinement have been made in order to obtain an accurate description of these resonances (Pascolini *et al* 1969).

One of the most recent of these, by Mikoshiba *et al* (1971), obtained a very good overall fit to the data using a rotational collective model for ^{12}C, with deformation parameter $\beta = -0{\cdot}5$. It was also necessary to include a spin–spin term in the optical potential, with $V_{SS} \approx 0{\cdot}4$ MeV, consistent with previous estimates. This term is present when the target nucleus is virtually excited to the first 2^+ state. Some of their results for proton and neutron scattering are shown in figures 18 and 19. They also included a term in the potential proportional to L^2 which enables the state dependence of the potential to be taken into account. These calculations are complicated, and depend on many parameters, so it is not easy to be sure that the ambiguities have been thoroughly explored, but it does seem to be one of the most promising ways of investigating the spin–spin term.

There are some indications that the spin–spin term may significantly affect the strength functions; in particular it may be responsible for the systematic difference between the neutron S wave strength functions for odd and even nuclei around $A = 65$ (Rahman Kahn 1966). Additional evidence for the spin–spin term has been found from neutron S wave strength functions for isotopes of Nd, Sm and Ho by Newstead *et al* (1971). The magnitude of the spin–spin term may be calculated from the strength of the $\boldsymbol{\sigma}_i . \boldsymbol{\sigma}_j$ term in the nucleon–nucleon interaction, and this gives values around 0·1 MeV (Rahman Kahn 1966, Fisher *et al* 1966). A more recent calculation by Satchler (1971a) shows that substantially smaller values are obtained when core polarization is taken into account.

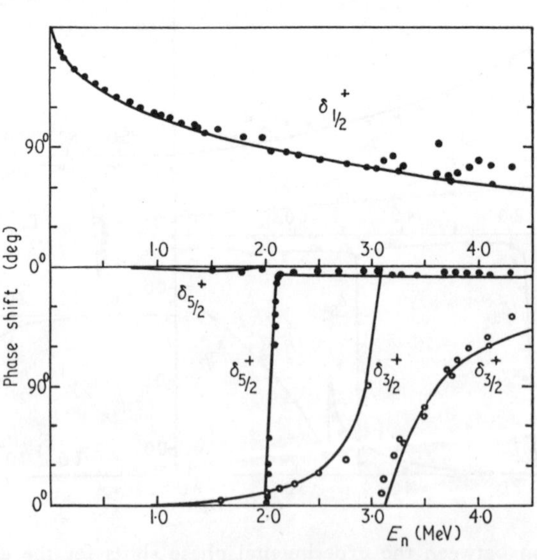

Figure 19. Comparison between the experimental phase shifts for the elastic scattering of neutrons by ^{12}C and coupled-channels calculations using the same model and parameters as the corresponding proton calculation shown in figure 18. From Mikoshiba *et al* (1971).

The next nuclear spin dependent potential in order of increasing complexity has a structure similar to the tensor force, and may be written

$$V_T(r) = V_T f_T(r)\{3(\boldsymbol{I}.\hat{\boldsymbol{r}})(\boldsymbol{\sigma}.\hat{\boldsymbol{r}}) - \boldsymbol{I}.\boldsymbol{\sigma}\}/2I. \tag{6.9}$$

The strength of this term may be found by comparing the total cross sections for polarized particles on polarized nuclei with the polarization axes parallel and antiparallel and in the beam direction and perpendicular to it (Fisher 1971). Estimates of V_T from existing data give values consistent with zero and with the small values obtained by Satchler (1971a).

The electrostatic interaction between an incident charged particle and a nucleus may easily be taken into account by adding to the optical potential the electrostatic potential due to a uniformly charged sphere. In addition, there are other small electromagnetic effects for both charged and uncharged particles due to the interaction of the magnetic moment of the incident particle with the electrostatic field of the nucleus and these can be represented by a potential of the form

$$V_{MM}(r) = \frac{\mu - a}{2}\left(\frac{\hbar}{mc}\right)^2 \frac{1}{r}\frac{dV_C(r)}{dr}\boldsymbol{L}.\boldsymbol{\sigma} \tag{6.10}$$

where μ is the magnetic moment of the incident nucleon, $a = \frac{1}{2}$ for protons and $a = 0$ for neutrons, and $V_C(r)$ is the electrostatic Coulomb potential. Apart from the radial dependence, which falls off much more slowly ($\sim r^{-3}$) at large distances, this is very similar to the spin–orbit potential, so in most circumstances any effects due to magnetic moment scattering are absorbed by that term. In the case of scattering at very small angles, however, such a term may have appreciable effects both on the differential cross section and on the polarization. Since the differential cross section for the elastic scattering of protons rises so rapidly towards small angles it is more convenient to investigate these magnetic moment effects in the differential cross section for neutron scattering and in the polarization for proton scattering. Several studies have shown that the magnetic moment effects are indeed observed, and that their magnitude is in accord with the theoretical calculations (Monahan and Elwyn 1964).

The incident nucleon may also be polarized by the Coulomb field of the target, and the induced electric dipole moment can interact with the Coulomb field, giving an additional potential

$$V_{DM}(r) = -\alpha E^2/2 \tag{6.11}$$

where α is the electric polarizability and E the electric field. This potential is significant only outside the nucleus, where it becomes

$$V_{DM}(r) = -\alpha Z^2 e^2/2r^4. \tag{6.12}$$

In principle it should be possible to compare the results of calculations including this potential with experimental data on the elastic scattering of neutrons, and hence to determine the electric polarizability of the neutron. In practice, however, the effects are so small that they can readily be absorbed by small changes in the real part of the optical potential.

7. The elastic scattering of deuterons

Deuterons have several special features that complicate the way they are scattered by nuclei. They are very loosely bound, and are therefore easily broken up when they encounter the nuclear field. The scattering is thus rather sensitive to the nuclear structure and it is correspondingly more difficult to define overall optical potentials. The charge and mass centres of the deuteron are significantly separated, and this gives rise to forces tending to twist and break the deuteron even before it encounters the nuclear field. Finally the deuteron has spin one so that the polarization phenomena accompanying the scattering are much more complicated than for nucleons; to specify the polarization it is necessary to define not only the vector polarization as for particles of spin $\frac{1}{2}$, but also three tensor polarizations as well.

The possibility that the deuteron can be broken up by the Coulomb field alone must be carefully considered since it could take place well beyond the region where the optical potential is appreciable, and thus could not be represented by it. Calculations of the Coulomb breakup process have, however, shown that it is negligible except in the region of the nuclear field, and so it can be taken into account by the optical potential.

Discrete ambiguities are encountered in the optical model analysis of the scattering of deuterons and heavier particles. It is found that several potentials of depths around 50, 100, 150, ... MeV all give equally good fits. This may be understood quantum mechanically if one partial wave is considered. If the potential is progressively deepened, more and more waves are drawn into the nuclear region, and if a good fit is obtained with one depth, an equally good fit is obtained when the potential is deepened sufficiently to bring in just one extra half wave, since the asymptotic wavefunction is then unaltered. This is exact for one partial wave, but it turns out that the different partial waves keep nearly in step in this respect, so it applies to the whole angular distribution as well. Then there is phenomenologically a whole series of potentials giving very similar fits.

This ambiguity may be resolved by calculating the deuteron potential from the constituent neutron and proton potentials. Although it is subject to many uncertainties, the calculation is sufficiently accurate for this purpose. The simplest expression for the deuteron optical potential is just the sum of the constituent nucleon potentials evaluated at half the deuteron energy. A more accurate expression may be obtained by averaging this over the internal wavefunction $\phi_D(\boldsymbol{x})$ of the deuteron so that (Watanabe 1958)

$$V_D(r) = \int \phi_D{}^2(\boldsymbol{x}) \{V_n(\boldsymbol{r}+\boldsymbol{x}/2) + V_p(\boldsymbol{r}-\boldsymbol{x}/2)\}\, d\boldsymbol{x} \qquad (7.1)$$

where $\boldsymbol{x} = \boldsymbol{r}_n - \boldsymbol{r}_p$ is the separation between the neutron and proton in the deuteron, and $V_n(\boldsymbol{r}+\boldsymbol{x}/2)$ and $V_p(\boldsymbol{r}-\boldsymbol{x}/2)$ are the neutron and proton optical potentials, usually evaluated at half the deuteron energy. Calculations with this expression show that the correct deuteron potential has a depth of rather less than the sum of the constituent neutron and proton depths (ie about 100 MeV) and this value is used in all optical model analyses of deuteron elastic scattering.

Many optical model analyses of deuteron elastic scattering have now been made, and it has been found possible in most cases to obtain a good fit to the differential cross sections. The optimum form factors have a radius for the imaginary part greater than that of the real part. Some investigations have covered a range of nuclei at different energies, and formulae have been obtained expressing the potential as a function of the appropriate variables. Thus for example Perey and Perey (1963) analysed an extensive set of data and obtained the expression for the real part of the potential

$$U = 81 + 2{\cdot}0Z/A^{1/3} - 0{\cdot}22E \qquad (7.2)$$

where energies are in MeV, with form factor parameters $r_u = 1{\cdot}15$ fm, $a_u = 0{\cdot}81$ fm, $r_w = 1{\cdot}34$ fm and $a_w = 0{\cdot}68$ fm.

Such expressions are not as accurate as for nucleon scattering due to the greater sensitivity of deuteron scattering to the surface potential and hence to details of nuclear structure. Unfortunately it has not yet been found possible to turn this sensitivity to good account and use it to determine specific features of the nuclear surface region.

The polarization phenomena associated with the elastic scattering of deuterons are much more complicated than those of nucleons, because the description of the state of polarization of a beam of particles of spin one requires not only a vector polarization as for nucleons, but three independent tensor polarizations as well. There are also several possible spin dependent interactions that could be included in the deuteron optical potential, and an examination of the forms that can be constructed from the vectors $\boldsymbol{r}$, $\boldsymbol{L}$, $\boldsymbol{p}$ and $\boldsymbol{S}$, subject to general invariance and symmetry restrictions, gives the forms (Satchler 1960)

Spin–orbit $$V_S(r) = U_S h_s(r) \boldsymbol{L}.\boldsymbol{S} \tag{7.3}$$

Tensor 1 $$V_{T1}(r) = U_{T1}\{(\boldsymbol{S}.\boldsymbol{r})^2/r^2 - \tfrac{2}{3}\} f_{T1}(r) \tag{7.4}$$

Tensor 2 $$V_{T2}(r) = U_{T2}\{(\boldsymbol{S}.\boldsymbol{p})^2 - \tfrac{2}{3}p^2\} f_{T2}(r) \tag{7.5}$$

Tensor 3 $$V_{T3}(r) = U_{T3}\{(\boldsymbol{L}.\boldsymbol{S})^2 + \tfrac{1}{2}(\boldsymbol{L}.\boldsymbol{S}) - \tfrac{2}{3}L^2\} f_{T3}(r) \tag{7.6}$$

where the form factors are functions of r that are not yet determined. Using any of these forms for the deuteron spin dependent potential, the polarization produced in an unpolarized beam can be calculated, or alternatively the change in polarization of a beam already polarized.

It is in principle possible to investigate the presence of these forces, and to determine their magnitudes, by comparing the results of optical model calculations with the appropriate measurements. In practice, however, it is difficult to determine the polarization of a deuteron beam. It is somewhat easier, though still difficult, to produce a deuteron beam of known polarization, and then to measure the asymmetry of the elastically scattered deuterons, and some measurements have been made in this way.

Some of the earliest investigations of the deuteron spin dependent potential were made with 21·6 MeV deuterons on ^{40}Ca, and the results were analysed by Raynal (1963). He found that it is possible to obtain a good overall fit to the differential cross section and to the vector and tensor polarizations with a spin–orbit potential alone; there was no need to include a tensor potential. This is in accord with calculations of the magnitude of the tensor potential, which indicate that it is very small (Testoni and Gomes 1966).

The deuteron polarization has also been measured by Kossanyi-Demay and Mayer (1967) and their results for 21·6 MeV deuterons on ^{58}Ni and ^{60}Ni are shown in figure 20. An extensive analysis by Schwandt and Haeberli (1969) of the differential cross section and vector and tensor polarizations of 5 to 34 MeV deuterons by ^{40}Ca gave the following expressions for the deuteron potential

$$\left.\begin{array}{lll} U = 115 - 0{\cdot}3E, & r_u = 1{\cdot}05, & a_u = 0{\cdot}85 \\ W_s = 12{\cdot}3 - 0{\cdot}01E, & r_w = 1{\cdot}71 - 0{\cdot}01E, & a_w = 0{\cdot}48 + 0{\cdot}007E \\ U_s = 6{\cdot}5, & r_s = 0{\cdot}9, & a_s = 0{\cdot}6 \end{array}\right\} \tag{7.7}$$

where energies are in MeV and distances in fermis. Further discussion of deuteron elastic scattering together with a table of potentials obtained up to 1966 may be found in the review by Hodgson (1966).

Many detailed attempts have been made to calculate the deuteron optical potential from the constituent neutron and proton potentials (Testoni and Gomes 1966, Mukherjee 1968, Rihan and Sharaf 1969, Satpathy *et al* 1969, Bencze and Doleschall 1970, Samaddar *et al* 1970, Keaton 1971). The overall features can be obtained from the simple folding expression (7.1) but to improve on this requires much detailed formalism unsuitable for summary here.

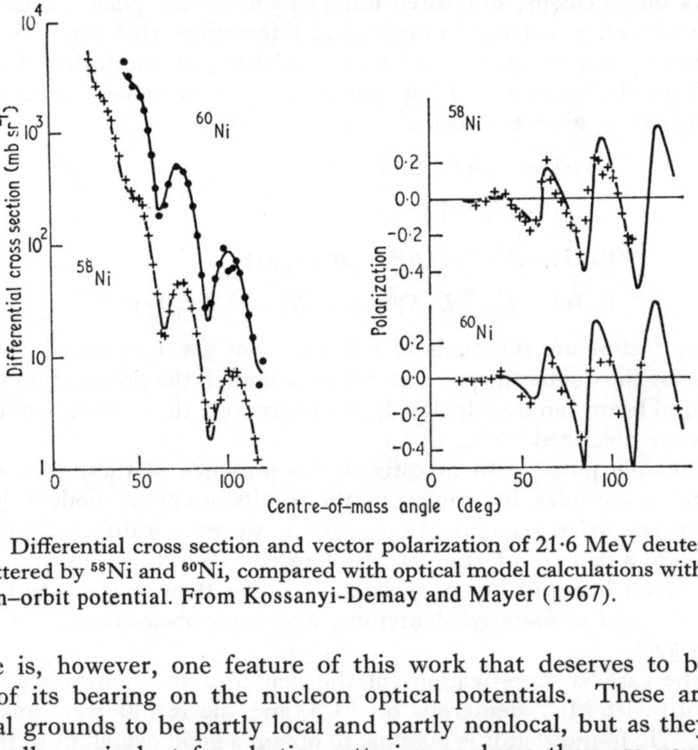

Figure 20. Differential cross section and vector polarization of 21·6 MeV deuterons elastically scattered by ^{58}Ni and ^{60}Ni, compared with optical model calculations with inclusion of a spin–orbit potential. From Kossanyi-Demay and Mayer (1967).

There is, however, one feature of this work that deserves to be mentioned because of its bearing on the nucleon optical potentials. These are known on theoretical grounds to be partly local and partly nonlocal, but as they are phenomenologically equivalent for elastic scattering analyses they are usually for convenience taken to be wholly local. There are, however, two ways of calculating the deuteron potential, even if this is taken to be wholly local as well. Thus either one can start from local nucleon potentials and hence obtain the local deuteron potential or one can start with the equivalent nonlocal nucleon potentials, calculate the corresponding nonlocal deuteron potential and then find its local equivalent. This has been done by Samaddar and Mukherjee (1971) and they find that these deuteron potentials have different energy dependences. It is thus possible in principle, by comparison with the phenomenological deuteron potentials, to find the extent to which the nucleon potentials are nonlocal, and the extent of their intrinsic energy dependence.

The calculation of the deuteron optical potential is made more difficult by the ease with which the deuteron may be broken up in the nuclear field. The effect of this process has been studied by Johnson and Soper (1970) using an adiabatic approximation, and they obtained a set of coupled equations for the elastic scattering and deuteron breakup wavefunctions. Solution of these equations shows that they account very well for the partial waves corresponding classically to reactions in the region of the nuclear surface, and this improves the agreement with the experimental data.

8. The elastic scattering of helions and tritons

Helions and tritons are more strongly bound than deuterons and so Coulomb breakup effects are negligible, and as they have spin $\frac{1}{2}$ the polarization phenomena are similar to those of nucleons. They interact strongly with the nuclear field and so their scattering is rather sensitive to the potential in the surface region, making it more difficult to find overall optical potentials.

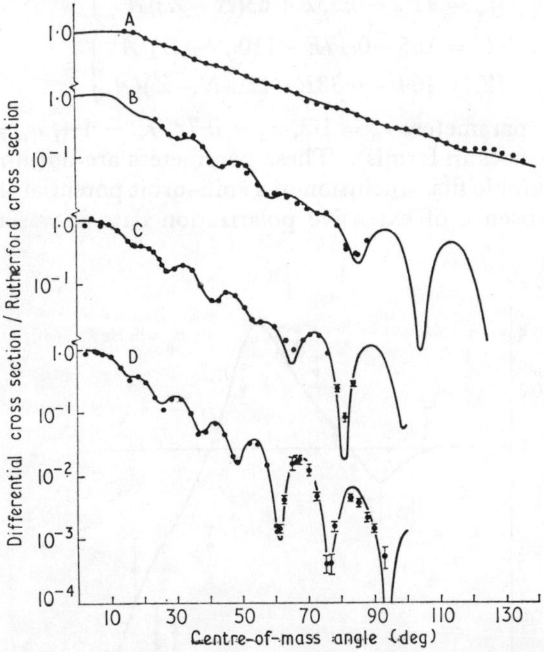

Figure 21. Differential cross sections for the elastic scattering of helions by ^{58}Ni compared with optical model calculations. A, E = 22 MeV (Los Alamos); B, E = 33 MeV (Argonne); C, E = 37·7 MeV (Gibson *et al*); D, E = 43·7 MeV (Gibson *et al*). From Gibson *et al* (1967).

The mass of the helion and triton ensures that at energies high enough for the effects of the nuclear field to be appreciable a substantial number of partial waves contribute to the interaction. It is then possible to describe many of the features of the elastic scattering and polarizations by diffraction models, or by parametrizing the elements of the scattering matrix by continuous functions of the orbital angular momentum quantum number (Springer and Harvey 1965, Fernandez and Blair 1970).

Many optical model analyses have also been made, and a typical study of the energy variation of the differential cross section is shown in figure 21. The potential depth ambiguity is particularly troublesome for helions and tritons because it is often found that there are many discrete potentials giving practically identical fits, and the calculations of the potential from the component nucleon–nucleus potential are not sufficiently accurate to show which must be chosen, as was the case for deuteron scattering. They do, however, indicate a value of the real potential around 150 MeV, so depths similar to this are most often used.

For a particular nucleus it is usually possible to find empirical formulae to represent the energy variation of the real and imaginary potentials when fixed form factors are used. An analysis of 20 MeV helion scattering by a range of nuclei gave potentials with a linear dependence on A (Gibson *et al* 1967). A particularly extensive analysis of helion and triton scattering has been made by Becchetti and Greenlees (1969) who find that it is quite well fitted by the potentials

$$\text{Helion} \quad \left.\begin{aligned} U &= 136{\cdot}4 - 0{\cdot}17E + 55(N-Z)/A \\ W_v &= 41{\cdot}3 - 0{\cdot}33E + 63(N-Z)/A \end{aligned}\right\} \tag{8.1}$$

$$\text{Triton} \quad \left.\begin{aligned} U &= 165 - 0{\cdot}17E - 110(N-Z)/A \\ W_v &= 46{\cdot}0 - 0{\cdot}33E - 110(N-Z)/A \end{aligned}\right\} \tag{8.2}$$

with the form factor parameters $r_u = 1{\cdot}2$, $a_u = 0{\cdot}72$, $r_w = 1{\cdot}4$, $a_w = 0{\cdot}86$ (energies are in MeV and distances in fermis). These parameters are not unique, and many other sets give comparable fits. Inclusion of a spin–orbit potential slightly improves the fit, but in the absence of extensive polarization data it was not used in the analysis.

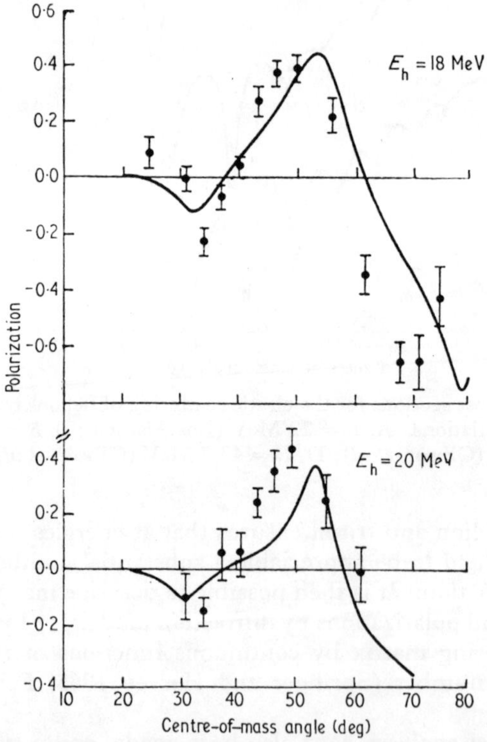

Figure 22. Polarization of 18 and 20 MeV helions elastically scattered by ^{12}C compared with optical model calculations with a spin–orbit potential $V_s = 4, 4{\cdot}5$, $r_s = 1{\cdot}0$, $a_s = 0{\cdot}5$. From McEver *et al* (1970).

Rather few measurements of the polarization of elastically scattered helions and tritons have been made. A recent example of such a measurement is shown in figure 22 together with optical model curves. Theoretical considerations indicate that the helion and triton spin–orbit potentials should be similar in form to the corresponding nucleon potential, with a strength three times smaller, and this is consistent with the available measurements.

The helion and triton optical potentials may be combined to give a general mass-three potential, just as the neutron and proton potentials can be combined in the nucleon potential. If the same form factors are used, the difference between the helion and triton potential depths indicates the presence of an isospin term similar to the corresponding term in the nucleon potential. The potentials also show a dependence on the symmetry parameter due to nuclear structure effects, and this may be determined by analysis of measurements on isotopic or isobaric sequences. In one of these investigations, comparison between the helion and triton optical potentials at 21 MeV shows that the imaginary part of the potential depends on the nuclear symmetry parameter, and that most of this is not of isospin origin (Urone *et al* 1971). The energy was not high enough to permit investigation of the real part of the isospin potential.

Further discussion of helion and triton elastic scattering together with a table of the potentials obtained up to 1968 may be found in the review by Hodgson (1968).

9. The elastic scattering of α particles

The α particle is a tightly bound spinless structure, and this simplifies the description of how it is scattered by nuclei. Its charge and mass ensure that a substantial number of partial waves contribute when the energy is above the Coulomb barrier, so that the diffraction and parametrized scattering amplitude

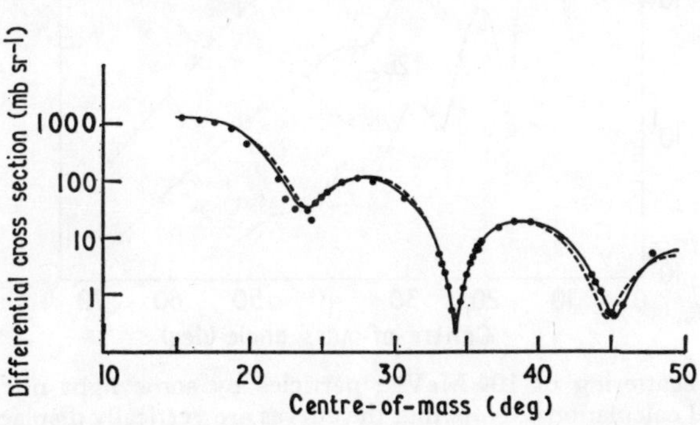

Figure 23. Differential cross sections for the elastic scattering of 42 MeV α particles by ^{44}Ca compared with: full line, optical model calculations with a four-parameter potential; broken line, the distribution given by the parameterization of the scattering amplitude. From Fernandez and Blair (1970).

models are very successful in many respects. This is shown by the comparison in figure 23 between the experimental data for the elastic scattering of 42 MeV α particles by calcium and the two types of theoretical description.

As for helions and tritons, the potential depth ambiguities constitute a serious difficulty. These have been extensively studied for α particle scattering and it is again found that several potentials give equally good fits within the range indicated by calculations from the nucleon–nucleon potential. It is not possible to resolve these ambiguities by coupled channels calculations (Eberhard and Robson 1971). There have been many optical model analyses of α particle scattering with both four-parameter (U, W, r_0, a) and six-parameter $(U, r_u, a_u, W, r_w, a_w)$ potentials. In many cases the adequate fit is obtained with a four-parameter potential, in which

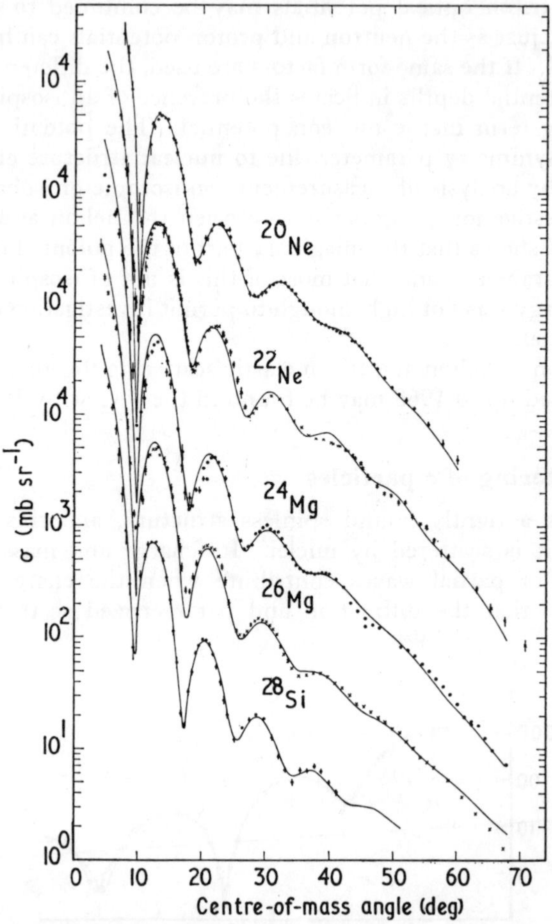

Figure 24. Elastic scattering of 104 MeV α particles by some light nuclei compared with optical model calculations. Note that the curves are vertically displaced from each other by a factor of 10. From Rebel (1971).

the form factors of the real and imaginary parts are the same, but if the data are extensive a significant improvement can usually be obtained with a six-parameter potential, and then it is found that the radius of the imaginary potential is optimally 20–30% greater than that of the real potential, typical values being $r_u = 1{\cdot}1$ fm, $r_w = 1{\cdot}4$ fm. An example of the precision that can now be attained is shown in figure 24.

The elastic scattering of α particles is particularly sensitive to the potential in the region of the nuclear surface, and may thus be used to study the radii of nuclei.

An indication of this sensitivity is shown by the cross sections for the elastic scattering of α particles by some calcium isotopes in the neighbourhood of a minimum (figure 25). The angle at which the minimum occurs can be measured to 0·1°, and hence a precise value of the nuclear radius can be found. If this is done for a series of isotopes, an accurate measure of the variation of the nuclear matter radius through the sequence is obtained, although the absolute value depends on the other parameters of the potential. Fernandez and Blair (1970) have used several different definitions of the nuclear radius, and find that they give very similar results for variations through isotopic sequences.

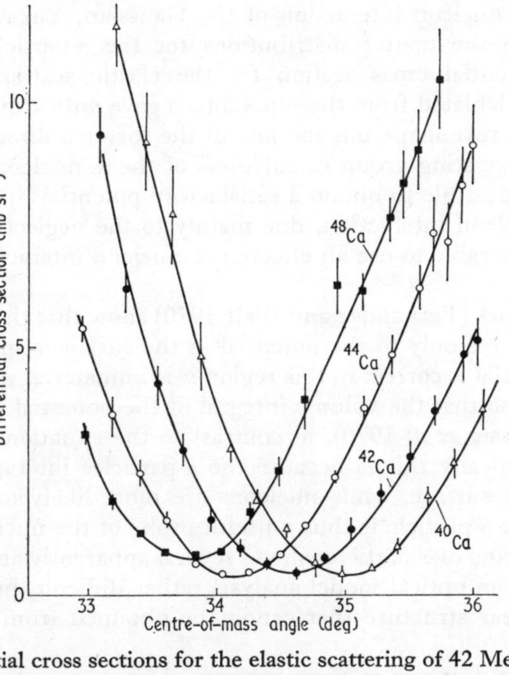

Figure 25. Differential cross sections for the elastic scattering of 42 MeV α particles by the Ca isotopes in the region of the minimum. From Fernandez and Blair (1970).

If these results for the nuclear matter radii are combined with the nuclear charge distributions obtained from electron elastic scattering then detailed information about the changes in neutron and proton density distributions from one nucleus to the next may be obtained. Some of this work has been reviewed by Jones (1970). The same surface sensitivity also makes it possible to determine the nuclear shape by analysing elastic and inelastic cross sections together by the coupled channels method. Some of this work is described in § 10.

Several attempts have been made to calculate the real part of the α particle potential from simpler interactions using folding procedures similar to those used for nucleons (see § 2). There are basically two ways of doing this: by a single folding from a nucleon–α potential or by a double folding from a nucleon–nucleon potential. The parameters of the nuclear matter distribution are treated as adjustable, and the real potential found in this way is supplemented by a phenomenological imaginary potential. The single folding procedure was used by Jackson and Kembhavi (1969) and by Morgan and Jackson (1969) using Gaussian and Yukawa forms for the effective interaction, and optimization of the two parameters. The Gaussian interaction was found to give the better results as it is sufficiently peaked to give the required behaviour in the surface region. The fits to the experimental data obtained with this model are comparable in quality to those found with the standard optical model and require fewer adjustable parameters. A more detailed analysis has been made by Lilley (1971) using the phenomenological α–nucleon potentials fitted by Satchler *et al* (1968) to the p–α and n–α phase shifts, and he finds that the calculated potentials are consistent with those obtained by phenomenological analyses of the data.

The double folding procedure has been used by Budzanowski *et al* (1970) with an effective nucleon–nucleon interaction of the Gaussian, Yukawa and McManus forms, together with the matter distributions for the α particle and the target nucleus. The differential cross section for the elastic scattering of 27·5 MeV α particles by ^{59}Co calculated from these potentials gave only a qualitative fit to the data, being of the correct shape but too low in the forward direction. This is not unexpected, since resonating group calculations of the α–nucleon interaction have shown that it is not possible to obtain a satisfactory potential by simple folding of the free nucleon–nucleon interaction, due mainly to the neglect of antisymmetrization. It is thus preferable to use an effective α–nucleon interaction fitted directly to the data.

This and other work (Fernandez and Blair 1970) show that the elastic scattering of α particles is sensitive only to the potential in the surface region. Provided the real part of the potential is correct in this region it is immaterial what value it has in the nuclear interior, so that the volume integral of the potential is not a physically useful quantity (Weisser *et al* 1970), in contrast to the situation for nucleon scattering (see §2). Physically this is because the α particles interact strongly in the region of the nuclear surface, while nucleons are more likely to penetrate to the nuclear interior. The α particle is thus a useful probe of the nuclear surface.

The elastic scattering of α particles shows several apparently anomalous features, some of which make an optical model analysis rather difficult, but which can give information on nuclear structure that cannot be obtained from the scattering of lighter particles.

The first of these features is the pronounced peak sometimes found in the extreme backward direction around 180°; this is sometimes called the 'glory effect' by analogy with a similar phenomenon in meteorology. It was at one time thought that this feature is not given by the optical model, but detailed calculations showed that it is indeed given automatically by the potentials fitted to the data at smaller angles. The width of the peak is in accord with that observed, but it is not possible to make a comparison of the magnitude because of the presence of the second anomaly described below. The glory effect is essentially due to the cumulative reinforcement of the contributions from several even partial waves with orbital angular momenta around that given by the classical input parameters. A typical cross section showing the glory effect is illustrated in figure 26.

The second anomaly is the rise in the cross section at backward angles that occurs for some nuclei but not for others. It is possible to fit the overall features of these distributions with the optical model by adjusting the parameters in each case, but they cannot all be fitted by the same potential. In the case of 22–29 MeV α particles, the effect is found to be strong for ^{36}A, ^{40}Ca and ^{39}K, weak for ^{42}Ca and ^{41}K, and absent for ^{40}A, ^{44}Ca and ^{48}Ca.

There were some attempts (Schmeing 1970) to explain this anomaly as a consequence of the α clustering that has been thought to occur in the surface region of nuclei (Wilkinson 1961) but a detailed calculation (Thompson 1971) showed that this hypothesis is untenable. A much more convincing explanation has been suggested by Eberhard (1970) who connects it with the level density in the residual nuclei in the open channels. It is a special characteristic of α particle interactions that they bring quite high angular momenta into the nucleus, and if final states of sufficiently high J are not available in the reaction channels the corresponding flux must be re-emitted into the elastic channel, thus raising the cross section in the backward direction. A detailed examination of the final state level densities showed that this effect is to be expected for just those nuclei where it is found to occur.

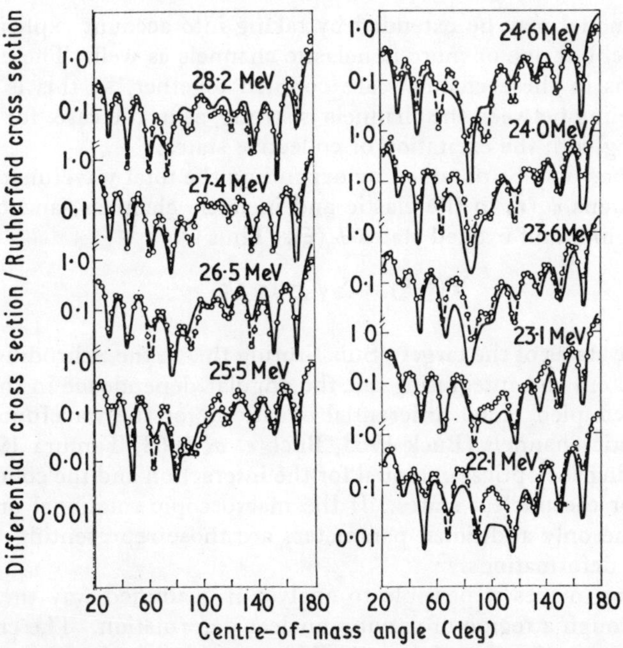

Figure 26. Differential cross sections for the elastic scattering of α particles at several energies from ^{39}K compared with optical model calculations: broken line, experimental curve; full line, optical model best fit. These show that the maximum in the cross section at 180° is very well given by the model. From Budzanowski *et al* (1968).

This effect may be brought within the optical model by allowing the imaginary potential to depend on the orbital angular momentum (Bisson *et al* 1970, Chatwin *et al* 1970); a suitable parametrization is

$$W_L(r) = W(r)/[1+\exp\{(L-L_c)/\Delta L\}] \tag{9.1}$$

where L_c is similar to the classical value for surface interactions. This expression ensures that for the higher partial waves the absorption potential is small, reflecting the physical fact that the scarcity of final states in the reaction channel prevents absorption taking place.

The third anomaly found in connection with the elastic scattering of α particles concerns the resonances that are sometimes observed in the excitation functions, particularly at back angles. These resonances are due to states in the compound nucleus of high J and they are particularly prominent because, as just mentioned, they sometimes have difficulty in decaying into the reaction channels. This distinguishes the situation from the case of lighter particles where the compound nuclear resonances are very numerous and readily decay into nonelastic channels. Several cases of these compound nucleus resonances have been found in the elastic scattering of α particles, and a well marked example is the broad maximum in the backward scattering of α particles by ^{39}K and ^{40}Ca around 24 MeV which may be fitted by modifying the reflection coefficient for $L = 10$ (Budzanowski *et al* 1967).

10. Inelastic scattering

The optical model may be extended by taking into account explicitly not only the elastic channel but one or more nonelastic channels as well. The equations for the wavefunctions in these channels are coupled together, so this is particularly useful if the coupling between the channels is strong, as in the case, for example, of inelastic scattering with the excitation of collective states.

These equations may be obtained by expanding the total wavefunction in terms of the wavefunctions $\psi_\alpha(\boldsymbol{r})$ in the elastic and inelastic channels, and of the target nucleus in its ground and excited states $\phi_\alpha(\boldsymbol{\xi})$. Thus

$$\Psi(\boldsymbol{r}, \boldsymbol{\xi}) = \sum_\alpha \psi_\alpha(\boldsymbol{r})\,\phi_\alpha(\boldsymbol{\xi}) \tag{10.1}$$

where α labels the states of the target. Substituting this in the Schrödinger equation for the whole system and integrating out the angular dependence in the usual way gives a series of coupled radial differential equations for the wavefunctions in the elastic and inelastic channels (Buck 1963, Buck *et al* 1963, Tamura 1965). These may be solved using the optical potential for the interaction and the collective model wavefunctions for the nuclear states. If the macroscopic rotational or vibrational model is used, the only additional parameters are those representing the static or dynamic nuclear deformations.

This formalism makes it possible to analyse in a unified way the differential cross sections through a region of varying nuclear deformation. The cross sections depend on the deformation, but they may all be accounted for by appropriate values of the deformation parameter, the optical potential remaining unchanged. This has been done for 17 MeV protons elastically scattered by some heavy nuclei (Perey 1964). A formal investigation of the effects of nuclear deformation in the optical potential has recently been made by Mackintosh (1971).

To first order, the inelastic cross sections for the one-step excitations are proportional to the square of the corresponding deformation parameter and thus provide an accurate way of determining them. In his pioneer work, Buck (1963) applied the coupled-channels formalism to fit the elastic and first inelastic cross sections with excitation of the lowest 2^+ states of several medium weight nuclei, and the β_2 deformation parameters he obtained were in good accord with those obtained by other methods. Analysis of the elastic and inelastic scattering of composite particles provides an even better way of determining the shapes of nuclei because of the sensitivity of the interaction to the nuclear surface. Several such analyses have been made with α particles, and have enabled the higher order deformations to be determined. A recent example of such an analysis, showing the sensitivity of the cross section to the β_4 deformation, is shown in figure 27. It is notable that the calculated cross sections for the excitation of the 4^+ state differ significantly for $\beta_4 = 0{\cdot}11$ and $\beta_4 = -0{\cdot}11$, so that the sign as well as the magnitude of β_4 can be obtained. The cross section is finite even for $\beta_4 = 0$ since the 4^+ state can be excited by a process proportional to $\beta_2{}^2$ that increases the spin by 2 acting twice, as well as by one proportional to β_4 that excites it directly.

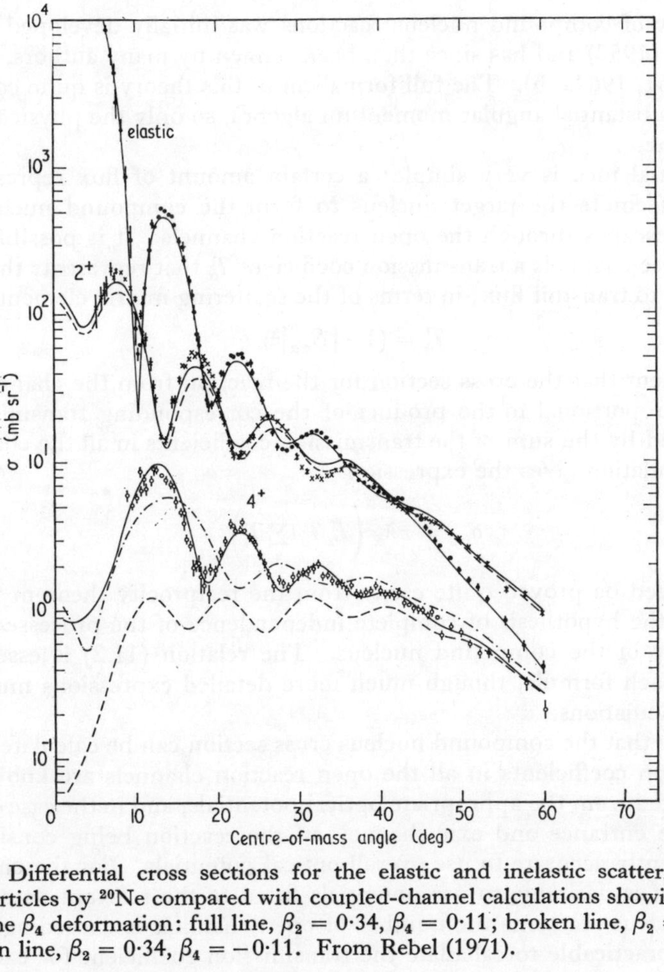

Figure 27. Differential cross sections for the elastic and inelastic scattering of 104 MeV α particles by ^{20}Ne compared with coupled-channel calculations showing the sensitivity to the β_4 deformation: full line, $\beta_2 = 0{\cdot}34$, $\beta_4 = 0{\cdot}11$; broken line, $\beta_2 = 0{\cdot}34$, $\beta_4 = 0{\cdot}0$; chain line, $\beta_2 = 0{\cdot}34$, $\beta_4 = -0{\cdot}11$. From Rebel (1971).

The coupled equations automatically give resonant cross sections at the energies of the collective excitations, just as the simple optical model gives them when the energy in the compound system corresponds to that of a single-particle state. In most cases, however, the resonances are so broadened by the imaginary part of the effective interaction that they are not observable. An important exception is the situation for some light nuclei at low energies, when most or all the nonelastic channels are closed. The most recent analysis of this type is that of Mikoshiba *et al* (1971) and as described in §6 this has yielded information on the spin–spin term in the optical potential.

11. Compound nucleus reactions

It was shown in § 4 that many elastic scattering cross sections, particularly at low energies, have an appreciable compound nucleus component, and this must be calculated and subtracted before an optical model analysis can be made. In such cases the compound nucleus calculation is essential for the optical model analysis, and as it relies very largely on optical model penetrabilities it is appropriate to describe it here.

The theory of compound nucleus reactions was initially developed by Hauser and Feshbach (1952) and has since then been refined by many authors, notably by Moldauer (1961, 1964a, b). The full formalism of this theory is quite complicated, and involves substantial angular momentum algebra, so only the physical ideas will be described here.

The essential idea is very simple: a certain amount of flux representing the incident beam enters the target nucleus to form the compound nucleus, and it subsequently escapes through the open reaction channels. It is possible to define for each of these channels a transmission coefficient T_α that represents the readiness of the channel to transmit flux; in terms of the scattering matrix element

$$T_\alpha = (1 - |S_{\alpha\alpha}|^2). \tag{11.1}$$

It is then evident that the cross section for the reaction from the channel α to the channel β is proportional to the product of the corresponding transmission coefficients, divided by the sum of the transmission coefficients in all the channels, and a detailed calculation gives the expression

$$\sigma_{\alpha\beta} = \pi \lambda\!\!\!^{-}{}_\alpha^2 \left(T_\alpha T_\beta / \sum_i T_i \right). \tag{11.2}$$

This may indeed be proved quite easily from the reciprocity theorem for nuclear reactions and the hypothesis of complete independence of the processes of formation and decay of the compound nucleus. The relation (11.2) is essentially the Hauser–Feshbach formula, though much more detailed expressions must be used in practical calculations.

This shows that the compound nucleus cross section can be calculated provided the transmission coefficients in all the open reaction channels are known. These can be calculated from the appropriate optical potentials, and in the case of channels other than the entrance and exit channels of the reaction being considered it is usually sufficiently accurate to use overall optical potentials. For the entrance and exit channels it is desirable to use potentials fitted to the relevant data. In many compound nucleus reactions the number of open channels is exceedingly large, so that it is not practicable to calculate the transmission coefficient for each of them individually. It is then sufficiently accurate to separate the denominator of (11.2) into two parts

$$\sum_i T_i = \sum_i{}' T_i + \sum_j \int \omega(E)\, T_j(E)\, \mathrm{d}E \tag{11.3}$$

where the first term refers to the important channels that can be treated individually and the second term is an integral over the regions where there are a large number of levels with level density $\omega(E)$ and transmission coefficient $T_j(E)$.

Several refinements to the Hauser–Feshbach calculation have been developed in recent years. The most important is the width fluctuation correction, which takes account of the partial breakdown of the hypothesis of the independence of the processes of formation and decay of the compound nucleus. This is done by multiplying the contribution to the cross section from each channel by the factor

$$W_{\alpha\beta} = \left\langle \frac{\Gamma_\alpha \Gamma_\beta}{\Gamma} \right\rangle \Big/ \frac{\langle\Gamma_\alpha\rangle\langle\Gamma_\beta\rangle}{\langle\Gamma\rangle} \tag{11.4}$$

where the Γ_i are widths and $\Gamma = \sum_i \Gamma_i$. For $\langle\Gamma_\alpha\rangle \ll D$, where D is the mean separation of the compound nucleus levels and the angular brackets indicate energy-averaging. The widths are related to the transmission coefficients by

$$T_\alpha \simeq \frac{2\pi\langle\Gamma_\alpha\rangle}{D}. \tag{11.5}$$

This width fluctuation correction factor may be calculated from the probability distribution of the nuclear levels. By using the Porter–Thomas distribution

$$P(x)\,\mathrm{d}x = (2\pi x)^{-1/2}\,\mathrm{e}^{-x/2}\,\mathrm{d}x \tag{11.6}$$

where $x = \Gamma/\langle\Gamma\rangle$, an explicit expression may be obtained for the width fluctuation correction, and this may be evaluated numerically from the values of the transmission coefficients. Several comparisons between compound nucleus calculations and experimental data show the necessity of including the width fluctuation correction (Hellström *et al* 1970).

There are some other corrections, notably the resonance interference correction (Moldauer 1964a,b), that are sometimes included in Hauser–Feshbach calculations, but they may not be significant in view of an unavoidable difficulty that arises in most cases from the presence of direct nonelastic reactions, which take flux directly from the elastic channels. This makes it incorrect to assume that all the nonelastic flux enters the compound nucleus, and if this is done the resulting compound nucleus cross sections will be too high. It is possible in principle to allow for this by computing the direct nonelastic flux and subtracting it from the total nonelastic flux to get the true flux entering the compound nucleus. Unfortunately this is nearly always impracticable since it requires a detailed theory of all the direct processes and also knowledge of all the corresponding spectroscopic factors, and this is seldom available. This effect may be allowed for in an approximate way by defining a reduction factor R

$$\sigma = \sigma_{\mathrm{DI}} + R\sigma_{\mathrm{CN}} \tag{11.7}$$

where σ_{DI} is the total direct nonelastic cross section and σ_{CN} the total compound nucleus cross section. If R is assumed to be the same in all channels, it may be determined from a comparison of theory and experiment in those reaction channels with no direct contribution and then used to analyse the data in the other channels, including the elastic channels. It is, however, unlikely that the reduction factor is the same in all channels, since the direct processes may be expected to take proportionately more flux from the higher partial waves. Nevertheless the approximation remains a useful one.

The analysis of an elastic scattering cross section with a substantial compound nucleus component must be made iteratively since the optical potential must be known in order to calculate the compound nucleus cross section. In practice, a suitable overall potential is first used to calculate the compound elastic cross sections, and with an assumed value of the reduction factor the shape elastic cross section is obtained. This is used as the input data for an optical model analysis in which the reduction factor as well as the usual optical model parameters are varied systematically to optimize the fit. With the resulting potential a new compound nucleus cross section is calculated and the whole process repeated until consistency is attained. The result of an analysis of the differential cross section for the elastic scattering of 5 MeV protons by ^{25}Mg, obtained in this way, is shown in figure 28.

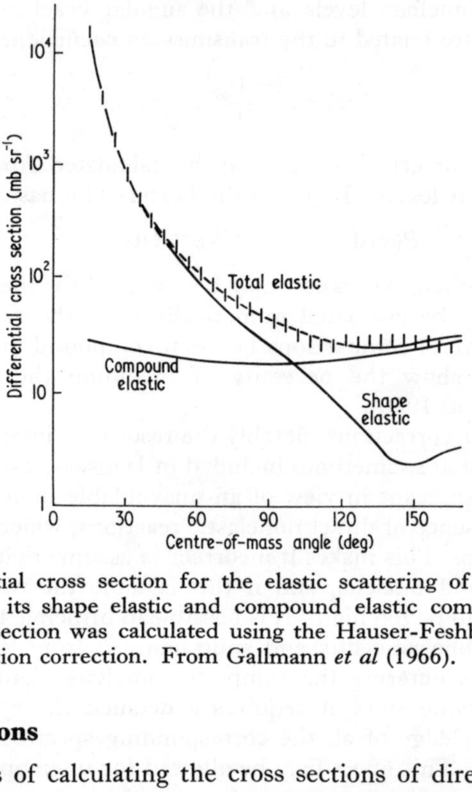

Figure 28. Differential cross section for the elastic scattering of 6 MeV protons by ^{25}Mg analysed into its shape elastic and compound elastic components. The compound elastic cross section was calculated using the Hauser-Feshbach theory including the width fluctuation correction. From Gallmann *et al* (1966).

12. Direct reactions

Many methods of calculating the cross sections of direct reactions have been developed over the last few decades and, as some of the more successful make use of wavefunctions calculated from optical potentials, it is appropriate to mention them here.

Typically, the cross section of a direct nuclear reaction is proportional to the square of a matrix element that may be written schematically in the form

$$M = \langle f|H|i\rangle \tag{12.1}$$

where $|i\rangle$ and $\langle f|$ are the wavefunctions of the initial and final systems and H the part of the hamiltonian responsible for the transition. Thus, for example, in the case of inelastic scattering, $|i\rangle$ is the product of the wavefunctions of the incident particle and the target nucleus in its ground state, and similarly for $|f\rangle$. In the case of a nucleon transfer reaction, such as a (d, p) stripping reaction, $|i\rangle$ is the product of the wavefunction of the incident deuteron, including its internal wavefunction, and that of the target nucleus and $|f\rangle$ represents the wavefunctions of the outgoing proton, the captured neutron and the residual nucleus. In all the expressions, the wavefunction representing the motion of the ingoing and outgoing particles can be calculated from the Schrödinger equation with the appropriate optical potential, and thus the optical model is an essential part of such theories.

There are several questions that arise concerning this use of optical model wavefunctions. One of the most important is how to justify the use of wavefunctions in the nuclear interior, when the potential from which they are calculated is obtained from elastic scattering that depends only on their asymptotic form in the region beyond the nucleus. There seems to be no way of knowing what the wavefunctions are in the nuclear interior, and yet it is just in this region that they contribute significantly to the direct reaction matrix element. Indeed in the case of deuterons and other composite particles it is difficult to see what meaning can be attached to the wavefunction in the nuclear interior. There is no easy answer to this question, but at least it can be said that the use of such wavefunctions often gives surprisingly good results in direct reaction calculations.

In other cases the distorted wave theory fails, and such situations can be dealt with in several ways. Thus, for example, it is sometimes possible to obtain agreement with experiment by an arbitrary modification of the distorted wave theory, such as imposing a radial cutoff on the overlap integral and ignoring contributions from smaller radii. This device is often successful, but it is unsatisfactory because it is arbitrary, and also because in some cases the cutoff radius is so large as to make the calculations unphysical. Thus the ^{208}Pb(p, d) cross-sections at 22 MeV can be fitted by the distorted wave theory only with a cutoff of 8·5 fm, which is well beyond the nuclear surface.

Another method of tackling failures of the distorted wave theory is to refine the theory itself, and there have been many attempts to do this, by taking into account processes that are normally ignored, such as the possibility of deuteron breakup in the case of (d, p) stripping. The formalism of such refined theory often gives a new prescription for the optical potentials describing the motion of the interacting particles. Using these potentials the old formalism can still be used but effects previously ignored are now included at least approximately.

A calculation of this type has been made for (d, p) stripping by Johnson and Soper (1970). They take into account the deuterons which have been broken up in the average field of the target but whose constituent nucleons continue to move together in a ^{3}S state with little relative momentum, and obtain an expression for the deuteron optical potential $U_{\mathrm{d}}(\boldsymbol{r})$ in terms of the neutron and proton optical potentials $U_{\mathrm{n}}(\boldsymbol{r})$ and $U_{\mathrm{p}}(\boldsymbol{r})$ at half the deuteron energy

$$U_{\mathrm{d}}(\boldsymbol{r}) = D_0^{-1} \int \{U_{\mathrm{n}}(\boldsymbol{r}+\tfrac{1}{2}\boldsymbol{s}) + U_{\mathrm{p}}(\boldsymbol{r}-\tfrac{1}{2}\boldsymbol{s})\} V_{\mathrm{np}}(s)\,\phi_{\mathrm{d}}(\boldsymbol{s})\,\mathrm{d}\boldsymbol{s} \tag{12.2}$$

where

$$D_0 = \int V_{\mathrm{np}}(\boldsymbol{s})\,\phi_{\mathrm{d}}(\boldsymbol{s})\,\mathrm{d}\boldsymbol{s} \tag{12.3}$$

$V_{\mathrm{np}}(\boldsymbol{s})$ is the neutron–proton interaction, $\phi_{\mathrm{d}}(\boldsymbol{s})$ the internal wavefunction of the deuteron in its ground state and $\boldsymbol{s}$ is the relative separation of the neutron and proton in the deuteron. In the limit of the zero-range neutron–proton interaction, this is just the sum of the neutron and proton potentials.

This deuteron potential differs from those found phenomenologically—in particular it has a larger surface diffuseness—and when used in the distorted wave theory it gives excellent results even for cases where the usual theory with optical potentials fitted to the elastic scattering fails. For the ^{208}Pb(p, d) reaction mentioned above, for example, it gives a good fit without a radial cutoff, and acceptable spectroscopic factors (Satchler, 1971b). Such an improvement has also been found for similar reactions on light nuclei (Johnson and Soper 1970). Further investigations are required, but already it seems likely that this prescription for the deuteron optical potential is a significant improvement on the usual distorted wave theory, and thus should be a useful tool in nuclear structure studies.

A second question concerns the effect of the nonlocality of the potential. As mentioned in § 1, the optical potential is partially nonlocal, though for convenience it is usual to work with the equivalent local potentials. While the wavefunctions corresponding to these potentials are by definition the same in the asymptotic region, they differ significantly in the nuclear interior. Indeed Perey (1962) has shown that to a good approximation

$$\psi_{\mathrm{NL}}(r) = \frac{\psi_{\mathrm{L}}(r)}{\{1+(2m\beta^2/\hbar^2)\,V(r)\}^{1/2}} \tag{12.4}$$

where $V(r)$ is the equivalent local potential and β the range of the nonlocality. The effect of this is to reduce $\psi_{\mathrm{L}}(r)$ by a factor that can be as much as 20% in the centre of the nucleus, falling to zero in the surface region. Using this expression the wavefunction can be corrected for the nonlocality, and calculations made with this correction show improved accord with the experimental data.

A final question regards the optical potential that should be used to describe the interaction of a particle with an excited nucleus, for this is needed whenever the residual nucleus is not in its ground state. This cannot be investigated experimentally, and at present there seems to be no strong theoretical reason for supposing that the optical potential depends at all strongly on the state of nuclear excitation, so it is usual to assume the dependence to be negligible. The error in so doing is certainly less than the other uncertainties in direct reaction calculations.

An attempt to calculate the potential due to an excited nucleus and to evaluate the effect of any difference from the ground-state potential has recently been made by Cooper *et al* (1971). They studied the ^{16}O(d, p)^{17}O* reaction and calculated the ^{17}O potential by adding the potential due to the valence neutron to the phenomenological optical potential. The potentials corresponding to the neutrons in the $1d_{5/2}$ and $2s_{1/2}$ state were calculated for this purpose using the Reid soft core potential and a Faddeev impulse approximation. The resulting ^{17}O potential gave a stripping cross section to the ground state similar to that obtained with a phenomenological ^{17}O optical potential, and reduced the cross section of the reaction to the excited state by a few per cent.

13. The future development of the optical model

The optical model has enjoyed impressive success over the last two decades, and it is worthwhile trying to outline the main lines of future development. It is convenient to begin with the model as applied to nucleon elastic scattering, and then to consider more complicated circumstances.

The optical model is able to give precision fits to elastic scattering differential cross sections, polarizations, reaction cross sections and (for neutrons) total cross sections, provided the phenomenological parameters are optimized for each nucleus at each energy. These particular potentials may be useful for some purposes, but it is obviously desirable to have an overall potential that fits the scattering from many nuclei at many energies with the same parameters. Such potentials have been found, subject to certain restrictions, but the corresponding fits are not quite as good, though still very satisfactory and useful for many applications. These overall phenomenological potentials have been progressively refined with the addition of isospin and other small terms, and include allowance for the energy dependence of the parameters.

These phenomenological potentials, both particular and overall, have several disadvantages. In general they are overparametrized, especially at low energies, so that many different potentials can be found to give equally good fits to the data. This poses the question of the physical content of the potentials, and raises doubts about the reliability of the distorted waves that are frequently calculated from them and used in reaction calculations.

It is of course to be expected that the particular potentials give better fits than the overall ones because they can take account of effects connected with the particular structure of the nucleus concerned, though at the expense of the generality of the model. The question to be faced is whether to use particular potentials for the sake of accuracy and accept the loss of generality and the large number of parameters, or whether to use overall potentials and try to refine them with the addition of more small terms, in both cases accepting the uncertainties connected with the ambiguities.

The way out of this impasse is to strengthen the physical basis of the potential by calculating it from the known nuclear characteristics and the nucleon–nucleon interaction. Ideally this could eliminate the ambiguities and also give the required sensitivity to details of nuclear structure without many arbitrary parameters. There are formidable difficulties in carrying out this programme, but the successes that have already been attained show that it is worth pursuing.

In its simplest form this approach gives the real part of the optical potential as an overlap integral of the nuclear density distribution and the nucleon–nucleon interaction. Together with a phenomenological imaginary potential this is able to give a fair account of the data for medium and heavy nuclei at energies of tens of MeV where some of the effects that are neglected, especially antisymmetrization and core polarization, are small. Further progress is being made by calculating these effects, and adding them to the first-order terms already known. Such calculations naturally require more or less detailed nuclear structure information, and thus provide a link between optical model calculations and the work on the nuclear charge and density distributions and on the nucleon–nucleon interaction. The aim of this work is to provide a perfectly general way of calculating the optical potential for any nucleus, including the spin–orbit, isospin and spin–spin terms and the energy dependence. It is not expressed in the usual Saxon–Woods form, as each term will have its own radial dependence, which will vary from one nucleus to the next depending on the differences in nuclear structure. Thus the simplicity of the old optical model is lost, but with a corresponding gain in physical reliability.

These remarks apply to the real part of the optical potential. The imaginary part poses a more complicated problem, and although many detailed attempts have been made to calculate it, a satisfactory way of doing so is still far away. This is only to be expected, since this part of the potential takes account of all the non-elastic processes, and these usually include many different types of reactions. Fortunately from one point of view, the elastic scattering is insensitive to the form chosen for the imaginary part of the potential, provided the parameters are optimized. This part of the potential will continue to be treated phenomenologically for the foreseeable future, and it may be possible to establish reliable expressions giving the variation of the parameters of this potential as a function of energy and from nucleus to nucleus.

A similar programme of work can be envisaged for the optical potentials for composite particles, with the additional complications due to their internal structure. Such potentials can in principle be calculated either directly from the nucleon–nucleon interaction and the structures of the nucleus and the incident particles or by making use of the nucleon–nucleus potentials as an intermediate step. The latter approach has been used most frequently so far, and significant progress has been made.

A further account of many of the questions discussed here, together with more extensive references, may be found in the review by Hodgson (1971).

Acknowledgment

I gratefully acknowledge the permission of the Clarendon Press to reproduce some figures from my book *Nuclear Reactions and Nuclear Structure*.

References

Afnan I R and Tang Y C 1970 *Nucl. Phys.* A **141** 653
Bang J and Zimanyi J 1969 *Nucl. Phys.* A **139** 534
Barschall H H 1952 *Phys. Rev.* **86** 431
Batty C J and Greenlees G W 1969 *Nucl. Phys.* A **133** 673
Batty C J and Tschalär C 1970 *Nucl. Phys.* A **143** 151
Becchetti F D and Greenlees G W 1969 *Annual Report of the J H Williams Laboratory of Nuclear Physics, University of Minnesota* 116
Bencze Gy and Doleschall P 1970 *Phys. Lett.* **32B** 539
Bethe H A 1956 *Phys. Rev.* **103** 1353
Bisson A E, Eberhard K A and Davis R H 1970 *Phys. Rev.* C **1** 539
Boyd R N *et al* 1971 *Nucl. Phys.* A **162** 497
Brown G E, Gunn J H and Gould P 1963 *Nucl. Phys.* **46** 598
Brueckner K A 1954 *Phys. Rev.* **96** 508
see also Brueckner K A, Eden R J and Francis N C 1955 *Phys. Rev.* **99** 76
—— 1955 *Phys. Rev.* **100** 891
Brueckner K A and Levinson C A 1955 *Phys. Rev.* **97** 1344
Brueckner K A, Lockett A M and Rotenberg M 1961 *Phys. Rev.* **121** 255
Buck B 1963 *Phys. Rev.* **130** 712
Buck B, Stamp A P and Hodgson P E 1963 *Phil. Mag.* **8** 1805
Budzanowski A *et al* 1967 *Proc.* 1966 *Gatlinburg Int. Nucl. Phys. Conf.* ed R L Becker *et al* (New York: Academic Press) p81
Budzanowski A *et al* 1969 *Nucl. Phys.* A **126** 369
Budzanowski A, Dudek A, Grotowski K and Strzalkowski A 1970 *Phys. Lett.* **32B** 431
Chatwin R A, Eck J S, Robson D and Richter A 1970 *Phys. Rev.* C **1** 795
Cohen B L 1965 *Am. J. Phys.* **33** 1011
Cole B J, Huby R and Mines J R 1970 *Phys. Lett.* **33B** 320
Cooper M D, Lerner G M and Redish E C 1971 *Bull. Am. Phys. Soc.* **16** 99
Dudek A and Hodgson P E 1972 *J. Phys., Paris* (in press)
Eberhard K A 1970 *Phys. Lett.* **33B** 343
Eberhard K A and Robson D 1971 *Phys. Rev.* C **3** 149
Elton L R B 1968 *Phys. Lett.* **26B** 689
Faraggi H *et al* 1971 *Ann. Phys.* **66** 905
Fernandez B and Blair J S 1970 *Phys. Rev.* C **1** 523

Fernbach S, Serber R and Taylor T B 1949 *Phys. Rev.* **75** 1352
Feshbach H 1958a *A. Rev. Nucl. Sci.* **8** 49
—— 1958b *Ann. Phys.* **5** 357
—— 1962 *Ann. Phys.* **19** 287
Feshbach H, Porter C E and Weisskopf V F 1954 *Phys. Rev.* **96** 448
Fisher T R 1971 *Phys. Lett.* **35B** 573
Fisher T R *et al* 1966 *Phys. Rev. Lett* **17** 36
Fortune H T, Gray T J, Trost W and Fletcher N R 1969 *Phys. Rev.* **179** 1033
Fortune H T and Vincent C M 1969 *Phys. Rev.* **185** 1401
Gallmann A *et al* 1966 *Nucl. Phys.* **88** 654
Gibson E F *et al* 1967 *Phys. Rev.* **155** 1194
Glasgow D W and Foster D G 1971 *Phys. Rev.* C **3** 604
Goldberger M L 1948 *Phys. Rev.* **74** 1269
Greenlees G W *et al* 1970a *Phys. Rev.* C **2** 1063
Greenlees G W, Makofske W and Pyle G J 1970b *Phys. Rev.* C **1** 1145
Greenlees G W and Pyle G J 1966 *Phys. Rev.* **149** 836
Greenlees G W, Pyle G J and Tang Y C 1968 *Phys. Lett.* **26B** 658
—— 1969 *Phys. Rev.* **171** 1115
Greenlees G W and Tang Y C 1971 *Phys. Lett.* **34B** 359
Grimm R C, McCarthy I E and Storer R G 1971 *Nucl. Phys.* A **166** 330
Hauser W and Feshbach H 1952 *Phys. Rev.* **87** 366
Häusser O, von Brentano P and Mayer-Kuckuk T 1964 *Phys. Lett.* **12** 226
Hellström J, Dallimore P J and Davidson W F 1969 *Nucl. Phys.* A **132** 581
Hodgson P E 1961 *Phys. Rev. Lett* **6** 358
—— 1963 *The Optical Model of Elastic Scattering* (Oxford: Clarendon Press)
—— 1966 *Adv. Phys.* **15** 329
—— 1967 *Nucl. Phys.* A **103** 127
—— 1968 *Adv. Phys.* **17** 563
—— 1970 *Nucl. Phys.* A **150** 1
—— 1971 *Nuclear Reactions and Nuclear Structure* (Oxford: Clarendon Press)
Huby R 1970 *Phys. Lett.* **33B** 323
Huby R and Mines J R 1965 *Rev. Mod. Phys.* **37** 406
Jackson D F and Kembhavi V K 1969 *Phys. Rev.* **178** 1626
Johansson A, Svanberg U and Hodgson P E 1961 *Ark. Fys.* **19** 541
Johnson R C and Soper P J R 1970 *Phys. Rev.* C **1** 976
Jones G A 1970 *Rep. Prog. Phys.* **33** 645
Keaton P W 1971 *Los Alamos Report* LA-DC-12739
Kidwai H R and Rook J R 1971 *Nucl. Phys.* A **169** 417
Kossanyi-Demay P and Mayer B 1967 *J. Phys.* C: *Solid St. Phys.* **1** 137
Lane A M 1957 *Rev. Mod. Phys.* **29** 191
—— 1962 *Nucl. Phys.* **35** 676
Le Levier R E and Saxon D S 1952 *Phys. Rev.* **87** 40
Lilley J S 1971 *Phys. Rev.* C **3** 2229
McEver W S *et al* 1970 *Phys. Rev. Lett.* **24** 1123
Mackintosh R S 1971 *Nucl. Phys.* A **164** 398
Meldner H 1969 *Phys. Rev.* **178** 1815
Mikoshiba O, Terasawa T and Tanifuji M 1971 *Nucl. Phys.* A **168** 417
Millener D J and Hodgson P E 1971 *Phys. Lett.* **35B** 495
Moldauer P A 1961 *Phys. Rev.* **123** 968
see also Moldauer P A 1967 *Phys. Rev.* **157** 907
—— 1968 *Phys. Rev.* **171** 1164
—— 1969 *Phys. Rev.* **177** 1841
—— 1964a *Rev. Mod. Phys.* **36** 1079
—— 1964b *Phys. Rev.* B **135** 642
Monahan J E and Elwyn A J 1964 *Phys. Rev.* B **136** 1678
Morgan C G and Jackson W F 1969 *Phys. Rev.* **188** 1758
Mukherjee S 1968 *Nucl. Phys.* A **118** 423
Nagamine K, Uchida A and Kobayashi S 1970 *Nucl. Phys.* A **145** 203
Newstead C M, Delaroche J and Canvin B 1971 *Proc. Albany Conf.* (in press)

Owen L W and Satchler G R 1970 *Phys. Rev. Lett.* **25** 1721
Pascolini A, Pisent G and Zardi F 1969 *Nuovo Cim. Lett.* **1** 643
Perey C M and Perey F G 1963 *Phys. Rev.* **132** 755
—— 1968 *Phys. Lett.* **26B** 123
Perey C M, Perey F G, Dickens J K and Silva R J 1968 *Phys. Rev.* **175** 1460
Perey F G 1962 *Direct Interactions and Nuclear Reaction Mechanisms* ed E Clementel and C Villi (New York: Gordon and Breach) p125
—— 1963 *Phys. Rev.* **131** 745
—— 1964 *Argonne National Laboratory Report* ANL-6848 114
Perey F G and Buck B 1962 *Nucl. Phys.* **32** 353
Pyle G J and Greenlees G W 1969 *Phys. Rev.* **181** 1444
Rahman Kahn M Z 1966 *Nucl. Phys.* **76** 475
Raynal J 1963 *Phys. Lett.* **7** 281
Rebel H 1971 *Karlsruhe Cyclotron Laboratory Report* KFK-1397
Rihan T H and Sharaf M A 1969 *Nucl. Phys.* A **134** 369
Rost E 1968 *Phys. Lett.* **26B** 184
Samaddar S K and Mukherjee S 1971 *Nucl. Phys.* A (in press)
Samaddar S K, Satpathy R K and Mukherjee S 1970 *Nucl. Phys.* A **150** 655
Satchler G R 1960 *Nucl. Phys.* **21** 116
—— 1967 *Nucl. Phys.* A **92** 273
—— 1971a *Phys. Lett.* **34B** 37
—— 1971b *Phys. Rev.* C **4** 1485
Satchler G R and Haybron R M 1964 *Phys. Lett.* **11** 313
Satchler G R *et al* 1968 *Nucl. Phys.* A **112** 1
Satpathy R K, Samaddar S K and Mukherjee S 1969 *Nucl. Phys.* A **132** 276
Schiffer J P 1963 *Nucl. Phys.* **46** 246
Schmeing N C 1970 *Nucl. Phys.* A **142** 449
Schwandt P and Haeberli W 1969 *Nucl. Phys.* A **123** 401
Sinha B C and Edwards V R W 1970 *Phys. Lett.* **31B** 273
Springer A and Harvey B G 1965 *Phys. Lett* **14** 116
Tamura T 1965 *Rev. Mod. Phys.* **37** 679
Tang Y C 1969 *Proc. Bochun Int. Conf. Clustering Phenomena in Nuclei* (Vienna: International Atomic Energy Agency) p109
Testoni J and Gomes L C 1966 *Nucl. Phys.* **89** 288
Thompson W J 1971 *Particles and Nuclei* **2** 47
Urone P P, Put L W, Chang H H and Ridley B W 1971 *Nucl. Phys.* A **163** 225
Watanabe S 1958 *Nucl. Phys.* **8** 484
Weisser D C, Lilley J S, Hobbie R K and Greenlees G W 1970 *Phys. Rev.* C **2** 544
Wilkinson D H 1961 *Proc. Rutherford Jubilee Int. Conf.* ed J B Birks (Manchester: Heywood) p339
Wilmore D and Hodgson P E 1964 *Nucl. Phys.* **55** 673
Woolam P B, Griffiths R J, Grace J F and Lewis V E 1970, *Nucl. Phys.* A **154** 513
Youngblood D H, Kozub R L, Kenefick R A and Hiebert J C 1970 *Phys. Rev.* C **2** 477

NEANDC-198-U
INDC(NEA)5

International Nuclear Model Code Comparison Study of the Spherical Optical Model for Charged Particles

P. E. Hodgson
Nuclear Physics Laboratory, Oxford, U.K.

and

E. Sartori
NEA Data Bank, Gif-sur-Yvette, France

1. Introduction

Several intercomparisons /1/,/2/ of computer codes used to calculate nuclear cross sections show general agreement but significant differences in detail. Since these codes embody the same mathematical functions and the intercomparisons use the same input parameters one would expect if the codes are indeed identical, that the results would be the same to a high degree of accuracy, limited only by the rounding errors of the computer used. For most practical purposes the differences between the codes tested in previous intercomparisons are not important. It is however desirable to identify and, where necessary, correct the sources of these differences because they may well cause significant effects when the program is used in other domains of parameter space or when the code is used as a subroutine in a more extended calculation.

This report is concerned with simple optical model calculations but it is planned to extend it to other calculations. In Section 2 we summarize the preliminary intercomparison and tabulate the results of calculations of proton and alpha-particle elastic scattering made with several different codes. Section 3 summarizes the possible reasons for the discrepancies that are found, and Section 4 presents the results of selected intercomparisons in which the codes are altered to bring them into close conformity. These results can therefore serve as a standard.

The optical model formalism is given by Hodgson /3/.

2. Preliminary Intercomparison

A series of test calculations for protons and alpha-particles scattered at 5, 10, 15 and 20 MeV from carbon 12, cobalt 59 and lead 208 was specified as described in Appendix 1 to this report. This specification was sent to several users of spherical optical model codes and the results obtained for the reaction cross sections are given in Tables I and II. Some characteristics of the codes used are given in Appendix 2.

Examination of these tables shows, as expected, that all the codes used are essentially correct and their accuracy is more than adequate for practical purposes.

Closer examination, however, shows significant differences and these are the subject of the present report. In particular it is notable that the results obtained with JIB and SCAT-2 are nearly always in accord to three decimal places, and often to four. Other codes show quite marked deviations from the average results, such as ABACUS-2 for 5 MeV protons on carbon 12, SMOG for 5 MeV protons on lead 208 and SCAT for 5 MeV protons on cobalt 59. On the whole the consistency is better for the heavy nuclei than for the light and better for the alpha-particles than for protons; this is expected because these interactions are more dominated by the Coulomb field and hence less sensitive to the nuclear potential. This is satisfactory confirmation that the routines calculating the Coulomb functions are essentially correct.

3. Reasons for the Discrepancies

There are several possible causes for the discrepancies between the codes and these may conveniently be divided into two classes: (a) mathematical and (b) numerical:

(a) Mathematical Discrepancies

There are several mathematical parameters that control the numerical integration of the radial wave equation, in particular (i) the step length (ii) the number of partial waves and (iii) the matching radius. These parameters are usually set automatically within the code to values that are more than adequate to ensure the required accuracy. The step length may be chosen to correspond to about ten steps per unit of kr, the number of partial waves so that the contributions of the last two waves included in the calculation contribute less than $\delta = 10^{-5}$ to the scattering matrix element and the matching radius to exceed R + na where n is about 10. The values of these parameters in the programs studied in this exercise are given in Table III. Tests showed that in all cases these values are adequate, and so the observed discrepancies cannot be attributed to these choices.

(b) Numerical Discrepancies

There are two combinations of fundamental constants that enter the calculation, namely those used to obtain the wave number and the Coulomb parameter, and these are also included in Table III. The parameters specifying the potential must obviously be the same, but even here there is one possible source of discrepancy, namely rounding errors evaluating the potential depths from the formulae given; this applies only to the proton calculations.

4. Precision Calculations

The calculations were rerun with three codes altered so as to have exactly the same numerical and output parameters; the results are compared in Table IV. In Table V we also present the results of exactly similar calculations to those in Table IV but with the charge of the incident particle set to zero. This can serve as a standard to check neutron optical model calculations.

For the precision calculations the following parameters were used for C^{12}:

E	V_R	W_V	W_{SF}	a_I	l_{max}
5	53.4	0.0	10.6	0.51	9
10	51.8	0.0	9.3	0.51	10
15	50.2	0.6	8.1	0.51	12
20	48.6	1.7	6.8	0.51	13

The other parameters are energy and mass independent and are given below:

r_R = 1.17 fm
a_R = 0.75 fm
r_I = 1.32 fm
V_{SO} = 6.2 MeV
r_{SO} = 1.01 fm
a_{SO} = 0.75 fm
r_C = 1.25 fm

Wave number = 0.218732
Coulomb parameter = 0.157484
Total number of steps = 122
Partial wave cut-off criterion = 10^{-6}
Matching radius n $\approx$ 10

In addition, the SMOG program was run on two different computers using the same word length (32 bits) for storing data, namely IBM and VAX. These two computers store data internally in a different way. The results obtained from these calculations are also shown in Tables IVa, IVb, Va and Vb. The observed discrepancies give an indication of the maximum agreement one can expect in this kind of comparison. SMOG was then rerun in double precision on both computers and the agreement improved by three decimal figures. The results obtained on VAX in single precision appeared to be more accurate than those obtained with IBM. Table VI shows the average fractional deviation of the results obtained with an IBM as compared to VAX both in single and double precision for the proton as incident particle.

5. Conclusion

The comparisons made in this report show the accuracy that is presently attainable with optical model calculations. They also provide benchmarks that can be used to check any optical model program before it is used to analyse experimental data.

References

/1/ E. Sartori, Report on the International Nuclear Model Code Intercomparison, Coupled Channel Model Study, NEANDC-182A-INDC(NEA)3, 1984.

/2/ A. Prince, G. Reffo and E. Sartori, Report on the International Nuclear Model Code Intercomparison Spherical Optical and Statistical Model Study, NEANDC-152A-INDC(NEA)4, 1983.

/3/ P. E. Hodgson, The Optical Model of Elastic Scattering, Oxford University Press, 1963.

Acknowledgements

We wish to thank the participants for their precious contributions, without which this comparison would not have been possible. Sheila Greenstreet accepted the continuous changes in the Tables with a smile. We thank her for editing this report.

Table Ia - Incident Particle: Proton

Reaction Cross Section (m barn)

E(lab) (MeV)	SCAT	ABACUS-2	ELIESE-3	SMOG	JIB	SCAT-2
			C^{12}			
5	807.20	802.59	810.94	811.48	811.27	810.65
10	751.48	752.79	753.52	749.97	749.66	749.56
15	679.50	680.90	681.91	678.07	677.72	677.82
20	602.05	602.38	608.35	602.03	601.91	601.91
			Co^{59}			
5	371.83	372.41	372.49	372.45	372.45	372.48
10	939.99	940.56	941.95	942.17	941.74	941.39
15	1128.47	1128.6	1126.9	1126.72	1126.54	1126.86
20	1185.14	1186.1	1188.0	1188.34	1188.09	1187.88
			Pb^{208}			
5	0.0864	0.0925	0.0909	0.0478	0.091	0.0925
10	188.14	190.79	190.38	189.80	190.61	190.76
15	932.35	933.60	933.13	932.61	933.52	933.70
20	1446.69	1450.7	1449.9	1449.45	1450.34	1450.48

Table Ib - Incident Particle: Proton; Target C^{12}

Shape Elastic Differential Cross Section (m barn/sterad)

Angle Degrees (C.M.)	ABACUS-2	ELIESE-3	SMOG	JIB	SCAT-2
			E(lab) = 5 MeV		
10	33863.	33801.	33798.	33803.	33806.
50	85.21	85.66	85.39	85.57	85.56
90	5.802	5.578	5.39	5.389	5.435
130	15.97	15.61	15.70	15.57	15.73
170	21.06	22.10	22.59	22.66	22.47
			E(lab) = 10 MeV		
10	7642.	7634.	7628.	7634.	7633.
50	49.68	50.06	51.15	51.33	51.26
90	15.20	15.28	15.50	15.48	15.44
130	9.454	8.463	8.65	8.63	8.673
170	4.132	5.263	5.67	5.639	5.526
			E(lab) = 15 MeV		
10	3300.	3283.	3259.	3263.	3263.
50	40.47	40.70	42.05	42.16	42.07
90	18.56	18.17	18.08	18.02	18.03
130	3.157	2.734	2.79	2.783	2.796
170	3.353	4.550	4.64	4.663	4.637
			E(lab) = 20 MeV		
10	1984.	1973.	1943.	1945.	1946.
50	34.65	34.32	35.38	35.39	35.38
90	12.92	12.61	12.56	12.54	12.55
130	1.248	0.9899	1.01	1.006	1.017
170	1.507	2.091	2.22	2.216	2.214

Table Ic - Incident Particle : Proton: Target C^{12}

Polarizations

Angle Degrees (C.M.)	ABACUS-2	ELIESE-3	SMOG	JIB	SCAT-2
			E(lab) = 5 MeV		
10	- 0.0012	0.0008	0.001	0.0008	0.0008
50	0.0088	- 0.0470	- 0.046	- 0.0473	- 0.0482
90	- 0.4868	0.5144	0.516	0.5146	0.5236
130	- 0.4146	0.3415	0.340	0.3396	0.3462
170	- 0.1462	0.1174	0.116	0.1155	0.1183
			E(lab) = 10 MeV		
10	- 0.0075	0.0028	0.003	0.0028	0.0029
50	0.1666	- 0.2344	- 0.234	- 0.2350	- 0.2398
90	- 0.3910	0.2668	0.264	0.2613	0.2685
130	0.1081	- 0.0920	- 0.087	- 0.0848	- 0.0850
170	- 0.6035	0.4216	0.408	0.4091	0.4205
			E(lab) = 15 MeV		
10	- 0.0354	0.0172	0.017	0.0174	0.0178
50	0.1646	- 0.2610	- 0.249	- 0.2498	- 0.2550
90	- 0.3685	0.2873	0.292	0.2892	0.2957
130	0.1873	- 0.0710	- 0.077	- 0.0767	- 0.0737
170	- 0.5121	0.3473	0.344	0.3437	0.3477
			E(lab) = 20 MeV		
10	- 0.0073	0.0345	0.035	0.0353	0.0361
50	0.1266	- 0.1993	- 0.186	- 0.1879	- 0.1911
90	- 0.2911	0.2706	0.276	0.2748	0.2811
130	- 0.5642	0.6673	0.630	0.6329	0.6451
170	- 0.5427	0.3807	0.376	0.3755	0.3786

Table II - Incident Particle : Alpha

Reaction Cross Section (m barn)

E(lab) (MeV)	SCAT	ABACUS-2	ELIESE-3	SMOG	JIB	SCAT-2
			C^{12}			
5	504.58	505.03	504.85	504.57	504.96	505.00
10	820.79	821.25	821.40	821.13	821.16	821.19
15	886.75	887.19	887.25	886.93	887.15	887.16
20	907.18	908.21	908.26	907.91	908.19	908.21
			Co^{59}			
5	0.0430	0.0163	0.0159	0.0210	0.016	0.0163
10	252.95	248.59	248.52	248.03	248.54	248.58
15	847.91	848.94	848.74	848.44	848.92	848.94
20	1144.65	1145.7	1145.5	1145.16	1145.68	1145.71
			Pb^{208}			
15		0.0908	0.0795	0.0611	0.086	0.0908
20		84.775	84.610	84.204	84.75	84.779

Table III - Values of Parameters of Codes

Parameter Code	Total Number of Steps	Partial Wave Cut-off	Matching Radius n	Wave Number	Coulomb Parameter
SCAT			7	0.21729	
ABACUS-2	varying with energy ≤ 500	varying with energy 10^{-6} - 10^{-9}	parameter given in input 4, 5, 6 or 7	0.21873427	0.15748603
ELIESE-3	varying	varying	≈ 10	0.2187342	0.1574862
SMOG	150	10^{-4}	10	0.2187342	0.1574862
JIB	122	10^{-7}	≈ 13	0.218732	0.157484
SCAT-2	200	10^{-6}	≈ 13	0.21873427	0.15748603

Table IVa - Incident Particle: Proton; Target C^{12}

Shape Elastic Differential Cross Section (m barn/sterad)

Reaction Cross Section (m barn)

C^{12}					
Angle Degrees (C.M.)	SMOG (IBM-VAX) Double P.	SMOG (VAX) Single P.	SMOG (IBM) Single P.	JIB (VAX)	SCAT-2 (CDC)
E(lab) = 5 MeV					
σ_r	811.14	811.14	811.33	811.16	810.58
10	33809.	33809.	33802.	33805.	33813.
50	85.83	85.83	85.70	85.84	85.84
90	5.400	5.400	5.388	5.374	5.418
130	15.65	15.65	15.64	15.66	15.65
170	22.52	22.52	22.48	22.52	22.33
E(lab) = 10 MeV					
σ_r	749.62	749.62	749.80	749.59	749.47
10	7635.	7635.	7629.	7634.	7636.
50	51.31	51.32	51.29	51.51	51.43
90	15.45	15.45	15.46	15.44	15.40
130	8.657	8.657	8.662	8.633	8.678
170	5.609	5.609	5.626	5.605	5.488
E(lab) = 15 MeV					
σ_r	679.83	679.83	680.01	678.90	678.98
10	3269.	3269.	3267.	3263.	3265.
50	41.91	41.91	41.90	42.05	41.98
90	18.05	18.05	18.06	17.89	17.89
130	2.762	2.762	2.767	2.732	2.743
170	4.555	4.555	4.556	4.576	4.547
E(lab) = 20 MeV					
σ_r	604.19	604.19	604.23	601.79	601.78
10	1960.	1960.	1959.	1944.	1945.
50	35.55	35.55	35.55	35.43	35.43
90	12.69	12.69	12.69	12.51	12.52
130	1.044	1.044	1.046	0.999	1.010
170	2.136	2.136	2.139	2.209	2.209

Table IVb - Incident Particle: Proton; Target C^{12}

Polarization

Angle Degrees (C.M.)	SMOG (IBM-VAX) Double P.	SMOG (VAX) Single P.	SMOG (IBM) Single P.	JIB (VAX)	SCAT-2 (CDC)
E(lab) = 5 MeV					
10	0.0008	0.0008	0.0008	0.0008	0.0008
50	- 0.0464	- 0.0464	- 0.0459	- 0.0466	- 0.0475
90	0.5139	0.5139	0.5147	0.5142	0.5233
130	0.3395	0.3395	0.3390	0.3385	0.3455
170	0.1151	0.1151	0.1151	0.1148	0.1176
E(lab) = 10 MeV					
10	0.0028	0.0028	0.0028	0.0028	0.0029
50	- 0.2349	- 0.2349	- 0.2341	- 0.2347	- 0.2395
90	0.2621	0.2621	0.2638	0.2620	0.2693
130	- 0.0831	- 0.0831	- 0.0848	- 0.0835	- 0.0836
170	0.4103	0.4103	0.4096	0.4096	0.4212
E(lab) = 15 MeV					
10	0.0173	0.0173	0.0173	0.0174	0.0178
50	- 0.2516	- 0.2516	- 0.2506	- 0.2494	- 0.2543
90	0.2888	0.2888	0.2903	0.2898	0.2965
130	- 0.0710	- 0.0710	- 0.0718	- 0.0780	- 0.0748
170	0.3454	0.3455	0.3441	0.3420	0.3460
E(lab) = 20 MeV					
10	0.0349	0.0349	0.0349	0.0353	0.0361
50	- 0.1880	- 0.1880	- 0.1872	- 0.1877	- 0.1908
90	0.2725	0.2725	0.2739	0.2758	0.2819
130	0.6340	0.6340	0.6332	0.6336	0.6450
170	0.3813	0.3812	0.3824	0.3747	0.3776

Table Va - Incident Particle: Neutron; Target C^{12}

Shape Elastic Differential Cross Section (m barn/sterad)

Reaction Cross Section (m barn)

Angle Degrees (C.M.)	SMOG (IBM-VAX) Double P.	SMOG (VAX) Single P.	SMOG (IBM) Single P.	JIB (VAX)	SCAT-2 (CDC)
			E(lab) = 5 MeV		
σ_{SE}	526.00	526.00	525.76	525.86	
σ_r	1082.32	1082.32	1082.62	1082.25	1081.55
10	319.4	319.4	319.2	319.2	318.9
50	60.20	60.20	60.18	60.18	60.18
90	5.747	5.746	5.736	5.760	5.745
130	14.31	14.31	14.30	14.32	14.38
170	25.73	25.73	25.78	25.69	25.43
			E(lab = 10 MeV		
σ_{SE}	676.15	676.15	676.00	677.29	
σ_r	871.56	871.56	871.65	871.47	871.66
10	563.4	563.4	563.3	564.2	563.9
50	40.67	40.67	40.68	40.88	40.78
90	23.06	23.06	23.07	23.04	23.01
130	8.781	8.781	8.765	8.763	8.813
170	11.90	11.90	11.90	11.88	11.72
			E(lab) = 15 MeV		
σ_{SE}	859.24	859.24	859.15	857.27	
σ_r	748.53	748.53	748.52	747.62	747.60
10	885.2	885.2	885.1	882.4	881.4
50	42.77	42.77	42.77	42.84	42.79
90	22.37	22.37	22.37	22.19	22.21
130	3.251	3.251	3.251	3.203	3.217
170	6.386	6.387	6.381	6.420	6.370
			E(lab) = 20 MeV		
σ_{SE}	950.19	950.18	950.27	940.21	
σ_r	645.57	645.57	645.53	643.30	643.26
10	1152.	1152.	1152.	1136.	1135.
50	43.18	43.18	43.17	42.82	42.85
90	14.29	14.29	14.29	14.13	14.13
130	1.741	1.741	1.741	1.695	1.707
170	2.124	2.124	2.127	2.199	2.199

Table Vb - Incident Particle: Neutron; Target C^{12}

Polarization

Angle Degrees (C.M.)	SMOG (IBM-VAX) Double P.	SMOG (VAX) Single P.	SMOG (IBM) Single P.	JIB (VAX)	SCAT-2 (CDC)
E(lab) = 5 MeV					
10	- 0.0089	- 0.0089	- 0.0091	- 0.0089	- 0.0091
50	- 0.0681	- 0.0681	- 0.0681	- 0.0688	- 0.0699
90	0.1423	0.1423	0.1400	0.1420	0.1487
130	0.1619	0.1619	0.1614	0.1608	0.1641
170	0.1782	0.1783	0.1771	0.1780	0.1825
E(lab) = 10 MeV					
10	- 0.0028	- 0.0028	- 0.0032	- 0.0028	- 0.0030
50	- 0.2473	- 0.2473	- 0.2477	- 0.2470	- 0.2524
90	0.2013	0.2012	0.2000	0.2013	0.2064
130	- 0.1491	- 0.1491	- 0.1493	- 0.1496	- 0.1498
170	0.3526	0.3526	0.3534	0.3521	0.3603
E(lab) = 15 MeV					
10	- 0.0113	- 0.0113	- 0.0115	- 0.0111	- 0.0114
50	- 0.2197	- 0.2197	- 0.2197	- 0.2186	- 0.2230
90	0.2011	0.2011	0.2010	0.2019	0.2065
130	0.1933	0.1932	0.1941	0.1876	0.1956
170	0.3701	0.3700	0.3705	0.3669	0.3724
E(lab) = 20 MeV					
10	- 0.0101	- 0.0101	- 0.0103	- 0.0095	- 0.0097
50	- 0.1176	- 0.1176	- 0.1174	- 0.1195	- 0.1212
90	0.2190	0.2190	0.2194	0.2215	0.2269
130	0.6878	0.6878	0.6886	0.6831	0.6959
170	0.4429	0.4429	0.4408	0.4353	0.4383

Table VI - Comparison of SMOG on IBM and VAX for $p + C^{12}$

Average Fractional Deviation(*)

a) Single Precision (10^{-3})

Energy	Differential Cross Section	Polarization
5	1.4	7.5
10	1.2	56.
15	0.9	8.2
20	0.8	4.3

b) Double Precision (10^{-6})

Energy	Differential Cross Section	Polarization
5	1.2	2.7
10	1.9	53.
15	1.8	7.1
20	1.6	2.4

(*) The averages were calculated over 36 angles at intervals of 5 degrees. However, these averages depend on the choice of the anguler mesh.

Appendix 1

Definition of Optical Potential

(according to Becchetti & Greenlees, Phys. Rev. 182, 1190 (1969))

Expression	Validity Range	Explanation
$U_{\mathrm{opt}}(r) = -V_R f_R$		central real
$+\left(\frac{\hbar}{m_\pi c}\right)^2 \frac{V_{so}}{r}\left(\frac{d}{dr} f_{so}\right)\vec{l}\cdot\vec{\sigma}$		spin orbit
$+\frac{Zze^2}{2R_c}\left[3-\left(\frac{r}{R_c}\right)^2\right]$	for $r \le R_c$	Coulomb
$+\frac{Zze^2}{r}$	for $r \ge R_c$	
$-iW_v f_I$		imaginary volume
$+i4a_I W_{SF}\frac{d}{dr} f_I$		imaginary surface

$$\text{where } f_X = f(r, R_X, a_X) = [1+\exp(r-R_X)/a_X]^{-1}$$

$$R_X = r_X A^{1/3}$$

$$r_c = \text{Coulomb radius}$$

Note: Whenever a parameter is omitted, it is assumed that the corresponding potential is not considered.

Proton Parameters: Becchetti & Greenlees, Phys. Rev. 182, 1190 (1969)

$$V_R = 54.0 - 0.32E + 0.4Z/A^{1/3} + 24.0(N-Z)/A \text{ MeV}$$

$$r_R = 1.17 \text{ fm}$$

$$a_R = 0.75 \text{ fm}$$

$$W_V = 0.22E - 2.7 \text{ or zero, whichever is greater}$$

$$W_{SF} = 11.8 - 0.25E + 12.0(N-Z)/A \text{ MeV or zero, whichever is greater}$$

$$r_I = 1.32 \text{ fm}$$

$$a_I = 0.51 + 0.7(N-Z)/A \text{ fm}$$

$$V_{SO} = 6.2 \text{ MeV}$$

$$r_{SO} = 1.01 \text{ fm}$$

$$a_{SO} = 0.75 \text{ fm}$$

$$r_c = 1.25 \text{ fm}$$

Alpha Parameters: McFadden & Satchler, Nucl. Phys. 84, 177 (1966)

$$V_R = 164.7 \text{ MeV}$$
$$r_R = 1.442 \text{ fm}$$
$$a_R = 0.520 \text{ fm}$$
$$W_v = 22.4 \text{ MeV}$$
$$r_I = r_R$$
$$a_I = a_R$$
$$r_c = 1.25 \text{ fm}$$

Appendix 2

Computer Codes Used

Name	Reference/Author	Participant
SCAT	W.R. Smith, Comp. Phys. Comm. 1/1969/106 "Nuclear Penetrability and Phase Shift Subroutine"	M. Herman IBJ, Warsaw, Poland
ABACUS-2	E.H. Auerbach, BNL-6592 (1964)	B.V. Carlson M.P. Isidro R.A. Rego IEAV-CTA, San José dos Campos, Brazil
ELIESE-3	S. Igarasi, JAERI-1224 (1972) "A Program for Calculation of the Nuclear Cross-Sections by Using Local and Non-Local Optical Models and Statistical Model"	S.B. Garg Amar Sinha BARC, Trombay, India
SMOG	F. Fabbri, CEC/63-61-SO-04 (1963) Un Programma per il Calcolo dei Parametri di Interazione Neutrone-Nucleo con il Modello Ottico"	F. Fabbri ENEA, Bologna, Italy
JIB	F.G. Perey Oak Ridge National Laboratory Oak Ridge, USA	P.E. Hodgson Nuclear Physics Laboratory University of Oxford, U.K.
SCAT-2	O. Bersillon, CEA-N-2227 (1981) "Un programme de modèle optique spherique"	B.V. Carlson M.P. Isidro R.A. Rego IEAV-CTA, San José dos Campos, Brazil

The neutron optical potential

P E Hodgson

Nuclear Physics Laboratory, University of Oxford, Keble Road, Oxford OX1 3RH, UK

Abstract

The nucleon optical model is described, with particular reference to the analysis of neutron scattering and reaction cross sections. The central, spin–orbit and isospin terms in the potential are defined and the results of phenomenological determinations of their parameters reviewed. The methods used to determine the asymmetry and isospin terms are described in detail. The optical potential at negative energies may be determined from the properties of the single-particle states and thus enables its behaviour at both positive and negative energies to be studied. Some of the methods used to calculate the optical potential from the nucleon–nucleon interaction are described and the results compared with the phenomenological potential. The coupled-channels formalism enables both elastic and inelastic scattering from collective nuclei to be analysed simultaneously. Some analyses made in this way are reviewed, together with similar analyses of charge-exchange isobaric analogue state reactions.

This review was received in December 1983.

0034–4885/84/060613+42$09.00

Contents

1. Introduction

1.1. The nucleon optical model

The interaction between a nucleon and a nucleus is of basic importance for the study of nuclear reactions. It is essentially rather complicated, as it is the sum of all the interactions between the incident nucleon and all the nucleons comprising the nucleus, modulated by the interactions among these nucleons. If we represent the nucleon–nucleus interaction by a potential $V(r)$, a first approximation is obtained by summing all the nucleon–nucleon interactions over the nuclear density distribution $\rho(r)$, giving

$$V(r)=\sum_i \nu(|\mathbf{r}-\mathbf{r}_i|)\approx\int \rho(\mathbf{r}')\nu(|\mathbf{r}-\mathbf{r}'|)\,\mathrm{d}\mathbf{r}' \qquad (1.1)$$

where $\nu(|\mathbf{r}-\mathbf{r}'|)$ is the effective interaction between the incident nucleon and a nucleon in the nucleus. It is not the same as the free nucleon–nucleon interaction because of the presence of the other nucleons. The expression (1.1) can indeed be taken as the definition of $\nu(|\mathbf{r}-\mathbf{r}'|)$, when it becomes exact, and then we are faced with the problem of calculating $\nu(|\mathbf{r}-\mathbf{r}'|)$ from the free nucleon–nucleon interaction.

This simple treatment ignores any effects due to mesonic degrees of freedom, or to the possibility that the nucleons in the nucleus spend part of their time in an excited state, or to the quark structure of the nucleons themselves. So far, it has proved possible to understand nucleon interactions without taking any of these effects into account, although the possibility that they must be included to account for more detailed effects cannot be excluded. They will not be further considered here.

In spite of the essential complexity of the nucleon–nucleus interaction, it has proved possible to represent it by quite simple analytical expressions which enable many observable features of the nucleon reactions to be calculated to high precision. This article is essentially a review of the progress that has been made in this work. Attention will be concentrated on the interactions of neutrons with nuclei, but inevitably frequent reference will be made to proton interactions. Emphasis will be put on the areas that have recently been developed, with references to previous work in other areas.

The study of the nucleon–nucleus interaction has quite a long history. As long ago as 1935, Bethe calculated the scattering of nucleons by a purely real potential $V(r)$ and found marked resonances that are not observed experimentally. Le Levier and Saxon (1952) showed that these are damped if the potential is allowed to become complex, and that such potentials are able to reproduce quite well the differential cross sections for the elastic scattering of medium-energy protons by nuclei. This work was extended to neutron scattering by Feshbach *et al* (1954) who examined the total and reaction cross sections for the interaction of neutrons with nuclei over a large range of energies and nuclei. They noticed that these cross sections vary rather smoothly with energy and mass number, suggesting that the overall features of the interaction are determined by average nuclear properties and are insensitive to the detailed properties of the individual nuclei. This is of vital importance for the development of a simple theory of nuclear reactions.

Feshbach *et al* also used a complex potential and this is referred to as the optical model by analogy with the interaction of light with a medium that is both refractive and absorptive. Just as this can be treated by allowing the refractive index to become complex, so the scattering and absorption of nucleons by nuclei can be represented by a complex potential. In their calculations they use a spherically symmetric potential with the simple square-well form:

$$V(r) = U + \mathrm{i}W \qquad \text{for } r < R \qquad (1.2)$$
$$= 0 \qquad \text{for } r \geqslant R.$$

Taking $U = 42$ MeV, $W = 1.6$ MeV and $R = 1.45\ A^{1/3}$ fm they were able to reproduce very well the overall features of the measured total and reaction cross sections.

Subsequently it was realised that this success is partly due to the relatively poor energy resolution of the early measurements. Later improvements in experimental techniques showed that the cross sections of reactions passing through the compound nucleus fluctuate as the incident energy varies, but if the energy spread of the incident beam is greater than the mean width of the fluctuations an averaged cross section is obtained that generally varies quite smoothly with energy. Detailed analyses showed that this energy-averaged differential cross section is the sum of the compound-nucleus and direct reaction components; the former may be calculated using the theory of Hauser and Feshbach (1952) and subtracted from the measured cross sections. As the energy increases the contribution of the compound-nucleus processes falls rapidly and soon becomes negligible.

In precision analyses at low energies it is necessary to remove the compound-nucleus contribution to the data by a preliminary Hauser–Feshbach calculation. For most nuclei, the compound-nucleus contribution to the elastic scattering cross section is negligible for energies above about 10 or 15 MeV, so that the data may be analysed directly using the optical model. In the forward direction (<2° for the differential cross section) allowance must also be made for an appreciable contribution from magnetic moment scattering (Hogan and Seyler 1969).

In the following years very many analyses of neutron and proton interactions have been made with optical potentials of increasing sophistication. This has amply confirmed the original idea by showing that a vast range of data can be understood in a simple way. The optical potentials are able to give not only the cross sections and polarisations of nucleon elastic scattering, but also provide the wavefunctions needed in many reaction analyses.

The nucleon optical potential can be determined either by phenomenological analyses of experimental data or by a more fundamental calculation starting from the nucleon–nucleon interaction. The former uses rather simple physical arguments to establish the form of the potential and the approximate values of its parameters, and then relies on comparisons with experimental data to fix these parameters more precisely. This approach has been developed to a high degree of sophistication and many of the results are summarised below, especially in § 2. The fundamental calculation is much more difficult, and only recently has it been developed to the stage where it can give reliable quantitative results. This has now made it possible to understand some apparently anomalous results of the phenomenological analyses, thus unifying both into a coherent whole. This work is described in § 3.

The optical potentials obtained from analyses of elastic scattering are widely used in distorted wave analyses of inelastic scattering and nucleon transfer reactions. This work is not reviewed here because, in most cases, it does not provide additional information on the potentials themselves. There are, however, two exceptions to this, namely inelastic scattering from collective nuclei analysed by the coupled-channels theory and (p, n) reactions to the isobaric analogue state (IAS), and these reactions are considered in § 4.

Neutron data are generally of lower accuracy than proton data, so it is useful to ask whether it is possible to calculate neutron cross sections from potentials obtained from analyses of proton data, in particular from (p, p) and (p, n) IAS reactions. This question is considered in § 4.2.

The striking success of the optical model in providing simple potentials that enable a very wide range of data to be fitted to quite high accuracy should not obscure its essential limitations. Each of the nuclei is different, and many of their properties do not vary smoothly with the atomic number or any other parameter. These differences inevitably affect the cross sections, so that we have to choose between a global potential that gives quite a good overall fit to the data for a range of nuclei and a particular potential that fits one set of data to much higher accuracy. In some cases the effects of the structure of individual nuclei are very marked, and some of these are noted later on. There are already some indications (Perey and Perey 1968, Noro *et al* 1981) of fine structure in proton optical potentials, but no similar effects have yet been found in neutron optical potentials.

There is currently much interest in optical model analyses of proton scattering at several hundred MeV and many new features are being found both in the phenomenological analyses and in the nuclear matter calculations. All this work is being carried out on proton scattering, so although the results are certainly also true with appropriate modifications for neutron potentials, they are not included in the present review. The energy range of the data and analyses discussed here thus extends up to about 60 MeV. At negative energies the neutron optical potentials can be determined from the properties of single-particle states, and this is described in § 2.7.

There is also much interesting work at lower energies on the behaviour of the potential around the Fermi surface. Indications of this are found in studies of the energy variation of the potential, supplemented by the application of dispersion relations to the imaginary potentials. Once again this work has been carried out on proton data and so, although the results certainly apply also to neutron potentials, it is not described here. This work has been reviewed elsewhere (Hodgson 1983b).

The remainder of this section is devoted to general remarks about the radial variation of the optical potential, its spin–orbit, isospin and asymmetry terms and the methods of calculating the observable quantities from them.

Throughout this review, all energies are in MeV and all lengths in fm (10^{-13} cm).

1.2. The radial variation of the nucleon optical potential

In their pioneer work Feshbach *et al* used a square-well potential; this is obviously unphysical and in particular it has too high a reflectivity. A more realistic form can be obtained from simple arguments based on the form of the nucleon–nucleon interaction. This is known to be of short range, falling exponentially with increasing distance. If it is replaced by a delta function, then from (1.1)

$$V(\boldsymbol{r}) \approx \int \rho(\boldsymbol{r}')\, V_0 \delta(\boldsymbol{r}-\boldsymbol{r}')\, \mathrm{d}\boldsymbol{r}' = V_0 \rho(\boldsymbol{r}). \tag{1.3}$$

Thus to first approximation we expect the optical potential to have a radial variation that follows the nuclear density quite closely, with perhaps a somewhat greater radius reflecting the finite range of the nucleon–nucleon interaction. It is convenient to represent this by an appropriate analytical expression, and for this purpose the Saxon–Woods form is particularly convenient:

$$f(r)=\frac{1}{1+\exp[(r-R)/a]} \tag{1.4}$$

where R is the radius and a is the surface diffuseness parameter. Since nuclear radii are quite accurately proportional to $A^{1/3}$, it is usual to define $R=r_0A^{1/3}$, where r_0 is independent of A.

The nucleon–nucleon interaction is purely real, so this argument applies only to the real part of the nucleon optical potential. Feshbach *et al* assumed that the real and imaginary parts of the optical potential have the same form, and this is still widely followed. Since the imaginary potential takes into account in a global way all the non-elastic processes that remove flux from the elastic channel it is not possible to establish its form by simple physical arguments. To give it some additional flexibility, the imaginary potential of the form (1.4) is often supplemented by a surface-peaked potential of the radial derivative form:

$$g(r)=-4a\frac{\mathrm{d}f(r)}{\mathrm{d}r}=\frac{4\exp[(r-R)/a]}{\{1+\exp[(r-R)/a]\}^2} \tag{1.5}$$

where the factor of $4a$ is introduced to ensure that $g(R)=1$. Some additional flexibility is sometimes introduced by allowing the values of r_0 and a to be different for the real and imaginary parts of the potential; when this is done they will be distinguished by the subscripts u and w. This difference was first shown to be necessary by analyses of proton scattering (Hodgson 1961).

Approximate numerical values of the parameters of the optical potential can be obtained by quite elementary arguments. Since the potential is the extension to positive energies of the potential of the simple shell model, its depth U can be calculated from the Fermi gas model, using the relation between the depth and the radius. Adding the average nucleon separation energy of 8 MeV gives $U\approx 50$ MeV. The depth of the imaginary part of the potential can be estimated from semi-classical arguments to be about 10 MeV. Finally, knowledge of the nuclear matter distributions gives $r_0\approx 1.2$ fm, $a\approx 0.65$ fm.

These radial forms for the nucleon optical potential sufficed until quite recently, when evidence from proton elastic scattering indicated the need for a modified form at higher energies (Hodgson 1983b).

1.3. The energy variation of the optical potential

Early phenomenological analyses showed that, if the form factors are kept fixed, then the depths of the real and imaginary potentials depend on energy. That of the real part of the potential is essentially linear and decreases with increasing energy, with some evidence for a small quadratic term. A recent analysis of proton elastic scattering from −70 to 200 MeV gave (Bauer *et al* 1982)

$$U(E)=52.4-0.37E+0.0007E^2+24\frac{N-Z}{A}+\frac{0.4Z}{A^{1/3}}. \tag{1.6}$$

A similar energy dependence is expected for neutron scattering. The data on neutron elastic scattering are not yet sufficiently accurate to establish the presence of the quadratic term. The nuclear asymmetry term is discussed in § 2.4, and the last term is of Coulomb origin and so is present only for protons.

The overall energy variation of the potential is that of the Hartree–Fock field and is associated with the non-locality of the potential. It is possible to define an equivalent energy-independent non-local potential that gives the same scattering cross sections, but generally the inconvenience of using such a potential outweighs that of accepting an energy dependence, so such potentials are now seldom used. They are discussed further in § 2.2.

In addition to this overall energy variation, the potential changes in the vicinity of the Fermi surface in a way that can be represented either by a modified energy variation or by the addition of a small energy-dependent surface-peaked term to the real potential. Physically this is due to coupling to low-lying nuclear collective states. The Fermi surface anomaly may also be described by an energy-dependent effective mass (Hodgson 1983c).

The imaginary part of the potential increases in magnitude with the energy to take account of the increasing cross sections of the various non-elastic reactions. It is found empirically that at low energies the surface-peaked component dominates, and as the energy increases the absorption progressively spreads through the nuclear volume.

1.4. The spin–orbit term

An unpolarised beam of nucleons is generally polarised after interacting with a nucleus and this polarisation can be calculated from the optical potential if it includes a term that couples the orbital and spin angular momenta. By analogy with the corresponding term for atoms this is taken to have the form

$$V_{so}(r) = \left(\frac{\hbar}{m_\pi c}\right)^2 (U_s + \mathrm{i}W_s)\,\frac{1}{r}\,\frac{\mathrm{d}f_s(r)}{\mathrm{d}r}\,\boldsymbol{L}\cdot\boldsymbol{\sigma}. \tag{1.7}$$

The square of the pion Compton wavelength is a relic of the derivation of this term from the meson theory of nuclear forces; its numerical value is close to 2 fm^2. The form factor $f_s(r)$ has the Saxon–Woods analytical form (1.4) and usually a rather smaller value of the radius parameter.

This spin–orbit potential is the extension to positive energies of the term in the simple shell model potential that gives the spin–orbit splitting needed to account for the magic numbers.

1.5. The isospin and asymmetry terms

The neutron and proton potentials at the same energy for the same nucleus differ by an amount that can be attributed to an isospin term in the nucleon optical potential proportional to $(\boldsymbol{t}\cdot\boldsymbol{T})$, where $\boldsymbol{t}$ and $\boldsymbol{T}$ are the isospins of the incident nucleon and target nucleus, respectively (Lane 1962a,b). The origin of this term may be traced to the $\boldsymbol{\tau}_i\cdot\boldsymbol{\tau}_j$ terms in the nucleon–nucleon interaction. Such a term in the nucleon potential gives rise to terms in the neutron and proton potentials that are proportional to the nuclear asymmetry $(N-Z)$ and of opposite sign.

There are, however, other effects that can contribute to a difference between the neutron and proton potentials, associated in particular with the presence of a Coulomb electrostatic potential acting on the protons and with small but significant differences between the neutron and proton density distributions. The former can be calculated and allowed for, but the latter ensures that the asymmetry dependence of the potential has a nuclear structure as well as an isospin component, so that the isospin dependence of the potential cannot be obtained from analyses of neutron scattering alone. It is necessary to combine the results of neutron and proton analyses in the way discussed in § 2.4.

1.6. Optical model calculations

Once the optical potential has been defined, and suitable values of its parameters chosen, the observable quantities such as the differential cross section for elastic scattering, the associated polarisations and the total and reaction cross sections may be calculated by standard quantum-mechanical and numerical techniques, using well-tested computer programs. Since this is all described in detail elsewhere (Hodgson 1963b, 1971) it will not be discussed further here.

2. Phenomenological analyses of neutron scattering

2.1. Method of analysis

The experimental data for neutron scattering that can be analysed with the optical model include the differential cross section for elastic scattering, the associated polarisations and the total and reaction cross sections. Most analyses have concentrated on these data, and this section will be mainly devoted to a review of the results. There are, however, other observables that have received much less attention due to the difficulty of measuring them. These include the spin rotation parameter, which is a measure of the rotation of the direction of polarisation of the incident beam by the elastic scattering process. No measurements of this quantity have been made for neutron scattering, but corresponding work on proton scattering indicates that it is probably quite sensitive to the form of the optical potential, particularly in the nuclear interior (Schwandt 1982). At low energies it is possible to measure the S- and P-wave strength functions (Hodgson 1971, chap 8.5), and these have been studied extensively by the Duke group (Tabony and Seth 1968, Pineo *et al* 1974, Malan *et al* 1975, Ramavataram *et al* 1976).

The foregoing remarks refer to the scattering of neutrons by nuclei of spin zero. If the nucleus has non-zero spin, and particularly if the target nuclei are aligned or polarised, many new measurements can be made, such as those related to the final spins of the incident neutron and target nucleus, and the correlations between them. A few measurements of the scattering of neutrons by aligned nuclei have been made, and this has made it possible to study the spin–spin term in the optical potential that is proportional to $\boldsymbol{\sigma} \cdot \boldsymbol{I}$, where $\boldsymbol{I}$ is the target nucleus spin. Due to the great difficulty of these experiments little work has been done in this area in recent years, so it is sufficient to refer to a review already published (Hodgson 1971, chap 9.2) and to the work of Heeringa and Postma (1976, 1983) and Fasoli *et al* (1978).

The extended optical model makes it possible to treat elastic and inelastic scattering within the same theoretical formalism, and such analyses are reviewed in § 4.1.

Returning now to what may be called mainstream optical model analyses, we collect together the terms in the spherically symmetric optical potential discussed in § 1 and express the potential in the form:

$$V(r)=Uf_u(r)+\mathrm{i}W_Vf_w(r)+\mathrm{i}W_Dg(r)+\left(\frac{\hbar}{m_\pi c}\right)^2(U_s+\mathrm{i}W_s)\frac{1}{r}\frac{\mathrm{d}f_s(r)}{\mathrm{d}r}\boldsymbol{L}\cdot\boldsymbol{\sigma} \quad (2.1)$$

where the form factors $f_i(r)=\{1+\exp[(r-R_i)/a_i]\}^{-1}$, with $R_i=r_iA^{1/3}$ and $g(r)=-4a(\mathrm{d}/\mathrm{d}r)f_d(r)$.

The potential varies with energy, and it is usual to parametrise this variation by keeping the form factor parameters R and $\boldsymbol{a}$ constant, and allowing the potential depths U, W_V, W_D, U_S and W_S to depend on energy. There are now definite indications that this approximate way of handling the energy dependence is no longer adequate, particularly at high energies (Hodgson 1983b).

As mentioned in § 1.6, the observable quantities may be calculated by well-known quantum-mechanical and numerical techniques from a potential of this form. The analysis of a particular set of experimental data is usually carried out by making such a calculation with values of the parameters chosen on general grounds such as those mentioned in § 1.2 or taken from previous analyses of similar data. The results of this calculation are compared with the experimental data and a measure of the goodness-of-fit defined and evaluated. The parameters of the potential are then systematically varied to optimise the fit to the data; this can all be done automatically by the computer program (Hodgson 1963b, chap 4). The resulting optimum set of parameters define a particular potential that corresponds to a particular data set.

Very frequently, several data sets are measured at the same time corresponding to the scattering of neutrons of a particular energy by a range of target nuclei. Each set of data may be analysed individually in the way described above and it is usually found that the parameters are quite similar. Close examination of the sensitivity of the calculated observables to the parameters of the potential shows that there are correlated pairs of parameters that can be varied together over certain ranges of values without much affecting the values of the observables. In particular, variations of U and R_u that maintain the constancy of UR_u^2 have little effect on the observables and similarly for the product Wa_W. This suggests that it is useful to fix the form factor parameters to average values and then to repeat the whole analysis allowing only the potential depths to vary. Because of the flexibility of the parameters just mentioned, the fits to the data obtained in this way are very little inferior to those obtained when all the parameters are allowed to vary. It is then found that these depths are quite similar and vary linearly with the nuclear asymmetry parameter, so that it is convenient to express them in the form:

$$U=U_0-\frac{N-Z}{A}U_I. \quad (2.2)$$

Such analyses give what are called global potentials that fit a wide range of data for different nuclei. These may be combined with the results of similar analyses made at different energies to give even more general potentials. Since the potentials vary linearly with energy this can be included by writing

$$U=U_0-\epsilon E-\frac{N-Z}{A}U_I. \quad (2.3)$$

2.2. Global potentials

A large number of optical model analyses of neutron scattering have been made over the last few decades and the parameters obtained have been tabulated by Perey and Perey (1974, 1976). It should be remarked that some of these analyses are based on early data of lower statistical accuracy and extending over a limited angular range, and may or may not include polarisation data, or the corresponding total and reaction cross sections. Conclusions about the optimum global potential should therefore be obtained from a critical analysis of a few accurate and comprehensive data sets rather than by simply combining the results of a large number of analyses of very different quality.

There is one feature of many global analyses that deserves to be mentioned. Very frequently most of the nuclei studied are those with closed or nearly closed shells, and so the potential may apply to scattering from such nuclei only. Significantly different potentials may be appropriate for nuclei between closed shells, as will be shown in § 2.4 on the asymmetry potential, and in § 2.6 on the scattering from deformed nuclei.

The earliest global analysis of neutron scattering was made by Perey and Buck (1962) using a non-local potential of the form:

$$V(\boldsymbol{r}, \boldsymbol{r}') = V(p)H(|\boldsymbol{r}-\boldsymbol{r}'|)$$

with (2.4)

$$H(|\boldsymbol{r}-\boldsymbol{r}'|) = \frac{1}{\pi^{3/2}\beta^3}\exp\left[-\left(\frac{\boldsymbol{r}-\boldsymbol{r}'}{\beta}\right)^2\right]$$

and $\boldsymbol{p} = \frac{1}{2}(\boldsymbol{r}+\boldsymbol{r}')$, β being the range of the non-locality. The potential $V(p)$ varies with p in the same way as the corresponding local potential $V(r)$ depends on r. Perey and Buck found that the data on neutron scattering from 1–25 MeV can be fitted by this potential with $U = 71$ MeV, $W = 15$ MeV, $r_0 = 1.22$ fm, $a = 0.65$ fm, $\beta = 0.85$ fm, $U_s = 7.2$ MeV and $a_s = 0.47$ fm. As mentioned in the introduction, such non-local potentials are no longer used, as it is more convenient to use the equivalent local potential. Such a potential has been obtained by Wilmore and Hodgson (1964) from that of Perey and Buck, and is specified by the parameters

$$\begin{aligned}
U &= 47.01 - 0.267E - 0.0018E^2 \\
W_D &= 9.52 - 0.053E \\
U_s &= 0 \\
r_u &= 1.322 - 0.00076A + (4\times10^{-6})A^2 - (8\times10^{-9})A^3 \qquad (2.5)\\
r_W &= 1.266 - 0.00037A + (2\times10^{-6})A^2 - (4\times10^{-9})A^3 \\
a_u &= 0.66 \\
a_W &= 0.48.
\end{aligned}$$

This potential has no spin–orbit term and so is unable to give polarisations, and also the energy variation of the imaginary potential is rather unphysical. Nevertheless it is able to fit a wide range of neutron data with an accuracy that is usually better and seldom worse than other global sets that have been proposed subsequently. It remains useful for practical calculations of cross sections for applications such as reactor shielding, for which polarisations are not required.

The Perey–Buck potential is not able to fit the neutron S-wave strength function around $A=100$ (Buck and Perey 1962) but Moldauer (1962, 1963, 1964) showed that this can be corrected by allowing the absorbing potential to peak sharply just outside the nuclear surface. An extensive analysis of neutron data below 1 MeV led to the potential

$$\begin{array}{llll} U=46 & R_U=0.6+1.16A^{1/3} & a_u=0.62 & \\ W_G=14 & R_G=1.1+1.16A^{1/3} & b=0.5 & U_S=7 \end{array} \tag{2.6}$$

where the subscript G refers to the imaginary potential of surface-peaked Gaussian form, $\exp[-(r-R_G)^2/b^2]$. Other global neutron potentials have been proposed by Rosen *et al* (1965), Engelbrecht and Fiedeldey (1967), Becchetti and Greenlees (1969), Holmqvist and Wiedling (1972), Eder *et al* (1977) and Rapaport *et al* (1979a, b).

Analyses with both a surface and a volume imaginary potential show a tendency for the surface component to decrease and the volume component to increase with increasing energy. Some typical results for the coefficient of the energy-dependent term for both real and imaginary potentials are shown in table 1. It is notable that there are substantial differences between isotopes. Over a greater energy range (1–30 MeV) Rapaport *et al* (1980) found for the even tin isotopes

$$\begin{aligned} &U=54.2-0.32E-22\alpha \\ &\left.\begin{aligned} W_V&=0 \\ W_D&=3.0+0.51E-14\alpha \end{aligned}\right\} E<15\ \text{MeV.} \\ &\left.\begin{aligned} W_V&=-10.7+0.52E \\ W_D&=18.7-0.52E-16.2\alpha \end{aligned}\right\} E>15\ \text{MeV} \end{aligned} \tag{2.7}$$

where $\alpha=(N-Z)/A$. Similar results were obtained for the molybdenum isotopes by Rapaport *et al* (1979a, b).

Table 1. Energy dependence of neutron optical potentials (El Kadi *et al* 1982). $U=U_0-\epsilon_u E$, $W_D=W_{DO}-\epsilon_w E$, $W_V=W_{vo}+\epsilon_v E$.

Nucleus	Energy (MeV)	ϵ_u	ϵ_w	ϵ_v
^{54}Fe	8–14	0.340	0.326	0.314
^{56}Fe	8–14	0.448	0.221	0.249
^{63}Cu	8–14	0.470	0.195	0.361
^{65}Cu	8–14	0.564	0.049	0.250

It is not possible to obtain global potentials for light nuclei due to resonance effects.

It is now increasingly realised that there is no particular reason to try to develop global potentials of ever-increasing complexity and precision. The influences of nuclear structure on neutron scattering, particularly those associated with shell closure and deformation, are sufficiently large for some nuclei to ensure that there is no simple and accurate global potential. It is more useful to seek potentials that give good fits to the data for particular types of nuclei or regions of the periodic table.

It should also be noted that the optical potential cannot be determined to any desired degree of precision simply by using more precise experimental data. Even for one nucleus at a particular energy, it is found that very accurate data cannot be fitted within the experimental standard deviations without altering the shape of the potential in a way that cannot easily be fitted by simple analytical formulae. This has been shown by Kobos and Mackintosh (1979a, b) for proton elastic scattering and there is no reason to suppose that their conclusions are not also applicable to neutron scattering. Furthermore, when potentials are very accurately adjusted in this way to fit a particular set of data they are unlikely to vary in a smooth way from one nucleus to another, so that increased accuracy of fitting is paid for by reduced generality. The usefulness of optical potentials is precisely that, with similar parameters, they can give good overall fits to a wide range of data and the precision and the range have to be chosen in accord with the needs of a particular investigation.

2.3. *The spin–orbit potential*

The polarisations and asymmetries of elastically scattered nucleons and the splitting of bound single-particle states of the same orbital angular momentum into two components can be described by a spin–orbit potential of the form

$$V_{\mathrm{so}}(r)=\left(\frac{\hbar}{m_\pi c}\right)^2 (U_{\mathrm{s}}+\mathrm{i}W_{\mathrm{s}})\frac{1}{r}\frac{\mathrm{d}f(r)}{\mathrm{d}r}\boldsymbol{L}\cdot\boldsymbol{\sigma} \tag{2.8}$$

where $f(r)$ is the Saxon–Woods form factor. The polarisations and asymmetries are usually not measured as accurately as the differential cross sections, but this is offset by their greater sensitivity to the fundamental features of the interaction. Careful analysis of these data is therefore of great importance for an understanding of the nucleon–nucleus interaction.

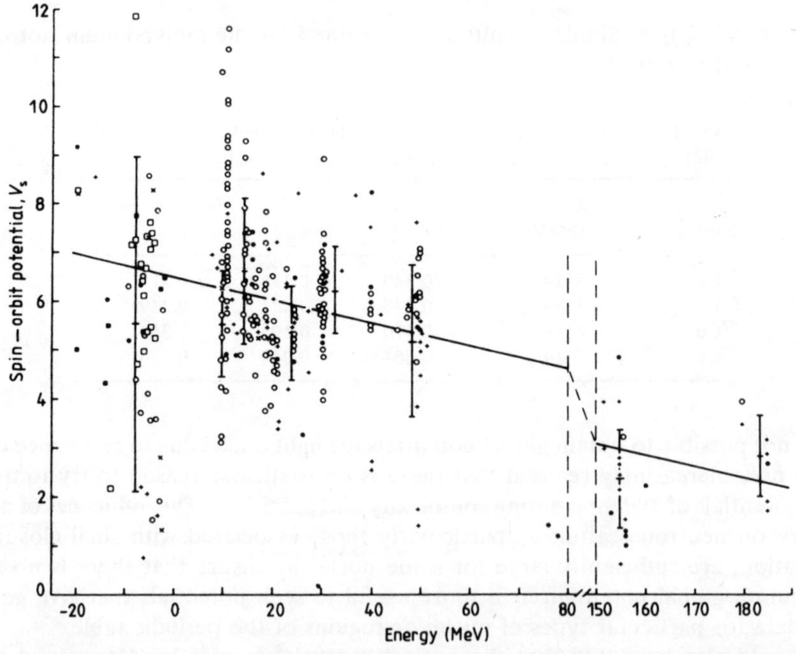

Figure 1. The energy variation of the spin–orbit potential (Cooper and Hodgson 1980). Light nuclei ($A<40$): ×, n; +, p. Medium nuclei ($40 \leqslant A \leqslant 100$): □, n; ○, p. Heavy nuclei ($A>100$): ■, n; ●, p.

The lack of very accurate data at low energies makes it difficult to investigate in detail the properties of the spin–orbit potential. It is usually found possible to reproduce the data with a potential of the above form with a radius parameter rather less than that of the real potential and a similar diffuseness parameter. With these form factor parameters the energy variation of the real spin–orbit potential is as shown in figure 1. This energy variation is much less than that of the real central potential and has the form (Cooper and Hodgson 1980):

$$U_s = (6.5 \pm 0.5) - (0.023 \pm 0.012)E. \tag{2.9}$$

There is no evidence of an anomaly around the Fermi surface and it is clear that it would be very difficult to establish its presence phenomenologically. There is little evidence at low energies for the presence of an imaginary spin–orbit potential (Floyd *et al* 1981, 1982, Tornow *et al* 1982).

More accurate values of the asymmetries or analysing powers are now becoming available and some recent measurements at 14 MeV are shown in figure 2. Analysis of these data showed that a small imaginary spin–orbit potential $W_s \approx 1$ MeV is necessary to optimise the fit and also that the parameters of the real spin–orbit potential ($U_s = 5.4$ MeV, $r_s = 1.025$, $a_s = 0.52$ fm) are very similar to those found in corresponding analyses of proton scattering ($U_s = 5.5$ MeV, $r_s = 1.025$, $a_s = 0.54$ fm) (Floyd *et al* 1981). Other analyses of neutron scattering gave $U_s = 5.5$ MeV, $r_s = 1.24$, $a_s = 0.66$ fm at 2.9 MeV (Galloway and Waheed 1979) and $U_s = 6.11$ MeV, $r_s = 1.02$, $a_s = 0.60$ fm for ^{54,56}Fe and ^{63,65}Cu at 8–14 MeV (El Kadi *et al* 1982). Preliminary analyses of neutron scattering from ^{40}Ca from 9.9–13.9 MeV, however, gave a different potential with $U_s = 3.2$ MeV, $r_s = 1.1$ and $a_s = 0.27$ fm (Tornow *et al* 1981, 1982).

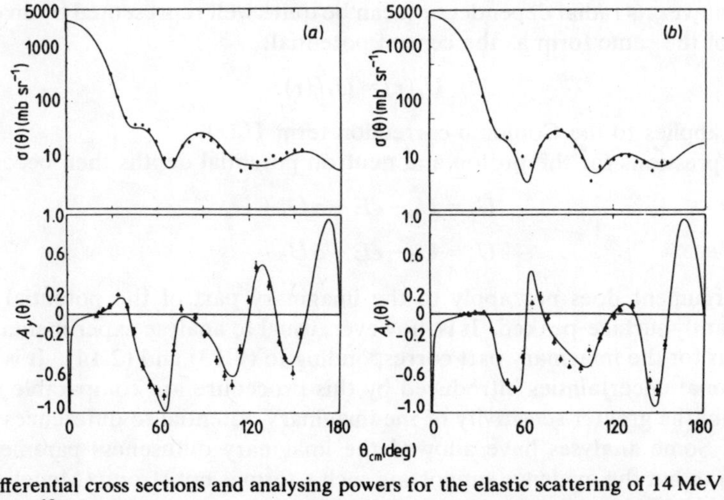

Figure 2. Differential cross sections and analysing powers for the elastic scattering of 14 MeV neutrons by (a) ^{54}Fe and (b) ^{65}Cu compared with optical model calculations using average geometry parameters (Floyd *et al* 1981).

The differential cross section in the far backward direction is particularly sensitive to the spin–orbit potential (Ramström and Göransson 1977) and gave optimum spin–orbit potentials of 6 MeV for Fe and Co and 4 MeV for In and Bi (with $r_s \approx 1.22$, $a_s \approx 0.67$ fm) at a neutron energy of 7.06 MeV.

At higher energies the volume integrals of the phenomenological spin–orbit potentials corresponding to the conventional Saxon–Woods parametrisation show an anomalous non-monotonic energy variation that is yet another symptom of the inadequacy of the Saxon–Woods parametrisation at higher energies (Schwandt 1982).

2.4. The isospin and asymmetry terms in the neutron optical potential

On the basis of microscopic arguments using the nucleon–nucleon potential, Lane (1962a, b) introduced a term proportional to $(\boldsymbol{t}\cdot\boldsymbol{T})V_I(r)$ in the nucleon–nucleus optical potential, where $\boldsymbol{t}$ and $\boldsymbol{T}$ are the isospin vectors of the incident nucleon and target nucleus, respectively. This gives terms proportional to $(N-Z)$ and of opposite signs in the proton and neutron potentials which then become

$$U_p(r) = (U_0 - \epsilon E)f(r) + \alpha U_I(r) + U_c(r) \tag{2.10}$$

$$U_n(r) = (U_0 - \epsilon E)f(r) - \alpha U_I(r) \tag{2.11}$$

where $\alpha = (N-Z)/A$ is the nuclear asymmetry and $U_I(r)$ is the isospin potential. The neutron and proton potentials must refer to the same energy in the nuclear interior, and since the protons are slowed down by the Coulomb repulsion and the nuclear potential depends on energy this implies a difference between the incident neutron and proton energies corresponding to the same energy in the interior. The term $U_c(r)$ is introduced into the expression for the proton potential to take account of this effect (Lane 1957). An isospin term can also be included in the imaginary potential (Satchler 1967). The isospin dependence of the nucleon optical potential has many applications in nuclear physics (Satchler 1969).

The isospin potential $\alpha U_I(r)$ is only a few per cent of the main central potential and so whatever its radial dependence it can be quite well represented by an equivalent potential of the same form as the central potential:

$$U_I(r) = U_I f(r). \tag{2.12}$$

The same applies to the Coulomb correction term $U_c(r)$.

The expressions for the proton and neutron potential depths then become

$$U_p = U_0 - \epsilon E + \alpha U_I + U_c \tag{2.13}$$

$$U_n = U_0 - \epsilon E - \alpha U_I. \tag{2.14}$$

This argument does not apply to the imaginary part of the potential since it is predominantly surface-peaked. It is, however, usual to analyse experimental data with expressions for the imaginary part corresponding to (2.13) and (2.14). It is likely that the additional uncertainties introduced by this procedure are comparable with those arising from the greater sensitivity of the imaginary potential to differences in nuclear structure. Some analyses have allowed the imaginary diffuseness parameter a_W to depend on α, but the evidence supporting such a term is not strong (Menet *et al* 1971, Zanelli *et al* 1981).

It might appear from the expression for U_n that it is possible to obtain U_I from analyses of neutron scattering alone, for example from the variation of U_n with α for a range of nuclei at the same energy. This is, however, not possible for the reasons explained below. The isospin potential can only be obtained by combining data for neutron and proton scattering. It is then necessary to know the value of the Coulomb correction term U_c and so the way of determining this will now be described.

In the case of proton scattering, the Coulomb potential reduces the energy of the incident proton so that it experiences the nuclear potential at an energy reduced by ΔE_c. Thus the comparison between neutron and proton potentials to determine the isospin term must be made at energies differing by ΔE_c or equivalently the term U_c must be included in the expression (2.13). The magnitude of U_c is thus

$$U_c = \Delta E_c\, \partial U/\partial E. \tag{2.15}$$

A reasonable estimate of ΔE_c is the average of the Coulomb potential over the nuclear volume and $\partial U/\partial E \approx 0.3$, giving (Perey 1963)

$$U_c = \frac{6e^2 Z}{5r_0 A^{1/3}} \times 0.3 = 0.4Z/A^{1/3}. \tag{2.16}$$

The magnitude of the Coulomb correction term may be determined experimentally from analyses of neutron and proton scattering from self-conjugate nuclei. For such nuclei the asymmetry parameter α is zero and so

$$U_p - U_n = U_c. \tag{2.17}$$

This method has been applied to several self-conjugate nuclei with the following results for the coefficient of $Z/A^{1/3}$: ^{16}O, 0.48 ± 0.20; ^{28}Si, 0.33 ± 0.12; ^{32}S, 0.45 ± 0.13; ^{40}Ca, 0.5 ± 0.07 (Rapaport 1977, 1980, 1982, Rapaport *et al* 1977); ^{28}Si, 0.37 ± 0.09 (De Vito *et al* 1983); ^{40}Ca, 0.32 ± 0.10 (Tornow *et al* 1982); ^{208}Pb, 0.44 (Austin 1983). In the case of ^{28}Si, the result may reflect coupled-channel effects not considered in the analysis. Excluding this nucleus, the average value is 0.46 ± 0.07 (Rapaport *et al* 1977), which is consistent with (2.16).

The Coulomb correction term has also been evaluated theoretically by Jeukenne *et al* (1974, 1976a, b, 1977a, b) and Mahaux (1979) who obtain a value for the integral of the Coulomb correction term about 30% greater than that corresponding to (2.16).

A similar analysis may be used to determine the imaginary part of the Coulomb correction term, with the results for ^{40}Ca (Rapaport 1980)

$$W_c = -[1 - (0.028 \pm 0.005)E](0.5 \pm 0.1)Z/A^{1/3} \tag{2.18}$$

and (Tornow *et al* 1982)

$$W_c = -(1 - 0.034E)0.55Z/A^{1/3} \tag{2.19}$$

and for ^{28}Si (De Leo and Sterrenberg 1980, De Vito *et al* 1983)

$$W_c = -[1 - (0.025 \pm 0.011)E](0.7 \pm 0.2)Z/A^{1/3}. \tag{2.20}$$

Physically, this term is due to several effects including Coulomb excitation, isospin impurities in the intermediate states, the effective closing of open channels due to the propagation of the intermediate projectile in a Coulomb potential, intermediate charge-exchange processes and the result of the smaller non-locality effect for protons than for neutrons when the local approximation is made (Osterfeld and Madsen 1981). Some of these effects give a larger absorption for protons, and others a smaller absorption. Detailed microscopic calculations give results consistent with the empirical determinations.

Once U_c is known, the value of the isospin potential can be determined from the neutron and proton optical potentials for a particular nucleus at a particular energy (Hodgson 1963a). If suitable data are not available at just the desired energies, results at other energies may be used, provided they are corrected using the known energy dependence of the potentials. The imaginary part of the isospin potential may be determined in essentially the same way.

The practical difficulty with such analyses is that suitable neutron and proton data are not always available for the same nucleus and the neutron data are frequently of lower precision. Several other methods of determining the isospin potential have therefore been used, but these involve particular assumptions about the parametrisation and the structure of the nuclei concerned that are seldom recognised and frequently difficult to justify. Some of these methods will now be critically examined.

The expressions (2.10) and (2.11) for the neutron and proton potentials relate to a particular nucleus, but it is possible to apply them to a range of nuclei and to identify the asymmetry dependence already noted in § 1.3 with the isospin dependence. This type of analysis may be applied to neutron and proton scattering separately. It is, however, open to two serious objections. In the first place, in the usual parametrisation the radii of nuclei of different A are assumed proportional to $A^{1/3}$. If, however, this is not quite correct, then the additional A dependence will be forced onto the potential depths U, the magnitude of the changes being determined by the need to maintain the constancy of UR^2. Since there are no A-dependent terms in (2.10) and (2.11) and since $(N-Z)/A$ increases more or less uniformly through the periodic table for the stable nuclei used in such analyses, this A dependence will appear as a spurious $(N-Z)/A$ dependence (Hodgson 1970, Rapaport *et al* 1979a, b). If this effect influences neutron and proton scattering in the same way, it may be represented by including in the expressions for U_p and U_n a nuclear structure term αU_{NS}:

$$U_p = U_0 - \epsilon E + \alpha U_{NS} + \alpha U_I + U_c \tag{2.21}$$

$$U_n = U_0 - \epsilon E + \alpha U_{NS} - \alpha U_I \tag{2.22}$$

and similar expressions for the imaginary potentials.

It might be preferable to include an A-dependent term explicitly, but few data are of sufficient precision to separate an A dependence from an asymmetry dependence. Most data have been analysed assuming an asymmetry dependence alone and so the above forms are retained. These analyses may be divided into three classes.

2.4.1. Particular analyses. The isospin potential may be obtained from proton and neutron analyses on the same nucleus at the same energy. Subtracting gives

$$U_p - U_n = 2\alpha U_I + U_c. \tag{2.23}$$

The Coulomb correction U_c is known, and hence U_I is determined.

Table 2. Particular analyses. Neutron and proton scattering by the same nuclei.

Energy (MeV)	Nuclei	r_u	U_I	W_I	Reference
11	^{51}V, ^{59}Co, ^{56}Fe, ^{181}Ta, ^{209}Bi	1.20–1.25	21.7 ± 0.6	12	Ferrer *et al* (1976)
11	Fe, ^{93}Nb, ^{120}Sn	1.20–1.25	22	14	Ferrer *et al* (1976)
11	^{51}V, ^{59}Co	1.20–1.25	18.5 ± 1.5	15 ± 2	Rapaport (1982)
14–24	^{59}Co, ^{208}Bi	1.25	25 ± 3	—	Picard (1964, 1965)
14.4	^{59}Co, ^{209}Bi	1.25	24 ± 3	—	Erramuspe (1967)
14.5, 17.3	Cu	1.25	≈18	—	Hodgson (1963a)

A variant of this method is to combine proton and neutron data at energies differing by the Coulomb displacement energy ΔE_c. This cancels out the effect of the Coulomb barrier, so

$$U_p(E+\Delta E_c) - U_n(E) = 2\alpha U_I. \quad (2.24)$$

The results of some analyses made in these ways are given in table 2.

2.4.2. Global analyses. If the results of phenomenological analyses of proton and neutron scattering at the same energy by several nuclei are combined, then plots of $U_p - U_c$ and U_n against α give straight lines with slopes $(U_{NS}+U_I)$ and $(U_{NS}-U_I)$, respectively. Subtraction gives a global value of U_I. The results of some analyses made in this way are summarised in tables 3 and 5. Corresponding results for the imaginary isospin potential are given in tables 4 and 6.

Table 3. Global analyses. Proton scattering by a range of nuclei: real potential.

Energy (MeV)	Nuclei	Number of nuclei	r_u	$U_I + U_{NS}$	Reference
9–22	Al–Au	35	1.25	27 ± 8	Perey (1963)
9–50	Fe–Pb	40	1.17	24 ± 3†	Becchetti and Greenlees (1969)
14.5	^{12}C–^{120}Sn	32	1.25	27.4	Rosen *et al* (1965)
17.8	S–In	8	1.25	0	Baugh *et al* (1966)
16.5, 18.6	Ti–Cu	9	1.25	9.3 ± 7.8	Kossanyi-Demay *et al* (1967)
18.6	Ti–Zr	10	1.25	17 ± 6	Kossanyi-Demay and de Swiniarski (1968)
18.6	Ti–Zr	10	1.10	16 ± 9	Kossanyi-Demay and de Swiniarski (1968)
30	Si–^{208}Pb	8	1.12	30	Satchler (1967)
30	^{40}Ca–^{208}Pb	8	1.25	17 ± 10	Barrett *et al* (1964)
30	^{40}Ca–^{208}Pb	8	1.20	23 ± 5	Greenlees and Pyle (1966)
40	^{12}C–^{208}Pb	6	1.18	26.4	Fricke *et al* (1967)
50	Mg–Cd	—	1.16	38 ± 8	Burge *et al* (1968)
61.4	Al–Pb	5	1.16	27.2	Fulmer *et al* (1967)

† In this analysis the imaginary diffuseness parameter depended on the asymmetry: $a' = 0.51 + 0.7(N-Z)/A$.

Table 4. Global analyses. Proton scattering by a range of nuclei: imaginary potential.

Energy (MeV)	Nuclei	r_w	a_w	$W_I + W_{NS}$	Reference
10–40	^{58}Ni–^{208}Pb	1.32	0.15 ± 0.7	12†	Becchetti and Greenlees (1969)
30–60	^{12}C–^{208}Pb	—	‡	15.5	Gross *et al* (1971)

† In this analysis the imaginary diffuseness parameter depended on the asymmetry, $a' = 0.51 + 0.7(N-Z)/A$.
‡ This analysis is based on proton total reaction cross sections. The isospin dependence is found in the surface part of the imaginary potential and the diffuseness parameter $a' = 0.74 - 0.008E + 1.0(N-Z)/A$.

2.4.3. Analyses of isotopic sequences. In the case of a range of isotopes the Coulomb correction term U_c is almost the same for all nuclei in the isotopic sequence, and if it is further assumed that the nuclear structure terms are also the same then for any two isotopes:

$$U_p(1) - U_p(2) \approx 2U_I(\alpha_1 - \alpha_2) \quad (2.25)$$

Table 5. Global analyses. Neutron scattering by a range of nuclei: real potential.

Energy (MeV)	Nuclei	Number of nuclei	r_u	$U_I - U_{NS}$	Reference
3–24	Fe–Pb	29	1.20	9 ± 2†	Becchetti and Greenlees (1969)
6–8	Al–Bi	13	1.21	50 ± 10	Holmqvist and Wiedling (1968)
7–26	^{40}Ca–^{208}Pb	6	1.198	22.7	Rapaport *et al* (1979a, b)
8	Al–Bi	22	1.22	50 ± 10	Holmqvist and Wiedling (1972)
8–14	$^{54,56}Fe$, $^{63,65}Cu$	4	1.175	15.3	El Kadi *et al* (1982)
8–24	^{54}Fe–^{208}Pb	9	1.165	23	El Kadi *et al* (1982)
8–24	^{54}Fe–^{208}Pb	9	1.173	16.4	El Kadi *et al* (1982)
9	Ni, Fe, Co	3	1.20	22 ± 2	Velkley *et al* (1974)
9	Ni, Fe, Co	3	1.20	1.4 ± 0.2†	Brandenberger and Schrils (1976)
11	Mg–Bi	21	1.27	22.5	Ferrer *et al* (1977)
11	^{51}V–^{59}Co	5	1.285	23.8	Ferrer *et al* (1977)
11	^{93}Nb–^{120}Sn	3	1.235	24.2	Ferrer *et al* (1977)
11	^{165}Ho–^{209}Bi	5	1.247	24.0	Ferrer *et al* (1977)
14	^{56}Fe–^{208}Pb	—	1.25	0	Perey‡
90	^{12}C–^{238}U	6	1.25	38	Hodgson (1960)

† Surface form factor.
‡ Unpublished work quoted by Satchler (1969).

Table 6. Global analyses. Neutron scattering by a range of nuclei: imaginary potential.

Energy (MeV)	Nuclei	Number of nuclei	r_w	$W_I - W_{NS}$	Reference
2–24	Fe–Pb	9	1.26	12†	Becchetti and Greenlees (1969)
7–26	^{40}Ca–^{208}Pb	6	1.295	12.8†‡	Rapaport *et al* (1979b)
8–14	$^{54,56}Fe$, $^{63,65}Cu$	4	1.282	24†	El Kadi *et al* (1982)
8–24	^{54}Fe–^{208}Pb	9	1.272	16†	El Kadi *et al* (1982)
8–24	^{54}Fe–^{208}Pb	9	1.282	19.6†	El Kadi *et al* (1982)
9	Ni, Fe, Co	3	1.20	0†	Velkley *et al* (1974)
11	Mg–Bi	21	1.27	2.03†	Ferrer *et al* (1977)
11	^{51}V–^{59}Co	4	1.285	12.2†	Ferrer *et al* (1977)
11	^{93}Nb–^{120}Sn	3	1.248	11.3†	Ferrer *et al* (1977)
11	^{165}Ho–^{209}Bi	5	1.247	11.9†	Ferrer *et al* (1977)

† Surface form.
‡ For $E \lesssim 15$ MeV; 10.4 MeV for $E \gtrsim 15$ MeV.

and

$$U_n(1) - U_n(2) \approx 2U_I(\alpha_1 - \alpha_2). \tag{2.26}$$

Thus in this special case U_I can apparently be obtained from proton or neutron scattering alone. The values of the isospin potential obtained in this way are, however, not reliable. Firstly, few nuclei are available in each isotopic sequence, reducing the statistical accuracy. Secondly, the structure differences between the isotopes produce effects that are often comparable in magnitude to the isospin term, so that they cannot be neglected. As an example of this, the analysis of Ball *et al* (1964) of the zirconium isotopes with $A = 90$, 91, 92, 94 and 96 gave $U_I \approx 28 \pm 6$ MeV overall. However, the data for the three lighter isotopes gave essentially the same proton potential, corre-

Table 7. Proton scattering by isotopic sequences: real potential.

Energy (MeV)	Nuclei	r_u	$U_I + U_{NS}$	Reference
9–12	^{42,44,48}Ca	1.25	20 ± 5	Marinov *et al* (1966)
9.8	^{112,116,120,124}Sn	1.25	3 ± 5	Durisch and Gould (1965)†
10	^{58,64}Ni, ^{56,58}Fe	1.25	23	Benveniste *et al* (1964)
12.29	^{63,65}Cu	1.28	23	Thomas *et al* (1964)
12.29	^{63,65}Cu	1.31	26.5	Thomas *et al* (1964)
12.7	^{90,92,94}Zr	1.25	≈0	Dickens *et al* (1968)
16	$^{110-116}$Cd	1.20	16	Makofske *et al* (1967, 1968)
	$^{112-124}$Sn			
16	^{126,128,130}Te	1.20	27	Makofske *et al* (1968)
17.8	^{58,60}Ni	1.25	23 ± 6	Baugh *et al* (1965)
19	^{90,92,94}Zr	1.20	≈0	Stautberg and Kraushaar (1966)
22.5	^{90,91,92,94,96}Zr	1.26	28.8 ± 6‡	Ball *et al* (1964)
30.3, 40	^{58,60}Ni	1.16	~23	Thomas and Burge (1969)
50	^{64,66,68}Zn	1.12	≈0	Calderbank *et al* (1967)

† Reanalysis by Sood (1965) gives 18 ± 0.5.
‡ Average value for all isotopes. If ^{96}Zr is omitted, $U_I + U_{NS} \approx 0$.

Table 8. Proton scattering by isotopic sequences: imaginary potential.

Energy (MeV)	Nuclei	r_w	$W_I + W_{NS}$	Reference
10	^{56,58}Fe, ^{56,64}Ni	1.25	40†	Benveniste *et al* (1964)
12	^{42,44,48}Ca	1.25	43†	Marinov *et al* (1966)
12.29	^{63,65}Cu	1.31	9‡	Thomas *et al* (1964)
12.29	^{63,65}Cu	1.31	21†	Thomas *et al* (1964)
16	$^{110-116}$Cd	1.20	23†	Makofske *et al* (1967, 1968)
	$^{112-124}$Sn			
16	^{126,128,130}Te	1.20	44†	Makofske *et al* (1968)
19	^{90,92,94}Zr	1.20	41†	Stautberg and Kraushaar (1966)
30.3, 40	^{58,60}Ni	1.16	15.5†	Thomas and Burge (1969)

† Surface form.
‡ Volume form.

sponding to $U_I = 0 \pm 30$ MeV. A similar situation was found for the zinc isotopes by Calderbank *et al* (1967) and a detailed study of the difficulties of determining the isospin potential from pairs of isotopes has been made by Thomas and Burge (1969). The results of analyses of such data are given in tables 7–10. As expected, they show considerable variation, so little weight can be attached to individual values.

The average results for all three methods are summarised in table 11. It is difficult to attach meaningful uncertainties to these numbers, so those given can only be considered as estimates. It is notable that the average value of U_{NS} is small but this does not justify its omission in particular analyses.

Since the nuclear asymmetry $(N-Z)/A$ is rather small, the value of the isospin potential depends on the difference between two rather similar potentials. Each of these is affected by details of nuclear structure and is subject to experimental uncertainties, so the resulting value of the isospin potential is not of high accuracy. It is thus justifiable to neglect second-order effects in its determination. In particular, various

Table 9. Neutron scattering by isotopic or isobaric sequences: real potential.

Energy (MeV)	Nuclei	Number of nuclei	r_u	$U_I - U_{NS}$	Reference
7–26	$^{92,96,98,100}Mo$	4	1.19	22 ± 4†	Rapaport *et al* (1979a)
8–14	$^{54,56}Fe$	2	1.165	20.3 ± 3†‡	Delaroche *et al* (1982)
8–14	$^{63,65}Cu$	2	1.165	24.7 ± 4†‡	Delaroche *et al* (1982)
11	^{92}Zr, ^{92}Mo	2	1.193	22.7§	Kulkarni and Rapaport (1980)
11	$^{92,96,98,100}Mo$	4	1.235	24.2†	Ferrer *et al* (1977)
11	$^{116-124}Sn$	5	1.2	16 ± 2†	Finlay *et al* (1980)
11	$^{116-124}Sn$	5	1.2	21.60 ± 0.2†	Rapaport *et al* (1980)
14.2	Ni–Bi	41	1.26	26.6§	Dukarevich *et al* (1967)
14.2	Ni–Bi	41	1.26	17 ± 2†	Dukarevich *et al* (1967)
24	$^{16,18}O$	2	1.10	−11.25†	Grabmayr *et al* (1980)

† Isotopic sequences.
‡ Coupled-channels analysis.
§ Isobaric sequences.

Table 10. Neutron scattering by isotopic or isobaric sequences: imaginary potential.

Energy (MeV)	Nuclei	Number of nuclei	r_w	$W_I - W_{NS}$	Reference
7–26	$^{92,96,98,100}Mo$	4	1.23	−10 ± 2†	Rapaport *et al* (1979a)
8–14	$^{54,56}Fe$	2	1.261	14.5 ± 4.5†	Delaroche *et al* (1982)
8–14	$^{63,65}Cu$	2	1.261	14.1 ± 4.4†	Delaroche *et al* (1982)
11	^{92}Zr, ^{92}Mo	2	1.237	10.4†	Kulkarni and Rapaport (1980)
11	$^{92,96,98,100}Mo$	4	1.235	10.0†	Ferrer *et al* (1977)
11	$^{116-124}Sn$	5	1.25	12 ± 2	Finlay *et al* (1980)
11	$^{116-124}Sn$	5	1.30	14†	Rapaport *et al* (1980)
14.2	Ni–Bi	41	1.26	50 ± 25†	Dukarevich *et al* (1967)
14.2	Ni–Bi	41	1.26	26 ± 9†	Dukarevich *et al* (1967)
24	$^{16,18}O$	2	1.40	−10.85	Grabmayr *et al* (1980)

† Surface form.

Table 11. Summary of results.

Real potential				
Method	$U_I + U_{NS}$	$U_I - U_{NS}$	U_I	U_{NS}
Particular	—	—	24 ± 2	—
Global	24 ± 2	18 ± 4	21 ± 3	3 ± 3
Isotopic	24 ± 5	20 ± 6	22 ± 6	2 ± 6
Imaginary potential				
Method	$W_I + W_{NS}$	$W_I - W_{NS}$	W_I	W_{NS}
Global	12	12	12	0
Isotopic	30	10	20	10

analyses use slightly different values for the radius and surface diffuseness parameters but it is doubtful if it is worth trying to correct for the corresponding second-order charge in the isospin potential. The radius correction could be made quite simply using the constancy of UR^2, but it is not obvious how to correct for small differences in the surface diffuseness parameters. These difficulties are avoided by analysing volume integrals of the potential as described in the next section.

Becchetti and Greenlees (1969) analysed a considerable range of proton and neutron data and found a good overall fit with an isospin potential of 24 MeV and a Coulomb correction term of $0.4Z/A^{1/3}$, but effectively setting U_{NS} to zero. They do, however, note that their analysis gives a slight difference between the proton and neutron central strengths and say that this can be removed by a Coulomb correction term of $0.27Z/A^{1/3}$ instead of $0.4Z/A^{1/3}$. This difference of $0.13Z/A^{1/3}$ is thus attributable to the neglect of U_{NS}.

2.4.4. Energy dependence of isospin potential. Some global analyses of neutron elastic scattering have been made that included a linear energy dependence of the asymmetry term. Values of the coefficients ϵ_R and ϵ_I in the expressions $(V_1-\epsilon_R E)\alpha$ and $(W_1-\epsilon_I E)\alpha$ are given in table 12.

Table 12. Values of coefficient of linear energy dependence of isospin potential.

Reaction	E	Nucleus	ϵ_R	ϵ_I	Reference
		Theoretical analyses	0.1	—	Rook (1973)
		Theoretical analyses	0.17	—	Dabrowski and Haensel (1974)
		Theoretical analyses	0.1	—	Jeukenne *et al* (1977b)
(n, n)	7–26	^{40}Ca, ^{208}Pb	0.19	—	Rapaport *et al* (1979a, b)
(p, n)	25–45	^{48}Ca–^{208}Pb	0.18	—	Patterson *et al* (1976)
(n, n), p, p)	10–50	^{208}Pb	0.183 ± 0.008	—	De Vito *et al* (1981)
(n, n), (p, p)	10–40	^{208}Pb	—	0.178 ± 0.052	De Leo and Micheletti (1981)
(p, p)	100	^{58}Ni–^{208}Pb	0.12 ± 0.06	—	Kwiatkowski and Wall (1978)

2.5. The volume integrals

The insensitivity of the elastic scattering cross section to variations in the depth U and radius r_u of the real part of the potential that maintain the constancy of Ur_u^2 suggests that it is useful to adopt the volume integral of the potential as a relatively invariant function of the parameters, and to investigate its variation with energy and nuclear asymmetry parameter (Greenlees *et al* 1968, Becchetti and Greenlees 1969, Sinha *et al* 1972, Kailas and Gupta 1977, 1978, Kailas *et al* 1979).

We therefore define the volume integral per nucleon as

$$J=\frac{1}{A}\int V(r)\,\mathrm{d}\mathbf{r}. \tag{2.27}$$

For a potential of Saxon–Woods form

$$J=\frac{4\pi R^3 U}{3A}\left\{1+\left(\frac{\pi a}{R_u}\right)^2+\mathrm{O}\left[\left(\frac{a}{R_u}\right)^4\right]\right\} \tag{2.28}$$

$$\simeq\frac{4\pi r_u^3 U}{3}\left[1+\left(\frac{\pi a}{R_u}\right)^2\right] \qquad (\mathrm{MeV\ fm^3}).$$

The volume integrals may be analysed using expressions obtained by integrating (2.10) and (2.11), viz:

$$J_R^{\mathrm{p}} = J_0 - \epsilon J_E + \alpha J_{\mathrm{NS}} + \alpha J_I + J_{\mathrm{c}} \frac{Z}{A^{1/3}} \tag{2.29}$$

$$J_R^{\mathrm{n}} = J_0 - \epsilon J_E + \alpha J_{\mathrm{NS}} - \alpha J_I \tag{2.30}$$

in an obvious notation. These volume integrals are defined so that the J are independent of Z and A. Subtracting:

$$J_R^{\mathrm{p}} - J_R^{\mathrm{n}} = 2\alpha J_I + J_{\mathrm{c}} \frac{Z}{A^{1/3}}. \tag{2.31}$$

The energy dependence of the volume integrals has been determined by several analyses of proton and neutron scattering, and some results are collected together in table 13.

Table 13. Determination of energy dependence of J_R^{p} and J_R^{n}.

Nucleus	Energy range (MeV)	Neutron	Proton
^{16}O†	E_{p}: 20–50	—	$521 - 2.9E$
^{28}Si‡	E_{p}: 14–40	$484 - (2.9 \pm 0.2E)$	$498 - (2.9 \pm 0.2)E$
^{32}S‡	$E_{\mathrm{n}} = 11$–25; $E_{\mathrm{p}} = 15$–35	$480 - (2.9 \pm 0.2)E$	$500 - (2.9 \pm 0.2)E$
^{40}Ca‡	$E_{\mathrm{n}} = 11$–40; E_{p}: 26–76	$473 - (2.9 \pm 0.2)E$	$498 - (2.9 \pm 0.2)E$
^{208}Pb‡	$E_{\mathrm{n}} = 5$–25; $E_{\mathrm{p}} = 10$–80	$(392 \pm 5) - (2.2 \pm 0.36)E$	$(490 \pm 8) - (2.65 \pm 0.36)E$

† Pignanelli *et al* (1981).
‡ Rapaport (1982).

The value of J_{c} may be obtained as before from analyses of scattering by $N = Z$ nuclei. The results of several of these analyses are given in table 14; an average value (omitting ^{28}Si) is

$$J_{\mathrm{c}} = 4.25 \pm 0.52 \tag{2.32}$$

corresponding to $U_{\mathrm{c}} = (0.45 \pm 0.05) Z/A^{1/3}$.

Table 14. Determinations of J_{c} and U_{c} from $N = Z$ nuclei (Rapaport 1982).

Nucleus	Energy range (MeV)	$J_{\mathrm{c}}Z/A^{1/3}$	J_{c}	U_{c}
^{16}O	E_{p}: 20–50; E_{n}: 24	14 ± 6	4.4 ± 1.9	0.48 ± 0.20
^{28}Si	E_{p}: 17–40: E_{n}: 11, 20, 26	14 ± 5	3.0 ± 1.1	0.33 ± 0.12
^{32}S	E_{p}: 15–35; E_{n}: 11, 20, 26	20 ± 6	4.0 ± 1.2	0.45 ± 0.13
^{40}Ca	E_{p}: 26–76; E_{n}: 11, 20, 26, 30, 40	25 ± 3.8	4.3 ± 0.6	0.5 ± 0.07

The isospin volume integral may be obtained from (2.31) and the results for ^{208}Pb in table 13 combined with (2.32) give

$$J_I = 95 \pm 27. \tag{2.33}$$

The values of J_0 and J_{NS} can be obtained from studies of the dependence of J_R on the asymmetry parameter α. A study of neutron scattering at 11 MeV by Ferrer *et al*

(1977) gave

$$J_R^n = (443 \pm 10)[1 - (0.7 \pm 0.18)\alpha] \tag{2.34}$$

and for neutrons at 25 MeV

$$J_R^n = (390 \pm 14)[1 - (0.6 \pm 0.2)\alpha]. \tag{2.35}$$

These may be combined using the energy dependence $2.9E_n$ to give at 11 MeV

$$J_R^n = (437 \pm 8)[1 - (0.65 \pm 0.16)\alpha]. \tag{2.36}$$

Hence

$$J_0 = 469 \pm 8 \qquad J_{NS} = -191 \pm 95. \tag{2.37}$$

It is notable that in this case the nuclear structure term is not negligible. The volume integral J_s of the spin–orbit potential has been determined by Begum *et al* (1979) for 16.1 MeV neutrons and they find a decrease with A approximately fitted by

$$J_s = 17 - 0.057A. \tag{2.38}$$

The volume integral per nucleon of the imaginary part of the potential shows little global variation with energy or target nucleus. Studies for neutrons at 11, 24 and 26 MeV gave (Ferrer *et al* 1977, Rapaport *et al* 1978, Rapaport 1982)

$$J_I^n = (105 \pm 5)[1 - (2 \pm 0.3)\alpha]. \tag{2.39}$$

2.6. *Scattering by deformed nuclei*

A global optical potential that has been fitted to experimental data relating to spherical and near-spherical nuclei does not always give acceptable fits to other nuclei. For

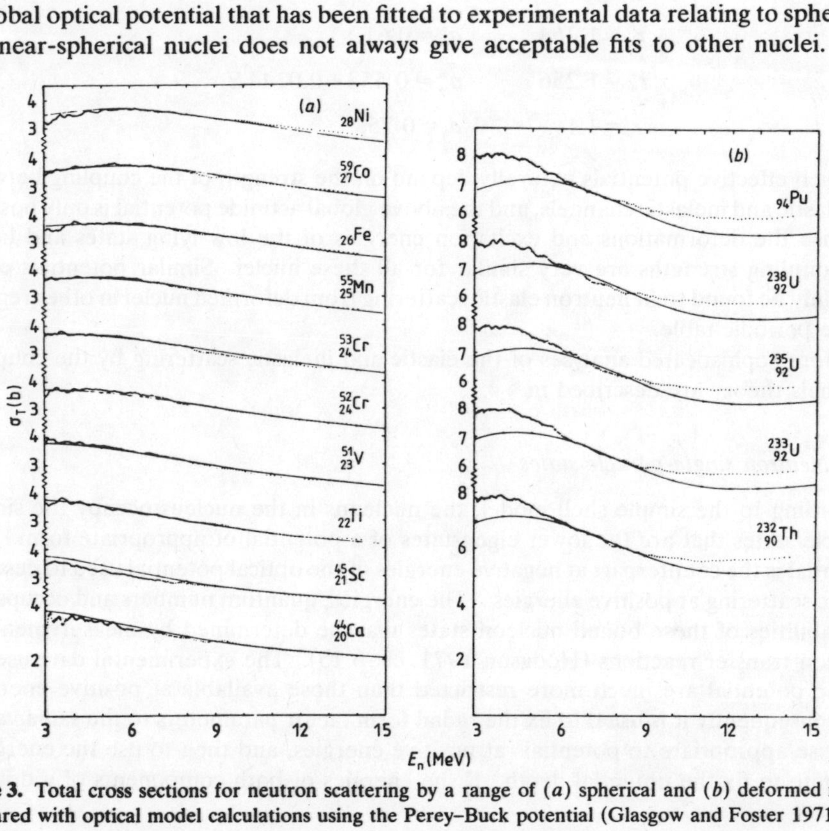

Figure 3. Total cross sections for neutron scattering by a range of (*a*) spherical and (*b*) deformed nuclei compared with optical model calculations using the Perey–Buck potential (Glasgow and Foster 1971).

example, figure 3 shows a comparison between the total cross sections for neutron interactions as a function of energy for (*a*) a group of spherical and near-spherical nuclei and (*b*) a group of deformed nuclei compared with the results of optical model calculations with the global potential of Perey and Buck (1962). The fit is excellent for the former groups and poor for the latter.

The physical reason for this is clear. The deformed nuclei have low-lying collective states that are strongly coupled to the ground state. The corresponding inelastic scattering cross sections are large and this removes flux from the elastic channel. The result of this strong coupling between the elastic and inelastic channels is that the optical model with parameters fitted to the data for spherical nuclei no longer gives the correct elastic and reaction cross sections.

There are two ways to tackle such a situation: one can either alter the optical model parameters to fit the data, or one can develop a more sophisticated formalism that takes account of both the elastic and inelastic scattering together. Both of these ways have been followed. In the course of their search for a coupled-channels potential Madland and Young (1978) found a global potential that gives an excellent fit to the elastic scattering and total cross-section data for the actinide nuclei, as shown in figure 4. The parameters of this potential are

$$\begin{aligned}
&U=50.378-0.354E-27.073(N-Z)/A \\
&V_{so}=6.12 \\
&W_D=9.265-0.232E+0.03318E^2-12.666(N-Z)/A \\
&r_u=1.264 \qquad a_u=0.612 \\
&r_w=1.256 \qquad a_w=0.553+0.0144E \\
&r_s=1.1 \qquad a_s=0.75.
\end{aligned} \tag{2.40}$$

Such effective potentials naturally depend on the strength of the coupling between the elastic and inelastic channels, and the above global actinide potential is only possible because the deformations and excitation energies of the low-lying states and hence the coupling strengths are very similar for all these nuclei. Similar potentials could certainly be found to fit neutron elastic scattering from deformed nuclei in other regions of the periodic table.

More sophisticated analyses of the elastic and inelastic scattering by the coupled-channels theory are described in § 4.1.

2.7. Neutron single-particle states

According to the simple shell model, the nucleons in the nucleus occupy the single-particle states that are the lower eigenstates of a potential of appropriate form. This potential is the counterpart at negative energies of the optical potential used to describe elastic scattering at positive energies. The energies, quantum numbers and occupation probabilities of these bound nucleon states may be determined by measurements of nucleon transfer reactions (Hodgson 1971, chap 15). The experimental data used to fix the potential are much more restricted than those available at positive energies, and consequently it is usual to fix the radial form factor parameters to the same values as those appropriate to potentials at positive energies, and then to use the energy of the state to fix the potential depth. If the energies of both components of a doublet

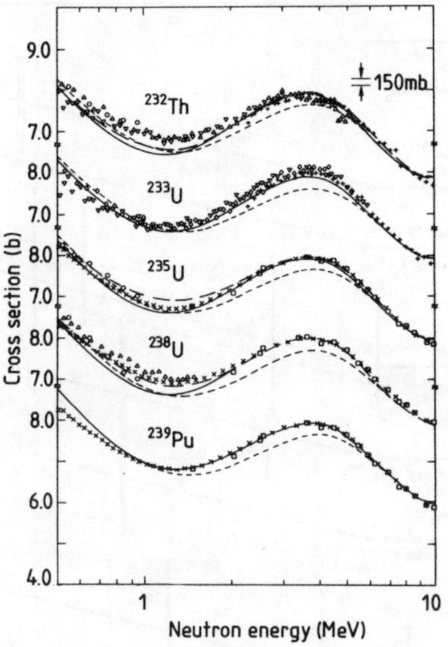

Figure 4. Neutron total cross sections compared with optical model calculations using the global potentials of Madland and Young (1978). The curves are (*a*) first iteration of spherical global optical potential (- - -), (*b*) second iteration of spherical global optical potential (——), (*c*) deformed global coupled-channels potential (— —). ∇, Green (1971); ○, Uttley (1966); +, Foster (1971); □, Shamu (1978); ×, Schwartz (1974); △, Whalen (1969), ◇, Poenitz (1978).

are known they can be used to calculate both the central and spin–orbit depths U and U_s; if only one is known it is usual to set U_s to an average value and use the known energy to determine U.

There is one respect in which the information obtained for the potential at negative energies is more detailed than that for positive energies, namely that the depth is determined for each quantum state separately. At positive energies it is usually neither possible nor desirable to fit each partial wave individually, so the potential refers to the weighted average of all the contributing states. The scattering potential is determined as a function of energy, whereas the corresponding bound state potential is determined only at its eigenenergies.

The results of several determinations of the potential at negative energies from the bound state data are shown in figure 5. It is evident that the depth decreases with increasing energy, as for the potentials at positive energies, and that there is a plateau at energies around the Fermi surface. Combining data from neutron and proton bound states, Bear and Hodgson (1978) found that the potential depth can be represented over a range of nuclei by the expressions

$$U = U_0 \qquad \text{for } -15 < E < 0 \quad \text{(surface states)}$$
$$U = U_0 - \beta(E+15) \qquad \text{for } E < -15 \quad \text{(deep states)} \tag{2.41}$$

for a potential of Saxon–Woods form (1.4) with $r_u = 1.236$ fm, $a_u = 0.62$ fm and $\beta = 0.51$.

Away from closed shell nuclei, the single-particle states are increasingly fragmented by the residual interactions. It is still possible to carry out analyses of the type just mentioned by using centroid energies defined by $E_c = \Sigma_i S^i_{JL} E_i$ where S_{JL} is the spectroscopic factor and the sum runs over all the fragments of the same NJL.

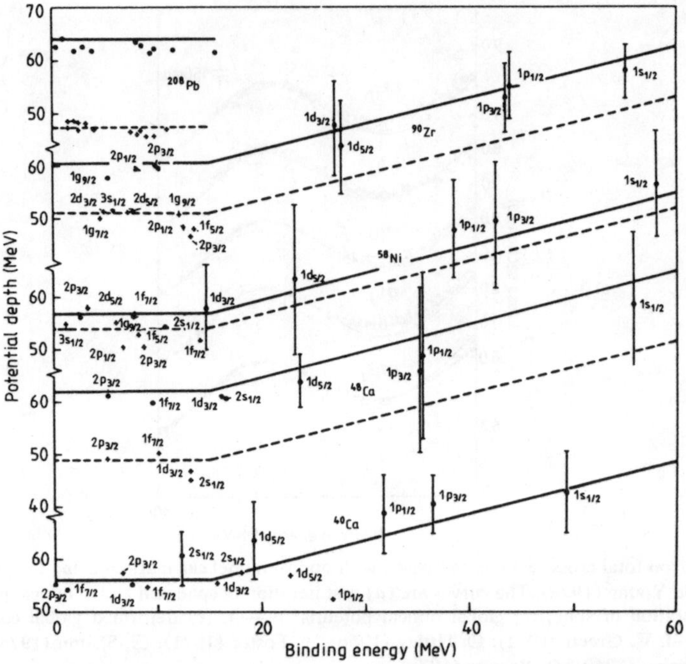

Figure 5. Nuclear potential depths for neutron and proton single-particle states of ^{40}Ca, ^{48}Ca, ^{58}Ni, ^{90}Zr and ^{208}Pb compared with the values given by equation (2.41) (Bear and Hodgson 1978).

The width of the fragment distribution Γ is related to the imaginary part of the bound state potential W by the relation $\Gamma = 2W$. There are few experimental determinations of fragment distributions and these are subject to uncertainties concerning whether all the fragments have been detected (Malaguti 1978). It is also possible to determine the widths of single-particle states by knock-out reactions at high energies. The widths of some proton states have been determined in this way from the cross sections for (p, 2p) reactions, but so far there are no corresponding results for neutrons.

3. The theory of the neutron optical potential

3.1. The real part of the potential

There have been many attempts to calculate the nucleon optical potential from more fundamental nuclear properties and these may be divided into the nuclear matter and the nuclear structure methods. It is possible to show quite generally that there is a potential that, when inserted into the one-body Schrödinger wave equation, gives the wavefunction in the elastic channel. The formalism of Feshbach gives the following expression for the optical potential:

$$U(r) = (\varphi_0|V|\varphi_0) + \left(\varphi_0\left|VQ\frac{1}{E-QHQ}QV\right|\varphi_0\right) \tag{3.1}$$

where φ_0 is the ground-state wavefunction of the target, V is the bare potential in the absence of target excitations, $P \equiv 1 - Q$ is the operator that projects out the ground-state part of the target wavefunction and the round brackets indicate integration over the internal coordinates of the target. The physical interpretation of this expression is that the first term represents direct transitions from the incident to the outgoing channel and the second term takes account of virtual nuclear excitations followed by de-excitations back into the elastic channel. Thus the first term gives most of the real part of the potential, with corrections from the second term, while the second term gives the imaginary part. This potential is highly complicated, non-local and strongly energy-dependent, so even if it could be calculated it would not be very useful. This suggests that an approximate calculation yielding a simpler but less accurate result may still be useful, and this is indeed the case. Many of the features of the optical potential can be calculated from the properties of nuclear matter and applied to finite nuclei by the local density approximation, i.e. by assuming that the properties of the part of the nucleus with a certain density are the same as those of infinite nuclear matter of that density. Other features of the potential, particularly those associated with the absorption processes, can only be found by taking into account the properties of the individual target nuclei.

The phenomenological finding that a simple optical potential of the form described in § 2 is able to fit quite well the elastic scattering from many nuclei at many energies itself suggests that one can obtain useful information by studying the limiting case of infinite nuclei, i.e. nuclear matter. The similarity between the radial dependence of the successful phenomenological potentials and that of the nuclear density distribution further suggests a connection between the two in terms of the local density approximation.

A first approximation to the real part of the potential is provided by the first term in (3.1), which is a simple folding of the nucleon–nucleon interaction $v(|\boldsymbol{r}-\boldsymbol{r}'|)$ and the nuclear density distribution $\rho(\boldsymbol{r})$:

$$V(r) \sim \int \rho(\boldsymbol{r}')v(|\boldsymbol{r}-\boldsymbol{r}'|)\,\mathrm{d}\boldsymbol{r}'. \tag{3.2}$$

This expression is quite good above about 500 MeV, but at lower energies it is progressively less accurate due to a number of effects including Pauli blocking, Fermi averaging, dispersion and antisymmetrisation. Qualitatively, the expression (3.2) yields several important features of the optical potential. In particular, the short-range nature of the nucleon–nucleon interaction ensures that the real part of the potential has the same radial shape as the nuclear density, namely a uniform variation in the interior and an exponential fall-off in the surface region. These general features are incorporated in the real part of the phenomenological optical potential. This gives quite good estimates of the radius and surface diffuseness of the potential and its depth can be estimated from the Fermi gas model. The imaginary part of the potential is similarly expected to have approximately the same radial extent as the real part and its strength can be estimated from the total absorption cross section by semi-classical arguments. The spin–orbit potential is taken to have the radial derivative form by analogy with the Thomas term in atoms and its strength obtained from the spin–orbit splitting of the nuclear shell model.

Thus, quite simple theoretical arguments suffice to give the main features of the optical potential, and more accurate values of the parameters may be obtained by comparison with experimental data.

3.2. The imaginary part of the potential

The imaginary part of the optical potential is much more sensitive to nuclear structure effects, as shown for example by the dependence of the reaction cross section, and hence the imaginary potential, on nuclear shell structure. It is found that these are lower than usual for the magic nuclei, and this is attributed to the lower number of low-lying excited states for such nuclei, and hence fewer reaction channels to remove the particle flux. It is thus desirable to express the imaginary part of the potential in terms of the low-lying nuclear excitations. This is provided formally by the theory of Feshbach, and detailed calculations have been made by Vinh Mau (1979), Bernard and Van Giai (1979), Osterfeld and Madsen (1981), Bouyssy *et al* (1981), Ngô (1982) and Fantoni *et al* (1982).

3.3. Nuclear matter calculations

In recent years there have been many calculations of the optical potential from the nucleon–nucleon potential, including those of Mahaux and his collaborators (Jeukenne *et al* 1974, 1976a, b, 1977a, b) and by Brieva and Rook (1977). This latter work has now been extended by Haider *et al* (1983) and some of their results will be summarised here.

The calculation starts from the Bethe–Goldstone equation for the relative motion of two nucleons in nuclear matter:

$$\Psi(\boldsymbol{r})=\Phi(\boldsymbol{r})-\int G(\boldsymbol{r},\boldsymbol{r}')\nu(|\boldsymbol{r}-\boldsymbol{r}'|)\Psi(\boldsymbol{r}')\,\mathrm{d}\boldsymbol{r}' \tag{3.3}$$

where $\Psi(\boldsymbol{r})$ is the correlated wavefunction, $\Phi(\boldsymbol{r})$ is the uncorrelated wavefunction and $\nu(\boldsymbol{r})$ is the nucleon–nucleon interaction. The Green function $G(\boldsymbol{r},\boldsymbol{r}')$ can be written

$$G(\boldsymbol{r},\boldsymbol{r}')=\frac{1}{(2\pi)^3}\int \exp[\mathrm{i}\boldsymbol{k}\cdot(\boldsymbol{r}-\boldsymbol{r}')]G_k(k',K,\theta)\,\mathrm{d}\boldsymbol{k}' \tag{3.4}$$

where K and k' are the relative momenta in the final and intermediate state of the nuclear pair and

$$G_k(k',K,\theta)=\frac{Q_p(k_1,k_2)}{e(k_1)+e(k_2)-W+\mathrm{i}\epsilon} \tag{3.5}$$

where k_1 and k_2 are the momenta of the two nucleons in the intermediate state; they are related to the centre of mass and relative coordinates by $\boldsymbol{K}=\boldsymbol{k}_1+\boldsymbol{k}_2$ and $\boldsymbol{k}=\frac{1}{2}(\boldsymbol{k}_1-\boldsymbol{k}_2)$. The starting energy $W=e(K)+e(k)$ and the energy denominator

$$e(q)=\frac{\hbar^2q^2}{2m}+U(K_\mathrm{F};q,E) \tag{3.6}$$

where the optical potential $U(K_\mathrm{F};q,E)$ is normally parametrised to depend quadratically on energy:

$$U(K_\mathrm{F};q,E)=V_0+V_1E+V_2E^2. \tag{3.7}$$

The Pauli operator $Q_p(\boldsymbol{k}_1,\boldsymbol{k}_2)$ restricts the interaction to those proceeding to the unoccupied final states, so that

$$\begin{aligned} Q_p(\boldsymbol{k}_1,\boldsymbol{k}_2|\boldsymbol{k}_1\boldsymbol{k}_2)&=|\boldsymbol{k}_1,\boldsymbol{k}_2) &&\text{if } k_1,k_2>K_\mathrm{F}\\ &=0 &&\text{otherwise} \end{aligned} \tag{3.8}$$

where K_F is the Fermi energy.

In order to solve the Bethe–Goldstone equation for the correlated wavefunction $\Psi(r)$ the Green function is first transformed to centre of mass and relative coordinates and then expanded in spherical harmonics:

$$G(k', k, \theta) = \sum_L G_L(k', k) Y_L^0(\theta). \tag{3.9}$$

The correlated and uncorrelated wavefunctions are similarly expanded and this enables the Bethe–Goldstone equation to be separated into equations for the wavefunctions for each partial wave:

$$U_{LL'}^{JS}(r) = j_L(kr)\delta_{LL'} - 4\pi \sum_{L''} \int_0^\infty G_L(r, r') V_{L'L''}^{JS}(r') U_{LL''}^{JS}(r') r'^2 \, \mathrm{d}r' \tag{3.10}$$

where $U_{LL'}^{JS}(r)$ and $V_{LL'}^{JS}(r)$ are, respectively, the radial wavefunction and the nucleon–nucleon interaction in the state $JSLL'$.

These integro-differential equations were solved by discretising the integral followed by matrix inversion for all contributing partial waves ($L \leqslant 5$) using the nucleon–nucleon interaction of Hamada and Johnston. The correlated wavefunction obtained in this way is then used to calculate the t-matrix element:

$$\langle t \rangle \equiv \langle ksm_s | t | ksm_s \rangle = \langle \Phi(k) | v | \Psi(k) \rangle \tag{3.11}$$

and hence the optical potential in infinite nuclear matter:

$$U(k, K_{\mathrm{F}}, E) = \sum_j n_<(j) \sum_{J,S} (2J+1) \sum_{m_s} \langle t \rangle \tag{3.12}$$

where j is the momentum of a particle in the Fermi sea and the summation runs over all occupied states.

Initially, the energy variation (3.7) of the optical potential is not known, so the calculation is started with reasonable values for the coefficients V_0, V_1 and V_2 and iterated until the final potential (3.12) has the same energy dependence. The results for the energy variation of the real potential for various values of the density show anomalous behaviour around $k \approx K_{\mathrm{F}}$, which may be connected with the similar anomaly found in phenomenological determinations at positive and negative energies (Brown *et al* 1963, Ahmad and Haider 1976, Bear and Hodgson 1978, Bauer *et al* 1982).

These results for nuclear matter must be applied to finite nuclei if quantitative comparisons are to be made. Eventually it is intended to do this using the prescription of Siemens (1970) but some preliminary results have been obtained using the local density approximation. These show that the potential can be separated into a part varying smoothly with energy and a part peaked around the Fermi surface. The former gives the familiar volume form for the real optical potential and the latter a surface-peaked term that becomes more important near the Fermi surface.

The real part of the potential for -10 MeV found from these calculations is shown in figure 6 and this is consistent with the determinations already mentioned. These calculations used proton and neutron densities calculated by the Hartree–Fock and single-particle potential methods. The energy variation of the depth of the equivalent real potential of volume form is compared in figure 7 with the calculations of Mahaux and Ngô (1981) from dispersion relations and with the phenomenological results of Bauer *et al* (1982).

It has been suggested (Perey 1963, Bertsch and Kuo 1968, Gyarmati *et al* 1981) that the anomaly around the Fermi surface is attributable to the coupling between the incident nucleons and collective states of the target nucleus. The present work shows

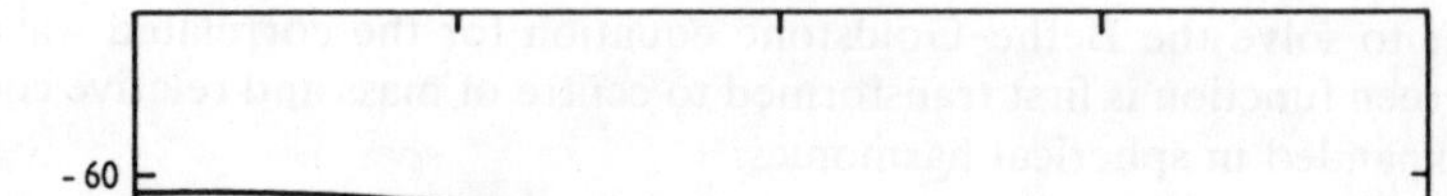

Figure 6. Radial variation of the real part of the nucleon optical potential at −10 MeV showing the separation into volume and surface-peaked components. The half-density radius of the nuclear matter distribution is indicated (Haider *et al* 1983).

that most of the effect is attributable to the presence of an absorbing potential, irrespective of the mechanism responsible for the absorption. Calculations are planned to see the effect of explicit inclusion of the coupling to collective states.

Another calculation of the optical potential has recently been made by Yamaguchi *et al* (1983). They start from the Hamada–Johnston and one-pion exchange (OPEC) nucleon–nucleon interactions and use essentially the same formulae to calculate the effective interaction in coordinate space. The central, spin–orbit and tensor parts of this interaction are then parametrised by sums of three Gaussian functions and the strength of the central potential is allowed to depend on energy and momentum. It is found that the resulting central potential has the energy dependence (for $K_F = 1.4\ \mathrm{fm}^{-1}$):

$$\begin{aligned} V(\mathrm{HJ}) &= 63.6 - 0.31E \\ V(\mathrm{OPEC}) &= 71.1 - 0.36E. \end{aligned} \tag{3.13}$$

The anomaly in the Fermi surface is found in these calculations, and examination of the behaviour of the phase shifts shows that it is mainly associated with the $^3S_1$ and $^3D_1$ partial waves. The results for the imaginary part of the potential shows a progression from surface-peaked to volume form with increasing energy.

These potentials are used to calculate the differential cross sections and polarisations for the elastic scattering of protons by ^{40}Ca for a range of energies, and good overall agreement is found, provided the imaginary potential is renormalised by a factor of 0.7–0.8; this is ascribed to the continuum level density in the region of the Fermi surface.

The isospin dependence of the microscopic optical potential has been studied by Dietrich *et al* (1983), who compared the results of nuclear matter calculations with

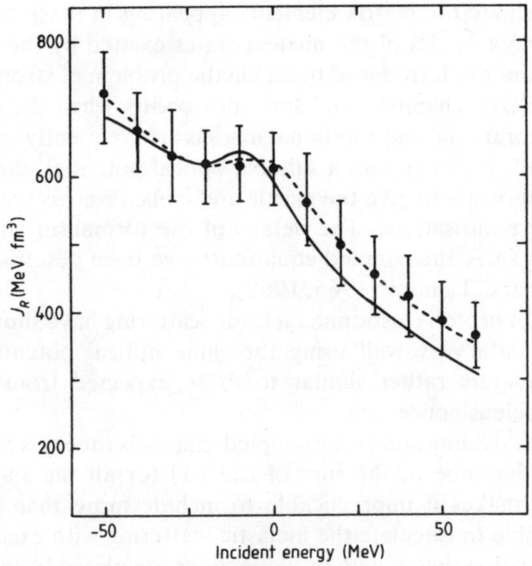

Figure 7. Energy variation of the volume integral of the real part of the nucleon optical potential for ^{40}Ca compared with the phenomenological results of Bauer *et al* (1982) (——) and with the calculations of Mahaux and Ngô (1981) (– – –) (Haider *et al* 1983). ●, empirical values.

data on neutron and proton elastic scattering by ^{208}Pb over a range of energies. They used the Yukawa parametrisation of the effective interaction of Brieva and Rook (1977, 1978a, b) and found good overall agreement. The calculation showed that proton scattering is much more sensitive than neutron scattering to the shape of the neutron density distribution.

4. Nuclear reactions

4.1. Inelastic scattering

In the optical model, the elastic channel is treated explicitly and all non-elastic processes are taken into account in a global way by the imaginary part of the potential. It is possible to extend this formalism so as to treat explicitly not only the elastic channel but also a few selected inelastic channels, leaving the remaining non-elastic processes to be taken into account by an appropriately reduced imaginary potential.

To do this, the total wavefunction is expanded as a sum of the wavefunctions in the elastic and the inelastic channels, each multiplied by the wavefunction of the corresponding nuclear state. When this is inserted into the Schrödinger equation the integration over the angular coordinates splits it into a set of coupled equations for the wavefunctions in the elastic and inelastic channels. The set of equations is truncated, leaving only the equations relating to the elastic channel and the few selected inelastic channels, which are usually those strongly coupled to the elastic channel. The truncated channels are taken into account by the imaginary part of the potential. The simple optical model corresponds to the situation when all the non-elastic channels are truncated.

In order to evaluate the matrix elements appearing in these coupled equations, it is necessary to have a model of the nuclear states excited by the inelastic scattering. Since these equations are introduced to tackle the problem of strong coupling between the elastic and inelastic channels, and since this occurs when the states are collective in character, the vibrational and rotational models are frequently used for this purpose. Using such a model, together with a suitable optical potential, the coupled equations can be solved numerically to give the elastic and inelastic cross sections, together with the corresponding polarisations. The details of the formalism and of the numerical techniques used to solve the coupled equations have been described by Buck (1963), Buck *et al* (1963) and Tamura (1965, 1969).

Many analyses of proton elastic and inelastic scattering have shown that it is possible to reproduce the data very well using the same optical potential parameters and deformation parameters rather similar to those expected from knowledge of the structure of the nucleus concerned.

The main practical limitation of the coupled-channels theory is that the computation time increases as the cube of the sum of $(2J+1)$ for all the states included in the calculation, which makes it impracticable to include more than a few states. It is, however, still possible to calculate the inelastic scattering with excitation of the higher states using perturbation theory with wavefunctions calculated from potentials adjusted to take account of the effects of the strong coupling (Hodgson 1983a).

After its initial success with proton scattering, the coupled-channels theory has been extensively applied to neutron scattering, with equally satisfactory results. As an example of some recent work we mention the analysis of scattering by actinide nuclei by Haouat *et al* (1981, 1982). They calculated the differential elastic and inelastic cross sections using the coupled-channels formalism using the rotational model with quadrupole and hexadecapole deformation to describe the nuclear excitations. Some of their results are compared with the experimental data in figure 8. At these low energies there are substantial compound-nucleus contributions to the cross sections and these were calculated using the Hauser–Feshbach formalism including the width fluctuation correction. In this calculation the radius and diffuseness parameters were given the same values for all nuclei, and the other parameters were adjusted to fit the s- and p-wave strength functions, potential scattering radius and total cross sections over a wide energy range. The deformation parameters were adjusted to fit the elastic and inelastic cross sections. Similar calculations have been made for actinide nuclei by Madland and Young (1978), for titanium by Smith *et al* (1978) and for oxygen by Grabmayr *et al* (1980).

Coupled-channels calculations have been made for the neodymium and samarium isotopes (Lagrange 1974, Lagrange *et al* 1975, Coope *et al* 1976, 1977, Shamu *et al* 1973, 1976, 1980, McEllistrem *et al* 1977) and good fits obtained to the total cross sections and their differences for isotopic pairs and also to the differential cross sections, using the vibrational and rotational models. This provides additional information concerning the structure of the nucleus; thus the vibrational model was found to be appropriate for ^{148}Sm and ^{152}Sm, and the rotational model for ^{154}Sm, which is in accord with other work. The results for ^{150}Sm are, however, quite well fitted assuming that it is a rotational nucleus, whereas other theoretical and experimental evidence indicates that it is neither a pure rotator nor a pure vibrator. The elastic and inelastic scattering of 3.4 MeV neutrons by four tungsten isotopes has been studied by Delaroche *et al* (1981) and used to extract information on their deformation parameters.

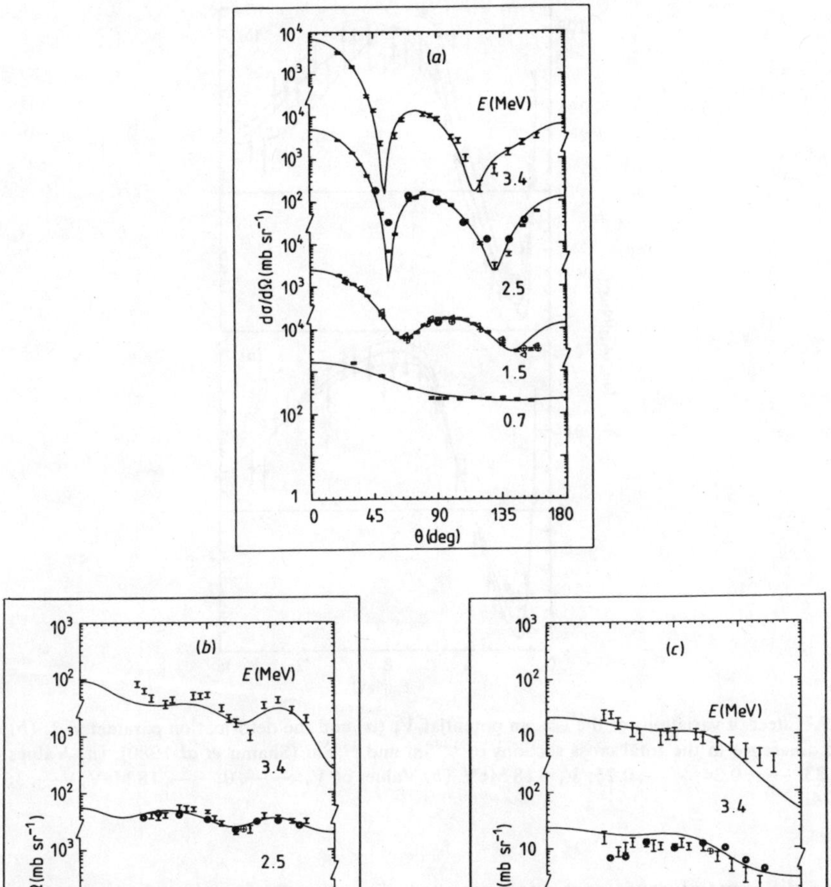

Figure 8. Neutron elastic and inelastic cross sections compared with coupled-channels calculations for ^{238}U (Haouat *et al* 1981). (*a*) (n_1n_0), (*b*) (n_1n_1), (*c*) (n_1n_2).

The total cross section for neutron scattering depends on the deformation of the ground state and not on the deformations of the excited states, so the differences between the total cross sections for neighbouring nuclei can be used to study the isospin dependence of the potential. As shown in figure 9, the comparison with experiment indicates that $V_1 \approx 25$ MeV for ^{148}Sm and ^{154}Sm, a result that is insensitive to the value chosen for the deformation parameter (Shamu *et al* 1976, 1980).

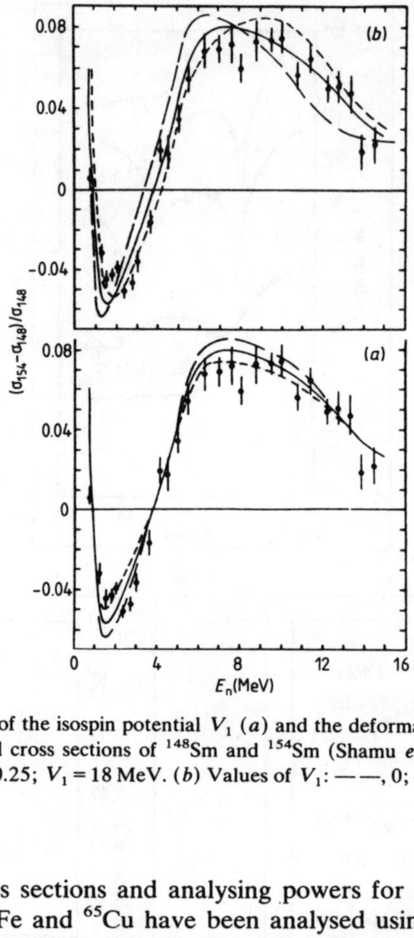

Figure 9. Effect of variations of the isospin potential V_1 (*a*) and the deformation parameter β_2 (*b*) on the relative difference in the total cross sections of ^{148}Sm and ^{154}Sm (Shamu *et al* 1980). (*a*) Values of β_2: ---, 0.23; ——, 0.24; — —, 0.25; $V_1 = 18$ MeV. (*b*) Values of V_1: — —, 0; ——, 18 MeV; ---, 36 MeV; $\beta_2 = 0.24$.

The differential cross sections and analysing powers for the elastic scattering of 10 MeV neutrons by ^{54}Fe and ^{65}Cu have been analysed using the coupled-channels formalism (Delaroche *et al* 1982, Floyd *et al* 1983). The parameters of the spin–orbit potential ($V_s = 5.3$ MeV, $r_s = 1.04$, 1.11 fm, $a_s = 0.464$ fm) are similar to those found in the corresponding spherical optical model analyses. There is an ambiguity between the values of the spin–orbit deformation δ_s and the strength of the imaginary part of the spin–orbit potential W_s, so that they cannot be determined independently using only the analysing powers for inelastic scattering. Coupled-channels calculations of neutron scattering from the even molybdenum isotopes were made by Rapaport *et al* (1979a, b), from ^{26}Mg by Tailor *et al* (1983) and from ^{28}Si by Böttcher *et al* (1983).

4.1.1. Isospin potential for deformed nuclei. If the methods of determining the isospin potential are applied to deformed nuclei, the coupling between the elastic and inelastic channels may so affect the elastic scattering, and hence the asymmetry dependence of the optical potential, that quite anomalous results are obtained for the isospin potential. If, however, the coupling is included in the analysis by solving the appropriate coupled-channels equations, then values of the isospin potential are obtained that are consistent with those found for spherical nuclei. This has been shown by Cereda *et al* (1982) in a series of studies of proton scattering from molybdenum, palladium and cadmium isotopes, and some of their results are shown in figure 10.

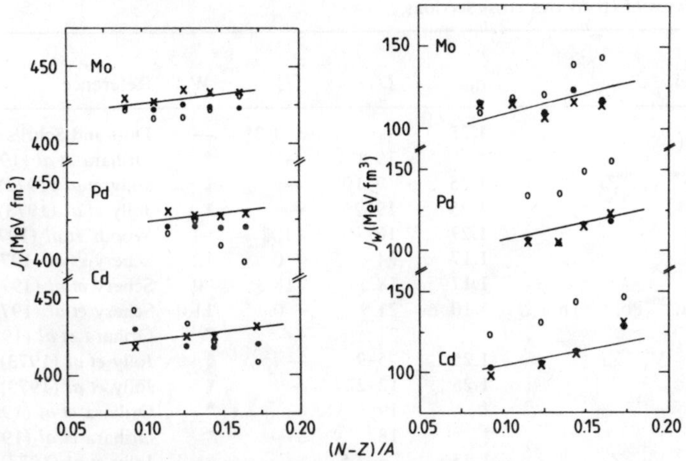

Figure 10. Asymmetry dependence of the volume integral per nucleon of the central real and imaginary parts of the optical potential used in optical model calculations (○), coupled-channels calculations with first-order vibrational model and coupling scheme $0_0^+–2_1^+–3_1^-$ (×) and second-order vibrational model with the coupling scheme $0_0^+–2_1^+–0_2^+–2_2^+–4_2^+$ (●). The full lines give the asymmetry dependence of the Becchetti–Greenlees proton potentials (Cereda *et al* 1982).

4.2. (p, n) isobaric analogue state reactions

The (p, n) reaction to the isobaric analogue state of the target depends directly on the isospin term in the optical potential. The reaction amplitude may be determined by solving the coupled Lane equations, or more simply from the first-order perturbation theory expression:

$$T_{\rm pn}=\int \chi_{\rm n}^{(-)*}(k_{\rm n},r)\frac{1}{2A}(2T_0)^{1/2}V_I(r)\chi_{\rm p}^{(+)}(k_{\rm p},r)\,{\rm d}\boldsymbol{r} \qquad (4.1)$$

where χ_n and χ_p are the neutron and proton distorted waves, T_0 is the isospin of the target nucleus and $V_I(r)$ is the isospin potential (Hodgson 1971, chap 19).

Although it is very direct, this method has the disadvantage that the cross section depends on both the real and imaginary parts of the isospin potential (Satchler *et al* 1964). In principle it might be thought possible to vary these to optimise the fit to the experimental cross section, but this is found to be impracticable (Schery *et al* 1974a, b). There are also significant effects due to Coulomb mixing and core polarisation (Lovas *et al* 1981).

Another complication of these analyses is the possible contribution of two-step (p, d, n) reactions (Madsen *et al* 1972). The contribution from such processes can be calculated explicitly and included in the analysis. Rickertsen and Kunz (1973) found that this greatly improves the consistency of the values obtained for the isospin term in the effective nucleon–nucleon interaction. Thus, analyses using the one-step process alone gave values from 15–33 MeV, whereas including the two-step processes gave values clustering around 29 MeV.

Several analyses of (p, n) IAS cross sections have been made using purely real or complex potentials and the results are given in table 15.

Coupled-channels analyses of the (p, n) reactions to the ground and first excited analogue states have been made by Orihara *et al* (1981) and they also find that the

Table 15. Analyses of (p, n) IAS cross sections.

Energy (MeV)	Nuclei	r_0	U_I^v	U_I^s	W_I	Reference
5	^{88}Sr	1.25	—	1.3†	—	Dutt and Schrils (1968)
20	^{56}Fe	‡	19.8	—	*	Orihara *et al* (1981)
22	^{27}Al, ^{51}V, ^{90}Zr	1.25	7–10	—	§	Jolly *et al* (1973)
		1.25	19–26	—	‡	Jolly *et al* (1973)
25–39	^{208}Pb	1.27	10	110‖	—	Woods *et al* (1972)
25.8	^{208}Pb	1.17	24	0	12	Schery *et al* (1974a)
25.8	^{208}Pb	1.17	18.5	18.3	0	Schery *et al* (1974a)
25.8	^{197}Au, ^{208}Pb, ^{232}Th	1.10	21.9	0	11.0	Schery *et al* (1974a)
28	^{56}Fe	‡	21	—	*	Orihara *et al* (1981)
30	^{27}Al, ^{51}V, ^{90}Zr	1.25	5–9	—	§	Jolly *et al* (1973)
		1.25	12–22	—	¶	Jolly *et al* (1973)
32	^{56}Fe	‡	19	—	*	Orihara *et al* (1981)
35	^{56}Fe	‡	18	—	*	Orihara *et al* (1981)
40	^{27}Al, ^{51}V, ^{90}Zr	1.25	6–7.5	—	§	Jolly *et al* (1973)
40	^{27}Al, ^{51}V, ^{90}Zr	1.25	14–18	—	¶	Jolly *et al* (1973)

* $W_I = 2U_I$.
† Surface peaked form factor (see note added in proof to Dutt *et al* (1969)).
‡ $r_0 = 2.69 - 0.40\,A^{1/3}$.
§ $W_1 = \frac{1}{2}V_1$, $r_I = 1.35$.
‖ Gaussian form factor.
¶ $W_1 = 1.5\,V_1$, $r_I = 1.29$.

strength of the two-step reactions diminishes with increasing energy (Madsen *et al* 1972). This work gives not only the isovector potential but also the isovector deformation parameter.

An imaginary isospin potential has been included in the analysis in several different ways. The simplest is to use a phenomenological term similar to isoscalar imaginary potential that is usually taken to have the volume form.

Schery *et al* (1978) have used the forward-scattering amplitude approximation (Sinha 1975) to calculate the imaginary isospin potential:

$$W_I(r) = -\frac{A}{4(N-Z)}\hbar v_{\rm p}(r)[\rho_{\rm n}(r) - \rho_{\rm p}(r)][\bar{\sigma}_{\rm np}(r) - \bar{\sigma}_{\rm pp}(r)] \qquad (4.2)$$

where $v_{\rm p}(r)$ is the velocity of the incident proton, $\rho_{\rm n}$ and $\rho_{\rm p}$ are the point neutron and proton matter densities and $\bar{\sigma}_{\rm np}$ and $\bar{\sigma}_{\rm pp}$ are nucleon–nucleon total cross sections averaged in the local Fermi gas model approximation to exclude scattering to states forbidden by the Pauli exclusion principle. Some additional calculations were made, taking account of the finite range of the nucleon–nucleon interaction by replacing $\rho_{\rm n}$ and $\rho_{\rm p}$ with effective densities with RMS ranges increased by the RMS range of the finite-range interaction. The real isospin potential was also calculated from the matter distributions using

$$U_I(r) = -\frac{A}{N-Z}\int v_\tau(|\boldsymbol{r}' - \boldsymbol{r}|)[\rho_{\rm n}(r') - \rho_{\rm p}(r')]\,{\rm d}\boldsymbol{r}' \qquad (4.3)$$

where $v_\tau(|\boldsymbol{r}' - \boldsymbol{r}|)$ is the isospin component of the effective nucleon–nucleon interaction. This expression has been found by Schery *et al* (1974b, 1976) to fit the data quite well.

A series of calculations was made with (*a*) real potential only, (*b*) complex potential without and (*c*) with finite range correction, and (*d*) with a phenomenological imaginary potential, and compared with (p, n) data at 35.3 MeV for ^{112,116,124}Sn at 45 MeV for ^{48}Ca, at 32 MeV for ^{58}Ni and at 25.8, 35 and 45 MeV for ^{208}Pb. On the whole, the best overall fits were found with the potential (*c*).

Carlson *et al* (1975) fixed the shape of $U_I(r)$ and the ratio U/W to the values found by Becchetti and Greenlees (1969) and then adjusted U_I to optimise the fit to the (p, n) data. This gave $U_I = 7.7 + 2.34A^{1/3}$, $r_I = 2.69 - 6.4A^{1/3}$ and $a_I = -0.71 + 0.46A^{1/3}$.

Several measurements have now been made of the analysing power in (p, n) reactions, and these make it possible to investigate the presence of a spin–orbit term in the isospin potential, corresponding to the $(\boldsymbol{L}\cdot\boldsymbol{\sigma})(\boldsymbol{T}\cdot\boldsymbol{t})$ term in the nucleon–nucleus interaction (Moss *et al* 1972, Gosset *et al* 1976). Although there are theoretical reasons for expecting such a term to be present (Satchler 1969) phenomenological analyses of (p, n) analysing powers do not yet require the inclusion of such a term.

4.2.1. The consistency of the Lane model. The introduction of the isospin potential unifies neutron and proton elastic scattering and (p, n) IAS reactions, so that it is possible to calculate the cross sections of all three interactions from the same nucleon potential. This unification can be tested and utilised in several ways.

A detailed test of the consistency of the Lane model was made by Byrd *et al* (1979, 1981, 1983), who searched for a simultaneous fit to the data on (p, p), (n, n) and (p, n) reactions with ^{9}Be, together with the analysing powers A_{pp} and A_{pn}, using a parametrisation of the isospin potential that included a volume real term, a surface imaginary term and a spin–orbit term. They obtained a parameter set that fits the data for neutron and proton energies from 10–18 MeV. While the parameters of the isoscalar potential had their usual energy variation, those of the isospin term were practically independent of energy. The optimum isospin spin–orbit strength was 1 MeV, but this term could be set to zero without appreciably affecting the fit; there is thus no evidence for a $(\boldsymbol{L}\cdot\boldsymbol{\sigma})(\boldsymbol{T}\cdot\boldsymbol{t})$ term in the optical potential. This is in accord with the results of previous analyses (Moss *et al* 1972, Gossett *et al* 1976). Calculations were also made with a surface-peaked real isospin term, but these did not fit the (p, n) analysing powers.

It should be pointed out that the success of this simultaneous analysis of (p, p), (n, n) and (p, n) cross sections does not necessarily imply that the isospin potential determined from analysis of the data for one or two of these cross sections will predict the remaining cross section. It is indeed quite likely that the parameters found in this work absorb the smaller effects described elsewhere that prevent such predictions being made, at least to the accuracy found in the overall analysis.

Another test of the consistency of the Lane model was made for ^{208}Pb by De Vito *et al* (1983). They measured the elastic scattering of neutrons and protons and used the optical potentials obtained from the data to calculate the (p, n) IAS cross section. This underestimated the (p, n) cross section, which they attribute to an incorrect Coulomb correction, or possibly to two-step reactions.

If the Lane model formalism provides a unified account of (n, n), (p, p) and (p, n) IAS interactions, then it should be possible to calculate the cross section of any one of these interactions from potentials obtained by analyses of the other two. In particular, it is interesting to see whether accurate (n, n) cross sections can be obtained from analysis of (p, p) and (p, n) IAS cross sections. The usual method of doing this is to obtain the optical model potential by analysis of (p, p) and then the isospin term from the (p, n) IAS data, both at the same energy. To avoid the ambiguity between the real and imaginary parts of the isospin potential it is assumed that $W_I = \frac{1}{2}U_I$, and that both have the same form factors. Use of the expressions (2.10) and (2.11) then gives the neutron potential at energy $E_n = E_p - \Delta_c$, and hence the (n, n) cross sections for comparison with the experimental data.

Several calculations have been made in this way, with generally satisfactory results (Carlson *et al* 1975, Schery *et al* 1974a, b, Patterson *et al* 1976, Fitzgerald *et al* 1977, 1978, Rapaport *et al* 1979a, b). These analyses all used the distorted wave theory to calculate the (p, n) cross sections. A more consistent analysis using the coupled Lane equations was made by Hansen (1982) and Hansen *et al* (1983) for some heavy nuclei and the calculated neutron cross sections fitted the data better than the corresponding distorted wave calculations, and almost as well as global neutron potentials.

5. Applications of the neutron optical potential

The neutron optical potential has many applications, both practical and theoretical. It is an essential tool for the economical description of the elastic and inelastic scattering of neutrons by nuclei that is needed for the design of nuclear reactors and their shielding. Low-energy neutrons make possible the chain reaction that is the basis of the fission reactor and the neutron optical potential describes how they interact with uranium and other actinide nuclei. The energy produced by the fusion reactors of the future will appear as fast neutrons, so a detailed understanding of their interactions with the surrounding material is essential for their design.

The nucleon–nucleus interaction occupies a central position in nuclear physics. It may be understood theoretically in terms of the nucleon–nucleon interaction and the structure of nuclei, and these in turn reflect the properties of the quarks and other sub-nuclear particles. The same folding model that can be used to obtain the nucleon–nucleus interactions from the nucleon–nucleon interaction can be used to calculate the potentials between deuterons, tritions, helions and alpha-particles and nuclei. These potentials are essential for the analysis of the interactions of such particles with nuclei and enable many of the properties of the nuclei to be obtained from the experimental data. In particular, the analysis of inelastic scattering yields important information on the collective nuclear excitations and the single- and multi-nucleon transfer reactions on the single-particle aspects of nuclear structure.

The same methods, with appropriate refinements, can be used to calculate the heavy-ion potentials that describe the high-energy interactions of one nucleus with another. This has, in recent years, opened up a whole new field of nuclear physics and has increased our knowledge of the properties of nuclei at high excitations.

Acknowledgments

Most of this review was written during a visit to the Physics Department of the American University of Beirut in April 1982, and I am grateful to Professor P Mourad and Professor H Mavromatis for their kind hospitality and for providing peaceful and pleasant working conditions. I am also grateful to Miss G M Field for help in the preparation of the review.

References

Ahmad I and Haider W 1976 *J. Phys. G: Nucl. Phys.* **2** L157
Austin S M 1983 *Michigan Preprint* 363
Austin S M, DeVito R P, Sterrenburg W and Berg U E P *Michigan Preprint*
Ball J B, Fulmer C B and Bassel R H 1964 *Phys. Rev.* **135** B706
Barrett R, Hill A D and Hodgson P E 1964 *Nucl. Phys.* **62** 133
Bauer M, Hernandez-Saldãna E, Hodgson P E and Quintanilla J 1982 *J. Phys. G: Nucl. Phys.* **8** 525
Baugh D J, Griffith J A R and Roman S 1965 *Proc. 2nd Int. Symp. on Polarisation Phenomena of Nucleons, Karlsruhe* (Basel: Birkhauser-Verlag) p 339
—— 1966 *Nucl. Phys.* **83** 481
Bear K and Hodgson P E 1978 *J. Phys. G: Nucl. Phys.* **4** L287
Becchetti F D and Greenlees G W 1969 *Phys. Rev.* **182** 1190
Begum A, Galloway R B and McNeil-Watson F K 1979 *Nucl. Phys.* A **333** 349
Benveniste J, Mitchell A C, Buck B and Fulmer C B 1964 *Phys. Rev.* **113** B323
Bernard V and Van Giai N 1979 *Nucl. Phys.* A **327** 397
Bertsch G F and Kuo T T S 1968 *Nucl. Phys.* A **112** 204
Bethe H A 1935 *Phys. Rev.* **47** 747
Böttcher J, Blank H, Finckh E, Forstner C, Jaumann W, Schall G, Scheuring H, Schneidereit U, Stauber K, Wiepert A, Tornow W and Woye E 1983 *J. Phys. G: Nucl. Phys.* **9** L65
Bouyssy A, Ngô H and Vinh Mau N 1981 *Nucl. Phys.* A **371** 173
Brandenberger J D and Schrils R 1976 *Phys. Rev.* C **13** 2559
Brieva F A and Rook J R 1977 *Nucl. Phys.* A **291** 299, 317
—— 1978a *Nucl. Phys.* A **297** 206
—— 1978b *Nucl. Phys.* A **307** 493
Brown G E, Gunn J H and Gould P 1963 *Nucl. Phys.* **46** 598
Buck B 1963 *Phys. Rev.* **130** 712
Buck B and Perey F G 1962 *Phys. Rev. Lett.* **8** 444
Buck B, Stamp A P and Hodgson P E 1963 *Phil. Mag.* **8** 1805
Burge E J, Calderbank M, Lewis V E, Rush A A and Thomas G L 1968 *Proc. Int. Conf. on Nuclear Structure, Tokyo, 1967* ed J Sanada *J. Phys. Soc. Japan Suppl.* **24** 684
Byrd R C, Floyd C E, Guss P P, Murphy K K and Walter R L 1983 *Nucl. Phys.* A **399** 94
Byrd R C, Walter R L and Contanch S R 1979 *Phys. Rev. Lett.* **43** 260
—— 1981 *Proc. 5th Int. Symp. on Polarisation Phenomena in Nuclear Physics, Santa Fe, 1980* (New York: AIP) p 1481
Calderbank M, Burge E J, Lewis V E, Smith D A and Ganguly N K 1967 *Nucl. Phys.* A **105** 601
Carlson J D, Zafiratos C D and Lind D A 1975 *Nucl. Phys.* A **249** 29
Cereda E, Pignanelli M, Micheletti S, Von Geramb H V, Harakeh M N, DeLeo R, D'Erasmo G and Pantaleo A 1982 *Phys. Rev.* C **26** 1941
Coope D F, Schell M C, Tripathi S N and McEllistrem M T 1976 *Phys. Rev. Lett.* **37** 1126
Coope D F, Tripathi S N, Schell M C, Weil J L and McEllistrem M T 1977 *Phys. Rev.* C **16** 2223
Cooper S G and Hodgson P E 1980 *J. Phys. G: Nucl. Phys.* **6** L21
Dabrowski J and Haensel P 1974 *Can. J. Phys.* **52** 1768
Delaroche J P, El Kadi S M, Guss P P, Floyd C E and Walter R L 1982 *Nucl. Phys.* A **390** 541
Delaroche J P, Haouat G, Lachkar J, Patin Y, Sigaud J and Chardine J 1981 *Phys. Rev.* C **23** 136
De Leo R and Micheletti S 1981 *Phys. Lett.* **98B** 233
De Leo R and Sterrenberg W A 1980 *KVI Rep.* 8
De Vito R P, Austin S M, Berg U E P, De Leo R and Sterrenberg W A 1983a *Phys. Rev.* C **28** 2530
—— 1983b *Michigan Preprint* MSUCL-424
De Vito R P, Austin S M, Sterrenberg W and Berg U E P 1981 *Phys. Rev. Lett.* **47** 628
Dickens J K, Eichler E and Satchler G R 1968 *Phys. Rev.* **168** 1355
Dietrich F S, Finlay R W, Mellema S, Randers-Pehrson G and Petrovich F 1983 *Phys. Rev. Lett.* **51** 1629
Dukarevich Y U V, Dyumin A N and Kaminhev D M 1967 *Nucl. Phys.* A **92** 433
Durisch J E and Gould P 1965 *Phys. Rev.* **137** B906
Dutt G C, Gabbard F, Schrils R and Egan J J 1969 *Phys. Rev.* **180** 1222
Dutt G C and Schrils R 1968 *Phys. Rev.* **175** 1413
Eder G, Leeb H and Oberhummer H 1977 *J. Phys. G: Nucl. Phys.* **3** L127

El Kadi S M, Nelson C E, Purser F O, Walter R L, Beyerle A, Gould C R and Seagondollar L W 1982 *Nucl. Phys.* A **390** 509
Engelbrecht C A and Fiedeldey H 1967 *Ann. Phys., NY* **42** 262
Erramuspe H J 1967 *Nucl. Phys.* A **105** 569
Fantoni S, Friman B L and Pandharipande V R 1982 *Nucl. Phys.* A **386** 1
Fasoli U, Galeazzi G, Pavan P, Toniolo D, Zago G and Zannoni R 1978 *Nucl. Phys.* A **311** 368
Ferrer J C, Carlson J D and Rapaport J 1976 *Phys. Lett.* **62B** 399
—— 1977 *Nucl. Phys.* A **275** 325
Feshbach H, Porter C E and Weisskopf V F 1954 *Phys. Rev.* **96** 448
Finlay R W, Rapaport J, Hadizadeh M H, Mirzaa M and Bainum D E 1980 *Nucl. Phys.* A **338** 45
Fitzgerald D H, Lilley J S, Poppe C H and Grimes S M 1978 *Phys. Rev.* C **18** 1207
Fitzgerald D H, Greenlees G W, Lilley J S, Poppe C H, Grimes S M and Wong C 1977 *Phys. Rev.* C **16** 2181
Floyd C E, Guss P P, Byrd R C, Murphy K, Walter R L and Delaroche J P 1983 *Phys. Rev.* C **28** 1498
Floyd C E, Guss P P, Murphy K, Byrd R C, Tungate G, Wender S A, Walter R L and Clegg T B 1981 *Phys. Rev. Lett.* **47** 1042
Floyd C E, Guss P P, Murphy K, Howell C R, Byrd R C, Tungate G, Wender S A, Walter R L and Clegg T B 1982 *Phys. Rev.* C **25** 1682
Fricke M P, Gross E E, Morton B J and Zucker A 1967 *Phys. Rev.* **156** 1207
Fulmer C B, Ball J B, Scott A and Whiten M L 1967 *Phys. Lett.* **24B** 505
Galloway R B and Waheed A 1979 *Nucl. Phys.* A **318** 173
Glasgow D W and Foster D G 1971 *Phys. Rev.* C **3** 604
Gossett J, Mayer B and Escudié J L 1976 *Phys. Rev.* C **14** 878
Grabmayr P, Rapaport J and Finlay R W 1980 *Nucl. Phys.* A **350** 167
Greenlees G W and Pyle G J 1966 *Phys. Rev.* **149** 836
Greenlees G W, Pyle G J and Tang Y C 1968 *Phys. Rev.* **171** 1115
Gross E E, Malanify J J, Menet J J H and Zucker A 1971 *Bull. Am. Phys. Soc.* **16** 99
Gymarti B, Lovas R G, Vertse T and Hodgson P E 1981 *J. Phys. G: Nucl. Phys.* **7** L209
Haider W, Kobos A M, Yamaguchi N and Rook J R 1983 *Oxford Preprint*
Hansen L F 1982 *NEANDC* 158 'U' (Paris) p 116
Hansen L F, Grimes S M, Poppe C H and Wong C 1983 *Phys. Rev.* C **28** 1934
Haouat G, Lachkar J, Lagrange Ch, Jary J, Sigaud J and Patin Y 1982 *Nucl. Sci. Engng* **81** 491
Haouat G, Patin Y, Lachkar J, Sigaud J, Jary J and Lagrange Ch 1981 *NEANDC* (E)21 'L', INDC (FR) 41/L33
Hauser W and Feshbach H 1952 *Phys. Rev.* **87** 366
Heeringa W and Postma H 1976 *Phys. Lett.* **61B** 350
—— 1983 *Phys. Rev.* C **27** 2012
Hodgson P E 1960 *Nucl. Phys.* **21** 21
—— 1961 *Phys. Rev. Lett.* **6** 358
—— 1963a *Phys. Lett.* **6** 75
—— 1963b *The Optical Model of Elastic Scattering* (Oxford: Clarendon)
—— 1970 *Nucl. Phys.* A **150** 1
—— 1971 *Nuclear Reactions and Nuclear Structure* (Oxford: Clarendon)
—— 1983a *Notas de Fisica* **6** 218
—— 1983b *Proc. Int. Symp. on Nuclear Physics at Large Tandem Accelerators, Legnaro, Padova*
—— 1983c *Contemp. Phys.* **24** 491
Hogan W S and Seyler R G 1969 *Phys. Rev.* **177** 1706
Holmqvist B and Wiedling T 1968 *Phys. Lett.* **26B** 620
—— 1972 *Nucl. Phys.* A **188** 24
Jeukenne J P, LeJeune A and Mahaux C 1974 *Phys. Rev.* C **10** 1391
—— 1976a *Phys. Rep.* **25C** 83
—— 1976b *Proc. Int. Conf. on Interactions of Neutrons by Nuclei* (Oak Ridge, Tenn.: US ERDA Technical Information Center) p 451
—— 1977a *Phys. Rev.* C **15** 10
—— 1977b *Phys. Rev.* C **16** 80
Jolly R K, Amos T M, Galonsky A, Hinrich S R and St Onge R 1973 *Phys. Rev.* C **7** 1903
Kailas S and Gupta S K 1977 *Phys. Lett.* **71B** 271
—— 1978 *Phys. Rev.* C **17** 2236
Kailas S, Mehta M K, Gupta S K, Viyogi Y P and Ganguly N K 1979 *Phys. Rev.* C **20** 1272

Kobos A M and Mackintosh R S 1979a *Ann. Phys., NY* **123** 296
—— 1979b *J. Phys. G: Nucl. Phys.* **5** 97
Kossanyi-Demay P and de Swiniarski R 1968 *Nucl. Phys.* A **108** 577
Kossanyi-Demay P, de Swiniarski R and Glashausser C 1967 *Nucl. Phys.* A **94** 513
Kulkarni V and Rapaport J 1980 *Can. J. Phys.* **58** 146
Kwiatkowski K and Wall N S 1978 *Nucl. Phys.* A **301** 349
Lagrange Ch 1974 *J. Physique Lett.* **35** 111
Lagrange Ch, Shamu R E, Burrows T, Glasgow G P, Hardie G and McDaniel F D 1975 *Phys. Lett.* **58B** 293
Lane A M 1957 *Rev. Mod. Phys.* **29** 191
—— 1962a *Phys. Rev. Lett.* **8** 171
—— 1962b *Nucl. Phys.* **35** 676
Le Levier R E and Saxon D S 1952 *Phys. Rev.* **87** 40
Lovas R G, Brown B A and Hodgson P E 1981 *Nucl. Phys.* A **357** 205
McEllistrem M T, Shamu R E, Lachkar J, Haouat G, Lagrange Ch, Patin Y, Sigaud J and Cocu F 1977 *Phys. Rev.* C **15** 927
Madland D G and Young P G 1978 *Proc. Int. Conf. on Neutron Physics and Nuclear Data* (OECD Nuclear Energy Agency) p 349
Madsen V A, Stamp M J, Brown V R, Anderson J D, Hansen L, Wong C and Wesolowski J J 1972 *Phys. Rev. Lett.* **28** 629
Mahaux C 1979 *Microscopic Optical Potentials. Lecture Notes in Physics No 89* ed H V von Geramb (Berlin: Springer-Verlag)
Mahaux C and Ngô H 1981 *Phys. Lett.* **100B** 285
Makofske W, Savin W W, Ogata H and Kruse T H 1968 *Phys. Rev.* **174** 1429
Makofske W, Slagowitz M, Savin W, Ogata H, Kruse T H and Tamura T 1967 *Phys. Lett.* **25B** 322
Malaguti F 1978 *Nucl. Phys.* A **308** 125
Malan J G, Pineo W F E, Divadeenam M, Choi B H, Bilpuch E G and Newson H W 1975 *Ann. Phys., NY* **89** 284
Marinov A, Lee L L and Schiffer J D 1966 *Phys. Rev.* **145** 852
Menet J J H, Gross E E, Malanify J J and Zucker A A 1971 *Phys. Rev.* C **4** 1114
Moldauer P A 1962 *Phys. Rev. Lett.* **9** 17
—— 1963 *Nucl. Phys.* **47** 65
—— 1964 *Phys. Rev.* B **135** 642
Moss J M, Brassard C, Vyse R and Gosset J 1972 *Phys. Rev.* C **6** 1698
Ngô H 1982 *Nucl. Phys.* A **390** 198
Noro T, Sakaguchi H, Nakamura M, Hatamaka K, Ohtani F, Sakamoto H and Kobayashi S 1981 *Nucl. Phys.* A **366** 189
Orihara H, Murakami T, Nishihara S, Nakagawa T, Maeda K, Miura K and Ohnuma H 1981 *Phys. Lett.* **106B** 171
Osterfeld F and Madsen V A 1981 *Phys. Rev.* C **24** 2468
Patterson D M, Doering R R and Galonsky A 1976 *Nucl. Phys.* A **263** 261
Perey C M and Perey F G 1968 *Phys. Lett.* **26B** 123
—— 1974 *Atom. Data Nucl. Data Tables* **13** 293
—— 1976 *Atom. Data Nucl. Data Tables* **17** 1
Perey F G 1963 *Phys. Rev.* **131** 745
Perey F G and Buck B 1962 *Nucl. Phys.* **32** 353
Picard J 1964 *Thesis* Paris
—— 1965 *Nucl. Phys.* **68** 153
Pignanelli M, von Geramb H V and De Leo R 1981 *Phys. Rev.* C **24** 369
Pineo W F E, Divadeenam M, Bilpuch E G, Seth K K and Newson H W 1974 *Ann. Phys., NY* **84** 165
Ramavataram S, Bergeron J, Divadeenam M and Newson H W 1976 *Ann. Phys., NY* **97** 245
Ramström E and Göransson P A 1977 *Nucl. Phys.* A **284** 461
Rapaport J 1977 *Phys. Lett.* **70B** 141
—— 1980 *Phys. Lett.* **92B** 233
—— 1982 *Phys. Rep.* **87** 27
Rapaport J, Carlson J D, Bainum D, Cheema T S and Finlay R W 1977 *Nucl. Phys.* A **286** 232
Rapaport J, Cheema T S, Bainum D, Finlay R W and Carlson J D 1978 *Nucl. Phys.* A **296** 95
—— 1979a *Nucl. Phys.* A **313** 1
Rapaport J, Kulkarni V and Finlay R W 1979b *Nucl. Phys.* A **330** 15

Rapaport J, Mirzaa M, Hadizadeh H, Bainum D E and Finlay R W 1980 *Nucl. Phys.* A **341** 56
Rickertsen L D and Kunz P D 1973 *Phys. Lett.* **47B** 11
Rook J R 1973 *Nucl. Phys.* A **222** 596
Rosen L, Beery J G, Goldhaber A S and Auerbach E H 1965 *Ann. Phys., NY* **34** 96
Satchler G R 1967 *Nucl. Phys.* A **91** 75, A **92** 273
—— 1969 *Isospin in Nuclear Physics* (Amsterdam: North-Holland) p 389
Satchler G R, Drisko R M and Bassel R H 1964 *Phys. Rev.* **136** B637
Schery S F, Austin S M, Galonsky A, Young L E and Berg U E P 1978 *Phys. Lett.* **79B** 30
Schery S F, Lind D A, Fielding H W and Zafiratos C D 1974a *Nucl. Phys.* A **234** 109
Schery S F, Lind D A and Wieman H 1976 *Phys. Rev.* C **14** 1800
Schery S F, Lind D A and Zafiratos C D 1974b *Phys. Rev.* C **9** 416
Schwandt P 1982 *Invited Lecture at ICUF Workshop, Bloomington*
Shamu R E, Bernstein E M, Blondin D, Ramirez J J and Rochau G 1973 *Phys. Lett.* **45B** 241
Shamu R E, Bernstein E M, Ramirez J J and Lagrange Ch 1980 *Phys. Rev.* C **22** 1857
Shamu R E, Lagrange Ch, Bernstein E M, Ramirez J J, Tamura T and Wong C Y 1976 *Phys. Lett.* **61B** 29
Siemens P J 1970 *Nucl. Phys.* A **141** 225
Sinha B 1975 *Phys. Rep.* **20** 1
Sinha B C, Edwards V R W, Burge E J and Tait W H 1972 *Nucl. Phys.* A **183** 401
Smith A B, Guenther P, Moldauer P and Whalen J 1978 *Nucl. Phys.* A **307** 224
Sood P C 1965 *Phys. Lett.* **19** 52
Stautberg M M and Kraushaar J J 1966 *Phys. Rev.* **151** 969
Tabony R H and Seth K K 1968 *Ann. Phys., NY* **46** 401
Tailor R C, Rapaport J, Finlay R W and Randers-Pehrson G 1983 *Nucl. Phys.* A **401** 237
Tamura T 1965 *Rev. Mod. Phys.* **37** 679
—— 1969 *Ann. Rev. Nucl. Sci.* **19** 99
Thomas G L and Burge E J 1969 *Nucl. Phys.* A **128** 545
Thomas R N, Burge E J and Hodgson P E 1964 *Nucl. Phys.* **52** 93
Tornow W *et al* 1981 *Proc. 5th Int. Symp. on Polarisation Phenomena in Nuclear Physics, Santa Fe 1980* (New York: AIP) p 404
—— 1982 *Nucl. Phys.* A **385** 373
Velkley D E, Glasgow D W, Brandenberger J D, McEllistrem M T, Manthuruthil J C and Poirier C P 1974 *Phys. Rev.* C **9** 2181
Vinh Mau N 1979 *Lecture Notes in Physics No 89* ed H V v Geramb (Berlin: Springer-Verlag) p 40
Wilmore D and Hodgson P E 1964 *Nucl. Phys.* **55** 673
Woods T J, Whitten C A and Igo G J 1972 *Nucl. Phys.* A **198** 542
Yamaguchi N, Nagata S and Matsuda T 1983 *Prog. Theor. Phys.* **70** 459
Zanelli C I, Urone P P, Romero J L, Brady F P, Johnson M L, Needham G A, Ullmann J L and Johnson D L 1981 *Phys. Rev.* C **23** 1015

MICROSCOPIC OPTICAL POTENTIALS
Summary of Hamburg Conference*

P. E. Hodgson
Nuclear Physics Laboratory, Oxford

1. Introduction

This meeting has been a timely one, marked by significant advances in the accuracy of the experimental data, in the phenomenological analyses and in the methods of calculating the microscopic potentials.

This summary begins with some general remarks about what we mean by an optical potential, how it is defined and the ways it can be determined, either microscopically from the nucleon-nucleon interactions, or by phenomenological analysis of the experimental data, or in some intermediate way.

The next Section 3 is devoted to the microscopic potentials themselves, both global for a range of nuclei and particular when the properties of particular nuclei are introduced into the calculation. Section 4 considers the ways of adding extra theoretical or phenomenological effective potentials to the standard optical potential to account for a recognised physical effect. Sections 5, 6 and 7 are devoted to new experimental data and phenomenological analyses of nucleon, light and heavy ion interactions, and Section 8 to some concluding remarks.

2. Definition of the Optical Potential

By an optical potential we mean a potential that represents the interaction between a nucleon or group of nucleons and a nucleus. When inserted into the Schrödinger equation it gives the differential cross-section and polarisation for elastic scattering, the reaction cross-section and some other less important observable quantities.

These optical potentials can be obtained in several ways. Most fundamentally, they can be calculated from the nucleon-nucleon interaction; this is difficult and has only recently been brought to the stage of quantitative success. Alternatively they may be found phenomenologically by postulating a form of potential and adjusting its parameters to optimize the fit to the experimental data. Such phenomenological analyses may be put on a firmer physical basis by using additional information from nuclear models.

Optical potentials may also be classified by the range of experimental data they are designed to fit. Global potentials give good

*Published in *Microscopic Optical Potentials*, Ed. H.V. von Geramb, Lecture Notes in Physics, No. 89 (Springer Verlag, 1979).

overall fits to the scattering from many nuclei over a range of energies. More precise fits can be obtained for particular interactions either by incorporating nuclear structure and nuclear reaction information into the calculation of the potential or by adjusting the parameters of a phenomenological potential to fit particular sets of data.

There are many ambiguities in optical model potentials. These are familiar in phenomenological analyses, where it is often found that several potentials fit the same data equally well. It is usually thought that one of these is the 'physical' potential, namely the one that is given by a microscopic calculation. It is important to identify the physical potential, for this can be used with more confidence in situations different from those from which it was obtained.

It should however be noticed that the 'physical' potential may have a form that is significantly different from any of the phenomenological potentials. Furthermore, a potential is of its nature a theoretical construct, so care is necessary in describing it as 'physical'. It is possible that even among microscopic potentials there are ambiguities in the sense that different types of calculations could conceivably give different potentials that nevertheless give equally good fits to the data.

This possibility follows most clearly from the different ways of expressing the potentials. We know from general considerations that the 'physical' is non-local, and this non-locality can be expressed in different ways. How can we compare these potentials with each other? It is of course possible to define local potentials that are in some specified sense equivalent to the corresponding non-local potential. But this at once opens up further possibilities of ambiguity. The local potential necessarily carries less information than the non-local potential and potentials equivalent in one sense may not be equivalent in another, and then which do we consider the more 'physical'?

These considerations concerning the ambiguities in the optical potentials suggest that it can be misleading to compare a theoretical microscopic potential with a phenomenological potential, since they may differ significantly from each other and yet both be consistent with experiment. It is therefore preferable to test theoretical microscopic potentials by comparing them directly with the experimental data.

The relations between some of the topics discussed at this meeting are summarised in the figure.

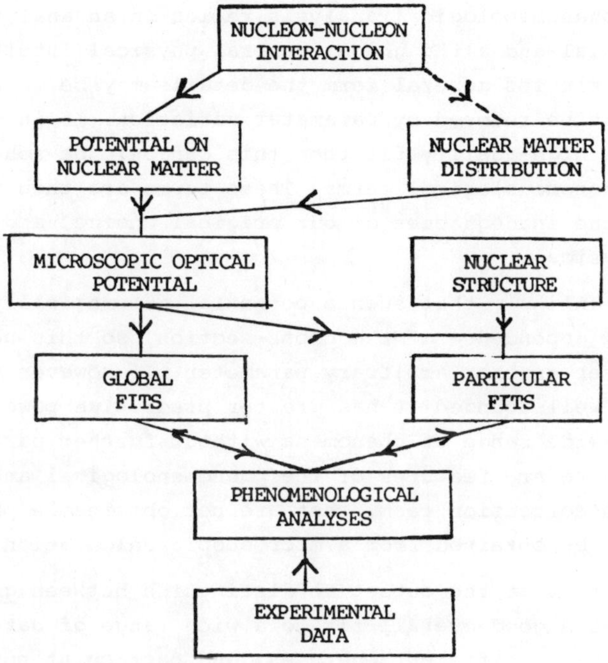

3. Microscopic Optical Potentials

By a microscopic optical potential we mean a potential calculated from the nucleon-nucleon interaction and some nuclear properties, usually the density distribution. Such potentials are distinguished from phenomenological potentials which are obtained by direct fitting to experimental elastic scattering data. These phenomenological potentials usually fit the data much better than microscopic potentials, so what is the use of the microscopic potentials?

We are interested in microscopic potentials primarily because we want to unify our understanding of nuclear physics. This understanding develops in several stages: for example if we find a bump in a cross-section we can fit it with some analytical curve. Then perhaps we can show that the bump is a true resonance, and analysing it with the Breit-Wigner theory can give its energy, width and quantum numbers. Later on we might be able to show that these properties follow automatically from a potential that also correlates other nuclear properties. And finally we may succeed in calculating that potential from the nucleon-nucleon interaction.

There is another reason for calculating microscopic optical potentials. Every phenomenological analysis relies on an analytical form for the potential and although our general physical intuition is sufficient to fix its general form the details may be wrong, and in a way that cannot be removed by parameter variation. If in such a case we insist on a high-quality fit then this can only be achieved by adding more phenomenological terms. These terms are then present solely to remove the inadequacies of our original choice, and correspond to nothing physical.

It is very unlikely that such a potential automatically gives the correct energy dependence of the cross-section, so this has to be rectified by yet another arbitrary parameter. If however the potential is physically well-founded it has greater predictive power and can account for a wide range of phenomena without further parametrisation. Furthermore there are features of the phenomenological analysis, such as the Coulomb correction term, that are not obtainable phenomenologically but must be obtained from a microscopic calculation.

It is important at the outset to distinguish between global potentials that give a good overall fit to a wide range of data for different nuclei at different energies, and particular potentials that give a precise fit to a narrow range of data. Both types of potentials have their value: the global potentials are useful for calculations that are relatively insensitive to the details of the potentials used, while the particular potentials are essential for detailed studies of specific reactions.

The calculation of microscopic optical potentials is fraught with many difficulties, and it is only in the last few years that for nucleons they have been developed to a degree of precision that approaches and in some cases even exceeds the phenomenological analyses.

These new methods have not yet been extended to light and heavy ions and so this review concentrates on nucleon potentials, with a briefer account of the phenomenological and semi-phenomenological analyses of the scattering of heavier particles that were also presented at the meeting.

Global Potentials

At this meeting we have heard of the nuclear matter approaches from Mahaux (1978) and Brieva (1978). Physically they are very similar: they start from nucleon-nucleon interactions that are equivalent in many respects and then solve the Bethe-Goldstone equation to give the correlated nucleon wave function in co-ordinate space and hence the mass operator and the t-matrix elements in momentum space. The principal difference arises in the application to finite nuclei: Mahaux applies the local density approximation directly to the mass operator whereas Brieva transforms the t-matrix to a co-ordinate representation before applying the local density approximation; this enables him to improve the geometrical aspect of the calculation and make explicit the range of the force. Both methods give complex energy-dependent microscopic potentials that already give a fair fit to the experimental data, a fit that is greatly improved by slight parameter adjustment. Mahaux has calculated many detailed features of the potential, including the volume integrals, the isospin terms and the Coulomb correction. The method of Brieva easily gives the spin dependence of the potential, and this is found to be similar to the phenomenological form. Both methods can be extended by improving the mathematical approximations, studying the convergence properties and evaluating higher order terms.

These nuclear matter calculations use information on nuclear structure in the form of nuclear density distributions. It is not certain that these contain all the nuclear structure information required even for global potentials, but it seems likely to be a good approximation. In principle it is possible to calculate the required nuclear structure information from the same nucleon-nucleon interaction, but this is generally regarded as a separate and very difficult problem. For the present purpose it is simpler and more accurate to use nuclear density distributions obtained either from Hartree-Fock calculations (Negele, 1970; Sick et al., 1975) or from the summation of squares of single-particle wave functions (Hodgson, 1976, Malaguti et al., 1978). The nuclear matter calculations are in general not too sensitive to the precise form of the density distribution, provided it has the correct RMS radius. An exception to this is those potentials that are sensitive to particular features of the density distribution, for example heavy ion potentials that are sensitive to its extreme tail.

The work described by Mahaux and by Brieva marks an important stage in the development of calculations of microscopic optical potentials from the nucleon-nucleon interaction: for the first time it is possible to obtain potentials that fit the data almost as well as phenomenological potentials with many adjustable parameters. Much work remains to be done to exploit this success. In particular, the spin-orbit and isospin terms can be studied with more realistic form factors obtained from the microscopic calculations. It is now possible to provide a much firmer physical basis for phenomenological calculations, and this should make it possible to obtain potentials that may be extrapolated in energy and from one nucleus to another with much greater confidence than before.

An alternative method of calculating the optical potential was described by McCarthy (1978) in the context of his studies of atomic potentials. The principal differences from the nuclear matter approach are that the momentum distribution of the struck particle is given by the momentum-space wave function, not by the Thomas-Fermi approximation, and the imaginary potential starts at the physical threshold for Q-space. The spherical averaging is carried out only in the projection operator Q and the multi-dimensional integration makes it possible to perform co-ordinate transformations exactly. For incident momentum k, the optical potential depends on K, p and $\underline{K}.\underline{p}$ so that a non-local potential in co-ordinate space is non-spherical in momentum space.

A study of the three-body Bethe-Faddeev equations and single-particle potentials in nuclei was presented by Zabolitzky (1978). A comparison of the single-particle potential felt by a real A+1st nucleon added to an A-nucleon system with the potential felt by one of the A nucleons in the system when virtually excited shows that the former quantity is of interest in optical model potential calculations while the latter is relevant for the calculation of the ground-state properties of the A-nucleon system.

The general theory of the optical potential was reviewed by Weigel (1978). He used the procedure of Bell and Squires to obtain the generalised optical potential from the Green's function formalism of many-particle theory. This approach includes the Pauli principle and the potential can easily be divided into an energy-independent folding term and an energy-dependent resonant term. In this formalism it is also easy to establish the connection with theories of nuclear structure and to calculate the contributions of inelastic and reaction processes to the imaginary potentials. The analytic properties of this potential are uniquely defined so that dispersion relations can be obtained. At higher energies, where particle-particle correlations in nuclear matter are important, the main part of the optical potential can be related to properties of nuclear ground state correlations. Numerical results were presented for the energy dependence of the optical potential and for the calculation of the optical potential from dispersion relations.

465

A study of the kaon-nucleus optical potential was presented by Barrett (1978). To first order, the potential is proportional to the nuclear matter density. The K^+-N interaction is rather weak and can be used to study $\rho_m(r)$, the K^--N interaction is considerably stronger and is needed for the energies and widths of states in k^- atoms. The measured shifts are analysed to give values of the scattering length. Calculations of the potential to second order were also presented.

A microscopic method of calculating the imaginary part of the inelastic scattering interaction was described by Madsen, Osterfeld and Wambach (1978). This calculation is based on the expression

$$\text{Im } F(r,r') = \text{Im} \sum_N \langle\psi_f|V|\psi_N\rangle\langle\psi_N|V|\psi_i\rangle G_N(r,r')$$

The imaginary part of the projectile Green's function $G_N(r,r')$ is always negative, so the terms add constructively for elastic scattering, whereas for inelastic scattering they need not have the same sign and may be random or destructive. This expression is evaluated for inelastic scattering by assuming (1) that the initial and final nuclear states are related by creation and annihilation operators according to the expression

$$\psi_f = \sum_{\bar{m}\underline{n}} A_{\bar{m}\underline{n}} a^+_{\bar{m}} a_{\underline{n}} \psi_i$$

where $\bar{m}$ refers to an unoccupied particle state and $\underline{n}$ to an occupied hole state based on the target nucleus. This is true for TDA and RPA collective states as well as for simple single-particle transitions. Also, (2) the intermediate states are taken as simple particle-hole doorway states based on the ground state, and (3) the identity of the projectile and target nucleons is neglected.

Evaluation of the expression then shows that there is indeed destructive interference but that it occurs in a systematic way so that the imaginary form factor is related to the strength of the real form factor and is sharply peaked at or beyond the nuclear surface due to the subtraction of the intermediate hole-scattering terms in the nuclear interior. This result provides a microscopic understanding of the phenomenological collective model prescription that relates the inelastic form factor to the derivative of the optical potential.

The odd-even dependence of the potential has been studied by Tang (1978) using the resonating group formalism. He shows that the anti-symmetrisation effects that arise when the Pauli principle is applied are represented by various nucleon-exchange terms in the exchange-normalisation and exchange-Hamiltonian kernel functions. The effects of the corresponding integral term in the Schrödinger equation can be represented by an equivalent local energy-dependent potential with a Majorana exchange term whose range decreases as the mass difference of the interacting nuclei increases. It is found that the amount of odd-even L-dependence is quite strong for the interaction of two S-shell nuclei, but is weak for α + ^{16}O and for n + ^{40}Ca. Exchange effects are also important for α + ^{6}Li, ^{12}C + ^{13}C, ^{12}C + ^{16}O and ^{28}Si + ^{16}O scattering (Bachelier et al., 1972; von Oertzen, 1970; Baye, 1976; Dehnhard et al., 1978).

This analysis leads one to expect that there are no odd-even effects when the masses of the interacting particles are widely different. However Kondo et al. (1975) have fitted the scattering of alpha-particles by ^{40}Ca using an L-dependent potential. Fits to this system over a wide range of energies have also been obtained with parity-independent potentials by Michel and Vanderpoorten (1977) and by Gubler et al. (1978). This shows that it is sometimes possible to fit the same data with L-dependent and with L-independent potentials, thus highlighting a pitfall in the phenomenological determination of L-dependent potentials. It should, however, be noted that it is possible to observe the effects of very small transfer amplitudes whenever there is high cancellation and this often occurs for backward scattering at high energies (Greenlees et al., 1972).

Explicit calculations were made for the h + α system at 60 MeV to study the importance of the various nucleon exchange terms, and these showed that the one-particle exchange and the core-exchange (three-particle exchange) terms are important, while the two-particle exchange term is negligible. Examination of the phase-shifts shows that the one-particle exchange effective potential is essentially an attractive Wigner potential, while the three-particle exchange effective potential is essentially a Majorana potential. In the case of the α + ^{16}O system at 18 MeV it is sufficient to include the one and two-particle exchange terms.

In general, it is found that the one-particle and core-exchange are the most important exchange terms, the latter being significant only when the nuclei are similar in mass. The effective local potentials corresponding to these two terms are of the Majorana and Wigner types respectively. Thus in phenomenological analyses the effective potential must in general include an odd-even L-dependent or parity-dependent term. However its effect is small when the interacting nuclei are dissimilar in mass, and this explains why it is frequently possible to get the data quite well with an L-independent potential. Further discussions of L-dependent potentials are summarised in the next section.

Ioannides and Johnson (1978) have analysed the propagation of energetic deuterons through nuclear matter, and have shown that the binding energy of the deuteron in nuclear matter depends strongly on the relative orientation of the deuteron spin and centre-of-mass momentum, when both the Pauli exclusion principle and the tensor force component of the neutron-proton interaction are included in the calculation. As a result, the tensor component of the duetron-nucleus optical potential is sensitive to the high momentum and tensor components of the neutron-proton interaction. In principle this could be used to obtain more information about the neutron-proton interaction, but at present the necessary experimental data on the elastic scattering of energetic deuterons is lacking.

Particular Potentials

Particular microscopic potentials can be calculated by making use of information on the structure of the nucleus concerned. If the nuclear matter distribution is used, it can be improved by adjusting for each state the parameters of the potential used to calculate the wavefunction so as to fit experimental binding energies and the model-independent charge distribution. Other information can also be included, such as the excitation energies and other properties of the excited states. The nuclear matter and nuclear structure methods of calculating microscopic optical potentials were compared by Vinh Mau (1978). They are complementary to each other because the former neglects the contributions of the collective states, which is only important at low energies, while the later uses a weak effective nucleon-nucleon interaction which is valid at higher energies.

Detailed calculations with the Green function formalism show that the strength of the imaginary potential calculated with the nuclear structure method depends strongly on the choice of effective interaction. The method was applied to calculate the n-^{40}Ca potential at low energy (Vinh Mau and Bouyssy, 1976). This work used information on the energies and collective properties of the low-lying states to give the contributions to the absorbing potential from each of the individual excitations. It is found that the mean square radii of both the real and the imaginary parts of the optical potential are well represented for energies less than 50 MeV. The volume integrals are too small for the real potential, but for the Reichstein-Tang nucleon-nucleon interaction the imaginary potential is in accord with the empirical value, and their variation with energy is well represented. These potentials have been used by Ngô to calculate cross-sections, and it is found that at 30.3 MeV the angular positions of the oscillations are correct but that the potential has to be renormalised to reproduce the amplitudes at large angles. Calculations of the n-α and α-^{40}Ca potentials were also presented.

A microscopic analysis of p-^{40}Ca elastic scattering at 40 MeV was made by Ngô et al. (1978). They obtained the non-local, energy-dependent complex potential, as well as the local equivalent potential, for several nucleon-nucleon interactions. This potential is in good overall agreement with the experimental data, but the fit is greatly improved by a 10 % increase in the real potential. The effects of particle-vibration coupling on the Hartree-Fock potential have been studied for ^{208}Pb by Van Giai and Bernard (1978). They include the effects of long-range particle-hole correlations and calculate the real and imaginary parts of the optical potential by a formalism in which the Hartree-Fock term and the correction terms are derived from a Skyrme effective interaction. For each L-multipole all the most important collective states are included, and in their calculations for ^{208}Pb they took account of the natural parity states corresponding to L=0,2,3,4,5 for isoscalar modes and L=0,1,2 for isovector modes. The resulting potential has a volume form for the real part and a surface-peaked form for the imaginary part, and strengths rather less than those found phenomenologically. These potentials are being used to calculate elastic scattering, and preliminary results are encouraging.

A shell model description of the optical model potential was presented by Micklinghoff (1978), who solved the Bethe-Goldstone integral equation for a finite system using an approximation in the framework of the collective model and the hole line expansion. This allows the absorption due to collective excitation to be included in the optical potential.

4. Effective Optical Potentials

There are several cases where the global potential does not fit the elastic scattering data at all well, for a reason that is physically evident. It is then possible either to make a more detailed calculation that includes this effect explicitly, or to devise a new term in the optical potential that reproduces its effects on the elastic scattering (Hodgson, 1974). Such effective potentials can be obtained either theoretically or phenomenologically. Notable examples of this are provided by the Coulomb excitation of low-lying collective states in heavy ion scattering and the coupling to giant resonances and to specific reaction channels (von Geramb, 1972, 1973).

The long-range absorption potential that gives the effect of the Coulomb excitation of low-lying collective states in heavy ion scattering has been studied by Baltz et al. (1978). In many of these reactions, especially at energies around the Coulomb barrier, it is found that the coupling to low-lying collective states is so strong that the elastic scattering cross-section is considerably perturbed from its usual Fresnel form and cannot be fitted by a standard optical potential. It was found possible to account for the coupling by adding a phenomenological absorbing potential falling off as $r^{-6.3}$ (Thorn et al., 1977). The coupling can be taken into account explicitly by a full coupled-channels calculation but this is very lengthy due to the long range of the Coulomb forces and so efforts have been made to calculate a potential that, when added to standard optical potential, gives the same effect as the coupling. This has been done by Love, Terasawa and Satchler (1977) using the Feshbach projection operator formalism. Their potential is almost entirely imaginary and has a long range, falling off as r^{-5} at large distances.

A more exact expression was obtained by Baltz et al. (1977, 1978) using Coulomb-distorted scattering states and a Coulomb-distorted Green's function. Their potential has a different radial dependence from that of Love et al., and is L-dependent, but the contribution of greatest range again varies as r^{-5}. They obtained an analytical closed-form expression for the differential elastic scattering cross-section that is valid below the Coulomb barrier where the absorption is weak.

Baltz et al. also calculated for the case of 50 MeV alpha-particles on ^{154}Sm the component of the optical potential due to the direct rotational coupling of the ground and first 2^+ excited state. This potential is highly L-dependent, repulsive in the low partial waves, increasing in magnitude to the surface, changing sign and becoming attractive, and then decreasing in magnitude for high partial waves. The empirical optical model component obtained as the difference between the optical model and coupled-channels parameters is also L-dependent and repulsive. The calculated imaginary potential is L-dependent and absorptive for all partial waves, while the empirical imaginary potential is of small magnitude but relatively diffuse in its small absorption outside the nuclear surface. Similar results were obtained for 60 MeV ^{16}O on ^{40}Ca. In both cases the L-dependence of the imaginary potential reflects the L-window of a direct reaction in the presence of a strongly absorptive background potential; flux is lost from the elastic channel into the inelastic channel primarily in the surface partial waves. This shows that it may well be insufficiently accurate to use an L-independent potential for heavy ion reactions in which a substantial amount of flux goes into direct channels.

Calculations of the effects of coupling to excited states have also been made by Frahn and Hill (1978) who show that Frahn's generalised Fresnel model for very heavy ion scattering can describe dynamic polarisation effects due to the Coulomb excitation of low-lying collective states. They use the Coulomb-distorted eikonal approximation to calculate the reflection functions corresponding to the potentials of Love et al. and of Baltz et al., and the resulting explicit formulae are in good accord with the experimental elastic scattering data. Measurements of dynamic polarisation effects in heavy ion scattering provide a sensitive way of determining B(EL) transition probabilities and deformations.

Several adiabatic and dynamic polarisation effects in sub-Coulomb elastic scattering were considered by Baur, Rösel and Trautmann (1978). The mutual electric polarisation of two nuclei can be interpreted as transitions to excited intermediate states, characterised by an adiabacity parameter $\xi_{if} = \eta_i - \eta_f$, where η_i and η_f are the Coulomb parameters. If $\xi_{if} > 1$, only virtual excitations are likely to occur, and the most important intermediate states are the strongly collective EL transitions to the giant dipole states. In this case the effects on the elastic scattering can be represented by a real potential. If however $\xi_{if} < 1$ then flux can be removed from the elastic channel due to the excitation of low-lying collective states. In this case the cross-section can be reliably calculated by the semiclassical coupled-channels method. Some explicit calculations for ^{40}Ar on ^{238}U at 340 MeV showed that the effect of reorientation in the second state and the coupling to the 4^+ state have little effect on the elastic scattering for angles less than the grazing angle.

L-dependence

The elastic scattering of particles by nuclei is fixed by the complex phase-shifts δ_L corresponding to several values of the orbital angular momentum L from zero to L max, the highest value of L for which the phase-shift contributes significantly to observed quantities. These phase-shifts can in principle be found by analysing the cross-section, but in practice this is only possible if very few partial waves contribute significantly. Solving the radial wave equation for each value of L gives the phase-shifts for a particular potential V(r). It is obviously remarkable if a potential of a particular analytic form gives all the phase-shifts correctly even if it has several adjustable parameters and δ_L is a smooth function of L, as is often found to be the case. In practice when there are many partial waves some errors in one may compensate some errors in another, so even if the fit is very good it does not necessarily mean that all the phase-shifts are correct. In principle any cross-section, and thus any set of phase-shifts, can be fitted by an L-independent potential, but this could be highly singular and not a smooth function of L. The high precision of the fits to alpha-particle cross-sections obtained with the spline method by Michel and Vanderpoorten shows that in this case an L-independent potential is adequate. In other cases it is necessary to assume some L-dependence. There is, of course, no guarantee that this L-independence will have a simple analytic form as a function of L, but this is the simplest initial hypothesis, and it seems to work quite well.

There is no need to consider L-dependent potentials for most nuclei as the scattering can be fitted very well by L-independent potentials. Even in these cases, however, it is not certain that there is not some L-dependence that is masked by parameter variation. The case for trying the effect of an L-dependent potential is stronger if no amount of parameter variation is able to give a fit to the data. There does still remain the possibility that a potential with a different analytical form would be able to fit the data without recourse to L-dependence. Every form of potential gives some L-dependence of the phase-shifts and it may be surmised that if this is not correct the potential has to be supplemented by an explicit L-dependence.

Leaving aside these considerations, and remaining in the frame of Saxon-Woods potential, it is interesting to see if data that cannot be fitted with an L-independent potential can be fitted with the addition of a simple L-dependent term. Mackintosh and Kobos (1978) have described their analyses using a potential with an L-dependent term of the form

$$V = f(\ell^2, L^2, \Delta^2)\{U^L\, g(r, R_R, a_R) + i\, W^L\, g(r, R_I, a_I)\}$$

where f is the standard Saxon-Woods form and $g(r,R,a) = -4a(d/dr)f(r,R,a)$. They found that very precise fits could be obtained to many sets of data, including some, such as p + ^{16}O, that cannot be well-fitted by a standard optical potential. These potentials are valuable for distorted wave calculations of reaction processes. The L-dependent terms are similar in all cases and have the following properties: (1) The potential for $L < L_c$ is substantially different from that for $L > L_c$, where L_c defines the transition region, (2) the new terms are always repulsive and absorptive for low partial waves, (3) the repulsive real term is almost always narrowly peaked within the nuclear surface, (4) the absorptive term is always peaked in the far surface and is usually narrow, (5) the L-dependent terms are particularly strong for energies around 20-40 MeV, and (6) the parameters of the L-dependent potential vary much more smoothly with energy than those of the L-independent potentials.

Another series of calculations by Mackintosh (1973, 1974) and Mackintosh and Kobos (1976) showed that the data can also be fitted by explicit inclusion of the coupling to (p,d) pick-up channels. This suggests that the L-dependence simulates the effect of this coupling, although the L-dependent term is much smaller than the part of the optical potential generated by the pick-up couplings. Some similar work has been carried out by Coulter and Satchler (1977).

The spline interpolation method is being used to obtain model-independent potentials, and this shows the connection between the L-dependent potentials and the corresponding L-independent potentials.

A special case of L-dependence is the odd-even staggering of the phase-shifts that may be represented by a $(-)^L$ term in the imaginary potential (Gelbke, Bock and Richter, 1974; Robson, 1971). In the case of elastic scattering of particles of nearly equal mass this is due to the constructive and destructive interference of the direct and transfer amplitudes; the matrix element of the transfer process changes sign depending on whether the orbital angular momentum L is even or odd, and this produces an odd-even staggering of the scattering S-matrix elements. However the staggering in the L-dependent model comes from the nuclear size resonances whereas in the transfer process it is due to the symmetry of the reaction and the bound state wave function of the exchanged particle. The energy dependence of the odd-even staggering is given naturally by the exchange model, whereas it can be reproduced with an L-dependent potential only by a special choice of parameters in each case. As described in the previous section, Tang

(1978) has used the resonating group formalism to show that such terms come from antisymmetrisation of the particle wave functions.

An L-dependent absorbing potential has been frequently used in analyses of heavy-ion scattering, and it can be justified physically by simple qualitative considerations concerning energy and angular momentum conservation in the various reaction channels (Hodgson, 1978).

5. Nucleon Analyses

Three detailed semi-phenomenological analyses of a wide range of nuclear data were described by Giannini (1978), Tarrats and Escudié (1978) and Leeb and Eder (1978). They obtain formulae for the potentials appropriate for many nuclei over a range of energies. It would be useful to make a critical comparison of these formulae, and also to study the simplifications that can be obtained by using information from microscopic optical potentials.

The microscopic optical potentials should make it possible to extract improved isospin potentials from the experimental data, in particular by providing more realistic form factors. At present it is usual to assume that the isospin potential has the same form factor as the bulk of the real potential, and to express the potential depth in the form

$$V_{p,n} = V_0 \pm V_1 \frac{N-Z}{A}$$

The values of V_0 and V_1 can then be obtained by analysing scattering data for a range of nuclei. For a particular choice for the dependence of the radius parameter on A, the asymmetry potential V_1 is in general partly geometrical and partly isospin, and it is important to separate these. For bound states this may be done using the energies of the $T_<$ and $T_>$ states, and for scattering by comparing the scattering of protons and neutrons from the same target. This comparison must be made for energies differing by the Coulomb energy difference Δ_C so that the compound states are analogues. An alternative method of separating the geometrical and isospin dependence was presented by Leeb and Eder (1978). It is also necessary to take into account the Coulomb-energy correction term V_C (sometimes taken as 0.4 $Z/A^{1/3}$. As Jeukenne et al. (1977) have shown, this cannot be obtained phenomenologically but must be obtained from a microscopic calculation.

A survey of the application of the folding model to elastic and inelastic scattering was presented by Petrovich (1978). He showed that with a proper choice of the effective interaction and inclusion of exchange it is possible to obtain a good overall fit to the experimental data.

It is found that precision fits to analysing powers give rather small values of the spin-orbit radius parameters, around 0.9 fm, and this is consistent with some analyses of bound state data. This does not seem to be consistent with the microscopic calculations of Rook and Brieva (1978) who found that the nucleon spin-orbit term is quite sharply peaked at the maximum of the derivative of the nuclear density. This merits further study.

Roman (1978) suggested that the low value of the spin-orbit diffuseness parameter for helions compared with the normal value for tritons may be attributable to a corresponding difference for nucleons. Reanalysis of neutron polarisation data showed that it can be quite well fitted by a potential with abnormally low spin-orbit diffuseness parameter. However the neutron polarisation data is rather imprecise, so it does not determine the parameter accurately. Improved neutron polarisation data is thus very desirable.

Another intriguing result was obtained by Bendiscioli et al. (1976). They found that the differential cross-sections and anti-symmetries for 36.2 MeV proton scattering at small angles around 1-7° from several nuclei differs markedly from the optical model values. An independent measurement is called for and if the result is confirmed its theoretical implications require examination.

A new technique for obtaining ratios of cross-sections by scattering protons from mixed targets was described by Austin (1978). These cross-section ratios are free of many of the uncertainties of absolute cross-sections, and thus may provide useful data for testing optical potentials. Some results were presented for 30.3 MeV proton elastic scattering from the calcium isotopes ^{40}Ca, ^{42}Ca and ^{48}Ca. These are well fitted by a standard optical potential, and give differences between the neutron RMS radii of the isotopes that are in accord with analyses of 800 MeV proton scattering data.

Extensive experimental data on the elastic scattering of 10-40 MeV protons by a range of light nuclei at closely-spaced energies was presented by the Milano-Bari group (Fabrici et al., 1978). It is found that the differential cross-sections for deformed nuclei like ^{12}C, ^{24}Mg and ^{28}Si can be quite well fitted by standard optical potentials, while those for spherical nuclei like ^{16}O, ^{40}Ar and ^{40}Ca are much harder to fit, even with non-standard potentials. These nuclei also show an anomalous peak in the differential cross-section for energies from 30 to 36 MeV. The magnitude of this peak is for many nuclei inversely proportional to the quadrupole deformation parameters β_2, suggesting that the anomaly is connected with collective excitation. The peak values of the cross-sections for excitation of 2_1^+ states are similarly proportional to $(R\beta_2)^2$ times the corresponding elastic peak cross-section. A phase-shift analysis shows that the partial waves involved in the enhanced cross-sections are centred on the grazing partial wave with L=kR. These features provide a severe test of microscopic optical potentials, which should be able to fit data for both spherical and deformed nuclei. It is important to test a calculated potential by comparing it with an extensive range of data; comparison with one cross-section could well be misleading.

6. Light-Ion Analyses

An impressive amount of precise data on the elastic scattering of deuterons, helions and alpha-particles by nuclei was presented by several speakers, together with detailed phenomenological analyses. This provides some knowledge of the interactions, which remain to be understood in a fully microscopic way.

The deuteron work was described by Djaloeis (1978), and referred to several nuclei at energies from 59 to 130 MeV. Data for 33 MeV helion scattering by a range of nuclei was presented by Karban (1978), together with the rather surprising result that the best optical model fits required a spin-orbit term with the abnormally-small diffuseness parameter of 0.2 fm. No satisfactory explanation of this has yet been proposed. Similar analyses of triton scattering gave a normal value for this parameter, and Roman (1978) suggested that the difference between the helion and triton potentials may be traceable to corresponding differences between neutron and proton potentials. This has been discussed in the previous section.

An analysis of the scattering of 104 MeV alpha-particles by ^{40}Ca and ^{48}Ca was presented by Gils et al. (1978). They used a potential consisting of the standard optical model term augmented by a Fourier-Bessel expansion to give greater flexibility and to remove the unwanted correlations between the different regions of the potential that are implicit in the usual analytic forms. The fits obtained with this potential are substantially better than those found with standard analyses using Saxon-Woods and Saxon-Woods-squared potentials. The analysis was repeated using a folding model potential obtained from an assumed nuclear density distribution and an effective alpha-nucleon interaction with a Gaussian radial dependence and a density-dependent factor to give the required saturation effects. The parameters of the interaction were first fixed by requiring a fit to the ^{40}Ca data assuming a known density distribution, and then determining the parameters of the neutron density distribution of ^{40}Ca from a fit to the ^{48}Ca data. This gives the following results for the RMS radii differences of ^{40}Ca and ^{48}Ca: $r_m(48)-r_m(40) = 0.12 \pm 0.07$ fm, and $r_n(48)-r_p(48) = 0.08 \pm 0.10$ fm.

The results of an analysis of the elastic scattering of 40-142 MeV alpha-particles by ^{90}Zr were presented by Put (1978). He found that the data above 80 MeV is well fitted by an optical potential with constant form factors and depths that vary smoothly with energy. A different potential is needed for lower energies.

The uniqueness and shape of the alpha-particle optical potential was studied by Wiktor et al. (1978) by using a Saxon-Woods potential raised to a variable power to analyse a wide range of experimental data from 120-172.5 MeV and from ^{12}c to ^{64}Ni. It is found that the best form of the potential depends on the target mass and projectile energy.

The validity of various folding models for the analysis of 104 MeV alpha-particle and 156 MeV ^{6}Li scattering by ^{40}Ca and ^{48}Ca has been studied by Majka, Gils and Rebel (1978). The folding models used were the simple double folding, and single-folding with target and projectile folding. The density-dependence of the nucleon-nucleon interaction was included in the sudden and in the intermediate approximations. The potentials used in the single folding were obtained in a semi-microscopic way including one-nucleon exchange effects and the density-dependence of the nucleon-nucleon interaction. In each case the real potential obtained from the folding model was supplemented by a phenomenological imaginary potential, and during the fitting process the real potential was normalised and the imaginary potential optimised. All the models give quite good fits, the best corresponding to projectile folding, and it is found that antisymmetrisation effects are of minor importance compared with the density-dependence of the nucleon-nucleon interaction. With the intermediate approximation for the density-dependence it is not necessary to normalise the real potential. Calculations were also made with various phenomenological form factors for the real part of the potentiala and it was found that the Saxon-Woods potential gives a poor fit, the Saxon-Woods-squared a much better fit and the modified Fermi parametrisation the best fit of all.

The results of an extensive analysis of alpha-particle bound and scattering data with a hard-core potential were presented by Gridnev et al. (1978) (Baz et al. 1977). The hard-core takes account of the repulsion due to the Pauli principle, and gives a total potential with a surface pocket that sustains unbound alpha-cluster states. It is found that with a particular choice of the parameters of the potential the energies and widths of many resonances in the $^{12}C + \alpha$ system are well reproduced, and similarly for the $^{16}O + \alpha$ system. The same potentials also give good overall fits to the alpha-particle elastic scattering cross-sections over a range of energies.

D.F. Jackson (1978) summarised some of her recent work on the theory of alpha-decay, and showed that the half-life of this process is very sensitive to the barrier penetrability and hence to the alpha-particle optical potential. The restriction that such potentials must be consistent with alpha-decay should be useful in resolving at least some of the ambiguities among the potentials that fit elastic scattering. Since the potentials may be calculated from a folding model, this should lead to improved nuclear matter distributions.

7. Heavy-Ion Analyses

After reviewing the folding model calculations of the optical potential, Brink (1978) presented a method for calculating the imaginary part of the heavy-ion optical potential based on Feynman's path integral method. This provides a natural link between a complete quantal theory and the semi-classical approximations. The method consists in equating two expressions for the scattering amplitude, one in terms of an optical potential and the other in terms of a detailed coupled-channels description. This gives an expression for the optical potential in terms of the time evolution operation $U[r(t),t_1,t_2]$,

$$\int_{t_0}^{t_i} V_{opt}(r(t))dt = i\hbar \ln[\langle 0|U[r(t),t_1,t_2]|0\rangle]$$

The right hand side of this equation is approximated by the leading term in time-dependent perturbation theory, and separating out the imaginary part of the potential gives

$$\int_{-\infty}^{\infty} \mathrm{Im}\, V_{opt}[r(t)]dt = -\frac{1}{2\hbar} \sum_{n\neq 0} \left| \int_{-\infty}^{\infty} dt\langle n|v(r(t),\xi)|0\rangle \exp[i(\varepsilon_n-\varepsilon_o)t\hbar] \right|^2$$

This expression can be used to calculate the long-range absorption potential studied by Love et al. (1977) and by Baltz et al. (1977). The path integral method can also be used to find a relation between the imaginary part of the optical potential describing the scattering of heavy ions in excited states and the friction coefficient for deep inelastic scattering.

The elastic scattering and vector and tensor analysing powers of 14.22 MeV ^{6}Li and ^{7}Li ions elastically scattered by ^{58}Ni have been measured by Tungate and Fick (1978). The data can be fitted quite well by spin-orbit potentials based on the folding model, and some of the marked differences between ^{6}Li and ^{7}Li scattering can be attributed to the differences in shape of the two projectiles.

The theoretical basis of the folding model of the optical potential was reviewed by Love (1978) and applied to calculate heavy-ion potentials. Both single- and double-folding methods were used, with realistic internucleon interactions based on G-matrices in a double-folding model. Many different internucleon interactions were used, and evaluated by comparing with data on heavy-ion scattering. It is usual in folding-model calculations to optimise the normalisation of the potential, and this provides a way of taking into account to some extent some of the higher-order corrections that are present in a more fully microscopic theory. While it is now possible to devise prescriptions for folding model calculations that fit the data quite well, progress at a more fundamental level can only come from the explicit introduction of these higher-order effects.

An extensive review of molecular states in heavy ion potentials was contributed by Terlecki et al. (1978). Particular attention was devoted to the conditions for the existance and excitation of nuclear molecules, and the calculation of heavy-ion potentials. A theoretical analysis using a rotating co-ordinate system clarifies the excitation mechanisms of molecular collective and single-particle states. A relatively new development is the application of parastatistics to nucleus-nucleus scattering. This is relevant to the scattering of nuclei with integral spin: when they are well separated they are two bosons and yet when they interact strongly they form a compound fermion system. To take account of this the relative wave function is written with a variable parameter that allows a continuous transition from Bose statistics to Fermi statistics. Analysis of the cross-section for ^{12}C-^{12}C elastic scattering shows that this parameter depends linearly on the incident energy. It can be used to simulate effects due to nuclear contributions to the potential, and from the internal structure and Coulomb excitation of the nuclei.

The folding model has also been used by Sinha (1978) to calculate the heavy-ion potential. The first-order real potential is obtained by folding the two nuclear densities with a Skyrme interaction, and higher order terms are also estimated.

Reinhard and Goeke (1978) described their work on the definition of the collective Hamiltonian for heavy-ion scattering. Using the concepts of the adiabatic time-dependent Hartree-Fock theory, they extend the collective path $|q\rangle$ to a dynamic path $|q,p\rangle$ and hence obtain the Hamiltonian function $k(q,p)$. This function is quantised by subtracting terms in $\hbar^2$, giving collective Hamiltonian identical to the one obtained in a Gaussian-overlap-expansion of the generator co-ordinate method.

8. Conclusion

The present time is a very lively and exciting period in the development of the optical model. Microscopic theories of the nucleon optical potential have now been developed to the stage where they can give without parameter adjustment potentials that are in good overall agreement with the experimental data. The way is now open to study particular features of the potential and to extend the microscopic theory to light and heavy ions.

Phenomenological anylses of the experimental data have greatly improved, and the spline technique now gives very precise fits. This shows that the usual Saxon-Woods form factor is in some respects inadequate, which had already been indicated by microscopic calculations. We can look forward to convergence of two approaches. It should be possible to use this as the basis of extensive and accurate global potentials that fit the data for many nuclei over a range of energies.

Work in the area of semi-phenomenological theories, particularly the folding model, is proving very successful and should facilitate the unification of the microscopic and phenomenological theories.

These developments in the theory of the optical potential have shown the need for more extensive and accurate experimental data in several areas, and this should stimulate further experimental work. Measurements of elastic scattering may seem somewhat tame, but in fact the attainment of high precision cross-sections and polarisations over a wide range of angles and energies presents a challenging technical problem.

In the brief compass of this summary it has not been possible to do justice to all the valuable and interesting contributions to the meeting. In particular, little attempt has been made to summarise several excellent papers that were themselves summaries of some specialised topic.

I thank all those who kindly made available copies of their papers and also Professor H.V. von Geramb and his colleagues for organising such a valuable and stimulating meeting.

REFERENCES

S.M. Austin, This Meeting, 1978.

D. Bachelier, M. Bernas, J.L. Boyard, H.L. Harney, J.C. Jourdain, P. Radvanyi, M. Roy-Stephen and R. De Vries, Nucl. Phys. A195,361,1972.

A.J. Baltz, N.K. Glendenning, S.K. Kauffmann and K. Pruess, This Meeting, 1987.

A.J. Baltz, S.K. Kauffmann, N.K. Glendenning and K. Pruess, Phys. Rev. Lett. 40,22,1978; Symposium on Nuclear Direct Reaction Mechanisms, Hosei University, Tokyo, 5-10 September 1977.

R.C. Barrett, This Meeting, 1978.

G. Baur, F. Rösel and D. Trautmann, This Meeting, 1978; Phys. Rev. C17, 2256,1978.

D. Baye, Nucl. Phys. A272,445,1976.

A.I. Baz, V.Z. Goldberg, N.Z. Darwisch, K.A. Gridnev, V.M. Semjonov and E.F. Hefter, Lett. Nuovo Cim. 18,227,1977.

G. Bendiscioli, E. Lodi-Rizzini, A. Rotondi, M.L. Stanga and A. Venaglioni, Lecture Notes in Physics Vol. 55 (Pavia Conference), Edited by S. Boffi and G. Passatore, Springer-Verlag, 1976.

F.A. Brieva, This Meeting, 1978. See also Brieva and Rook, 1977, 1978.

F.A. Brieva and J.R. Rook, Nucl. Phys. A291,299,317,1977; A297,206,1978.

D.M. Brink, This Meeting, 1978.

P.W. Coulter and G.R. Satchler, Nucl. Phys. A293,269,1977.

D. Dehnhard, V. Shkolnik and M.A. Franey, Phys. Rev. Lett. 40,1549,1978.

A. Djaloeis, This Meeting, 1978.

E. Fabrici, S. Micheletti, M. Pignanelli, F. Resmini, R. De Leo, G. D'Erasmo and A. Pantaleo, This Meeting, 1978.

W.E. Frahn, Preprint 1978.

W.E. Frahn and T.F. Hill, Preprint 1978.

H.V. von Geramb, Nucl. Phys. A199,545,1973; Nuovo Cim. Lett. 5,333,1972.

C.K. Gelbke, R. Bock and A. Richter, Phys. Rev. C9,852,1974.

M. Giannini, This Meeting, 1978.

H.J. Gils, E. Friedman, H. Rebel and Z. Majka, This Meeting, 1978.

G.W. Greenlees, W. Makofske, Y.C. Tang and D.R. Thompson, Phys. Rev. C6, 2057,1972.

K.A. Gridnev, V.M. Semjonov, V.B. Subbotin, E.F. Hefter and H.V. von Geramb, This Meeting, 1978.

H.P. Gubler, U. Kiebele, H.O. Meyer, G.R. Plattner and I. Sick, Phys. Lett. 74B, 202, 1978.

P.E. Hodgson, Nature 249,412,1974; Lecture Notes in Physics Vol. 55 (Pavia Conference), Edited by S. Boffi and G. Passatore, Springer-Verlag 1976, p. 88; Nuclear Heavy-Ion Reactions Ch.3.4 (Oxford) 1978.

A.A. Ioannides and R.C. Johnson, Phys. Rev. C17,1331,1978; This Meeting, 1978.

D.F. Jackson, This Meeting, 1978.

J.P. Jeukenne, A. Lejeune and C. Mahaux, Phys. Rep. 25C,83,1976; Phys. Rev. C15,10,1977.

O. Karban, This Meeting, 1978.

Y. Kondo, S. Nagaka, S. Ohkubo and O. Tanimura, Prog. Theor. Phys. 53, 1006,1975.

H. Leeb and G. Eder, This Meeting, 1978.

W.G. Love, This Meeting, 1978.

W.G. Love, T. Terasawa and G. Satchler, Phys. Rev. Lett. 39,6,1977; Nucl. Phys. A291,183,1977.

R.S. Mackintosh, Phys. Lett. 44B,437,1973; Nucl. Phys. A230,195,1974; Phys. Lett. 59B,431,1975.

R.S. Mackintosh and A.M. Kobos, Phys. Lett. 62B,127,1976, This Meeting 1978.

V.A. Madsen, F. Osterfeld and J. Wambach, This Meeting, 1978.

C. Mahaux, This Meeting, 1978. See also Jeukenne, Lejeune and Mahaux, 1976, 1977.

Z. Majka, H.J. Gils and H. Rebel, This Meeting, 1978.

F. Malaguti, A. Uguzzoni, E. Verondini and P.E. Hodgson, Nucl. Phys. A297,287,1978.

I.E. McCarthy, This Meeting, 1978.

F. Michel and R. Vanderpoorten, Phys. Rev. C16,142,1977; This Meeting, 1978.

M. Micklinghoff, This Meeting, 1978.

J.W. Negele, Phys. Rev. C1,1260,1970.

H. Ngô, A. Bouyssy and N. Vinh Mau, This Meeting, 1978.

F. Petrovich, This Meeting, 1978.

L. Put, This Meeting, 1978.

P.G. Reinhard and K. Goeke, This Meeting, 1978.

D. Robson, Argonne National Laboratory Report 7837,239,1971.

S. Roman, This Meeting, 1978.

I. Sick, J.B. Bellicard, M. Bernheim, B. Frois, M. Huet, P.L. Leconte, J. Mougey, Phan Xuan-Ho, D. Royer and S. Turek, Phys. Rev. Lett. 35, 910, 1975.

B.K. Sinha, This Meeting, 1978.

Y.C. Tang, This Meeting, 1978.

A. Tarrats and J.L. Escudié, This Meeting, 1978.

G. Terlecki, D. Hahn, W. Scheid, R. Koennecke and W. Greiner, This Meeting, 1978.

C.E. Thorn, M.J. LeVine, J.J. Kolata, C. Flaum, P.D. Bond and J.C. Sens, Phys. Rev. Lett. 38,384,1977.

A. Tungate and D. Fick, This Meeting, 1978.

N. Van Giai and V. Bernard, This Meeting, 1978.

N. Vinh Mau and A. Bouyssy, Nucl. Phys. A257,189,1976.

W. von Oertzen, Nucl. Phys. A148,529,1970.

M. Weigel, This Meeting, 1978.

S. Wiktor, C. Mayer-Böricke, A. Kiss, M. Rogge and P. Turek, This Meeting, 1978.

J.G. Zabolitzky, This Meeting, 1978.

Nucleon removal and rearrangement energies

P E HODGSON

Nuclear Physics Laboratory, University of Oxford, Keble Road, Oxford, OX1 3RH

Abstract

The process of nucleon removal from a nucleus is analysed together with the associated nuclear rearrangement. Several possible definitions of the rearrangement energy are reviewed, and their connection with experimentally observable quantities is investigated.

This review was completed in January 1975.

Rep. Prog. Phys. 1975 **38** 847–902

Contents

1. Introduction

The removal of a nucleon from a nucleus is one of the most basic and important processes in nuclear physics, and numerous studies of pickup and the complementary stripping reactions have provided much of our knowledge of nuclear structure, particularly of its single-particle aspects.

This process can be visualized very simply, and the experimental results can be analysed with a model that gives a clear picture of nuclear structure. The nucleus can be envisaged as a number of nucleons moving in a simple one-body potential of the form

$$V(r)=Uf_1(r)+\left(\frac{\hbar}{m_\pi c}\right)^2 U_S \frac{1}{r}\frac{\mathrm{d}f_2}{\mathrm{d}r}\boldsymbol{L}\cdot\boldsymbol{\sigma} \tag{1.1}$$

where U and U_S are the central and spin–orbit potential depths, and the form factors $f_{1,2}(r)=\{1+\exp[(r-R_{1,2})/a_{1,2}]\}^{-1}$.The eigenvalues of this potential are the allowed nucleon states, and these are filled by the available nucleons starting from the lowest state and subject to the Pauli principle.

These nucleons can be removed individually by pickup reactions, and the energy balance gives the energy of the corresponding nucleon state. Similarly the stripping reaction deposits a nucleon in one of the unfilled states, and again the energetics give the energy of the state. It is found that the potentials U and U_S, together with the form factor parameters $R_{1,2}$ and $a_{1,2}$, can be chosen to reproduce these energies quite well, and furthermore the sum of the squares of the wavefunctions of the nucleons in all the occupied states is in good accord with the nuclear charge and matter distributions.

Nucleons can be removed from deeper states by high-energy knockout reactions like (p, 2p). The corresponding cross sections show broad maxima; the peak energies are identified with those of the single-particle states, and their widths are related to their lifetimes.

This simple picture is corroborated by detailed studies of the stripping and pickup reactions. The angular distributions of the outgoing particles in each case can be analysed by the distorted-wave theory to give the quantum numbers of the nuclear state, and these are in accord with the simple model, and the absolute magnitudes of the cross sections give the strengths of the single-particle states. We thus have a model that is able to account in a very direct way for a wide range of nuclear reaction and structure experiments (Wilkinson 1968).

The difficulties of this simple model become apparent as soon as the single-particle states are measured experimentally. It is not possible to remove a nucleon without disturbing the whole nucleus, so that the final state is a quantum mechanical superposition of many nuclear states. Thus we can write the wavefunction ψ_T of the target nucleus as a sum of products of the wavefunctions ψ_R^i of the residual nucleus and ϕ_i of the nucleon removed

$$\psi_T=\sum_i C_i\psi_R^i\phi_i \tag{1.2}$$

where C_i stands for coupling and fractional parentage coefficients.

We observe the states of the residual nucleus, and these are fixed by its internal dynamics, which are distinct from those of the initial nucleus and of the removal process. The latter cannot alter the energies of the states of the residual nucleus, but only their relative population probabilities, and in this way we can learn about the single-particle state in the initial nucleus. It is seen as a doorway state in the residual nucleus, and if looked at quickly by a fast reaction (with correspondingly poor energy resolution) it appears as a broad peak with long tails; if looked at slowly (with high energy resolution) it is found dissolved into a number of discrete fragments that are states of the residual nucleus.

This can also be expressed by saying that the single-particle states are split into a number of fragments by the residual interactions that are not included in the simple one-body potential. It is possible to recover the original simplicity to some extent by defining the centroid energy of a single-particle state as the average energy of the fragments, weighted by their spectroscopic factors determined from analyses of the reaction data. However, it is not possible to be sure that all the fragments have been found, and this introduces uncertainty into the analysis discussed in detail in §2.

The spreading of the single-particle states is essentially due to two types of forces: firstly the long-range or 'soft' interactions that can be accurately evaluated by perturbation theory; and secondly the short-range or 'hard' interactions, representable by a hard core potential, that cannot. For convenience these will be referred to as 'soft' and 'hard' interactions respectively. Physically, they correspond to nucleon–nucleon interactions in the tail and in the core region of the potential. As indicated in figure 1, the soft interactions spread the single-particle strength over a region of a few MeV around the main concentration of strength, while the hard interactions produce long tails extending many tens of MeV from the peak energy.

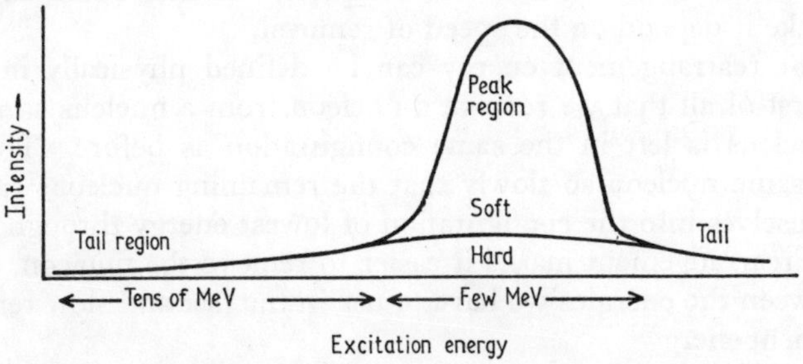

Figure 1. Idealized sketch of the division of single-particle strength into the peak region, mainly due to the soft (or longe-range) interactions, and the tail region, due to the hard (or short-range) interactions.

It might well be asked how sharp is the distinction made in figure 1 between the peak region and the long tails, for this is crucial to the subsequent discussion. Such estimates as have been made indicate that around 90% of the strength is in the peak region and only a few per cent in the tail. The studies of the variation of the centroid energies of the observed fragments from one nucleus to the next, described in §3, show a very smooth systematic behaviour, and this provides supporting evidence that they are experimentally well defined and that the widths of the peaks do not vary drastically. Thus, although it is not possible to make a completely sharp distinction between the peak and tail regions, there is good evidence that the experimental division between the fragments that are observable in nucleon transfer reactions and those that are not is sufficiently accurate for practical purposes.

The first-order term in the perturbation theory expansion for the energy of the single-particle state gives a good approximation to the peak energy. The second-order terms are complex, and not only do they give a small correction to the peak energy, but also spread the state around the peak energy, the amount of the spreading depending on the imaginary parts of the second-order terms. In this way perturbation theory to second order gives the envelope of the single-particle strength in the peak region. The energies and widths of the individual fragments can only be calculated by a more complete theory that gives a detailed model for the structure of each component. The high-energy tail is entirely due to the hard core or the short-range correlations, but the latter also contribute to the main peak in a way that is included in a renormalized Brueckner–Hartree–Fock (BHF) calculation. Thus the BHF theory, in which the self-consistency condition of the iteration process is arranged to cancel certain first-order diagrams only, includes all the effects of the soft interactions, and in addition, after renormalization, the effects of the hard interactions in the peak region as well. Thus the BHF theory essentially refers to the peak region, and to first order the BHF single-particle energy is the peak energy.

The theoretical picture of the nucleus is now more complicated than before. Instead of nucleons completely occupying the states up to the Fermi level, we have a whole series of particle–hole excitations that reduce the occupation probabilities below unity for the occupied states, and populate the states that are unoccupied on the simple picture. This more realistic situation may be described in terms of quasi-particles.

Associated with nucleon removal is the rearrangement of the nucleus left behind. In general it will not be left in its ground state, and so it subsequently loses energy in a variety of ways until it reaches a stable configuration. It may start to do this even before the process of nucleon removal is complete, and thus could affect the removal energy and make it depend on the speed of removal.

The nuclear rearrangement energy can be defined physically in a simple way. We imagine first of all that we remove a nucleon from a nucleus so rapidly that the rest of the nucleus is left in the same configuration as before. Then we imagine removing the same nucleon so slowly that the remaining nucleons can continuously rearrange themselves into the configuration of lowest energy throughout the removal process. This rearrangement makes it easier to remove the nucleon, and we call the difference between the energies we have to use in the fast and slow removal processes the rearrangement energy

$$R = E\,(\text{fast}) - E\,(\text{slow}). \tag{1.3}$$

This energy depends on the state from which the nucleon is removed.

Unfortunately difficulties arise when this simple concept is applied to real nuclei, and we try to measure rearrangement energies and to compare them with those calculated from theories of nuclear structure.

As will be shown in §4, a quantum mechanical analysis of the nucleon removal process leads to a definition of the rearrangement energy in terms of the distribution of single-particle strength. If the centroid energy is defined as the single-particle energy, this can be written

$$R = E\,(\text{single particle}) - E\,(\text{separation}) \tag{1.4}$$

where E (separation) is the energy needed to leave the nucleus in the state of lowest energy. Unfortunately this definition is practicable only for the ground state, so it is usual to *redefine* the rearrangement energy as

$$R' = E\,(\text{single particle}) - E\,(\text{peak}) \tag{1.5}$$

where E (peak) is the energy of the largest fragment of the single-particle strength if it is resolved into its components, and the maximum of the distribution function if it is not. These two definitions nearly always coincide for the ground state, since in this case the strongest fragment usually has the lowest energy, but for excited states they do not. The continuity with the original physical meaning can only be retained if the separation process is redefined to be such that it leaves the residual nucleus with the peak energy.

At this point it is useful to recall the basic requirements for any physical concept, namely that it should be precisely defined in a clear theoretical framework and also be measurable by a specified series of operations. There are of course usually mathematical difficulties in calculating the precise value of the quantity in any particular case, and practical difficulties in measuring it accurately, but it remains necessary that the theoretical and experimental definitions correspond, and that successive improvements in computational techniques on the one hand and experimental techniques on the other enable the corresponding theoretical and experimental values of the quantity to approach each other in any particular set of circumstances.

In practice these simple requirements are often difficult to satisfy, and it may happen, for example, that the definition of a particular physical quantity is altered to make it more general, or more easy to calculate, or more easy to measure. The same word is often retained after such changes, even if the connection with the original physical definition has vanished. This has happened with nuclear rearrangement energies.

Time-dependent rearrangement energies are discussed in §4, and it is shown that it is possible to define such rearrangement energies so that they can in principle be measured experimentally. In practice, however, there is no evidence for substantial time-dependent effects, and small effects would be very difficult to observe. All the available evidence is consistent with sudden removal, so at present the concept of a time-dependent rearrangement energy cannot be related to observable quantities.

The rearrangement energy can be calculated using the Hartree–Fock and Brueckner–Hartree–Fock theories, and some of these calculations and the associated formalism are described in §5. It is possible to distinguish formally between the rearrangement energy due to the change in the wavefunctions and that due to the change in the G matrix between the initial and final nuclei. Although these quantities cannot be distinguished experimentally, this gives a useful insight into the removal process. The BHF theory can be renormalized to give single-particle energies in agreement with the experimental peak energies; in the process the rearrangement energy is redefined so that it no longer corresponds to a physical quantity but instead becomes a measure of the success of the theory in taking account of higher-order processes.

The expression for the total nuclear binding energy in terms of the nucleon kinetic and potential energies can be used to define a total rearrangement energy as a function of the total removal energy, the calculated total kinetic energy, and the measured peak removal energies. Some analyses using various forms of this expression are discussed in §6, and some conclusions are presented in §7.

This brief summary shows that over the years the term 'nuclear rearrangement energy' has been applied in many different ways, causing some confusion. Whenever this term is encountered, it is important to see exactly how it is defined, and whether or not it corresponds to a real physical quantity.

It should perhaps be pointed out that a detailed treatment of the nucleon removal process requires highly technical analysis using the formalism of many-body theory, and no attempt is made here to describe this work in detail. The aim of this review is the more modest one of describing some of the work that has been done from the experimental point of view and showing how concepts may be defined to attain some understanding of what is going on. Where appropriate, reference is made to the detailed theoretical papers, and in them may be found a full account of the calculations whose results are summarized here. No attempt has been made to present a complete historical account of the subject, so many important early papers are not referred to.

Throughout this review, all energies are in MeV and all lengths in fm (10^{-13} cm).

2. Nucleon removal from a nucleus

It is useful to begin by describing what happens when a particle is removed from a nucleus, for example by a pickup or knockout reaction, and thus to define some of the concepts used.

For definiteness we consider the (d, h) and (p, 2p) reactions that remove a proton from a nucleus, though all the results are equally applicable to other nucleon removal reactions. The (d, h) reaction is indeed particularly suitable for nucleon removal studies because it usually has good angular momentum matching in the range of Q values of interest.

In the proton pickup reaction, for example ^{24}Mg(d, h)^{23}Na, the outgoing helions at a particular angle have an energy distribution with sharp peaks corresponding to the states of the final nucleus ^{23}Na, as shown in figure 2. The same peaks are of course found at all angles, but their intensities change, and analysis of the angular distributions gives the orbital angular momentum quantum number and the spectroscopic factor for each transition. As the incident deuteron energy changes, the same peaks are again found, but their intensity varies with incident energy corresponding to the changes in dynamical conditions. The cross section for this reaction may be written in the plane-wave Born approximation (Gross and Lipperheide 1970)

$$\frac{d^3\sigma}{dE_h d\Omega_h} = \frac{2\pi}{(2\pi\hbar)^3} \frac{m_d m_p p_h}{\hbar p_d} |M(p_h, p_d, p_1)|^2 P(\boldsymbol{p}, E) \tag{2.1}$$

where the suffices h and d refer to the outgoing helion and incident deuteron, and $\boldsymbol{p}$ is the momentum of the struck proton. The matrix element

$$M(p_h, p_d, p_1) = \left[W_h - W_d - \frac{m_h}{2m_d m_p}\left(\frac{p_d - 2p_1}{3}\right)^2\right] \times \iint \phi_d(\boldsymbol{r})\, \phi_h(\boldsymbol{x}, \boldsymbol{r}) \exp\left[\tfrac{1}{3}i\,(p_d - 2p)\,x\right] d\boldsymbol{x}\, d\boldsymbol{r} \tag{2.2}$$

is the amplitude for the decay of the helion into a deuteron and a virtual proton. In this expression E_h is the total relativistic energy of the outgoing helion, W_h and W_d are the (negative) binding energies of the helion and the deuteron, and $\phi_d(\boldsymbol{r})$ and $\phi_h(\boldsymbol{x}, \boldsymbol{r})$ are the internal wavefunctions of the incident deuteron and outgoing helion. The spectral function $P(\boldsymbol{p}, E)$ can be interpreted as the probability per unit energy interval of finding a proton of momentum $\boldsymbol{p}$ in the target nucleus of mass A and leaving an energy $E = W_A^{(0)} - W_1$ in the residual nucleus after the proton has been removed. This spectral function contains all the information on the energy level

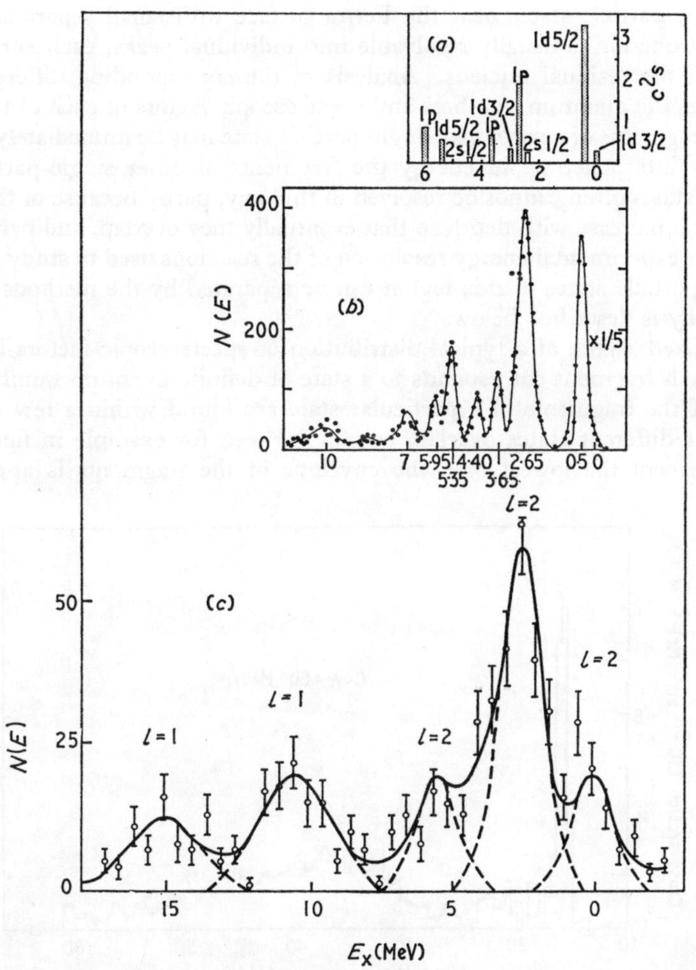

Figure 2. Comparison of excitation energy spectra for proton removal from ^{24}Mg by pickup and knockout reactions (after Wagner 1969): (*a*) ^{24}Mg (d, h) ^{22}Na, $E_d = 40$ MeV; (*b*) ^{24}Mg (d, h) ^{23}Na, $\theta_{lab} = 12°\ 6'$, $E_d = 82$ MeV; (*c*) ^{24}Mg (p, 2p) ^{23}Na, $\theta = 33°$.

spectrum of the residual nucleus and on the spectroscopic factors for transitions to its ground and excited states.

In practice the incident and outgoing nucleons are deflected and partially absorbed by interactions with other nucleons in the nucleus, and this may be taken into account using the distorted-wave Born approximation. This introduces certain technical complications, but the essential physics is contained in the above plane-wave expression. Indeed, at sufficiently high energies the main effect of these interactions is to reduce the flux of particles rather than to alter their momenta, so (2.1) may be used directly with an appropriate reduction factor to take account of the absorption (Gross and Lipperheide 1970, Kullander *et al* 1971).

The analysis of a set of experimental data gives the spectral function corresponding to the energy range of the data; this comprises the energies of the single-particle states of the residual nucleus together with their corresponding spectroscopic factors.

For single-particle states near the Fermi surface with small separation energies the spectral function is usually resolvable into individual peaks, each corresponding to a state in the residual nucleus. Analysis of the corresponding differential cross sections gives the quantum numbers and spectroscopic factors of each of these states, so that the fragments of a particular single-particle state may be immediately identified, even if they interpenetrate in energy the fragments of other single-particle states. The deeper states often cannot be resolved in this way, partly because of their natural widths, which increase with depth so that eventually they overlap, and partly because of the poorer experimental energy resolution of the reactions used to study this region. The single-particle states in this region can be separated by the methods of spectral function analysis described below.

An idealized sketch of a typical distribution of spectroscopic factors is shown in figure 3. Each fragment corresponds to a state of definite quantum numbers (*NJL*), and most of the fragments of a particular state are found within a few MeV. The fragments of different states overlap, as may be seen for example in figure 2. For each state except the lowest one, the envelope of the fragments is approximately

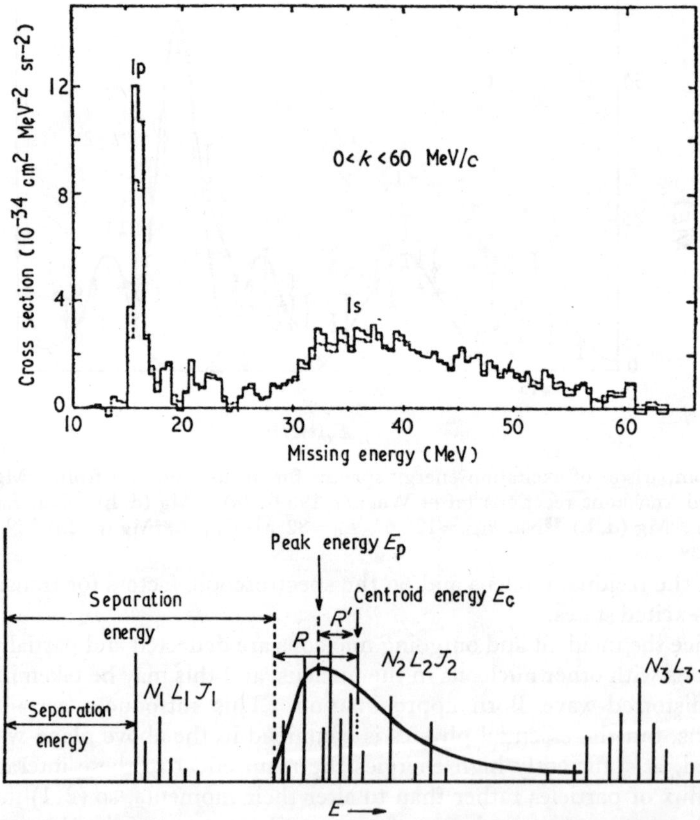

Figure 3. Spectroscopic factors for nucleon removal as a function of the excitation energy of the residual nucleus. The envelope of the spectroscopic amplitudes is sketched for one of the states, and the peak energy, the centroid energy, the separation energy, and the rearrangement energies R and R' are shown. For comparison the missing-energy spectrum for the reaction ^{12}C (e, e′p) is shown (after Bernheim *et al* 1974a).

Lorentzian in the region of the peak, and in each case the short-range correlations shift a few fragments to much higher energies, maybe tens of MeV, above the main concentration of strength, and even into the continuum. Although the total strength in these faraway fragments is only a few per cent of the total (Clement and Perez 1973, 1974), they are important for two reasons: firstly, if the mean energy of the state is defined as an energy-weighted centroid of the fragments,

$$\epsilon = \sum_i \mathscr{S}_i E_i \Big/ \sum_i \mathscr{S}_i \tag{2.3}$$

where the $\mathscr{S}_i$ are the spectroscopic factors of the fragments, then the value obtained from experiment depends quite critically on how many of these fragments are found. Secondly, the energy spread may be so great that fragments of the (NJL) state interpenetrate those of the $(N+1, JL)$ state. The principal quantum number can be determined from the number of nodes in the spatial wavefunction, which can be found from the Fourier transform of the momentum wavefunction. However, with present techniques this can only just be done for strong states, and an attempt to do the same for small fragments would encounter extreme experimental difficulties. The separation of fragments of different principal quantum numbers would be even more difficult if they were broadened and showed significant overlap. It is thus at present impossible in practice to determine accurately the centroid of all the fragments of a particular state. Thus we have to be content with the centroid of the states in the region of greatest concentration of strength; this not very precise quantity is referred to as the 'local centroid'. If the distribution is symmetrical in this region, this is the same as the 'peak energy'. All experimental determinations of centroid energies give this quantity and not the real centroid.

It is also useful to define the separation energy as the energy of the fragment of a particular state that has the lowest energy. This is a well defined quantity for the ground state, since most of the strength is usually concentrated in the fragment of lowest energy; but for higher states the lowest component may be a very small fragment, and so for these states the separation energy has no useful meaning. This definition is a generalization of the usual definition of separation energy as the energy required to raise one nucleon to the continuum energy; when applied to excited states, the energy of excitation is added to the separation energy of the lowest state to give the separation energies appropriate to the higher states, as shown in figure 3.

In §4 it is shown that the nuclear rearrangement energy can be chosen in two ways. The first leads to the definition

$$R = E_c - E_s \tag{2.4}$$

where E_c is the centroid energy and E_s the separation energy. The second leads to

$$R' = E_c - E_p \tag{2.5}$$

where E_p is the peak energy. For the lowest state only the peak energy is the same as the separation energy, so for this state $R \equiv R'$. For excited states the two energies, both referred to as rearrangement energies, are quite distinct.

The pickup reaction may be used to determine the fragment distribution with good resolution for states in the region of the Fermi surface, but if the incident energy is raised to examine the deeper states, the individual peaks in the missing-energy spectrum become increasingly difficult to resolve. At higher energies the (d, h) reaction proceeds mainly by a higher-order process that takes place in two stages: firstly the deuteron breaks up into a proton and a neutron, and then one of these combines with a highly correlated pair of nucleons to form the emitted helion (Strobel and von Geramb 1973). Above the threshold for nucleon emission this gives a continuous spectrum that obscures the discrete peaks due to the simple pickup process. There is some recent evidence (Sakai *et al* 1974) that the (h, α) reaction, which is not so likely to proceed by such a breakup process, is able to remove particles from deeply lying states in a way that allows their energy to be determined experimentally.

Similar results are obtained with the (p, d) reaction, and figure 4 shows a comparison between the cross sections for nucleon removal from ^{16}O by the (p, 2p) and (p, d) reactions. The (p, d) reaction reveals more detailed structure at low excitation energies, but has a much higher background due to breakup processes at the higher energies.

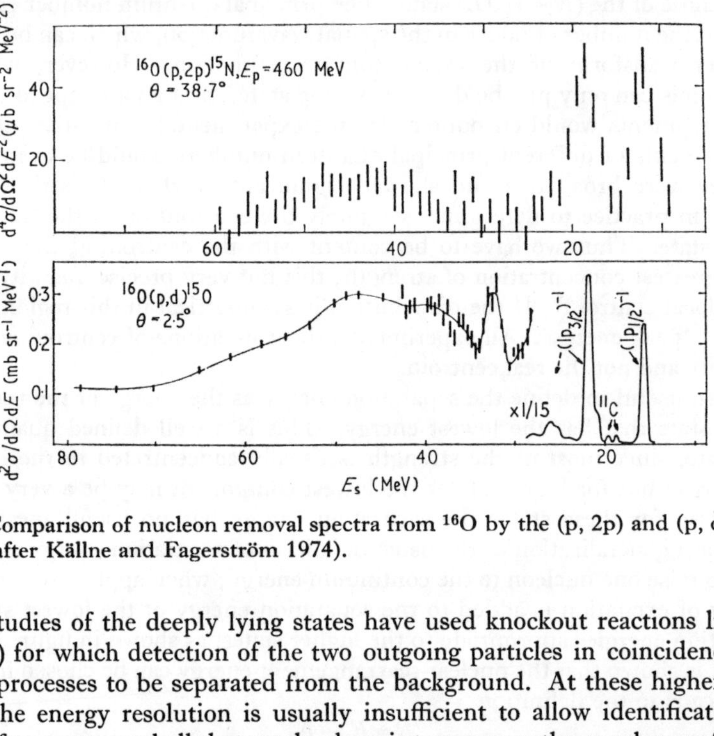

Figure 4. Comparison of nucleon removal spectra from ^{16}O by the (p, 2p) and (p, d) reactions (after Källne and Fagerström 1974).

Most studies of the deeply lying states have used knockout reactions like (p, 2p) and (e, e′p) for which detection of the two outgoing particles in coincidence enables the direct processes to be separated from the background. At these higher energies, however, the energy resolution is usually insufficient to allow identification of the individual fragments, and all that can be done is to measure the envelopes of the peaks as shown in figure 2. This leads to inaccuracies when fragments of the different peaks interpenetrate. The averaged characters of the (p, 2p) and (e, e′p) data require special techniques for their analysis, and these will now be described. In the knockout reactions, for example $^{24}Mg(p, 2p)^{23}Na$, the emission of two particles removes a kinematic constraint, and this increases the complexity of the data. The reaction goes to the same final states of ^{23}Na as the (d, h) reaction, and, by fixing two angles, the cross section for a particular incident energy may be obtained as a function of the excitation energy of the final nucleus. When the dynamical factors are removed, this gives the same spectroscopic factors as before. The distribution for the reaction $^{24}Mg(p, 2p)\,^{23}Na$ is given in figure 2(*c*), and comparison of the two distributions shows the correspondence of the peaks, and, incidentally, the typically poorer resolution of the (p, 2p) data.

The earlier studies of knockout reactions used protons, but this has the disadvantage that the interpretation of the (p, 2p) data is greatly complicated by the high probability that the protons will undergo multiple scattering before or after the quasi-elastic interaction. In this respect the (e, e′p) reaction is much better, but it has the disadvantage of absolute cross sections lower by four or five orders of magnitude. With the advent of high-intensity electron linear accelerators this difficulty has been largely overcome, so many of the most accurate proton removal experiments are now made with the latter reaction. The (e, e′p) reaction also has the advantage that the Born approximation may be used in place of the impulse approximation. The steady improvement in the experimental determinations of nucleon removal energies has recently been reviewed by Wagner (1973).

Historically the formalism of quasi-elastic knockout reactions has been developed for the (p, 2p) reaction, and this will be continued here, remembering that all the formal results are equally applicable to the (e, e′p) reaction. The formalism has been reviewed in detail by Jacob and Maris (1966, 1973), and only a brief summary will be given here.

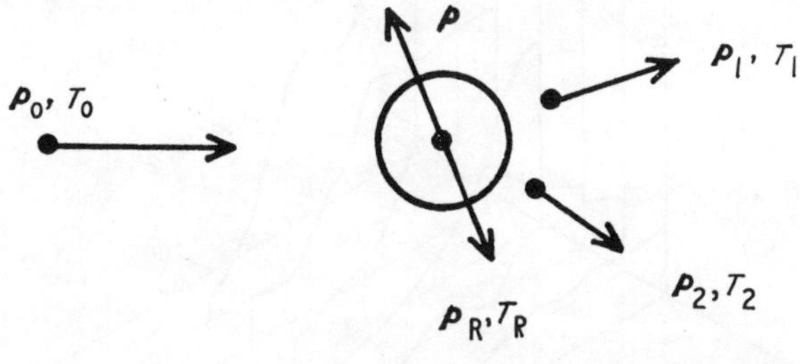

Figure 5. Energies and momenta in a (p, 2p) knockout reaction.

Let ($\boldsymbol{p}_0$, T_0) be the momentum and kinetic energy of the incident particle, and ($\boldsymbol{p}_1$, T_1), ($\boldsymbol{p}_2$, T_2) and ($\boldsymbol{p}_R$, T_R) the corresponding quantities for the two outgoing particles and the recoiling residual nucleus respectively (figure 5). Then energy and momentum conservation for the reaction A(p, 2p) B gives

$$T_0 = T_1 + T_2 + T_R + E^* - Q \tag{2.6}$$

and

$$\boldsymbol{p}_0 = \boldsymbol{p}_1 + \boldsymbol{p}_2 + \boldsymbol{p}_R \tag{2.7}$$

where E^* is the excitation energy of the residual nucleus and $Q = E_A - E_B$ is the Q value of the reaction, E_A and E_B being the ground-state energies of A and B.

At high energies the incident particle collides with only one target nucleon, and if this has momentum $\boldsymbol{p}$ and it is assumed that the outgoing particles do not interact with the nucleus, then

$$\boldsymbol{p}_0 + \boldsymbol{p} = \boldsymbol{p}_1 + \boldsymbol{p}_2 \tag{2.8}$$

so that $\boldsymbol{p} = -\boldsymbol{p}_R$. Thus the interaction suddenly removes a nucleon from the nucleus, and the energy required to do this is

$$T_0 - (T_1 + T_2 + T_R) = E^* - Q \equiv E. \tag{2.9}$$

This energy is called the *removal* energy.

These relations make it possible to obtain both $\boldsymbol{p}$ and E by measuring T_0, T_1, T_2 and the angles of emission of the outgoing particles. Then at high energies, where the plane-wave impulse approximation is valid, the cross section for the reaction is given by

$$\frac{d^6\sigma}{dE_1 d\Omega_1 dE_2 d\Omega_2} = \frac{2\pi}{(2\pi\hbar)^6} \frac{E_0 E_1 E_2 p_1 p_2}{\hbar p_0 c^6} |M(p_1 p_2, p_0 p)|^2 \, P(\boldsymbol{p}, E) \tag{2.10}$$

where $M(p_1 p_2, p_0 p)$ is the off-shell two-body scattering matrix element for the proton–proton scattering in the nucleus, and $P(\boldsymbol{p}, E)$ is the same spectral function as before. If the kinematics of the reaction are chosen so that the proton–proton interaction is nearly on the energy shell, the value of the matrix element obtained from analyses of proton–proton scattering can be used to obtain the required spectral function. A typical result, for the (p, 2p) reaction at 385 MeV on Sn, is shown in figure 6. A more accurate analysis may be made using the distorted-wave formalism.

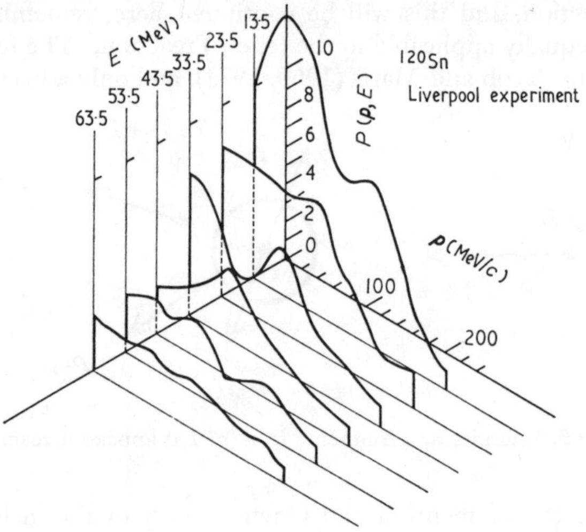

Figure 6. Probability or spectral function for single-particle removal as a function of the momentum $\boldsymbol{p}$ of the proton removed and of the energy E of the final state (after Koltun 1972).

Providing the removal is sufficiently rapid for there to be no appreciable interaction between the outgoing particles and the nucleus, the spectral function $P(\boldsymbol{p}, E)$ obtained in this way is independent of the incident energy. At lower energies this may not be so, and this situation is discussed in §4.

Formal expressions for $P(\boldsymbol{p}, E)$ may be obtained by defining $a(\lambda)$ as the removal operator for a particle in the state λ in the nucleus $|A\rangle$. Then a_i, the probability amplitude for leaving the residual nucleus $|B\rangle$ in the state i, is given by (Koltun 1972, 1974, Dieperink *et al* 1972)

$$a(\lambda)|A\rangle = \sum_i a_i |B_i\rangle \tag{2.11}$$

so that

$$a_i = \langle B_i | a(\lambda) | A\rangle. \tag{2.12}$$

The probability $P(\lambda, E)$ that removal of a particle from the state λ in $|A\rangle$ leaves the residual nucleus with excitation energy E is then

$$P(\lambda, E) = \sum_i |a_i|^2 \, \delta(E - E_i). \tag{2.13}$$

This may also be written

$$P(\lambda, E) = \langle \mathrm{A} | a^{\dagger}(\lambda)\, \delta(E-H)\, a(\lambda) | \,\mathrm{A}\rangle \tag{2.14}$$

where H is the nuclear Hamiltonian.

The distorted-wave analysis gives a more complicated expression in terms of the non-diagonal spectral function (Gross and Lipperheide 1970)

$$P(\lambda, \lambda', E) = \langle \mathrm{A} | a^{\dagger}(\lambda')\, \delta(E-H)\, a(\lambda) | \,\mathrm{A}\rangle. \tag{2.15}$$

The off-diagonal terms appear to be small, and in what follows this will be assumed to be so.

In principle, if the orbital wavefunctions $\phi_{\lambda}(\boldsymbol{p})$ are known, it is possible to obtain the $P(\lambda, E)$ from the off-diagonal spectral function in the momentum representation

$$P(\lambda, E) = \iint \phi_{\lambda}^{*}(\boldsymbol{p}')\, P(\boldsymbol{p}, \boldsymbol{p}', E)\, \phi_{\lambda}(\boldsymbol{p})\, \mathrm{d}\boldsymbol{p}\, \mathrm{d}\boldsymbol{p}'. \tag{2.16}$$

In practice it is easier to analyse the experimental $P(\boldsymbol{p}, E)$ into the components coming from states of different λ by fitting it by a finite set of oscillator functions

$$P(\boldsymbol{p}, E) = \sum_{\lambda\lambda'} \phi_{\lambda}^{*}(\boldsymbol{p})\, \phi_{\lambda'}(\boldsymbol{p})\, P(\lambda, \lambda', E) \approx \sum_{\lambda} |\phi_{\lambda}(\boldsymbol{p})|^{2}\, P(\lambda, E). \tag{2.17}$$

The results of such an analysis are given in figure 7 and show how it is possible to assign quantum numbers to each state and to obtain the distributions of their strengths as a function of the removal energy from the target nucleus.

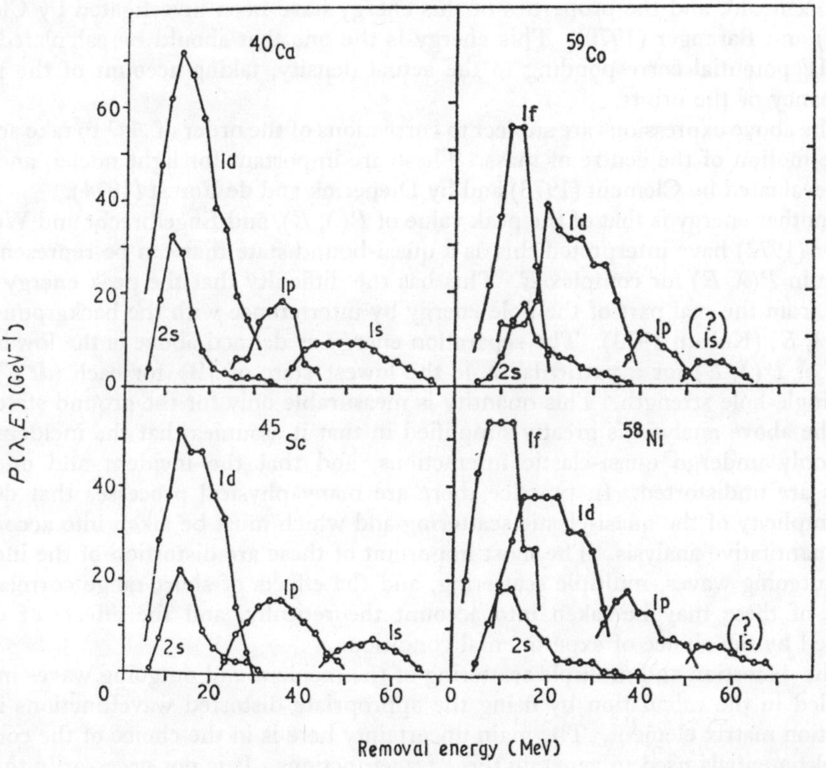

Figure 7. Orbital probability functions for several nuclei (after James *et al* 1969).

The momentum distributions of the struck nucleus in each state may be obtained by integrating the spectral function over E. It is useful to define the centroid energies (or mean removal energies) of these distributions by the relations (Koltun 1972, 1974, Shakin and Da Providencia 1971, Dieperink *et al* 1972)

$$E_\lambda = \epsilon(\lambda)/n(\lambda) \tag{2.18}$$

where

$$\epsilon(\lambda) = -\int_{-\infty}^{\infty} EP(\lambda, E)\,\mathrm{d}E = \langle \mathrm{A}|a^\dagger(\lambda)\,[a(\lambda), H]|\mathrm{A}\rangle \tag{2.19}$$

and

$$n(\lambda) = \int_{-\infty}^{\infty} P(\lambda, E)\,\mathrm{d}E = \langle \mathrm{A}|a^\dagger(\lambda)\,a(\lambda)|\,\mathrm{A}\rangle. \tag{2.20}$$

This definition has been quite widely used, although it is subject to serious uncertainty in practice as $P(\lambda, E)$ must be known for all E and $\lambda\,(K)$.

Inserting a complete set of intermediate states gives the usual definition of the single-particle energy as the centroid of the fragments

$$E_\lambda = \sum_i \mathscr{S}_i^\lambda\,(E_0^{\mathrm{A}} - E_i^{\mathrm{A}-1})\Big/\sum_i \mathscr{S}_i^\lambda \tag{2.21}$$

where the spectroscopic factors

$$\mathscr{S}_i^\lambda = |\langle \mathrm{B}_i|\,a(\lambda)|\,\mathrm{A}\rangle|^2. \tag{2.22}$$

It is possible to combine the particle removal and particle addition spectra to give a mean centroid, and the properties of this energy have been investigated by Clement (1969) and Baranger (1970). This energy is the one that should be calculated from the HF potential corresponding to the actual density, taking account of the partial occupancy of the orbits.

The above expressions are subject to corrections of the order of A^{-1} to take account of the motion of the centre of mass. These are important for light nuclei, and have been evaluated by Clement (1973) and by Dieperink and de Forest (1974).

Another energy is that of the peak value of $P(\lambda, E)$, and Engelbrecht and Weidenmuller (1972) have interpreted this as a quasi-bound state that can be represented as a pole in $P(\lambda, E)$ for complex E. This has the difficulty that the peak energy could differ from the real part of the pole energy by interference with the background part of $P(\lambda, E)$ (Koltun 1973). The separation energy as defined above is the lower end-point of $P(\lambda, E)$ for a particular λ, ie the lowest state of $|\mathrm{B}\rangle$ for each $(J^\pi T)$ with any single-hole strength. This quantity is measurable only for the ground state.

The above analysis is greatly simplified in that it assumes that the incident protons only undergo quasi-elastic interactions, and that the incident and outgoing waves are undistorted. In practice there are many physical processes that destroy the simplicity of the quasi-elastic scattering and which must be taken into account in any quantitative analysis. The most important of these are distortion of the incident and outgoing waves, multiple scattering, and the effects of short-range correlations. Some of these may be taken into account theoretically, and the effects of others reduced by the choice of experimental conditions.

The distortion and multiple scattering of the incident and outgoing waves may be included in the calculation by using the appropriate distorted wavefunctions in the transition matrix element. The main uncertainty here is in the choice of the complex optical potentials used to generate these wavefunctions. It is not necessarily the case that the best potentials for this purpose are those obtained from an analysis of the corresponding elastic scattering cross sections; indeed it may be preferable to use potentials derived directly from the nucleon–nucleon interaction.

Since the mean free path for the interaction of the incident proton in the (p, 2p) reaction is generally less than the nuclear radius, there is an appreciable likelihood that unless it interacts in the nuclear surface it will be scattered by another nucleon before the quasi-elastic interaction. Furthermore the two protons from this interaction may themselves be scattered before they emerge from the nucleus. Such multiple scattering naturally upsets the clean kinematics of the quasi-elastic process, broadens the sharp energy and angular peaks, and smears the spectral function, especially if one or more of the additional nucleons involved in the multiple scattering themselves escape from the nucleus. However, the angular distribution of the emitted protons usually still shows marked correlations due to the quasi-elastic process, in particular the 90° difference between their angles of emission, but this peak is broadened and superposed on a background due to interactions where multiple scattering has taken place. Thus providing the quasi-elastic peak is not too weak or broad it may be separated from the multiple-scattering events. Multiple scattering still has to be taken into account because it reduces the magnitude of the quasi-elastic peak, and this is included in the distorted-wave calculation through the imaginary part of the optical potential. Thus both the distortion and the multiple-scattering effects may be included in the theoretical analysis, to an accuracy that depends on the reliability of the optical potentials used.

Even if measurements are confined to the region of the quasi-elastic peak, the multiple-scattering effects are important. For example, in an (e, e′p) reaction the outgoing proton may be scattered and so lose energy to the nucleus, so that the low-energy part of the spectrum is reduced by an energy- and momentum-dependent factor that in practice may vary in the range 0·3–0·7. This transfers part of the lower-energy region of the spectral function to higher energies. The effect depends on the size of the nucleus, but even in ^{16}O the average absorption is about 50%, so that the tail contains at least half the strength of the spectrum. It is possible with distorted-wave calculations to take this absorption into account for the low-energy peaks, but the tail is a far more difficult problem. The situation for (p, 2p) is similar but more severe, with reduction to 10% or less, so that the tail has an order of magnitude more weight than the low-energy end of the spectrum.

The short-range correlations pose a much more difficult problem, since they are completely interwoven with the dynamics of the nuclear states and imply a basic deviation from the simple picture of knocking a proton out of a single-particle state in the target nucleus. The physical basis of this effect is the possibility that the incident proton may collide with the target proton just when the latter is interacting strongly with another target nucleon. This will destroy the quasi-elastic nature of the interaction, and since the two correlated target nucleons are likely to be ejected from the nucleus with high energies, such events will also contribute to the high-energy tail of the spectrum.

Another way of looking at this is to expand the wavefunction of the ejected proton in terms of the nuclear shell model single-particle wavefunctions. There is a high overlap with the low-lying bound states, but the wound in the wavefunction due to the short-range correlations introduces an appreciable component of higher-energy states. There is thus still a large probability that the residual nucleus will be left in one of its low-lying states after the proton has been ejected, but nevertheless an appreciable probability that it will be left in a highly excited state (Brueckner 1959).

The effects of all these processes on the spectrum may be illustrated by a schematic diagram due to Maris (1974) given in figure 8. In this diagram the first line (*a*) corresponds to the extreme single-particle model with infinitely sharp states and strengths corresponding to the numbers of protons in the shells concerned. The second line (*b*) shows the effects of including the soft residual interactions: the single-particle states are each split into several fragments, each with a finite width, and those at higher excitation energies are significantly broadened. This is the situation considered in the first part of this section. If now the short-range correlations are included, it becomes quite likely that the incident particle interacts with a highly correlated pair, and the corresponding high-momentum components can result in the emission

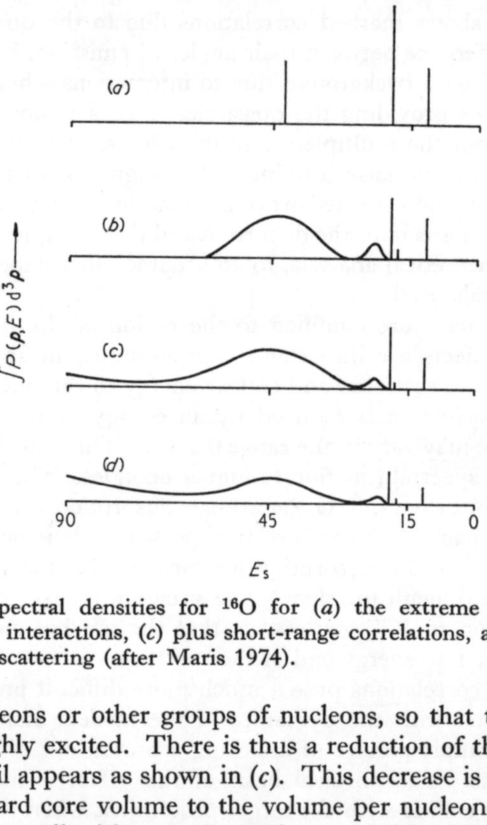

Figure 8. Schematic spectral densities for ^{16}O for (*a*) the extreme single-particle model, (*b*) with soft interactions, (*c*) plus short-range correlations, and (*d*) plus distortion by multiple scattering (after Maris 1974).

of two or more nucleons or other groups of nucleons, so that the residual $(A-1)$-particle system is highly excited. There is thus a reduction of the low-energy peaks, and a high-energy tail appears as shown in (*c*). This decrease is given approximately by the ratio of the hard core volume to the volume per nucleon; this is about 10%, which compares quite well with an estimate from many-body theory of about 15% (Brueckner and Goldman 1959).

The effect of multiple scattering is also to transfer some of the lower-energy part of the spectral function to higher energies, and the result of this is shown in (*d*). The combined effect of short-range correlations and multiple scattering is thus to give a spectral function that is qualitatively different from that assumed in the simplified quasi-elastic analysis outlined above.

The fragmentation of single-particle states may also be studied by nucleon transfer reactions in heavy nuclei. Such nuclei have high level densities at relatively low excitation energies, so it is possible to measure their strength functions from the nucleon transfer cross sections. Such an investigation has been made by Back *et al*

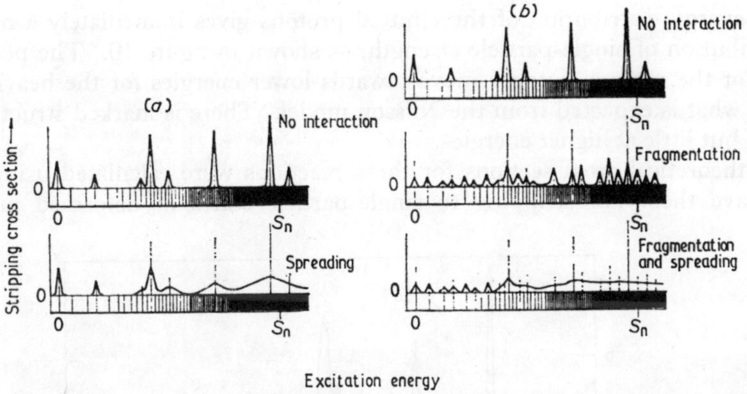

Figure 9. Sketch illustrating (*a*) the gradual spreading of the single-particle states over compound levels, and (*b*) the fragmentation of the single-particle states resulting from strong configuration mixing. The effect of combined fragmentation and spreading is also shown. The total level density is illustrated below each graph (after Back *et al* 1974).

(1974) for (d, p) reactions on some rare earth nuclei. They distinguish between the 'spreading' of the single-particle states (due to weak interactions) and the 'fragmentation' due to strong configuration mixing, as shown in figure 9. The fragmentation can result from a coupling of the single-particle states to vibrational degrees of freedom or from coupling between one- and three-quasi-particle states. The spreading width is expected to increase with excitation energy, while the fragmentation width remains almost constant.

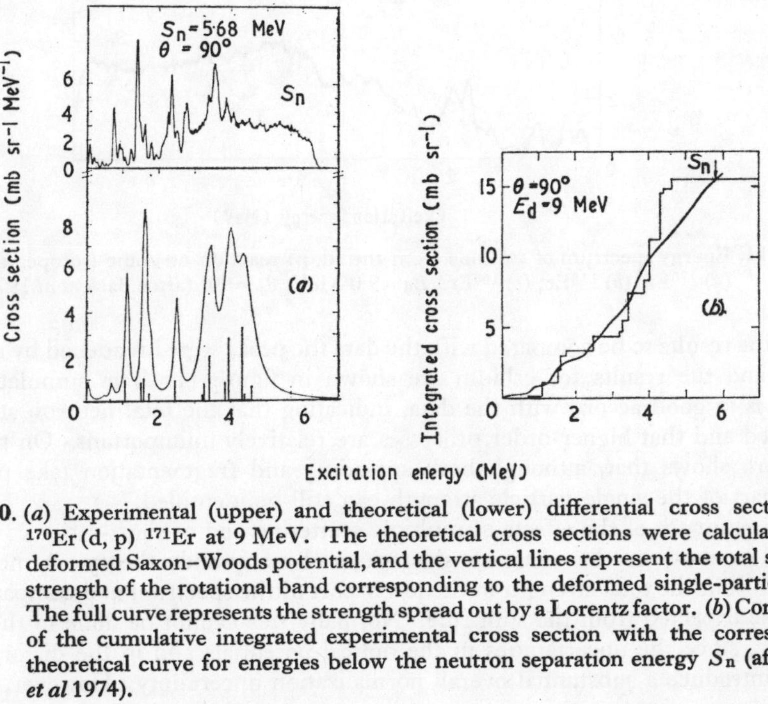

Figure 10. (*a*) Experimental (upper) and theoretical (lower) differential cross sections for ^{170}Er (d, p) ^{171}Er at 9 MeV. The theoretical cross sections were calculated in a deformed Saxon–Woods potential, and the vertical lines represent the total stripping strength of the rotational band corresponding to the deformed single-particle state. The full curve represents the strength spread out by a Lorentz factor. (*b*) Comparison of the cumulative integrated experimental cross section with the corresponding theoretical curve for energies below the neutron separation energy S_n (after Back *et al* 1974).

The energy distribution of the emitted protons gives immediately a measure of the distribution of single-particle strength, as shown in figure 10. The peaks in the spectra for the erbium isotopes move towards lower energies for the heavier nuclei, which is what is expected from the Nilsson model. There is marked structure below 40 MeV, but little at higher energies.

The theoretical cross sections for these reactions were calculated using the distorted-wave theory of stripping to single-particle states in deformed nuclei. To

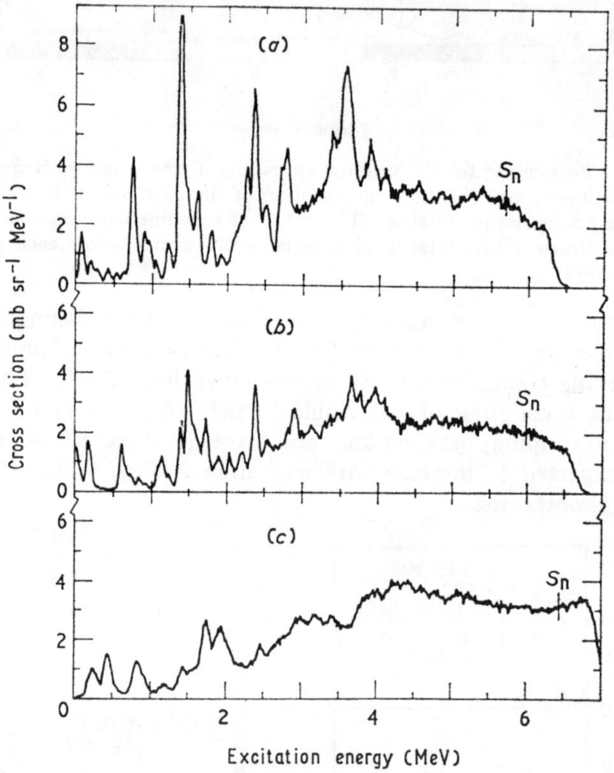

Figure 11. Energy spectrum of protons from the (d, p) reaction on some isotopes of erbium: (*a*) ^{170}Er; (*b*) ^{168}Er; (*c*) ^{166}Er. $E_d = 9{\cdot}0$ MeV, $\theta_p = 90°$ (after Back *et al* 1974).

enable the results to be compared with the data the peaks were broadened by a Lorentz factor, and the results for erbium are shown in figure 11. The cumulative cross section is in good accord with the data, indicating that the total neutron strength is preserved and that higher-order processes are relatively unimportant. On the whole this work shows that, although both spreading and fragmentation take place, the major part of the single-particle strength can still be identified.

The estimates of the effects of multiple scattering and of short-range correlations referred to above are based on considerations of many-body theory. A more direct way is to sum the total strength in the lower part of the spectrum and compare it with the value expected from the sum rule. Normally this cannot be done to the desired accuracy, since the uncertainties in the optical potentials and in the distorted-wave theory introduce a substantial overall normalization uncertainty. However, Clement and Perez (1973, 1974) have recently shown that these uncertainties may be reduced by applying the j-dependent sum rules to stripping and pickup reactions on the same nucleus. In the particular case of the $1f_{7/2}$ state of ^{45}Sc they find at least 90% of the strength in the peaks in the lower end of the spectrum. In this case the combined effects of multiple scattering and short-range correlations therefore amount to less than 10%. This is substantially less than the values estimated from many-body theory for the knockout reactions, so it seems that the stripping and pickup reactions are better for determining the spectral function, though it may also be that the magnitude of the effect depends quite strongly on the state and nucleus concerned.

The effects due to multiple scattering thus depend strongly on the way the proton is removed from the nucleus, and can be minimized by an appropriate choice of experimental conditions. The effect of the short-range correlations is different in that it is always present and remains the same whatever the experimental conditions.

A study of the effects of short-range correlations on nucleon pickup reactions has been made by Blum *et al* (1973) for the reaction $^{16}O(p, d)\,^{15}O$ from 25 to 45 MeV. They modified the wavefunction of the transferred neutron by a Jastrow correlation function and found that this introduced more structure into the angular distributions and improved the agreement with the experimental data. Since the short-range correlations reduce the wavefunction near the origin, the normalization condition necessitates a corresponding increase at larger distances and hence larger calculated cross sections. This reduces the spectroscopic factors extracted by comparison with the experimental data and brings them nearer to the shell model values. However, the overall fits are still not particularly good, and it is likely that the effects of short-range correlations are lost in the uncertainties in the optical potentials, particularly for so light a nucleus.

There is yet another effect that modifies the measured spectral function: if the lifetime of the state is comparable with or less than the reaction time, then it can begin to decay before it is properly formed. In this case the state is broadened, and it is not possible to separate the reaction mechanism from the structure of the target nucleus, for this determines how the hole state will decay. These time-dependent effects are considered in §4.

It is very difficult to calculate the hole spectral function with any accuracy, but there have been preliminary attempts that reproduce some of the general features of the observed spectra. Becker (1973) has calculated the distribution of hole strengths after particle removal using the Brueckner shell model, and the results for the $(1s_{1/2})^{-1}$ strength distribution between 35 and 60 MeV in ^{15}O are shown in figure 12. He used the Hamada–Johnston interaction and removed the spurious states due to motion of the centre of mass. As shown in the figure, this spectrum is very similar to the experimental (p, 2p) cross section obtained by Tyrén *et al* (1966).

Extensive calculations have been made by Lipperheide and collaborators. They used a Green function method with a mass operator determined up to the second order in the shell model space of bound and continuous single-particle states to calculate the separation energy spectra in the $^{12}C(p, 2p)$ and $^{16}O(p, 2p)$ reactions (Gross and Lipperheide 1970, Wille *et al* 1971, Wille and Lipperheide 1972). Some of their results are compared in figure 13 with subsequent microscopic calculations by Fritsch *et al* (1972, 1973) of the hole spectrum of ^{11}C based on the continuum shell model with residual interactions and including all correlations among 1h and 1p2h configurations in the residual ^{11}B system. There are substantial differences between the two calculations, indicating that the coupling among the various 1p2h configura-

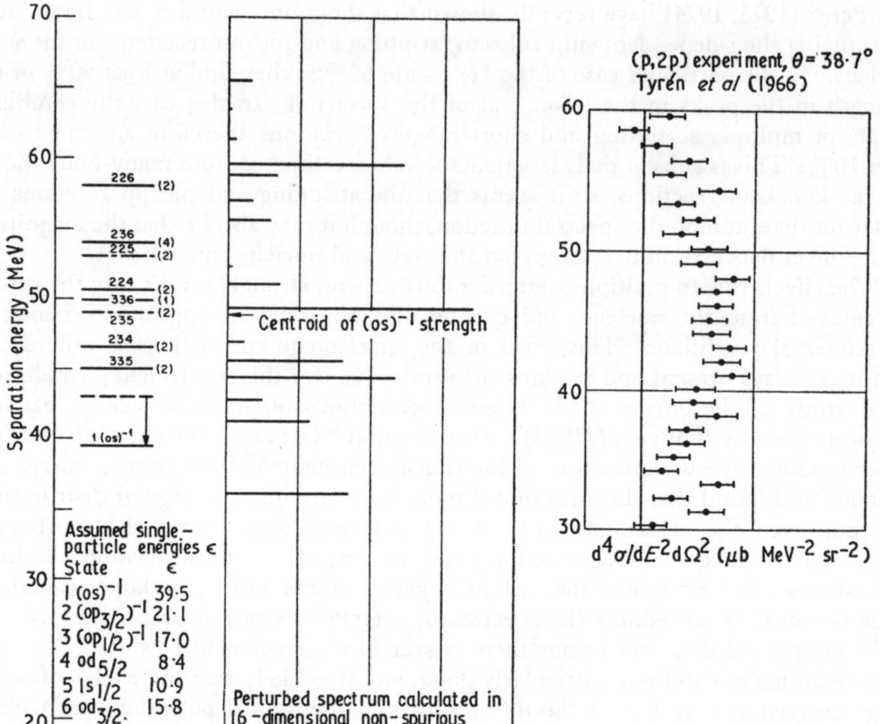

Figure 12. Energy spectrum and single-particle strength of $s_{1/2}$ states in ^{15}O obtained by mixing $(1s)^{-1}$ and quasi-degenerate $(1p)^{-2}$ (sd) states in a Brueckner shell model calculation. The experimental data of Tyrén *et al* (1966) for final states in ^{15}N have been shifted by a Coulomb energy for comparison with the calculated spectrum in ^{15}O (after Becker 1973).

tions is rather strong, so that no peak in figure 13(*a*) can be associated with a particular 1p2h state. The calculations showed that the narrow peaks of figure 13(*b*) gradually merge into the broad structure of figure 13(*a*) as the coupling between the 1p2h configurations is increased from zero to its full value.

The method of Gross and Lipperheide has been applied by Faessler *et al* (1972) to the ^{16}O (d, h) excitation spectrum. The reaction mechanism is described by the distorted-wave theory, and the model takes into account explicitly the effect of residual interactions due to the short-range correlations and the decay of the residual nucleus by particle emission. They included the higher-order effects of the residual interaction on the self-energy operator, and some of their results with the Lemmer–Shakin potential for the residual interaction are shown in figure 14. The fragmentation of the single-particle states is due to coupling to 1p2h (one particle, two holes), 2p3h and higher configurations. For example, the $1s_{1/2}$ single-hole configuration can couple to nineteen 1p2h configurations, giving twenty peaks in the spectral function, and so on for the higher-order configurations. The calculated spectral function has peaks near the 1p2h configuration energies and is a continuous function of energy when the residual nucleus can decay into the continuum by particle emission.

These calculations reproduce some of the overall features of the experimental excitation spectra, but are not yet at the stage where a detailed comparison with experiment is useful. Indeed this work of Faessler *et al* shows that the perturbation expansion used to calculate the separation energies converges only for energies less than 20 MeV. For higher energies a new approach is needed. Thus in principle it

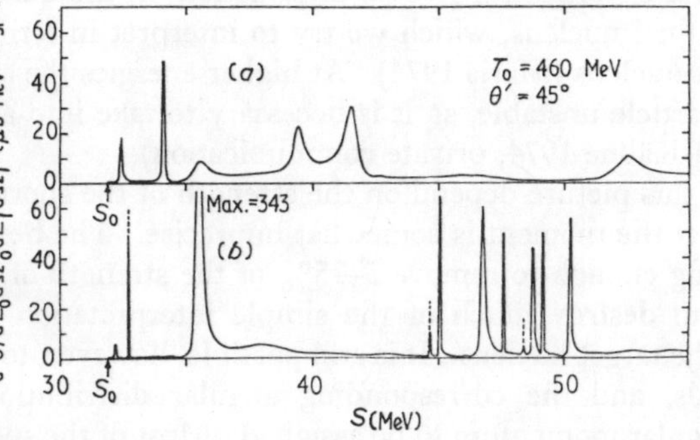

Figure 13. Cross section for the reaction ^{12}C (p, 2p)^{11}B as a function of the proton separation energy *S*, calculated (*a*) in the framework of the continuum shell model, and (*b*) from a second-order mass operator. The energy S_0 indicates the continuum threshold of the spectrum (after Fritsch *et al* 1973).

is possible to calculate the spectral function for comparison with experiment, but so far this has not yet been done with sufficient accuracy for useful comparisons to be made.

If these calculations of the spectral function attain sufficient reliability, they could be used to find the energies and strengths of the small fragments of single-particle strength that are far away from the peak energy, and hence a relation between the real centroid and the peak centroid could be obtained. However, the difficulty of distinguishing between states of different principal quantum number would still remain.

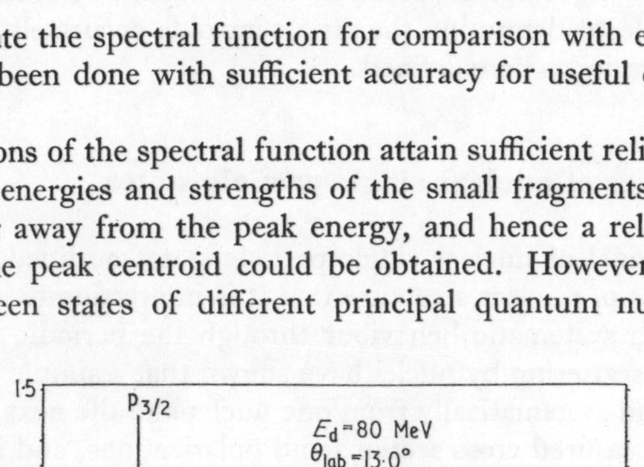

Figure 14. Excitation spectrum for the reaction ^{16}O (d, h)^{15}N calculated using the Lemmer–Shakin potential (after Faessler *et al* 1973a).

The presence of the short-range correlations has an important bearing on the interpretation of the results of proton removal experiments. The energies of the outgoing particles correspond to the states of the residual nucleus, and in the absence of short-range correlations the energy centroids can be interpreted as single-particle states in the target nucleus. If strong short-range correlations are present, however, much of the single-particle strength is shifted to high energies, so that the centroid energy becomes ill-defined. In this case the experimental data provide information only about the energies, lifetimes and momentum distributions of the hole states, so that proton removal experiments give us information on the quasi-stationary single-hole states in the final nucleus, which we try to interpret in terms of single-particle states in the target nucleus (Maris 1974). At higher energies the states in the residual nucleus may be particle unstable, so it is necessary to take into account the breakup channels as well (J Källne 1974, private communication).

The details of this picture depend on the strength of the short-range correlations, and the evidence at the moment is somewhat imprecise. The best estimates indicate that they are strong enough to remove 5–15% of the strength of the main peak, and this is sufficient to destroy much of the simple interpretation in terms of single-particle states in the target nucleus. It is still possible, however, to define and measure the peak centroids, and the corresponding angular distributions usually allow a definite orbital angular momentum to be assigned. Most of the spectroscopic strength is still concentrated in these peaks, and their systematic behaviour from nucleus to nucleus shows that they have a definite physical meaning. To this extent they certainly still give information about the single-particle states of the target nucleus. The short-range correlations blur this picture to some extent, and in particular the small fraction of single-particle strength that is spread in an unknown way over a large energy range makes it impossible to determine the true centroids accurately, and with them the rearrangement energies as here defined.

3. The systematics of nuclear single-particle states

The simple model of nuclear single-particle states outlined in §1 agrees so well with many features of nuclear structure that it is interesting to see to what extent its predictions show a systematic behaviour through the periodic table. Many studies of nucleon elastic scattering by nuclei have shown that a simple optical potential that varies smoothly and systematically from one nucleus to the next is able to give a very good account of measured cross sections and polarizations, and it is therefore natural to see how successful such a potential is in correlating the energies of the bound single-particle states (Wyatt *et al* 1960).

Such a study cannot be concerned with the details of the fragmentation of the single-particle states, for these are peculiar to each nucleus and would not be expected to show any systematic behaviour. It is only the centroid energies that might show such behaviour, though even for them it is possible that the differences in structure from one nucleus to the next are sufficient to destroy it.

There are important differences between the states of bound nucleons and the scattering of free nucleons. For bound nucleons the potential fixes the energy of each state with a particular set of quantum numbers, while for free nucleons many states (partial waves) contribute at each energy, and except at low energies it is not practicable to separate them. It is thus possible to study the potential corresponding to each bound state in a way that is not possible for free nucleons. On the other hand the data to be fitted are much more restricted as they consist of a number of centroid energies for the single-particle states of each nucleus, compared with the extensive cross sections and polarizations as a function of incident energy and scattering angle that are available for scattered nucleons.

The bound-state data can be analysed in two basically different ways, depending on the aspect of the potential that one wants to study. These are firstly to analyse together all the single-particle states in a particular nucleus, and secondly to analyse the energies of a single state in a range of nuclei. Ultimately of course one requires a potential that will account for all the data, so this division is only of methodological convenience. The former method has so far been applied mainly to medium and heavy nuclei, for only in these cases are enough states available in a particular nucleus to make the analysis worthwhile. The latter method has been applied mainly to lighter nuclei where data are available over greater ranges of adjacent nuclei. The results of such analyses will now be described.

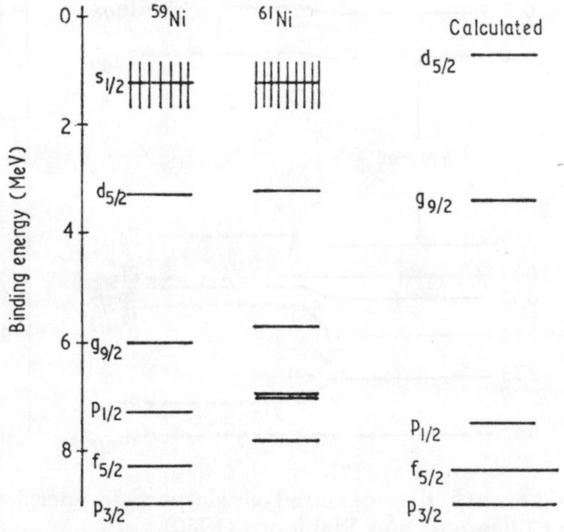

Figure 15. Comparison between measured single-particle energies in ^{59}Ni and ^{61}Ni and those calculated by Brown *et al* (1963).

One of the first analyses of a series of states in a particular nucleus was made by Brown *et al* (1963), and some of their results are shown in figure 15. They used a potential of the form

$$V(r) = Vf(r) + \left(\frac{\hbar}{m_\pi c}\right)^2 V_s \frac{1}{r}\frac{df}{dr} \boldsymbol{L}.\boldsymbol{\sigma} \qquad (3.1)$$

where $f(r) = \{1 + \exp[(r-R)/a]\}^{-1}$ is the Saxon–Woods form factor, and found that the calculated energies are more widely spaced than those found experimentally. This can be corrected by allowing the potential to depend on energy, or, equivalently, by giving the bound particle an effective mass m^* greater than its real mass. The best overall fit was obtained with m^*/m slightly greater than unity. A similar result had earlier been found by Blomqvist and Wahlborn (1960), whose results for ^{208}Pb are shown in figure 16.

Of all the nuclei, most single-particle states have been measured by adding and removing nucleons from ^{208}Pb, and several extensive studies of this nucleus have been made. Rost (1968) optimized the parameters of the standard Saxon–Woods potential (3.1) to fit the states of ^{208}Pb, and obtained a good overall fit. His potentials were used by Elton (1968) to calculate the wavefunctions and hence the neutron and proton density distributions, and the electron elastic scattering cross sections calculated from the latter were found to be in good accord with the experimental data.

There are, however, considerable differences between Rost's proton and neutron potentials: in particular the neutron radius parameter of 1·35 fm is substantially greater than the corresponding proton radius parameter of 1·27 fm, while the spin–orbit radii were 1·35 fm for neutrons and 1·25 fm for protons. The data were therefore re-analysed by Batty and Greenlees (1969) to see if an equally good fit could be

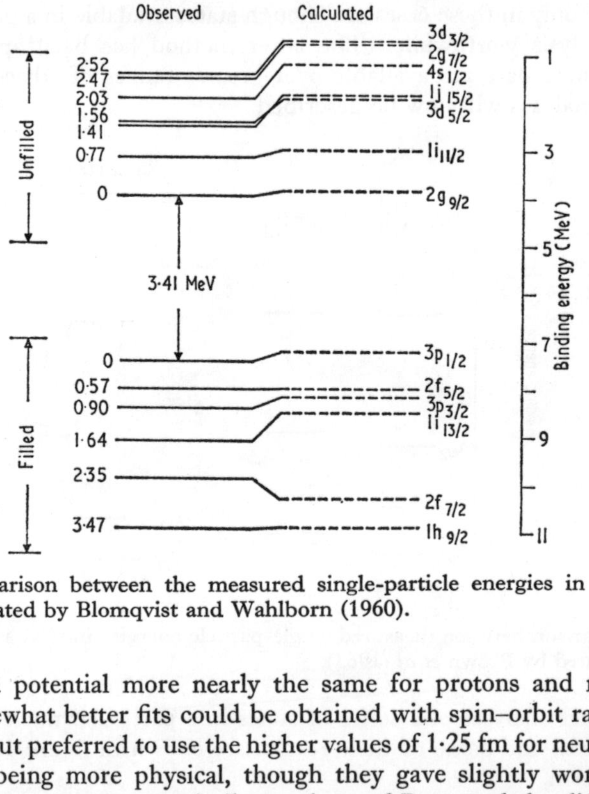

Figure 16. Comparison between the measured single-particle energies in ^{208}Pb and those calculated by Blomqvist and Wahlborn (1960).

obtained with a potential more nearly the same for protons and neutrons. They found that somewhat better fits could be obtained with spin–orbit radius parameters around 0·9 fm, but preferred to use the higher values of 1·25 fm for neutrons and 1·1 fm for protons as being more physical, though they gave slightly worse fits. On the whole their parameters are very similar to those of Rost, and the difference between the best neutron and proton potentials remains. They calculated the proton density distributions and found that they gave cross sections for electron elastic scattering in excellent agreement with the data at 115 and 250 MeV. They also analysed the single-particle states in ^{40}Ca and ^{48}Ca in the same way, and their results for these nuclei as well as for ^{208}Pb are shown in figure 17. The corresponding charge distributions again agreed with the electron elastic scattering, and the difference between the charge distributions of ^{40}Ca and ^{48}Ca was qualitatively reproduced. Similar analyses of the calcium data were made by Gibson and Van Oostrum (1967), Swift and Elton (1966) and Elton and Swift (1967). The success of these analyses suggests that the conclusions of previous analyses giving theoretical energy levels more widely spaced than the experimental ones, and interpreted as requiring an effective mass greater than unity, depend on the particular form factors chosen, and are not justified by the data. There are, however, good theoretical reasons for supposing that the conclusion itself is correct.

The systematic behaviour of particular single-particle states from nucleus to nucleus was first analysed by Cohen (1965) using his own data on (d, p) and (d, t) reactions on a series of nuclei. He found that it is necessary to allow the potential depth to depend on the nuclear asymmetry parameter $(N-Z)/A$ and on the total quantum number of the major shell, and then it is able to give a good qualitative account of the trends from nucleus to nucleus.

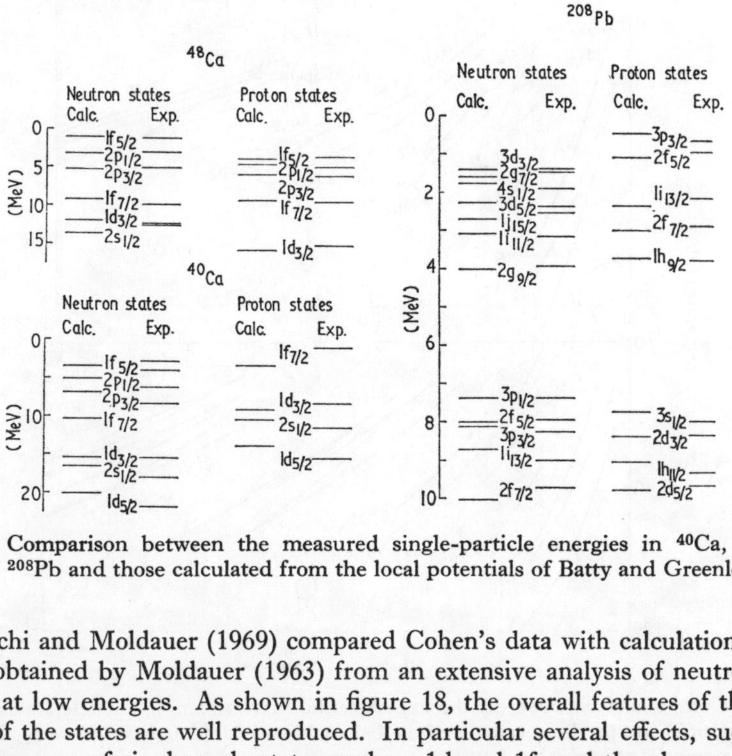

Figure 17. Comparison between the measured single-particle energies in ^{40}Ca, ^{48}Ca and ^{208}Pb and those calculated from the local potentials of Batty and Greenlees (1969).

Takeuchi and Moldauer (1969) compared Cohen's data with calculations using a potential obtained by Moldauer (1963) from an extensive analysis of neutron elastic scattering at low energies. As shown in figure 18, the overall features of the energy variation of the states are well reproduced. In particular several effects, such as the relative steepness of single-node states such as 1d and 1f, and the abnormally large spin–orbit doublet splitting in cases where the $j=l-\frac{1}{2}$ level is a particle state and the $j=l+\frac{1}{2}$ level is a hole state, for which special theoretical explanations were at one time sought, are shown to arise naturally from the potential.

Alongside these calculations with local potentials, several studies were made with non-local potentials of the form suggested by nuclear matter calculations. These were applied both to particular nuclei and to series of nuclei across the periodic table.

The calculations by Meldner and Süssmann (1963) were based on a non-local potential of the form successfully applied by Perey and Buck (1962) to account for the elastic scattering of neutrons by nuclei, viz

$$V_{\rm NL}(\boldsymbol{r}, \boldsymbol{r}') = V\left(\tfrac{1}{2}(\boldsymbol{r}+\boldsymbol{r}')\right)\frac{1}{\pi^{3/2}\beta^3}\exp\left[-\left(\frac{\boldsymbol{r}-\boldsymbol{r}'}{\beta}\right)^2\right] \qquad (3.2)$$

where β is the range of the non-locality.

Instead of solving the non-local wave equation, Meldner and Süssmann used an equivalent local potential $V_L(r)$ related to $V_{NL}(r)$ by an approximate expression derived by Perey and Buck, viz

$$V_{NL}(r) = V_L(r) \exp\left(\frac{m\beta^2}{2\hbar^2}(E - V_L(r))\right). \tag{3.3}$$

To determine the non-local potential they assumed that it has the same form factor as the local potential and used (3.3) to obtain the non-local depth V_{NL}. The resulting

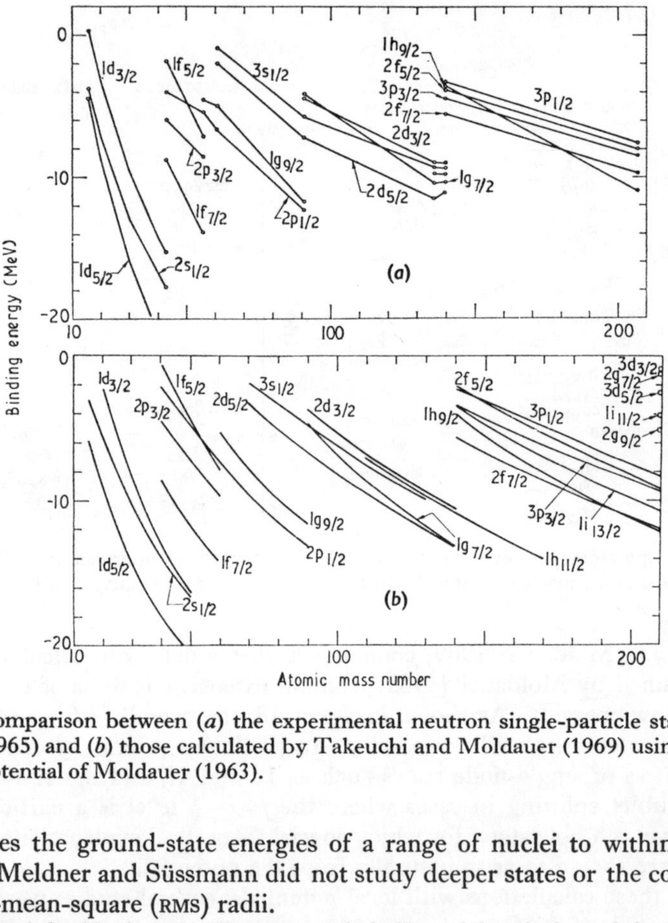

Figure 18. Comparison between (*a*) the experimental neutron single-particle states of Cohen (1965) and (*b*) those calculated by Takeuchi and Moldauer (1969) using the neutron potential of Moldauer (1963).

potential gives the ground-state energies of a range of nuclei to within about 1 or 2 MeV, but Meldner and Süssmann did not study deeper states or the corresponding nuclear root-mean-square (RMS) radii.

Similar work by Koshel (1966) gave good shell model ordering of the bound states with a non-local potential with parameters similar to the optimum values for scattering, and fitted the energies of the bound states quite well. However, later calculations by Grimm *et al* (1971) using the full Perey–Buck potential (3.2) and solving the non-local wave equation exactly showed that it is not sufficiently accurate to assume that the local and non-local potentials have the same form factors; this is indeed to be expected, since the major difference between the local and non-local potentials occurs in the region of the nuclear surface. Further calculations with (3.2) showed that it does not fit the binding energies of the last neutron in several nuclei.

The equivalent potential has been obtained by Batty and Greenlees (1969) in its full r-dependent form and used to fit the proton levels in ^{40}Ca, ^{48}Ca and ^{208}Pb. They found that on the whole the energies of the less tightly bound single-particle states and the RMS proton radii are rather less well fitted by the non-local than by the corresponding local potential. All these comparisons indicate that the Perey–Buck potential is not useful for bound states.

A detailed investigation was made by Meldner (1969) using a non-local density-dependent potential of the form

$$V_{\mathrm{n,p}}(\boldsymbol{r}, \boldsymbol{r}') = V \exp\left(-\frac{y}{a}\right)\left(\frac{y}{a}\right)^{-1}\left[1-\left(\frac{\rho_{\mathrm{n,p}}(x)}{\rho_1}\right)^{2/3}\right]\left(1-\frac{V_{\mathrm{s}}}{x}\,\boldsymbol{L}.\boldsymbol{s}\frac{\mathrm{d}}{\mathrm{d}x}\right)\rho(x) \qquad (3.4)$$

where $x = \frac{1}{2}(\boldsymbol{r}+\boldsymbol{r}')$, $y = |\boldsymbol{r}-\boldsymbol{r}'|$,

$$\rho_{\mathrm{n,p}} = \frac{\rho_{\mathrm{p,n}} + \tau\rho_{\mathrm{n,p}}}{1+\tau} \qquad (3.5)$$

and $\rho_{\mathrm{n,p}}$ are the neutron and proton density distributions obtained by summing the squares of the appropriate single-particle wavefunctions over all occupied orbits. By optimizing the five parameters V, a, ρ_1, V_{s} and τ Meldner obtained a satisfactory overall fit to the charge density distributions, including isotope and isomer shifts, the energies of 1s proton states, the total binding energies, and the shell model spin assignments for a range of nuclei throughout the periodic table. He did not, however, investigate how well his potential fits other single-particle states, so it is difficult to evaluate its usefulness in the present context.

Further work by Grimm *et al* (1971) used a non-local potential of the form

$$V(\boldsymbol{r}, \boldsymbol{r}') = (V(r)\, V(r'))^{1/2}\, \frac{1}{\pi^{3/2}\gamma^3} \exp\left(-\frac{(\boldsymbol{r}-\boldsymbol{r}')^2}{\gamma^2}\right) \qquad (3.6)$$

where $V(r)$ is the Saxon–Woods local potential. They found that it cannot reproduce the observed energies of $1s_{1/2}$ states for light nuclei. This potential can be adjusted to give the correct binding energies for the s and p states in ^{12}C, together with the total RMS radius of the charge distribution, but the RMS radii of the s and p state distributions are out by about 0·2 fm. They concluded that the use of a non-local potential does not remove the need for energy-dependent potential parameters, but that it is possible to distinguish between the energy dependence due to non-local effects and the intrinsic energy dependence of the nuclear potentials.

The same potential was used by Janiszewski (1972) to fit the single-particle states of the doubly closed shell nuclei ^{12}C, ^{16}O, ^{40}Ca, ^{90}Zr and ^{208}Pb. Optimization of the parameters gave a potential that reproduces the measured single-particle energies to within a few MeV, together with the RMS radii of the charge distributions.

Substantial improvement was made by including a density-dependent non-locality of the form $\gamma(V(r))^{1/2}$ in place of γ in (3.6). Even for this potential, however, the differences between the calculated and measured single-particle energies is often several MeV. Thus it seems that the use of the non-local potential does not provide an easy way of describing the energies of single-particle states, and in view of the substantial time taken even by a fast computer to solve the non-local wave equation it is probably preferable to use local potentials.

The variation of single-particle energies from nucleus to nucleus has been studied by Millener and Hodgson (1973) and by Malaguti and Hodgson (1973), who analysed the available data in the ranges of nuclei with $35 \leqslant A \leqslant 65$ and $A \leqslant 35$ respectively. They used a potential of the standard form

$$V(r) = V f_1(r) + V_s \left(\frac{\hbar}{m_\pi c}\right)^2 \frac{1}{r} \frac{\mathrm{d} f_2(r)}{\mathrm{d}r} \boldsymbol{L}.\boldsymbol{\sigma} \tag{3.7}$$

with Saxon–Woods form factors and fixed the values of the form factor parameters to the values $R_1 = 1{\cdot}10A^{1/3} + 0{\cdot}15$, $R_2 = 1{\cdot}10A^{1/3}$, $a_1 = 0{\cdot}52$ and $a_2 = 0{\cdot}65$ fm. The values of R_1 and a_1 come from the Hartree–Fock calculations of Dover and Van Giai (1972), and those of R_2 and a_2 from many analyses of the polarization of protons elastically scattered by nuclei. Any deficiencies in this parametrization are thrown on the potential depth V, and this has a sufficiently flexible parametrization to absorb any nuclear structure effects that may be present.

The potential depth V depended on the nuclear structure according to the expression

$$V = V_0 - 4V_1 (\boldsymbol{t}.\boldsymbol{T})/A + \gamma A + 0{\cdot}4Z/A^{1/3}. \tag{3.8}$$

The isospin term $4V_1(\boldsymbol{t}.\boldsymbol{T})/A$, where $\boldsymbol{t}$ and $\boldsymbol{T}$ are the isospins of the incident nucleon and target nucleus respectively, gives the dependence on the asymmetry parameter $(N-Z)/A$ found by Cohen. It is a good approximation to assume that this isospin potential has the same radial dependence as the main part of the potential, even if it is really partly surface-peaked (Satchler *et al* 1964).

The behaviour of each state (NLJ) was studied separately, leaving open the possible state dependence of the potential. It is then natural to allow the potential to depend on A, and this also absorbs any inadequacies in the parameterization of the nuclear radius parameter R as a function of A. Since the energy of a bound state is fixed by $V(r)$, unlike the situation for the scattering phase shifts, the potential V cannot without redundancy depend on both A and E; the potential was therefore allowed to have no energy dependence, and the A dependence was assumed linear for simplicity. The final term $0{\cdot}4Z/A^{1/3}$ arises from the change of energy due to the Coulomb potential, and its value is taken from analyses of proton elastic scattering (Perey 1963).

The form of the potential in a particular reaction depends on whether a proton or a neutron is being picked up or stripped, and on the isospin of the final state. Since isospin is conserved in nuclear reactions, proton stripping can go to final states with $T_Z = T - \frac{1}{2}$ and isospin $(\boldsymbol{T} + \frac{1}{2}) = T \pm \frac{1}{2}$, and these are referred to as $T_>(\equiv T + \frac{1}{2})$ and $T_<(\equiv T - \frac{1}{2})$ states respectively. Neutron stripping, on the other hand, can go only to $T_>$ states, since $T_Z = T + \frac{1}{2}$ and $T \geqslant T_Z$ so that the isospin can be $(T + \frac{1}{2})$ but not $(T - \frac{1}{2})$. The converse applies to pickup reactions. Assuming pure isosopin for all states, the expressions for the potential depths obtained by stripping reactions for particle states are (Bayman 1966)

$$V_{\mathrm{n}}^{>} = V_0 - \frac{N-Z}{A} V_1 + \gamma A \tag{3.9}$$

$$V_{\mathrm{p}}^{<} = V_0 + \frac{N-Z+2}{A} V_1 + \gamma A + \frac{0{\cdot}4Z}{A^{1/3}} \tag{3.10}$$

$$V_{\mathrm{p}}^{>} = V_0 - \frac{N-Z}{A} V_1 + \gamma A + \frac{0{\cdot}4Z}{A^{1/3}} \tag{3.11}$$

and the corresponding expressions for pickup reactions for hole states are

$$V_{\mathrm{p}}^{>}=V_0+\frac{N-Z+3}{A}V_1+\gamma A+\frac{0{\cdot}4(Z-1)}{A^{1/3}} \tag{3.12}$$

$$V_{\mathrm{n}}^{<}=V_0-\frac{N-Z-1}{A}V_1+\gamma A \tag{3.13}$$

$$V_{\mathrm{n}}^{>}=V_0+\frac{N-Z+3}{A}V_1+\gamma A \tag{3.14}$$

where (N, Z, A) refer to the target nucleus.

The experimental data on the difference between the energies of the $J=L\pm\frac{1}{2}$ states were not sufficiently accurate to allow a systematic study of the depth of the spin–orbit potential V_{S}, so this was fixed to an average value of 7 MeV.

The optimum values of the parameters V_0, V_1 and γ found by least-squares fitting to the potential depths corresponding to measured centroid energies are given for several states in table 1. In the last two columns σ_{B} is the RMS deviation of the experimental from the calculated binding energies, and N is the number of states on which

Table 1. Values of the parameters of the potential fitted to the experimental energies of several sets of single-particle states in the region $35\leqslant A\leqslant 65$

2s hole states	V_0	V_1	γ	σ_{B}	N
Proton $T_>$	51·4	36·4	−0·130	0·22	27
Neutron $T_<$	52·4	33·0	−0·131	0·23	13
Neutron $T_>$	54·6	26·5	−0·130	0·36	9
Proton $T_>$ + neutron $T_<$	52·6	31·3	−0·139	0·27	40
2p particle states					
Neutron $T_>$	58·9	32·0	−0·196	0·26	17
Proton $T_<$	53·2	30·7	−0·090	0·32	12
Neutron $T_>$ + proton $T_<$	57·2	30·6	−0·164	0·38	29
$1d_{3/2}$ hole states					
Proton $T_>$	48·0	39·8	−0·074	0·50	21
Neutron $T_<$	49·7	28·1	−0·073	0·34	14
Neutron $T_<$ + proton $T_>$	49·8	28·5	−0·075	0·52	35

the analysis in each line is based. The overall goodness of fit to the experimental data, shown graphically in figure 19, confirms the adequacy of the assumed linear dependence on A.

This work shows that the energies of the single-particle states for nuclei in the range $35\leqslant A\leqslant 65$ are given to within a few hundred keV by a simple potential, so that they vary in a very systematic way from nucleus to nucleus, irrespective of the complexity of the underlying nuclear structure. This potential was found also to give nuclear charge distributions with mean-square radii in agreement with those found from analyses of electron elastic scattering.

The analyses of the single-particle states in the region of $A\leqslant 35$ (Malaguti and Hodgson 1973) gave similar results, though not unexpectedly the deviations between the parametrized potentials and the data were on average somewhat larger. In the case of the 2s and 1d states it is possible to follow them through the whole region from $A=15$ to $A=70$, and as shown in figure 20 a smooth behaviour is found only

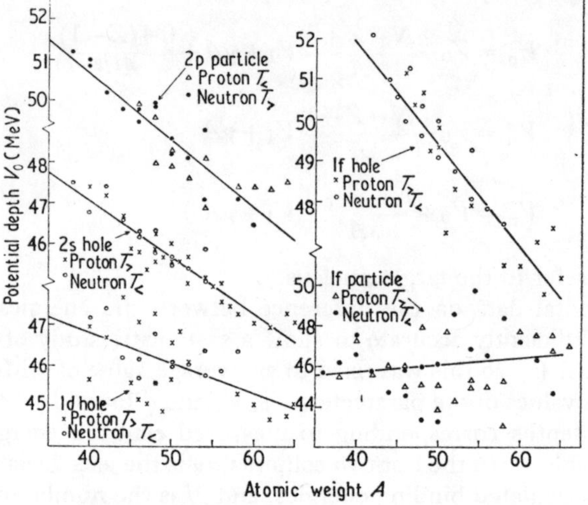

Figure 19. Nuclear potentials for single-particle states as a function of A. The points are fitted to the experimental binding energies, and the straight lines are the best fits to the neutron and proton data combined. In all cases the spin and isospin dependence has been removed for clarity (after Millener and Hodgson 1973).

if the single-particle energy is defined as the sum of the particle and hole potentials, weighted by the corresponding occupation probabilities.

The binding energies of the centroids of single-particle states in the $f_{7/2}$ shell region can also be fitted by the expressions of Bansal and French (1964)

$$\begin{aligned} E_B^{n<} &= E_B^n\,(^{40}\mathrm{Ca}) + (A-40)\,\bar{E} - \tfrac{1}{4}\overline{\Delta E}\,(N-Z+2) \\ E_B^{p>} &= E_B^p\,(^{40}\mathrm{Ca}) + (A-40)\,\bar{E} + \tfrac{1}{4}\,\overline{\Delta E}\,(N-Z) + e_C(Z-20) \\ E_B^{n>} &= E_B^n\,(^{40}\mathrm{Ca}) + (A-40)\,\bar{E} + \tfrac{1}{4}\,\overline{\Delta E}\,(N-Z) \end{aligned} \qquad (3.15)$$

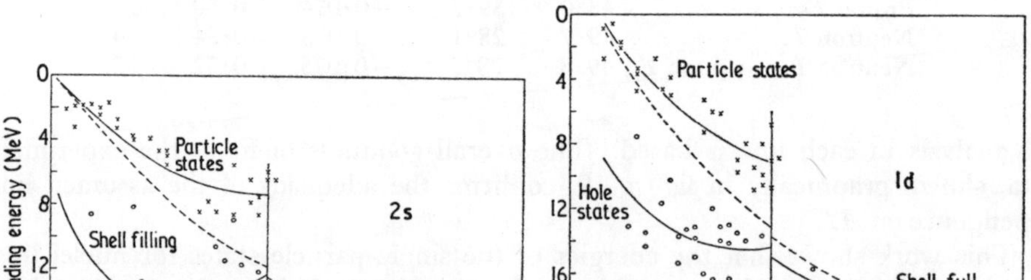

Figure 20. Nucleon binding energies for 2s and 1d particle (×) and hole (○) states as a function of A. The particle and hole curves were obtained from a least-squares fitting of the parametrized potentials to the experimental data, and the broken curves are the sum of the particle and hole curves, weighted by the corresponding occupation probabilities. The isospin and Coulomb dependence has been removed for clarity (after Malaguti and Hodgson 1973).

Table 2. Parameters of the Bansal–French formulae for 2s hole states

State	$E_B^n(^{40}Ca)$	$E_B^p(^{40}Ca)$	$\overline{\Delta E}$	$\bar{E}$	e_C	RMS deviation
Neutron $T_<$	15·89	—	2·01	0·168	—	0·20
Proton $T_>$	—	10·60	1·90	0·150	−0·4	0·25
Neutron $T_>$	17·57	—	1·81	0·143	—	0·33
Neutron $T_<$ + proton $T_>$	15·81	10·56	1·90	0·158	−0·41	0·24

where E_B^n (^{40}Ca) and E_B^p (^{40}Ca) are the binding energies of 2s neutrons and protons in ^{40}Ca, $\bar{E}$ and $\overline{\Delta E}$ are the scalar and isovector parts of the space–spin monopole interaction, and e_C is the Coulomb correction. The results of such a fit to the data on 2s hole states are given in table 2, where it is assumed that the 2s–1f$_{7/2}$ and 2s–2p interactions are the same. The quality of the fit is very similar to that obtained previously with a potential.

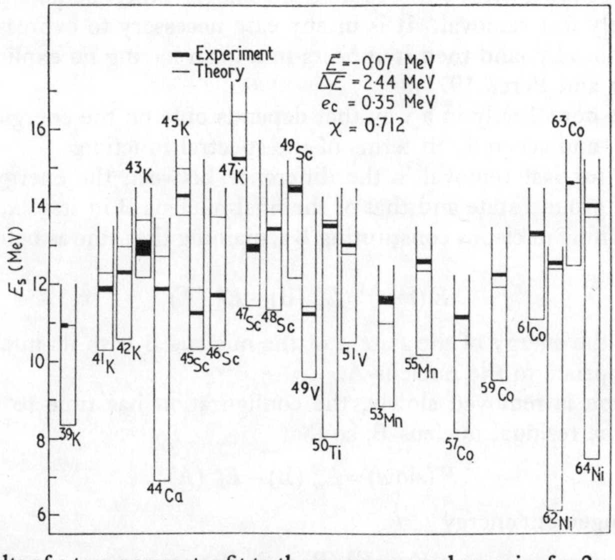

Figure 21. Results of a two-parameter fit to the mean removal energies for 2s$_{1/2}$ protons obtained with the method of Bansal and French (after Wenneis 1971).

The results of a recent analysis of 2s proton separation energies using the Bansal–French formalism and adjusting $\bar{E}$ and $\overline{\Delta E}$ to fit the neutron and Coulomb separation energies are shown in figure 21. The values of $\bar{E}$ and $\overline{\Delta E}$ agree quite well with the spin and isospin averages of the diagonal particle–hole matrix elements attained from shell model calculations with Talmi fits or from realistic forces (Wenneis 1971).

4. The physical basis of nuclear rearrangement energies

The concept of a nuclear rearrangement energy can be defined in a direct physical way by considering the time dependence of the nucleon removal process. This will be done first of all by assuming that the final state is stable, and subsequently the modifications introduced by the finite lifetime of the final state will be considered.

If a particle is removed instantaneously from a nucleus, it has gone before the rest of the nucleus knows what is happening, and so the remaining nucleons are left in the same configuration as they were before the removal. If, however, the nucleon is removed very slowly, the nucleus is able to readjust itself at each stage of the removal process, which therefore requires less energy than instantaneous removal. We can thus define the time-dependent rearrangement energy $R(t)$ as the difference between the energy required for instantaneous removal and that required for removal in a time t:

$$R(t)=E(0)-E(t). \tag{4.1}$$

In the limit of infinitely slow removal

$$R(\infty)=E(0)-E(\infty)\equiv E(\text{fast})-E(\text{slow}). \tag{4.2}$$

This is a clear concept physically, though it is not possible to realize it practically except as a limiting case. Nucleons cannot be removed infinitely slowly from a nucleus, and, as shown below, most nucleon removal experiments correspond effectively to infinitely fast removal. It is in any case necessary to express the definition quantum mechanically, and then it appears in a form having no explicit time dependence (Meldner and Perez 1971).

This may be done firstly in a way that depends only on the energies of the initial and final nuclei, and secondly in terms of the spectral function.

The energy for fast removal is the difference between the energy of the initial nucleus A in its ground state and that of the final nucleus B in state α, the configuration of the remaining nucleons constituting B remaining the same as before the removal process. Thus

$$E\,(\text{fast})=E_\alpha^{\text{A}}\,(\text{B})-E_0^{\text{A}}\,(\text{A}) \tag{4.3}$$

where $E_\alpha^{\text{A}}(\text{B})$ is the energy of the state α of the nucleus B with its nucleons in a configuration appropriate to the nucleus A.

If the nucleon is removed slowly, the configuration has time to change to that appropriate to the residual nucleus B, so that

$$E\,(\text{slow})=E_\alpha^{\text{B}}\,(\text{B})-E_0^{\text{A}}\,(\text{A}). \tag{4.4}$$

Thus the rearrangement energy

$$R_\alpha=E_\alpha^{\text{A}}\,(\text{B})-E_\alpha^{\text{B}}\,(\text{B}). \tag{4.5}$$

This expression will be used in §5 to evaluate the rearrangement energy in the framework of the BHF theory.

The energy for fast removal is just the single-particle energy

$$\epsilon_\alpha\equiv E\,(\text{fast})=E_\alpha^{\text{A}}\,(\text{B})-E_0^{\text{A}}\,(\text{A}) \tag{4.6}$$

and that for slow removal is the separation energy

$$S_\alpha\equiv E\,(\text{slow})=E_\alpha^{\text{B}}\,(\text{B})-E_0^{\text{A}}\,(\text{A}). \tag{4.7}$$

So the rearrangement energy can also be written

$$R_\alpha=\epsilon_\alpha-S_\alpha \tag{4.8}$$

or, in words,

$$R_\alpha=E\,(\text{single particle})-E\,(\text{separation}). \tag{4.9}$$

The relation between these energies is shown in figure 22. It should be noted that in the definition (4.6) the sign is different from that usually found in the literature; this choice ensures that R, ϵ and S are all positive quantities.

To obtain the rearrangement energy in terms of experimental quantities it is necessary to take account of the fragmentation of the single-particle strength. If the single-particle strength is concentrated in just one state, the separation energy must be independent of the reaction time, as the energy of the state is invariant. Thus in this particular case the time-dependent rearrangement energy is zero. If it is fragmented, however, it is possible in principle for the degree of excitation of the

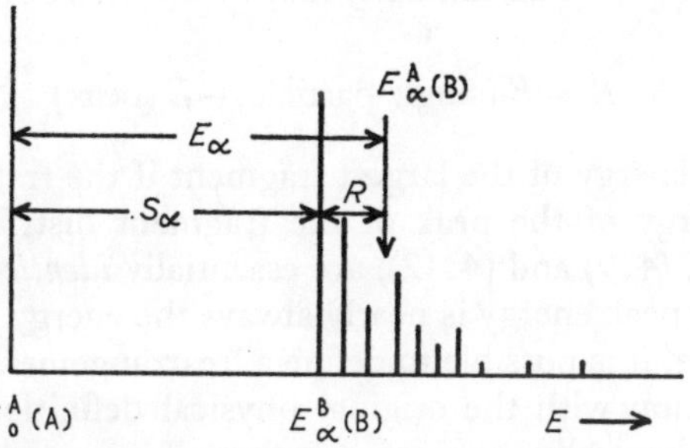

Figure 22. The relation between the separation energy S_α, the single-particle energy E_α, and the rearrangement energy R.

fragments to depend on the reaction time (Brueckner *et al* 1972). If the nucleon is removed instantaneously, the energy uncertainty is maximal; thus, provided there is sufficient energy available, all the fragments are excited with their full spectroscopic amplitudes, so that

$$E\,(\text{fast}) = \sum_i |\langle B_i|a|A\rangle|^2 E_i(B) - E_0(A) \tag{4.10}$$

where $|\langle B_i|a|A\rangle|^2$ is the spectroscopic factor for removing a nucleon from A leaving the nucleus B in the ith state with energy $E_i(B)$, and the summation is over the continuum as well as over the bound states of the final nucleus. If there is insufficient energy available, some of the higher states will not be excited, and $E_\alpha(B)$ will depend on the incident energy; this kinematic energy dependence of $E_\alpha(B)$ and of the associated rearrangement energy is of little physical interest.

Thus the rearrangement energy

$$R = \sum_i |\langle B_i|a|A\rangle|^2 E_i(B) - E_0(B). \tag{4.11}$$

This definition satisfies the requirements of a physical concept in that it is clearly defined theoretically, apart from the difficulties connected with the faraway fragments. The definition of rearrangement energy is now expressed in terms of the fragment distribution.

This definition of rearrangement energy applies only to the ground state, but it is interesting to see if it can be extended to other states as well. Considered as a time-dependent process, this raises difficulties, since infinitely slow removal automatically leaves the nucleus in its ground state. However, we can imagine, at least in principle, that we can remove the nucleon in a time that is long compared with the nuclear rearrangement time and yet short compared with the lifetime of the excited state concerned. In each case it must be considered whether this is possible in practice, and this will be done later; here we are concerned with the conceptual implications of this definition. To avoid the difficulties connected with the time dependence of the process we can work entirely in terms of the fragment distribution.

If we consider the fragment distribution for an excited state (cf figure 3) we could define its rearrangement energy by the same expression (4.6) as for the ground state. However, the separation energy is now the energy of the lowest fragment of the state, and, as this may be a very small fragment, it is neither experimentally observable with certainty (as there may always be a small fragment of lower energy that has escaped detection) nor theoretically very interesting (since it may correspond to a rare and complicated configuration). This difficulty may be avoided be *redefining* the rearrangement energy as

$$R' = E\,(\text{single particle}) - E\,(\text{peak}) \tag{4.12}$$

where E (peak) is the energy of the largest fragment if the fragments are individually identified, or the energy of the peak of the fragment distribution if they are not. These two definitions, (4.9) and (4.12), are essentially *identical* for the ground state, since for this state the peak energy is nearly always the energy of the lowest fragment. Thus, by a redefinition, it is possible to define a 'rearrangement energy' for an excited state, but the connection with the original physical definition has been lost. Considered quantum mechanically, one can say that, if a particle is removed infinitely slowly from a nucleus, the energy is precisely defined; and by suitable arrangement of the mode of removal, the nucleus can be left in any one of the fragments of the single-particle state that we specify, and among these is the strongest or peak state. However, as this cannot in practice be done experimentally, this remains a notional concept.

If the nucleon is removed neither infinitely rapidly nor infinitely slowly it is possible that most of the fragments are excited, but not to their full spectroscopic amplitudes. In this case the centroid energy, defined as the mean of the energies of the fragments, weighted by the squares of their excitation amplitudes, varies with the removal time and with the rearrangement energy.

It has sometimes been argued that a time dependence of the rearrangement energy is impossible in principle, since the coefficients of the energies in (4.3) are matrix elements with no time dependence. It is, however, still worthwhile to see if we could conceive an experiment to falsify this prediction, since general theoretical arguments have often been falsified by experience.

There was indeed at one time some experimental evidence that the centroid energies as determined by (d, h) and (p, 2p) reactions differed from each other, and this was interpreted as evidence for the time dependence of nuclear rearrangement energies (Wagner 1969). However, the expressions (2.1) and (2.10) for the cross sections of these reactions both contain the same spectral function; and so, if the dynamical terms are evaluated correctly, the data for both types of reactions must give the same rearrangement energy. The essential physical process is proton removal from nucleus A, leaving nucleus B, and for the same energy transfer the rearrangement effects must be the same, however the proton is removed (Gross and Lipperheide 1970). Later experimental work did indeed give the same centroid energies for both reactions (Wagner 1973).

It is nevertheless interesting to see if there is any other experimental evidence for a dependence of the rearrangement energy on the reaction or on the incident energy. This may be done by varying the energy of a pickup reaction and seeing if the relative population of the fragments of the single-particle states changes. This can be done over quite an appreciable range of energies, and some peak or 'local centroid' energies obtained from (e, e′p), (p, 2p) and (d, h) reactions are given in table 3.

Table 3. Proton local centroid energies as a function of incident energy

Energy (MeV) and reaction	^{12}C 1s	1p	^{27}Al 1p	^{40}Ca 1p	1d	2s	^{48}Sc 2s	^{59}Co 2s	^{58}Ni 2s
52 (d, h)	—	16·8	—	—	12·2±1	10·8±0·1	11·9	—	11·2
82 (p, 2p)	—	—	—	—	—	11·6±0·8	—	—	11·2
150–185 (p, 2p)	34±1·5	16±1·2	30·5±1·9	36·8±4·2	—	10·9±0·2	12±1	12·1±0·3	10·8±0·3
387 (p, 2p)	36±4	16±2	~31	34±4	—	12±1	12·5±1	13·5±1	11·3±1
460 (p, 2p)	34±2	14·7±0·8	—	—	—	11·6±0·5	—	13·9±0·8	—
500 (e, e′p)	38±0·8	17·5±0·4	—	—	14·9±0·8	11·2±0·3	—	—	14·7±0·5
600 (e, e′p)	33±6	14±6	32±4	32±6	—	—	—	—	—
750 (e, e′p)	30±10	11·5±6·5	—	35·3±0·5	14·6±1·5	13·6±0·4	—	—	—
1000 (p, 2p)	38±12	18±8	—	—	—	—	—	—	—

References: Garron *et al* (1962), Tyrén *et al* (1966), Amaldi (1967), Hinterberger *et al* (1968), Jackson (1969), James (1969), Wagner *et al* (1969), Hiramatsu (1970), Landaud *et al* (1971), Kullander *et al* (1971), Campos Venuti *et al* (1973), Bernheim *et al* (1974b), Doll (1974), Nakamura *et al* (1974).

Some spectroscopic factors are plotted as a function of incident energy in figure 23. In neither case is there any convincing evidence of energy variation. It is likely that the removal would have to be effected much more slowly before any evidence of rearrangement became apparent, but there is no physical process by which this can be done. At least one of the incident and outgoing particles must be charged, and thus must have an energy comparable with that of the Coulomb barrier, and this is probably already too high for slow removal.

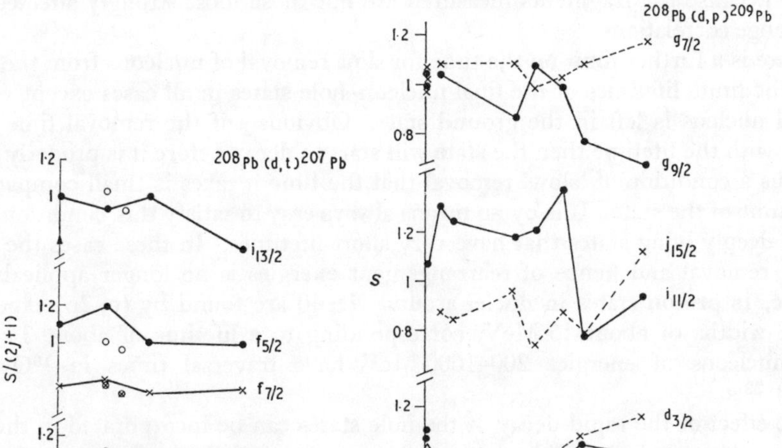

Figure 23. Neutron spectroscopic factors in lead (after Macfarlane 1969).

It thus seems that there is no evidence for a gross variation of removal energies with reaction time, and little prospect of such evidence being obtained.

The detection of small variations of removal energies with reaction times raises another type of difficulty. The data in table 3 and figure 23 refer to measurements and analyses made by different groups in different conditions, and this no doubt accounts for most of the fluctuations about the mean values. These fluctuations could be removed by careful experiments under controlled conditions and by analysing the data in a uniform way. This could give removal energies or spectroscopic factors that varied slowly and smoothly with incident energy and hence with removal time.

It would, however, be impossible to attribute this variation to the rearrangement energy, because such variations could also be due to defects in the energy dependence of the optical potentials used to describe the wavefunctions of the interacting particles, or in the theories used to extract them from the experimental data. Indeed, it is usual to *test* the theories of the nucleon transfer process used to extract the spectroscopic factors from the experimental data by *requiring* that they give spectroscopic factors independent of incident energy.

There is thus at present no evidence that nucleons are ever removed from nuclei with speeds less than those sufficient to excite all the fragments to their full spectroscopic amplitude, so that all existing data correspond to the fast limit.

The incident nucleon energies corresponding to fast removals may be estimated by requiring that the velocity of the incident particle be large compared with the average velocity of the struck nucleon, taking account of the high-velocity components due to short-range correlations. By making a Fourier transform of some BHF wavefunctions of Demos and Banerjee (1972), Brueckner *et al* (1972) found that about half the momenta correspond to energies above 100 MeV, so such energies are necessary for fast removal. The apparent lack of removal time dependence at lower energies may be because the fragments measured are not those most strongly affected by the short-range correlations.

There is a further limit on the time for slow removal of nucleons from the nucleus set by the finite lifetimes of the final nucleon–hole states in all cases except when the residual nucleus is left in the ground state. Obviously if the removal time is comparable with the lifetime then the state will start to decay before it is properly formed. It is thus a condition of 'slow' removal that the time it takes is small compared with the lifetime of the state. It is by no means always easy to satisfy this condition, especially for deeply lying states that have very short lifetimes. In these cases the concept of slow removal and hence of rearrangement energies is no longer applicable. For example, 1s proton states in nuclei around $A \simeq 40$ are found by (p, 2p) experiments to have widths of about 15 MeV, corresponding to a lifetime of about 3×10^{-23} s, while nucleons of energies 200–1000 MeV have traversal times in ^{16}O of 2·2–$1{\cdot}4 \times 10^{-23}$ s.

The effect of the rapid decay of the hole states can be incorporated in the (p, 2p) formalism by using a complex potential to generate the wavefunction of the hole state and adjusting the imaginary part to give the required width (Herscovitz *et al* 1968, Herscovitz 1971). There have been several recent calculations of the lifetimes of hole states and the effect of short lifetimes on nucleon removal spectra (Pittel and Austern 1972, 1974, Shanta 1973).

Although it seems unlikely that time-dependent rearrangement energies will be observed, the expression (4.6) could be used to calculate R if the energies and spectroscopic factors of all the fragments of a single-particle state were known. However, as shown in the last section, it is the peak energy $E_{\rm p}$ and not the centroid energy that is obtained experimentally, so (4.6) becomes, using (2.2),

$$R - R' = E_{\rm p} - E_0({\rm B}). \qquad (4.13)$$

For the ground state, the peak energy coincides with the separation energy, so that $R = R'$ and the centroid $E_{\rm c}'$ of the observed fragments gives a lower limit to the rearrangement energy:

$$R = R' \geqslant E_{\rm c}' - E_{\rm p} \quad \text{(ground state)}. \qquad (4.14)$$

For excited states the energy of the lowest fragment is difficult to obtain, but the energy E_L of the lowest one actually observed sets a lower limit to R:

$$R \geqslant E_c' - E_L \quad \text{(excited states).} \tag{4.15}$$

Some lower limits to rearrangement energies obtained in this way are given in table 4. It is notable that the energies for the $1d_{3/2}$ states are remarkably constant and substantially less than the two values for the $1d_{5/2}$ states. The greater rearrangement energies for the deeper $1d_{5/2}$ states are in accord with the trends of the calculations in the next section.

Table 4. Lower limits to nuclear rearrangement energies obtained from measured fragment distributions in neutron removal reactions

Nucleus	$2s_{1/2}$	$1d_{3/2}$	$1d_{5/2}$
^{36}Ar	0·94	0·32	3·05
^{40}Ar		0·43	
^{40}Ca	0·12		0·82
^{42}Ca	0·47	0·33	
^{44}Ca	0·30	0·46	
^{46}Ti	0·42	0·35	
^{48}Ti	0·06		
^{50}Ti	0·02		
^{50}Cr	1·43	0·42	
^{52}Cr		0·09	
^{63}Ni	0·59		
^{65}Ni	0·85		

5. Hartree–Fock and Brueckner–Hartree–Fock rearrangement energies

The process of nucleon removal can be analysed within the framework of the Hartree–Fock (HF) and Brueckner–Hartree–Fock (BHF) theories. This may be done using the expression (4.5) for the rearrangement energy in terms of the energies of the residual nucleus B:

$$R_\alpha = E_\alpha^A(B) - E_\alpha^B(B). \tag{5.1}$$

Using the expression for the total energy of a nucleus in terms of the kinetic energies T and the G matrices, this becomes in the notation of Faessler and collaborators (Faessler and Müther 1973)

$$R_\alpha = \sum_{\substack{i=1\\ i\neq\alpha}}^{A} (\langle\phi_i^A|T|\phi_i^A\rangle - \langle\phi_{i\alpha}^B|T|\phi_{i\alpha}^B\rangle) + \tfrac{1}{2}\sum_{\substack{i,j=1\\ i,j\neq\alpha}}^{A} (\langle\phi_i^A\phi_j^B|G_A|\phi_i^A\phi_j^A\rangle - \langle\phi_{i\alpha}^B\phi_{j\alpha}^B|G_B^\alpha|\phi_{i\alpha}^B\phi_{j\alpha}^B\rangle). \tag{5.2}$$

Subtracting the expression for the BHF single-particle energy ϵ_α of the state α in the nucleus A,

$$\epsilon_\alpha = -\langle\phi_\alpha^A|T|\phi_\alpha^A\rangle - \sum_{j=1}^{A}\langle\phi_\alpha^A\phi_j^A|G_A|\phi_\alpha^A\phi_j^A\rangle \tag{5.3}$$

this becomes

$$R_\alpha - \epsilon_\alpha = E_0^A(A) - E_\alpha^B(B) \tag{5.4}$$

which is the separation energy S_α.

Thus as before (4.8)

$$R_\alpha = \epsilon_\alpha - S_\alpha. \tag{5.5}$$

This expression is often used to *define* the rearrangement energy.

The total energy of the nucleus A may then be written

$$E_0^{\rm A}({\rm A}) = \tfrac{1}{2}\sum_i (\langle \phi_i^{\rm A}|T|\phi_i^{\rm A}\rangle + \epsilon_i). \tag{5.6}$$

It is useful to distinguish between the rearrangement energy due to the change in the wavefunctions from $\phi_i^{\rm A}$ to $\phi_i^{\rm B}$ and that due to the change in the G matrix from $G_{\rm A}$ to $G_{\rm B}$; these are referred to as the orbital and the Brueckner rearrangement energies respectively. The total rearrangement energy is not given by the sum of the orbital and Brueckner rearrangement energies, due to the strong coupling between the wavefunctions and the interaction.

The orbital rearrangement energy is obtained by putting $G_{\rm A} = G_{\rm B}^\alpha$ in (5.2), giving

$$R_\alpha^{\rm O} = \sum_{\substack{i=1\\ i\neq\alpha}}^{\rm A} (\langle \phi_i^{\rm A}|T|\phi_i^{\rm A}\rangle - \langle \phi_{i\alpha}^{\rm B}|T|\phi_{i\alpha}^{\rm B}\rangle) + \tfrac{1}{2}\sum_{\substack{i,j=1\\ i,j\neq\alpha}}^{\rm A} (\langle \phi_i^{\rm A}\phi_j^{\rm A}|G_{\rm A}|\phi_i^{\rm A}\phi_j^{\rm A}\rangle - \langle \phi_i^{\rm B}\phi_j^{\rm B}|G_{\rm A}|\phi_i^{\rm B}\phi_j^{\rm B}\rangle) \tag{5.7}$$

and the Brueckner rearrangement energy by putting all $\phi_i^{\rm A} = \phi_{i\alpha}^{\rm B}$, so that

$$R_\alpha^{\rm B} = \tfrac{1}{2}\sum_{\substack{i,j=1\\ i,j\neq\alpha}}^{\rm A} (\langle \phi_i^{\rm A}\phi_j^{\rm A}|\, G_{\rm A} - G_{\rm B}^\alpha|\, \phi_i^{\rm A}\phi_j^{\rm A}\rangle). \tag{5.8}$$

The Brueckner rearrangement energy arises from the difference between the G matrices corresponding to the A and B nuclei. These G matrices are obtained by solving the Bethe–Goldstone equation

$$G = V + V(Q/e)\,G \tag{5.9}$$

where the intermediate energy denominator $e = W - H_0$, and W is the starting energy for a particular nucleon–nucleon interaction V. The Pauli operator Q depends on the occupation probabilities of the single-particle states and the energy denominator e on the single-particle energies, and as these differ in the A and B nuclei, so do the effective interactions G obtained from the same interaction V. The resulting rearrangement energy was first calculated for nuclear matter and to some extent for finite nuclei by Brueckner and Goldman (1960) and by Köhler (1965, 1966).

Subtracting the Bethe–Goldstone equations for the A and B nuclei gives, after some operator algebra (Brown 1967),

$$G_{\rm A} - G_{\rm B} = -G_{\rm A}\left(\frac{Q_{\rm B}}{e_{\rm B}} - \frac{Q_{\rm A}}{e_{\rm A}}\right) G_{\rm B} \equiv -G_{\rm A} p_\alpha G_{\rm B}. \tag{5.10}$$

Iterating gives

$$G_{\rm A} - G_{\rm B} = -G_{\rm A} p_\alpha G_{\rm A} - G_{\rm A} p_\alpha G_{\rm A} p_\alpha G_{\rm A} - \ldots. \tag{5.11}$$

The Brueckner rearrangement energy may be further separated into a part due to the change in the starting energy and a part due to the change in the Pauli operator. Thus

$$R_\alpha^{\rm BS} = \tfrac{1}{2}\sum_{\substack{i,j=1\\ i,j\neq\alpha}}^{\rm A} \langle \phi_i^{\rm A}\phi_j^{\rm A}|\, G\,(W_{\rm A} = e_i^{\rm A} + e_j^{\rm A}) - G\,(W_{\rm B} = e_i^{\rm B} - e_j^{\rm B})|\, \phi_i^{\rm A}\phi_j^{\rm A}\rangle \tag{5.12}$$

and

$$R_{\alpha}^{\mathrm{BP}}=\frac{1}{2}\sum_{\substack{i,j=1\\ i,j\neq\alpha}}^{\mathrm{A}} \langle\phi_i^{\mathrm{A}}\,\phi_j^{\mathrm{A}}|\; G_{\mathrm{B}}\left(\frac{Q^{\mathrm{A}}-Q^{\mathrm{B}}}{e}\right) G_{\mathrm{A}}|\phi_i^{\mathrm{A}}\,\phi_j^{\mathrm{A}}\rangle. \tag{5.13}$$

As for the orbital and Brueckner rearrangement energies these terms are strongly coupled, so their sum is not the same as the Brueckner rearrangement energy.

The rearrangement energy is zero if $G_{\mathrm{A}}=G_{\mathrm{B}}^{\alpha}$ and all $\phi_i^{\mathrm{A}}=\phi_i^{\mathrm{B}}$. Under these conditions the separation and single-particle energies are identical; this is Koopmans' (1933) theorem. Detailed studies of the applicability of Koopmans' theorem for various interactions have been made by Köhler and Lin (1971). For infinite nuclear matter the Brueckner rearrangement energy has a finite value, but the orbital rearrangement energy is identically zero.

In atoms where the interactions really are weak the experimental separation energies agree with the calculated HF single-particle energies to a very high degree of accuracy (Slater 1955, Clementi 1964, Lindgren 1965, Rosen and Lindgren 1968), and the calculated absolute atomic binding energies agree with experiment to within less than 1% (Froman 1958, Clementi 1963).

This agreement, which amounts to an experimental verification of Koopmans' theorem, shows that in atoms the conditions for the exact validity of the theorem are well satisfied; that is, the wavefunctions and the interactions are not appreciably affected by the removal of an electron.

In nuclei the situation is different, because the wavefunctions and the interaction both change appreciably. The change in the interaction is the more important, and this is due to the renormalization of the effective force in the Brueckner reaction matrix that comes from the change in the occupation probabilities of the single-particle states due to the Pauli operator in the Bethe–Goldstone equation.

In atomic calculations the interaction is well known, but in nuclear calculations it is necessary to use either an interaction that has been derived, inevitably with some approximations and uncertainties, from a phenomenological nucleon–nucleon interaction, or an effective interaction whose parameters have been chosen to fit selected experimental data. Both approaches are evidently subject to uncertainties that are not present in the case of atoms.

There have been many calculations of nuclear rearrangement energies within the framework of the HF and BHF theories and using one of the interactions available from other work. The results of these calculations are difficult to evaluate for several reasons. Firstly it is always difficult to know how much importance to attach to a quantity calculated from a theory using an effective interaction obtained by fitting quite different data, and secondly there are no reliable experimental values of the rearrangement energy with which they can be compared. The best that can be done is to see how well the theory agrees with other data (not used to obtain the effective interaction) and to assume that the better the agreement here, the better are the corresponding rearrangement energies. A third difficulty is that the HF and BHF theories take no account of the hard core interactions responsible for the faraway fragments that make the greatest contribution to the rearrangement energies.

Over the last decade there have been very many HF and BHF calculations of the properties of nuclei, and some of the more recent interactions give a good overall agreement with the experimental data. The earlier calculations tended to give single-particle energies that were larger than the experimental peak separation energies (this is a valid comparison since neither take into account the hard core interactions or the faraway fragments), and charge densities that were too large at the centre of the nucleus and had mean-square radii that were too small. They also gave rearrangement energies of several MeV. Furthermore, since the single-particle energies differ substantially from the separation energies, the BHF wavefunctions are quite different from those used in the empirically very successful shell model calculations of nuclear structure and direct nuclear reactions. These difficulties have been substantially removed by the new density-dependent interactions, which decrease the absolute values of the single-particle energies and reduce the charge density at the centre of the nucleus. This work will now be reviewed in more detail.

The orbital rearrangement energy has been calculated by applying the HF theory to nuclei with A and $A-1$ nucleons, and using the resulting wavefunction in the above expression to evaluate R^{O}, putting $G_{\mathrm{A}}=G_{\mathrm{B}}^{\alpha}$. The early calculation of Faessler and Wolter (1969) used wavefunctions symmetric under time reversal, which implies that after a nucleon is removed from an even nucleus the hole state and its time-reversed state are both half-filled. The expression for R_{α}^{O} then includes a spurious term representing the interaction between these two 'half-holes'. This term was evaluated in subsequent calculations and found to be of the same order of magnitude as or larger than the orbital rearrangement energy (Müther *et al* 1972).

Faessler and Wolter also showed that to a good approximation the separation energy is given by the mean of the single-particle energies in the A and A − 1 nuclei, ie

$$E_{\alpha}\approx\tfrac{1}{2}\,(S_{\alpha}^{\mathrm{A}}+S_{\alpha}^{\mathrm{A}-1}). \tag{5.14}$$

This relation is unaffected by the restriction to time-reversal symmetric wavefunctions, since the same spurious term appears in the expressions for the total and for the single-particle energies. A comparison between some calculated values of these approximately equal quantities is given in table 5.

Table 5. Orbital rearrangement energies in ^{16}O showing the effect of different ways of treating time-reversal invariance, as described in the text (Faessler and Wolter 1969)

State	$R_{\mathrm{FW}}^{\mathrm{T}}$	R_{MGF}	$R_{\mathrm{MGF}}^{\mathrm{T}}$
$s_{1/2}$	3·4	0·94	0·80
$p_{3/2}$	1·0	0·08	0·06
$p_{1/2}$	2·1	0·02	0·01

The effect on the orbital rearrangement energy in ^{16}O of various ways of treating time-reversal invariance is shown in table 5. The calculation of $R_{\mathrm{FW}}^{\mathrm{T}}$ assumes time-reversal invariance for the wavefunction, keeping the spurious energy term in the calculation of the energies (Faessler and Wolter 1969). In the next column R_{MGF} shows the result of ignoring time-reversal invariance, while $R_{\mathrm{MGF}}^{\mathrm{T}}$ is calculated including time-reversal invariance of the wavefunction but omitting the spurious matrix element in the expressions for the final energy and the single-particle energy after the iteration process (Müther *et al* 1972). It is seen that $R_{\mathrm{MGF}}\approx R_{\mathrm{MGF}}^{\mathrm{T}}$, but both differ substantially from $R_{\mathrm{FW}}^{\mathrm{T}}$.

A further correction to the calculated single-particle energies must be made to allow for the fact that the self-consistent potential of a finite nucleus is not translationally invariant. This may be done to a good approximation by using the centre-of-mass corrected Hamiltonian (Davies and Becker 1971, Khadkikar and Kamble 1974)

$$H_{\mathrm{CM}}=H-\frac{1}{2mA}P^{2} \tag{5.15}$$

where P is the total momentum operator, in place of the Hamiltonian H. Unfortunately the resulting single-particle energies are not equal to the separation energies in the limit of equal single-particle wavefunctions. This condition is satisfied if the corrected single-particle energies are defined by (Müther *et al* 1972)

$$\epsilon_\alpha^{\mathrm{CM}} = \epsilon_\alpha + \langle \mathrm{A}-1 | T_{\mathrm{CM}} | \mathrm{A}-1 \rangle - \langle \mathrm{A} | T_{\mathrm{CM}} | \mathrm{A} \rangle. \tag{5.16}$$

Since the centre-of-mass motion may be allowed for to a good approximation by first carrying out the HF calculation with the Hamiltonian H and then subtracting the kinetic energy $\langle T_{\mathrm{CM}} \rangle$ of the centre-of-mass motion, the centre-of-mass contribution to the rearrangement energy is approximately given by

$$R_\alpha^{\mathrm{CM}} = \sum_{\substack{i<F \\ i\neq\alpha}} t_i \left(\frac{1}{\mathrm{A}-1} - \frac{1}{\mathrm{A}} \right) + \frac{1}{2} \sum_{\substack{i,k<F \\ i,k\neq\alpha}} \left\langle ik \left| \frac{\boldsymbol{P}_1 . \boldsymbol{P}_2}{m} \right| ik \right\rangle \left(\frac{1}{\mathrm{A}-1} - \frac{1}{\mathrm{A}} \right). \tag{5.17}$$

Some typical orbital rearrangement energies calculated with the Brink–Boeker force are given in table 6. It is notable, as might be expected, that these are greater for the

Table 6. Proton single-particle and orbital rearrangement energies for ^{40}Ca (Müther *et al* 1972)

State α	$S_\alpha^{\mathrm{A}\prime}$	$\frac{1}{2}(S_\alpha^{\mathrm{A}\prime} + S_\alpha^{\mathrm{B}\prime})$	ϵ_α'	$R_\alpha^{\mathrm{O}\prime}$	$R_\alpha^{\mathrm{CM}\prime}$
$1s_{1/2}$	67·51	66·89	66·97	0·54	0·88
$1p_{3/2}$	39·30	39·18	39·21	0·09	0·52
$1p_{1/2}$	39·29	39·05	39·05	0·24	0·68

low-lying states than for those near the Fermi surface. The spurious centre-of-mass contribution to the orbital rearrangement energy is at least comparable with, and often substantially greater than, the rearrangement energy itself. In this table the prime indicates that the centre-of-mass corrections have been applied, using equation (5.17). Similar results for the orbital rearrangement energies were obtained by Köhler and Lin (1971) using a Serber force.

The single-particle energies are rather insensitive to orbital rearrangement, so the smallness of the rearrangement energies does not imply that all such rearrangement effects are small. Other nuclear properties, such as quadrupole moments, show much more marked rearrangement effects (Müther *et al* 1972).

Several calculations have been made taking into account both the orbital and the Brueckner rearrangement energies (Bassichis and Strayer 1971, Köhler and Lin 1971, Becker 1970a, Faessler and Müther 1973).

The first term in expression (5.11) has been evaluated by Faessler and Müther (1973) with the additional assumption $e_{\mathrm{A}-1} = e_{\mathrm{A}}$, and some of their results are given in table 7. Their calculations show that the Brueckner rearrangement energy is about twice the orbital rearrangement energy for the $1s_{1/2}$ proton state of ^{16}O. Both are substantially less for states nearer the Fermi surface, the decrease being stronger for the orbital rearrangement energy.

Table 7. Rearrangement energies for ^{16}O (Faessler and Müther 1973)

State	Orbital	Pauli	Orbital plus Pauli	Pauli potential
$1s_{1/2}$	2·54	4·96	8·89	7·41
$1p_{3/2}$	0·41	1·52	2·99	1·73
$1p_{1/2}$	0·18	3·38	3·28	3·93

The energy denominator can also be expressed in terms of the A-nucleon system

$$e^{i,\alpha}_{A-1} \approx e^i_A - \langle i\alpha\,|G_A|\,i\alpha\rangle \tag{5.18}$$

and some of the resulting higher-order terms have been evaluated by Köhler (1966).

Several extensions of the BHF theory have been made to overcome the inadequacies of the earlier theories. Three such developments will be mentioned here: the renormalized BHF theory of Becker and colleagues; the Pauli BHF theory of Faessler and colleagues; and the many papers using density-dependent interactions. While the relation between these approaches is not yet clear, they all bear in some way on the problem of rearrangement energies.

The essential feature of Becker's renormalized BHF theory (Becker 1970a, b, Becker and Patterson 1971) is that certain higher-order diagrams that previously contributed to the rearrangement energy are *included* in the single-particle potential; in other words the single-particle potential is *redefined* to include the major part of what was previously called the rearrangement energy. The effect of this is more than a simple redefinition of terms, because now these higher-order diagrams are included in the self-consistent field and thus affect the calculated values of all observables. This renormalization, originally suggested by Brandow (1963), is much simpler and more elegant theoretically than the previous approach, where the contributions of the higher-order diagrams to each observable had to be calculated separately in a perturbative way. The effect of this renormalization is that the centroid energies are now very nearly equal to the separation energies, apart from the orbital rearrangement energy which is expected to be small for doubly closed shell nuclei. Thus Koopmans' theorem is now approximately satisfied, but the concept of a rearrangement energy related to physical observables has now disappeared. If we retain the definition of the rearrangement energy as the difference between the single-particle and separation energies, it has become a quantity that we hope to reduce to zero as we steadily refine the theory.

The modification of the self-consistent field inherent in the renormalized BHF theory is equivalent to taking into account the depletions of the normally occupied single-particle states that result from two-nucleon correlations. These depletions so weaken the self-consistent BHF field that the single-particle energies become equal to the separation energies. The renormalization also alters the form of the expression for the total energy in terms of the single-particle kinetic and potential energies so that, as shown in the next section, the (redefined) mean rearrangement energy found in this way is not distinguishable from zero. The incorporation of the higher-order diagrams in the self-consistent field gives a set of coupled equations for the occupation probabilities, so that occupation–rearrangement effects are calculated self-consistently. The renormalized BHF theory also gives nuclear density distributions and radii that are in better agreement with the experimental data for light nuclei (Becker *et al* 1974) and for heavy nuclei (Davies and McCarthy 1971).

The single-particle and separation energies are very similar in the renormalized BHF theory, and they are very closely related to the mean removal energies defined by (2.14) (Koltun 1974), but it still remains to be seen how well they compare with the experimental values. Such a comparison is shown in table 8 for ^{16}O, and it is seen to be generally satisfactory. Since the matter and charge distributions are also in good accord with the data, these calculations show a definite improvement over the unrenormalized ones.

An alternative and to some extent equivalent way of improving the BHF theory

Table 8. Results of renormalized BHF calculations for protons in ^{16}O. The rearrangement energies R are those calculated by Becker and Patterson. The last two columns give experimental separation energies found respectively by Tyrén *et al* (1966) from (p, 2p) reactions and from the energies of the eigenstates of ^{15}N and ^{15}O (Becker *et al* 1974)

State	ϵ	R	S	S_1(exp.)	S_1(exp.)
$1s_{1/2}$	~38·5 (45·5)†	13·7	31·8†	42±5	—
$1p_{3/2}$	~18·6	2·1	16·3	$19^{+0.5}_{-1}$	18·46
$1p_{1/2}$	~14·9	2·4	12·5	$12.4^{+0.3}_{-1}$	12·13

† Corrected for the centre-of-mass spuriosity in the $s_{1/2}$ hole state.

is by the use of density-dependent effective interactions. Many of these calculations have been made (Meldner and Shakin 1969, Negele 1970, Nemeth and Ripka 1973, Krieger and Moszkowski 1972, Vautherin and Brink 1972, Fai and Nemeth 1973), and on the whole they also give substantially improved overall agreement with the experimental single-particle energies and nuclear matter and charge distributions. Some results are shown in table 9, and it is notable that although the separation

Table 9. Single-particle, rearrangement and separation energies for proton orbitals in a self-consistent solution for ^{16}O (Meldner and Shakin 1969)

State	ϵ	R	S	S (exp.)
$1s_{1/2}$	64·3	26·2	38·1	—
$1p_{3/2}$	29·2	9·5	19·7	18·4
$1p_{1/2}$	14·6	2·3	12·3	12·1

energies agree well with the experimental values, the single-particle and rearrangement energies for the deeper states are substantially different from those in table 7.

The density dependence of the forces introduces a new term in the expression for the single-particle energy, and this is sometimes called the Pauli rearrangement potential (Brueckner and Goldman 1959)

$$V^{\mathrm{P}}_{\mathrm{HF}} = \tfrac{1}{2} \sum_{l,k<F} \langle lk \,|\partial V/\partial\rho|\, lk\rangle \qquad (5.19)$$

where V is the density-dependent potential. Faessler and Müther (1973) have shown that this term is equal to the Brueckner rearrangement energy plus a small term that is approximately equal to the orbital rearrangement energy. Their values of the Brueckner and orbital rearrangement energies and of the Pauli rearrangement potential are shown in table 7, and it is notable that the rearrangement energy for the Brueckner and orbital together (which is not the sum of the two effects acting separately) is approximately the same as the Pauli rearrangement potential. Since now this term is *included* in the calculation of the single-particle energy (instead of being defined as the difference between the single-particle and the separation energy), the use of density-dependent forces gives single-particle energies in good accord with the separation energies, and in practice with the peak energies.

The rearrangement effects in ^{16}O have been calculated by Müther *et al* (1973) treating the Pauli operator as well as the starting energy in the basis of a self-consistent HF method. The Pauli operator then becomes density-dependent, and they find that its use in BHF calculations gives the same result as the use of density-dependent forces. Some of their results are given in table 10. Column I gives the rearrangement due to the change of the starting energy only; column II has in addition the orbital

Table 10. Rearrangement energies in ^{16}O calculated by the self-consistent BHF method. The meaning of columns I–IV is given in the text (Müther *et al* 1973)

State	I	II	III	IV
$1s_{1/2}$	11·85	15·04	16·11	13·26
$1p_{3/2}$	9·19	10·13	10·62	9·91
$1p_{1/2}$	8·57	9·44	9·19	9·67

rearrangement; column III the total rearrangement in which both nuclei are treated fully self-consistently; and column IV shows the effect of Pauli and starting-energy rearrangement (Brueckner rearrangement) alone, neglecting the orbital rearrangement. Although all three rearrangement energies are coupled, a comparison of columns I–IV shows that the starting-energy rearrangement contributes predominantly to the total rearrangement.

Similar calculations were made for ^{40}Ca by Faessler *et al* (1973b). They found that the self-consistent Pauli operator decreases the absolute values of the single-particle energies by about 10%, and that the rearrangement term decreases the single-particle energy by a further 12–20%. This improves both the charge distribution and the agreement of the calculated single-particle energies with the measured peak energies. Examination of the rearrangement energies in tables 5–10 shows that: (*a*) rearrangement energies are greater for more deeply lying states, which is to be expected as removal from such states causes greater disturbance of the nucleus than removal from states near the Fermi surface; (*b*) rearrangement energies are greater for lighter nuclei than for heavier nuclei, which is to be expected as removal of a nucleon from a light nucleus causes a proportionately greater disturbance than removal from a heavier nucleus; and (*c*) orbital rearrangement energies are usually less than one-half the corresponding Brueckner rearrangement energies, and vary more rapidly with binding energy.

The effective density dependence of the nucleon–nucleon interaction in finite nuclei was taken into account microscopically by Tripathi *et al* (1973). They included the dependence of the Pauli operator on the self-consistent single-particle states within a BHF calculation, and called this a Pauli–Brueckner–Hartree–Fock (PBHF) calculation. Some of their results, given in table 11, show how this improvement together with the effects of rearrangement bring the calculated single-particle energies into satisfactory agreement with the experimental data.

Table 11. Energies of neutron and proton single-particle states, binding energy per nucleon, and RMS matter and charge radii for ^{16}O calculated by the BHF and PBHF theories, with and without rearrangement (Tripathi *et al* 1973)

^{16}O		BHF	PBHF without rearrangement	PBHF with rearrangement	Experiment
Proton states	$1s_{1/2}$	−50·60	−46·91	−40·94	−40±8
	$1p_{3/2}$	−20·98	−19·97	−18·02	−18·4
	$1p_{1/2}$	−15·55	−14·88	−12·56	−12·1
Neutron states	$1s_{1/2}$	−54·05	−50·26	−43·99	
	$1p_{3/2}$	−23·94	−22·85	−20·71	−21·9
	$1p_{1/2}$	−18·87	−18·14	−15·64	−15·7
B/A (MeV)		−5·36	−4·85	−4·89	−7·98
$\langle r_m^2\rangle^{1/2}$ (fm)		2·36	2·39	2·48	—
$\langle r_e^2\rangle^{1/2}$ (fm)		2·37	2·41	2·49	2·6±0·1

It is noteworthy that all these HF calculations refer to closed-shell nuclei, whereas the results described in §3 show that single-particle centroid energies and thus the rearrangement energies vary smoothly *for all nuclei* through the regions of the periodic table studied. This behaviour lacks theoretical interpretation, so that in the regions away from closed shells there is a conspicuous theoretical gap that remains to be filled.

The foregoing discussion has shown that single-particle and rearrangement energies can be defined in several different ways, in accord with different criteria of usefulness, and this is acceptable provided their differences, the separation energies, are in accord with the experimental data. The fundamental choice is that of the definition of the single-particle energies, for when this is fixed that of the rearrangement energies follows.

So far, three definitions have been used, so it is useful to review their relations to each other (Brueckner *et al* 1972, 1973). The first of these is the simplest and is the one used in the beginning of this section when the single-particle energy is defined in terms of the kinetic and potential energy by (5.3). This definition is essentially that of the centroids of the hole states, and gives a single-particle spectrum that is much more widely spaced than that of the separation energies, and hence rearrangement energies that increase to large values away from the Fermi surface. This difference in spacing implies an effective nucleon mass of one-half or less.

An improved definition of the single-particle states is obtained by summing the particle addition and removal spectra to give a combined centroid for the particle and hole states (Clement 1969, Baranger 1970):

$$\epsilon_2 = \sum_i |\langle \psi_i^{A-1}|a|\,\psi_0^A\rangle|^2\,(E_0^A - E_i^{A-1}) - \sum_i |\langle \psi_i^{A+1}|a^\dagger|\,\psi_0^A\rangle|^2\,(E_0^A - E_i^{A+1}). \quad (5.20)$$

This definition is symmetrical with respect to particles and holes and has the advantage that the corresponding potential varies smoothly with A through the nuclei where the state is filling (Millener *et al* 1973, Malaguti and Hodgson 1973), but is subject to some theoretical difficulties connected with the singular behaviour of the particle addition spectrum in the presence of a hard core potential (Dieperink *et al* 1972, Koltun 1972). The practical difficulty in using this definition is that it is rarely possible to obtain sufficient information on both the hole and the particle states for nuclei where both are important. For states far from the Fermi surface only one or the other is important, and the definition then reduces to the previous one.

A third definition of single-particle energies is that of the renormalized BHF theory of Becker and colleagues and of the various theories using density-dependent interactions. Since this effectively gives single-particle energies equal to the corresponding separation energies, the single-particle energy can be defined by (Brueckner and Goldman 1959)

$$\epsilon_3^\alpha = \delta\epsilon/\delta n_\alpha \quad (5.21)$$

where n_α is the occupation number of the level α. This is the definition of single-particle energies used in Landau's theory of fermions, and at the Fermi energy ϵ_3^α is the mean energy per particle, so that the Hugenholtz–Van Hove (1958) theorem is satisfied. The definition is closer to the concept of a local shell model potential (Brueckner *et al* 1973) and gives an effective mass ratio of unity (Brueckner *et al* 1972). According to this definition the total rearrangement energy is automatically zero, but to ensure that (5.21) is satisfied in practice it is necessary to use density-dependent forces.

With this definition of the single-particle energies they can be compared directly with the experimental data on centroid separation energies, apart from the difficulty due to faraway fragments. The rearrangement energy in the physical sense is less prominent in this framework, but it can always be calculated if necessary whenever sufficiently accurate experimental data are available to make this worthwhile.

6. Rearrangement energies obtained from total binding energies and peak energies

In the previous section it was shown that in the HF or BHF theory the total nuclear binding energy is given by (cf. 5.6) (Weisskopf 1957)

$$\mathscr{E}=\tfrac{1}{2}\sum_i (T_i+\epsilon_i)-T_{\mathrm{CM}} \tag{6.1}$$

where i runs over all normally occupied states, T_i is the kinetic energy of the ith particle, $\epsilon_i=T_i+V_i$ is its single-particle energy, and T_{CM} is the correction for the spurious motion of the centre of mass. This expression is indeed quite general, and as shown below can be obtained directly from a Hamiltonian restricted to two-body interactions with three-body interactions excluded.

Using the definition of the rearrangement energy as the difference between the centroid energy and the peak energy

$$R'=E_{\mathrm{c}}-E_{\mathrm{p}} \tag{6.2}$$

it is possible to obtain from (6.1) an expression for the mean rearrangement energy R', provided the single-particle energy ϵ_i is identified with the centroid energy E_{c}^i; this gives

$$A\bar{R}'=\sum_i (T_i-E_{\mathrm{p}}^i)-2\,(\mathscr{E}+T_{\mathrm{CM}}). \tag{6.3}$$

This expression has been used by Köhler (1966), Elton (1967) and Becker (1970b) to estimate $\bar{R}'$ from the experimental peak energies in nuclei for which data relating to all states are available. Köhler calculated the kinetic energies T_i using harmonic oscillator wavefunctions, while Elton used Saxon–Woods wavefunctions, and both authors obtained the total binding energy from mass tables. Elton's results, given in table 12, show that the mean rearrangement energy $\bar{R}$ is about 5–6 MeV, independent of A. The lower values for ^{28}Si and ^{32}S were interpreted as a result of their deformation. Becker repeated these calculations with improved experimental peak energies, using $\frac{3}{4}\hbar\omega$ for the centre-of-mass energy, as the potential energy of the centre of

Table 12. Mean nuclear rearrangement energies $\bar{R}'$ (Elton 1967)

	^{12}C	^{16}O	^{28}Si	^{32}S	^{40}Ca
ΣE_{p}^i†	−281	−407	−902	−995	−1144
$A^{-1}\,\Sigma E_{\mathrm{p}}^i$	−23·4	−25·4	−32·2	−31·0	−28·6
$A^{-1}\,\Sigma \langle T_i\rangle$	17·9	16·0	18·4	17·0	18·0
$A^{-1}\,\epsilon$	−7·7	−8·0	−8·4	−8·5	−8·6
$\mathscr{E}+2A^{-1}\,T_{\mathrm{CM}}$	9·9	6·6	3·0	3·0	6·5
$A^{-1}\,T_{\mathrm{CM}}$‡	1·9	1·2	0·7	0·6	0·4
$\bar{R}'$	6·1	4·2	1·6	1·8	5·8

† Includes the Coulomb energy.
‡ This term is the spurious energy of the centre of mass, given by $\frac{3}{2}\hbar\omega$.

mass does not enter spuriously into the expression for the total energy. All these estimates of mean rearrangement energies are given in table 13.

Table 13. Mean rearrangement energies $\bar{R}'$ (Becker 1970b)

^{12}C	^{16}O	^{28}Si	^{32}S	^{40}Ca	^{56}Ni	Reference
11	8	—	—	—	—	Köhler (1966)
6·1	4·2	1·6	1·8	5·4	—	Elton (1967)
8	5·4	2·3	2·4	6·1	—	Elton (1967)†
6–7	6·1–6·8	12·8	13·2	8·0	8·2	Becker (1970b) Renormalized
−0·2–(+0·5)	0·3–0·8	5·8	6·2	1·1	1·2	$\bar{R}'$ for $\bar{d}=0{\cdot}15$
0·145–0·165	0·16–0·17	0·31	0·33	0·18	0·18	$\bar{d}$ for $\bar{R}=0$

† Recalculated with $\frac{3}{4}\hbar\omega$ for the centre-of-mass energy.

Further more detailed calculations have been made using the renormalized BHF theory. As explained in the previous section, this involves a redefinition of the rearrangement energy, so that now it becomes a quantity whose value we expect to tend to zero as the experimental and theoretical work is progressively improved. This renormalization has the effect of taking into account the partial occupancy of the single-particle states, so that the expression for the total energy becomes (Becker 1970a, Da Providencia and Shakin 1971)

$$\mathscr{E}_1=\sum_i \epsilon_i(1-\tfrac{1}{2}\gamma_i)+\tfrac{1}{2}\sum_i t_i\,\gamma_i-T_{\rm CM} \tag{6.4}$$

where as before the summation runs only over the normally occupied states. It might be thought that in the present context the summation should also include the normally unoccupied states since they are now partially occupied; however, this would be equivalent to introducing higher-order corrections into an essentially first-order theory, and for consistency this is not done (Becker and Patterson 1971).

Becker wrote his expression in terms of the depletion factors d_i, which are related to the occupation probabilities γ_i by

$$1-d_i=\gamma_i. \tag{6.5}$$

Using (6.4) as before to obtain an expression for the mean rearrangement energy gives

$$A\bar{R}'(1-\tfrac{1}{2}\bar{\gamma})\equiv\sum_i \bar{R}'(1-\tfrac{1}{2}\gamma_i)=\sum_i E^i_{\rm p}(1-\tfrac{1}{2}\gamma_i)+\tfrac{1}{2}\sum_i t_i\gamma_i-(\mathscr{E}+T_{\rm CM})\equiv\mathscr{E}_1-\mathscr{E}. \tag{6.6}$$

The occupation probabilities γ_i are known in principle from pickup reactions (though in practice with insufficient accuracy) and may be calculated from the shell model, and the other quantities may be found as before.

There have been several calculations of the occupation probabilities of the filled orbits of doubly closed shell nuclei, and most of them give probabilities around 0·80–0·90 for all orbits (Gillet and Sanderson 1967, McCarthy and Davies 1970, Becker 1970b, Davies and McCarthy 1971, Barrett *et al* 1972, Becker *et al* 1974). At first this is rather surprising, since we might expect the lowest 1s state to be almost fully occupied, with the occupation probabilities of the higher states falling progressively towards the Fermi surface. Detailed calculations using the theory of nuclear matter show, however, that the expected decrease due to the tensor interaction is offset by the action of the hard core in depleting the lower states, with the net result that the occupation probability is similar for all the occupied orbits.

Expression (6.6) has been evaluated using the typical occupation probabilities given in tables 14 and 15, and the results are given in table 16. It is notable that the occupation probabilities are not very well known, and indeed some of the sets are

Table 14. Occupation probabilities for ^{16}O

State	γ_1	γ_2	γ_3	γ_4	γ_5
$1s_{1/2}$	0·85	0·873	0·832	0·81	0·86
$1p_{3/2}$	0·815	0·904	0·825	0·78	0·89
$1p_{1/2}$	0·83	0·906	0·822	0·78	0·88

References: γ_1, Gillet and Sanderson (1967); γ_2, γ_3, McCarthy and Davies (1970); γ_4, Davies and McCarthy (1971); γ_5, Barrett *et al* (1972).

Table 15. Occupation probabilities for ^{40}Ca

State	γ_1	γ_2	γ_3	γ_4
$1s_{1/2}$	0·85	0·8233	0·88	0·80
$1p_{3/2}$	0·81	0·8325	0·90	0·79
$1p_{1/2}$	0·86	0·8275	0·89	0·78
$1d_{5/2}$	0·80	0·8367	0·92	0·72
$2s_{1/2}$	0·83	0·8379	0·92	0·71
$1d_{3/2}$	0·82	0·8320	0·92	0·70

References: γ_1, Becker (1970b); γ_2, γ_3, γ_4, Davies and McCarthy (1971).

Table 16. Values of the mean rearrangement energies $\bar{R}'$ evaluated from (4·14) and (4·16)

Nucleus	Occupation probabilities	$\mathscr{E}_1$	$\mathscr{E}_2$	$\mathscr{E}$ (exp.)	$\bar{R}'$
^{16}O	γ_1	−113	−154	−127·6	1·6
	γ_2	−92	−116		4·0
	γ_3	−114	−156		1·5
	γ_4	−119	−172		0·9
	γ_5	−97	−125		3·4
^{40}Ca	γ_1	−336	−451	−342·0	0·24
	γ_2	−328	−437		0·59
	γ_3	−268	−324		3·40
	γ_4	−394	−570		−2·07

certainly too low because they are not in accord with the total number of nucleons required by the data on pickup and stripping reactions. Since the calculated energies are rather sensitive to these occupation probabilities, the estimates of the mean rearrangement energies are not significantly different from zero.

Essentially the same calculation has been made by Becker (1970b), who showed that the mean occupation probabilities required to give no rearrangement energy are very similar to those found from nuclear matter calculations.

While the accuracy of these comparisons can certainly be improved both experimentally and theoretically, the best estimates are consistent with (redefined) rearrangement energies of zero, and thus confirm that the renormalized BHF theory is able to give a good account of the experimental centroid energies.

A sum rule similar to (6.1) can be derived using only very general properties of the nuclear Hamiltonian (Koltun 1972, 1974). If the Hamiltonian is expressed in terms of interactions involving not more than two particles, ie

$$H = H_1 + H_2 \tag{6.7}$$

where, in the second quantization formalism,

$$H_1 = \sum_{\alpha\beta} \langle \alpha | H_1 | \beta \rangle \, a^\dagger(\alpha) \, a(\beta) \tag{6.8}$$

$$H_2 = \tfrac{1}{2} \sum_{\alpha\beta\delta\gamma} \langle \alpha\beta | H_2 | \gamma\delta \rangle \, a^\dagger(\alpha) \, a^\dagger(\beta) \, a(\gamma) \, a(\delta) \tag{6.9}$$

then the single-particle energy

$$\begin{aligned}\epsilon_\lambda = \langle \mathrm{A} | a^\dagger(\lambda) \, [a(\lambda), H] | \mathrm{A} \rangle = &\sum_\beta \langle \lambda | H_1 | \beta \rangle \langle \mathrm{A} | a^\dagger(\lambda) \, a(\beta) | \mathrm{A} \rangle \\ &+ \sum_{\beta\delta\gamma} \langle \lambda\beta | H_2 | \gamma\delta \rangle \langle \mathrm{A} | a^\dagger(\lambda) \, a^\dagger(\beta) \, a(\delta) \, a(\lambda) | \mathrm{A} \rangle. \end{aligned} \tag{6.10}$$

Summing over all states λ gives

$$\sum_\lambda \epsilon_\lambda = \langle \mathrm{A} | H_1 | \mathrm{A} \rangle + 2 \langle \mathrm{A} | H_2 | \mathrm{A} \rangle \tag{6.11}$$

and since the total energy of the nucleus

$$\mathscr{E} = \langle \mathrm{A} | H | \mathrm{A} \rangle = \langle \mathrm{A} | H_1 + H_2 | \mathrm{A} \rangle \tag{6.12}$$

then

$$\mathscr{E} = \tfrac{1}{2} \langle \mathrm{A} | H_1 | \mathrm{A} \rangle + \tfrac{1}{2} \sum_\lambda \epsilon_\lambda. \tag{6.13}$$

This expression reduces to (6.1) if all the occupation probabilities of the normally occupied orbits are set equal to unity, but in general this is not so. In practical calculations the spurious centre-of-mass kinetic energy must be subtracted as before.

This expression for the total nuclear energy has been evaluated directly from the spectral function $P(k, E)$ by Koltun (1972). Since only proton removal (p, 2p) data are available, it is necessary for an accurate calculation to obtain an expression for the total proton energy of the nucleus. This may be done by writing the Hamiltonian in a proton–neutron notation

$$H = H_1(\mathrm{p}) + H_1(\mathrm{n}) + H_2(\mathrm{pp}) + H_2(\mathrm{nn}) + H_2(\mathrm{pn}) \tag{6.14}$$

and defining the total proton energy by

$$\mathscr{E}_Z = \langle \mathrm{A} | H_1(\mathrm{p}) + H_2(\mathrm{pp}) + \tfrac{1}{2} H_2(\mathrm{pn}) | \mathrm{A} \rangle. \tag{6.15}$$

Then (6.13) becomes

$$\mathscr{E}_Z = \tfrac{1}{2} \int_{-\infty}^{\infty} \mathrm{d}E \int \mathrm{d}^3k \, (k^2/2m - E) \, P(k, E). \tag{6.16}$$

The spectral function is known experimentally for a limited range of k and E, so it is necessary to use the normalized expressions for the mean kinetic and removal energies per proton

$$T_\mathrm{m} = n^{-1} \int \mathrm{d}E \int \mathrm{d}^3k \, (k^2/2m) \, P(k, E) \tag{6.17}$$

$$E_\mathrm{m} = n^{-1} \int \mathrm{d}E \int \mathrm{d}^3k \, (-E) \, P(k, E) \tag{6.18}$$

where

$$n = \int \mathrm{d}E \int \mathrm{d}^3k \, P(k, E). \tag{6.19}$$

Then the mean energy per proton

$$(\mathscr{E}_Z/Z)_\mathrm{m} = \tfrac{1}{2} T_\mathrm{m} + \tfrac{1}{2} E_\mathrm{m}. \tag{6.20}$$

Koltun (1972) used the experimental data of James *et al* (1969) to evaluate this quantity for several nuclei, and the results are compared in table 17 with values obtained from the nuclear masses with the help of the semi-empirical mass formula.

Table 17. Total proton energies of several nuclei calculated from the (p, 2p) spectral function for removal and compared with the experimental data (Koltun 1972)

Nucleus	T_m	E_m	$(\mathscr{E}_Z/Z)_m$	$(\mathscr{E}_Z/Z)_{exp}$
^{12}C	14·1	−28·2	−7·05	−6·93
^{40}Ca	13·8	−28·0	−7·1	−6·73
^{120}Sn	14·4	−30·2	−7·9	−9·32
^{208}Pb	14·3	−27·7	−6·7	−8·22

While the agreement is as good as can be expected in view of the uncertainties in the experimental data, it is fortuitous for several reasons. In particular the spectral function is available only over a limited range of E and k, the effects of the short-range correlations leading to highly excited states with large energy weight are not contained in the calculated sum rule, and multiple-scattering effects should be taken into account more accurately than by just an overall constant reduction factor.

A more stringent test of the sum rule has been made by Bernheim *et al* (1974a) with 497 MeV electrons on ^{12}C. The use of electrons greatly reduces the corrections due to multiple scattering, and the experimental conditions enabled the ranges $0<E_m<74$ MeV and $0<k<300$ MeV/c to be covered. Their results are shown in table 18 before and after correction for the final-state interaction.

Table 18. Total proton energies of ^{12}C calculated from the ^{12}C (e, e′p) data of Bernheim *et al* (1974a)

Energy region	Experimental results			Corrected values		
	n	T_m	E_m	n	T_m	E_m
15–25	1·7	16·3	−16·9	2·6	18·4	−16·9
25–74	6·56	11·5	−38·7	1·1	13·4	−38·7
15–74	—	—	—	3·7	16·9	−23·4

The binding energy per proton calculated from the sum rule corrected for recoil energies

$$(\mathscr{E}_Z/Z)_m=\tfrac{1}{2}\,(T_m+E_m-m_pT_m/M_{A-1}) \tag{6.21}$$

gives $-4{\cdot}0\pm0{\cdot}5$ MeV compared with the value of $-6{\cdot}93$ MeV obtained from nuclear masses with Coulomb corrections. There is thus a marked discrepancy that could be due to faraway fragments or to the effect of three-body forces, since the sum rule is derived assuming that the Hamiltonian contains only two-body interactions. Bernheim *et al* estimate that the discrepancy would be accounted for if 5% of the events had an average separation energy of 150 MeV, but this seems very unlikely.

The measurement of the spectral function in this experiment was restricted to separation energies from 0 to 50 MeV and momenta less than 300 MeV/c, so it is important to see how this affects the sum rule. The range of energy and momenta is approximately the same as that of the occupied single-particle shell model wave-functions, which is essentially the domain of the HF theory. There are three ways of renormalizing the nucleon–nucleon interaction to take account of this restriction, and this makes it possible to see if the lack of balance in the sum rule can be attributed to the restrictions on the range of energy and momenta.

As mentioned in the previous section, these three ways are: (i) allowing the effective force to become density-dependent due to the short-range correlations; (ii) representing this density dependence by an effective three-body force; and (iii) the introduction of partial occupation probabilities. Faessler *et al* (1975b) have made numerical calculations of the corrections to the sum rule due to the restrictions on the energy and momenta of the ejected nucleon. The density dependence introduces a rearrangement correction (5.19), and this was evaluated using the modified delta interaction of Faessler *et al* (1975a). The three-body effective force of Skyrme used by Vautherin and Brink (1972) was used in the second series of calculations. The uncertainties in the occupation probabilities made it more convenient to use the sum rule itself to calculate the required mean occupation probabilities, and these were found to be consistent with the theoretical values. The results of the first two sets of calculations are shown in table 19, and it is clear that essentially all of the discrepancies in the

Table 19. Values of the sum-rule discrepancy Δ obtained from analyses of data on (e, e′p) reactions at 500 MeV (Mougey 1975), compared with calculations using density-dependent interactions $\Delta(\rho)$ and the equivalent three-body forces Δ (3-body) (Faessler *et al* 1975b)

Nucleus	n	$\langle E\rangle$	$\langle T\rangle$	E_Z/Z	Δ	$\Delta(\rho)$	Δ (3-body)
^{12}C	3·7	23·4	16·9	−6·93	−2·9±0·5	−2·15	−2·93
^{28}Si	11·1	24·0	17·0	−6·84	−2·9±0·6	−2·65	−3·61
^{40}Ca	16·2	27·8	16·6	−6·51	−0·7±0·5	−2·81	−3·81
^{58}Ni	25·1	25·0	18·8	−6·95	−3·7±0·7	−2·96†	−4·06†

† Calculation for ^{56}Ni. Other calculated values of Δ (3-body) are −3·18 (^{16}O) and −4·26 (^{208}Pb).

sum rule are accounted for, except in the case of calcium. Thus the explanation of the discrepancy in the original sum rule must be sought in missing strength at high relative momenta or in the presence of (real) three-body forces. The recent calculations of Yang (1974) strengthen the likelihood of the latter alternative, as he showed that real three-body forces make a substantial contribution to the binding energy of the triton.

This sum rule has been tested in a different way by Tripathi and Goldhammer (1972), who used it to define a relation between total shell model energies E and the single-particle energies ϵ_i. Thus for ^{16}O they obtain

$$E(^{16}\text{O}) = E(\text{s}^4) + \sum_j (2j+1)\,(E(\text{p}_j) + \epsilon(\text{p}_j)) + \text{Coulomb corrections.} \qquad (6.22)$$

Using experimental values for $E(\text{s}^4)$, $E(\text{p}_j)$ and $\epsilon(\text{p}_j)$ gives −122·4 MeV for this expression, compared with the known binding energy $E(^{16}\text{O}) = -127{\cdot}6$ MeV. The expression (6.23) might be expected to be an overestimate of the energy $E(^{16}\text{O})$, but in fact it is an underestimate, indicating a definite discrepancy that can be ascribed to three-body forces. It sets a lower limit to the importance of higher-order correlations in ^{16}O, which is consistent with that of three-body correlations in nuclear matter, about 1 MeV per nucleon (Rajaraman and Bethe 1967).

Thus on the whole one can conclude that at present there is no evidence that the sum rule is not satisfied, within the uncertainty connected with the evaluation of the three-body forces.

There is in principle another way of determining the nuclear rearrangement energy from the total binding energy and the peak energies for nucleon removal.

The total binding energy of a nucleus is the energy required to separate all its nucleons to an infinite distance from each other, and this energy is known accurately from the nuclear mass. The nucleons can in fact be removed one by one by the appropriate pickup and knockout reactions, and so it is possible to relate the experimental removal energies to the total binding energy and to the rearrangement that takes place after the removal of each nucleon.

Thus if we always remove the nucleon from the highest state, so as to leave the residual nucleus in the single-particle state that contains the ground state of the residual nucleus, then the total energy

$$\mathscr{E}(A)=E_{\mathrm{p}}^{A}+R_{A}+\mathscr{E}(A-1). \tag{6.23}$$

Repeating the nucleon removal until only one nucleon is left gives

$$\mathscr{E}(A)=\sum_{i=2}^{A-1}(E_{\mathrm{p}}^{i}+R_{i}) \tag{6.24}$$

so that the mean rearrangement energy $\bar{R}$ is given by

$$(A-1)\,\bar{R}=\mathscr{E}(A)-\sum_{i=1}^{A-1}E_{\mathrm{p}}^{i}. \tag{6.25}$$

This expression has not yet been used to find $\bar{R}$, as it requires the peak energies for all nuclei smaller than A, instead of all E_{p} for a particular nucleus, as was required previously. This $\bar{R}$ is not a mean rearrangement energy in a particular nucleus, but a mean of the ground-state rearrangement energies for all nuclei up to A.

7. Summary and conclusions

This survey shows that the present experimental data on reactions that add or subtract a nucleon from a nucleus can be analysed to give local centroid energies that are well defined and vary smoothly from nucleus to nucleus in a way that can be described by the eigenvalues of a simple local potential. The true centroid energies are not observable due to the impossibility of observing faraway fragments.

The physical definition of a rearrangement energy as the difference between the energies spent in removing a nucleon quickly and slowly from a nucleus has as yet no practical utility, as all known processes correspond to fast removal. This situation could change if experimental evidence were found for the dependence of the removal energy on the reaction time, though at present this seems very unlikely.

The rearrangement energy can be expressed in terms of nuclear energies before and after the removal process, and this can be calculated in the BHF theory. After renormalization, this theory gives single-particle energies that are comparable with the local centroid energies.

The expression for the total energy of the nucleus in terms of the sum of the kinetic and potential energies of the constituent nucleons together with the data on local centroid energies may be used to obtain rearrangement energies averaged over all the nuclear states. This gives evidence for a large effect from the faraway fragments or from three-body forces.

This account of nucleon removal and rearrangement energies is somewhat complicated by the effects of the unobservable faraway fragments of the single-particle strength. The available evidence, however, indicates that their effect does not destroy the simple picture of nuclear single-particle states. In principle their distribution could be calculated from a knowledge of short-range nucleon–nucleon correlations, and this would make possible a more complete account of nucleon removal processes.

Acknowledgments

I am grateful to many friends, particularly Bruce Barrett, Richard Becker, Claudio Ciofi degli Atti, Charles Clement, Amand Faessler, Klaus Goeke, Gerhard Jacob, Michael Kirson, Tony Lane, Alan MacKellar, Eivind Osnes and Ram Tripathy, who have commented on drafts of this paper and have shared with me their knowledge of nucleon removal and rearrangement energies.

References

AMALDI U 1967 *Proc. Int. School of Physics 'Enrico Fermi'* XXXVII (New York, London: Academic) p284

BACK B B, BANG J, BJØRNHOLM S, HATTULA J, KLEINHEINZ P and LEIN J R 1974 *Nucl. Phys.* A **222** 377

BANSAL R K and FRENCH J B 1964 *Phys. Lett.* **11** 145

BARANGER M 1970 *Nucl. Phys.* A **149** 225

BARRETT B R, HEWITT R G L and MCCARTHY R J 1972 *Nucl. Phys.* A **184** 13

BASSICHIS W H and STRAYER M R 1971 *Ann. Phys., NY* **66** 457

BATTY C J and GREENLEES G W 1969 *Nucl. Phys.* A **133** 673

BAYMAN B F 1966 *Isobaric Spin in Nuclear Physics* ed J D Fox and D Robson (New York: Academic) p503

BECKER R L 1970a *Phys. Rev. Lett.* **24** 400

—— 1970b *Phys. Lett.* **32B** 263

—— 1973 *The Nuclear Many-Body Problem* vol II, ed F Calogero and C Ciofi degli Atti (Bologna: Editrice Compositori) p205

BECKER R L, DAVIES K T R and PATTERSON M R 1974 *Phys. Rev.* C **9** 1221

BECKER R L and PATTERSON M R 1971 *Nucl. Phys.* A **178** 88

BERNHEIM M, BUSSIÈRE A, GILLEBERT A, MOUGEY J, PHAN XUAN HO, PRIOU M, ROYER D, SICK I and WAGNER G J 1974a *Phys. Rev. Lett.* **32** 898

BERNHEIM M, BUSSIÈRE A, GILLEBERT A, MOUGEY J, PRIOU M, ROYER D, SICK I and WAGNER G J 1974b *Proc. Int. Conf. on Nuclear Structure and Spectroscopy* vol II (Amsterdam: Scholar's Press) p412

BLOMQVIST J and WAHLBORN S 1960 *Ark. Fys.* **16** 545

BLUM B, DUNNWEBER W and HERING W R 1973 *Nucl. Phys.* A **206** 321

BRANDOW B H 1963 *Phys. Lett.* **4** 8

—— 1967 *Rev. Mod. Phys.* **39** 771

BROWN G E 1967 *Unified Theory of Nuclear Models and Forces* (Amsterdam: North-Holland) p203

BROWN G E, GUNN J H and GOULD P 1963 *Nucl. Phys.* **46** 598

BRUECKNER K A 1959 *Proc. Int. Conf. on the Nuclear Optical Model, Tallahassee* (Tallahassee: Florida State University) p143

BRUECKNER K A and GOLDMAN D T 1959 *Phys. Rev.* **116** 424

—— 1960 *Phys. Rev.* **117** 207

BRUECKNER K A, MELDNER H W and PEREZ J D 1972 *Phys. Rev.* C **6** 773

—— 1973 *Phys. Rev.* C **7** 537

CAMPOS VENUTI G, CORTELLESSA G, FARCHI G, FRULLANI S, GIORDANO R, SALVADORI P, TAKAMATSU K, CAPITANI G P and DE SANCTIS E 1973 *Nucl. Phys.* A **205** 628

CLEMENT C F 1969 *Phys. Lett.* **28B** 398

—— 1973 *Nucl. Phys.* A **213** 469

Clement C F and Perez S M 1973 *Nucl. Phys.* A **213** 510
—— 1974 *J. Phys.* A: *Math., Nucl. Gen.* **7** 1464
Clementi E 1963 *J. Chem. Phys.* **38** 1001
—— 1964 *J. Chem. Phys.* **41** 295, 303
Cohen B L 1965 *Am. J. Phys.* **33** 1011
Da Providencia J and Shakin C M 1971 *Phys. Rev.* C **4** 1560
Davies K T R and Becker R L 1971 *Nucl. Phys.* A **176** 1
Davies K T R and McCarthy R J 1971 *Phys. Rev.* C **4** 81
Demos J R and Banerjee M K 1972 *Phys. Rev.* C **5** 75
Dieperink A E L, Brussaard P J and Cusson R Y 1972 *Nucl. Phys.* A **180** 110
Dieperink A E L and de Forest T 1974 *Phys. Rev.* C **10** 543
Doll P 1974 *Thesis* University of Heidelberg
Dover C B and Van Giai N 1972 *Nucl. Phys* A **190** 373
Elton L R B 1967 *Phys. Lett.* **25B** 60
—— 1968 *Phys. Lett.* **26B** 689
Elton L R B and Swift A 1967 *Nucl. Phys.* A **94** 52
Englebrecht C A and Weidenmuller H A 1972 *Nucl. Phys.* A **184** 385
Faessler A, Galonska J E, Ehlers J W and Moszkowski S A 1972 *Nuovo Cim.* **11A** 63
Faessler A, Galonska J E, Goeke K and Moszkowski S A 1975a *Nucl. Phys.* A **239** 477
Faessler A, Krewald S and Wagner G J 1975b *Preprint*
Faessler A, Kusuno S and Strobel G L 1973a *Nucl. Phys.* A **203** 513
Faessler A, MacKellar A D and Tripathi R K 1973b *Nucl. Phys.* A **215** 525
Faessler A and Müther H 1973 *Z. Naturf.* **28a** 362
Faessler A and Wolter H H 1969 *Z. Phys.* **223** 192
Fai G and Nemeth J 1973 *Preprint*
Fritsch W, Lipperheide R and Wille U 1972 *Nucl. Phys.* A **198** 515
—— 1973 *Phys. Lett.* **45B** 103
Froman A 1958 *Phys. Rev.* **112** 870
Garron J P, Jacmart J C, Riou M, Ruhla C, Teillac J and Stranch K 1962 *Nucl. Phys.* **37** 126
Gibson B F and Van Oostrum K J 1967 *Nucl. Phys.* A **90** 159
Gillett V and Sanderson E A 1967 *Nucl. Phys.* A **91** 292
Grimm R C, McCarthy I E and Storer R G 1971 *Nucl. Phys.* A **166** 330
Gross D H E and Lipperheide R 1970 *Nucl. Phys.* A **150** 449
Herscovitz V E 1971 *Nucl. Phys.* A **161** 321
Herscovitz V E, Jacob G and Maris Th A J 1968 *Nucl. Phys.* A **109** 478
Hinterberger F, Mairle G, Schmidt-Rohr U, Turek P and Wagner G J 1968 *Nucl. Phys.* A **106** 161
Hiramatsu S 1970 *University of Tokyo, Institute for Nuclear Study Rep.* INS J-122
Hugenholtz N M and Van Hove L 1958 *Physica* **24** 363
Jackson D F 1969 *Symp. on Pickup and Quasi-free Knockout of Particles and Clusters, Maryland* (unpublished)
Jacob G and Maris Th A J 1966 *Rev. Mod. Phys.* **38** 121
—— 1973 *Rev. Mod. Phys.* **45** 6
James A N 1969 *Proc. Int. Conf. on Properties of Nuclear States, Montreal* (Montreal: Montreal UP) p515
James A N, Andrews P T, Kirkby P and Lowe B G 1969 *Nucl. Phys.* A **138** 145
Janiszewski J 1972 *MSc Thesis* Flinders University
Källne J and Fagerström B 1974 *High Energy Physics and Nuclear Structure, Uppsala* (Amsterdam, London: North-Holland; New York: Elsevier; Stockholm: Almqvist and Wiksell) p369
Khadkikar S B and Kamble V B 1974 *Nucl. Phys.* A **225** 352
Köhler H S 1965 *Phys. Rev.* **137B** 1145
—— 1966 *Nucl. Phys.* **88** 529
Köhler H S and Lin Y C 1971 *Nucl. Phys.* A **167** 305
Koltun D S 1972 *Phys. Rev. Lett.* **28** 182
—— 1973 *The Nuclear Many-Body Problem* vol II, ed F Calogero and C Ciofi degli Atti (Bologna: Editrice Compositori) p88
—— 1974 *Phys. Rev.* C **9** 484
Koopmans T 1933 *Physica* **1** 104

Koshel R D 1966 *Phys. Rev.* **144** 811
Krieger S J and Moszkowski S A 1972 *Phys. Rev.* C **5** 1990
Kullander S, Lemeilleur F, Renberg P U, Landaud G, Yonnet J, Fagerström B, Johansson A and Tibell G 1971 *Nucl. Phys.* A **173** 357
Landaud G, Yonnet J, Kullander S, Lemeilleur F, Renberg P U, Fagerström B, Johansson A and Tibell G 1971 *Nucl. Phys.* A **173** 337
Lindgren I 1965 *Phys. Lett.* **19** 382
McCarthy R J and Davies K T R 1970 *Phys. Rev.* C **1** 1644
Macfarlane M H 1969 *Proc. Int. Conf. on Properties of Nuclear States, Montreal* (Montreal: Montreal UP) p385
Malaguti F and Hodgson P E 1973 *Nucl. Phys.* A **215** 243
Maris Th A J 1974 *High Energy Physics and Nuclear Structure, Uppsala* (Amsterdam, London: North-Holland; New York: Elsevier; Stockholm: Almqvist and Wiksell)
Meldner H W 1969 *Phys. Rev.* **178** 1815
Meldner H W and Perez J D 1971 *Phys. Rev.* A **4** 1388
Meldner H W and Shakin C M 1969 *Phys. Rev. Lett.* **23** 1302
Meldner H W and Süssmann G 1963 *Phys. Lett.* **6** 353
Millener D J and Hodgson P E 1973 *Nucl. Phys.* A **209** 59
Millener D J, Malaguti F and Hodgson P E 1973 *Proc. Int. Conf. on Nuclear Structure, Munich* (Amsterdam: North-Holland) p60
Moldauer P A 1963 *Nucl. Phys.* **47** 65
Mougey J 1975 *Proc. 12th Int. Meeting in Nuclear Physics* (Paris: Gauthier-Villars)
Müther H, Faessler A and Goeke K 1973 *Nucl. Phys.* A **215** 213
Müther H, Goeke K and Faessler A 1972 *Z. Phys.* **253** 61
Nakamura K, Hiramatsu S, Kamae T, Muramatsu H, Izutsu N and Watase Y 1974 *Phys. Rev. Lett.* **33** 853
Negele J W 1970 *Phys. Rev.* C **1** 1260
Nemeth J and Ripka G 1973 *Preprint*
Perey F G 1963 *Phys. Rev.* **131** 745
Perey F G and Buck B 1962 *Nucl. Phys.* **32** 353
Pittel S and Austern N 1972 *Phys. Rev. Lett.* **29** 1403
—— 1974 *Nucl. Phys.* A **218** 221
Rajaraman R and Bethe H A 1967 *Rev. Mod. Phys.* **39** 745
Rosen A and Lindgren I 1968 *Phys. Rev.* **176** 114
Rost E 1968 *Phys. Lett.* **26B** 184
Sakai M, Sekiguchi M, Soga F, Hirao Y, Yagi K and Aoki Y 1974 *Phys. Lett.* **51B** 51
Satchler G R, Drisko R M and Bassel R H 1964 *Phys. Rev.* B **136** 637
Shakin C M and Da Providencia J 1971 *Phys. Rev. Lett.* **27** 1069
Shanta R 1973 *Nucl. Phys.* A **199** 624
Slater J C 1955 *Phys. Rev.* **98** 1039
Strobel G L and von Geramb H 1973 *Nucl. Phys.* A **210** 67
Swift A and Elton L R B 1966 *Phys. Rev. Lett.* **17** 484
Takeuchi K and Moldauer P 1969 *Phys. Lett.* **28B** 384
Tripathi R K, Faessler A and MacKellar A D 1973 *Phys. Rev.* C **8** 129
Tripathi R K and Goldhammer P 1972 *Phys. Rev.* C **6** 101
Tyrén H, Kullander S, Sundberg O, Ramachandran R, Isacsson P and Berggren T 1966 *Nucl. Phys.* **79** 321
Vautherin D and Brink D M 1972 *Phys. Rev.* C **5** 626
Wagner G J 1969 *Bull. Am. Phys. Soc. II* **14** 85
—— 1973 *Nuclear Structure Physics, Lecture Notes in Physics No.* 23, ed V Smilansky, I Talmi and H A Weidenmuller (Berlin: Springer) p16
Wagner G J, Mairle G, Schmidt-Rohr U and Turek P 1969 *Nucl. Phys.* A **125** 80
Weisskopf V F 1957 *Rev. Mod. Phys.* **29** 174
Wenneis S 1971 *Dissertation* University of Heidelberg
Wilkinson D H 1968 *Comm. Nucl. Particle Phys.* **2** 48
Wille U, Gross D H E and Lipperheide R 1971 *Phys. Rev.* C **4** 1070
Wille U and Lipperheide R 1972 *Nucl. Phys.* A **189** 113
Wyatt P J, Wills J G and Green A E S 1960 *Phys. Rev.* **119** 1031
Yang S N 1974 *Phys. Rev.* C **10** 2067

The Neutron Optical Model in the Actinide Region

P.E. Hodgson

Nuclear Physics Laboratory, Oxford.

Invited paper presented to the meeting on

'Fast Neutron Scattering on Actinide Nuclei'

Paris, November 1981

Abstract:

The usefulness of the simple optical model for calculating neutron cross-sections in the actinide region is assessed. Comparisons are made between many proposed sets of optical model parameters and the experimental data, and an optimum set is selected. The dependence of the potential on the incident energy and on the nuclear asymmetry and deformation is discussed, and several ways of improving this set are proposed. The usefulness of the simple optical model is compared with that of the coupled-channels method and a new way of calculating inelastic cross-sections is proposed.

1. Introduction

The optical model has been extensively used to analyse the elastic scattering of neutrons by nuclei, and several sets of parameters have been obtained that give good overall fits for a wide range of energies and nuclei.

In this review we are concerned with the actinide nuclei, and these pose particular problems due to their large deformations. These introduce a strong coupling between the elastic and inelastic channels, so that the elastic scattering is no longer given by the global potential fitted to the scattering from a range of spherical and near-spherical nuclei. It is however still possible to fit the elastic scattering from deformed nuclei by an optical potential, with suitably modified parameters.

The important practical question is whether this effective potential is simple in form and able to fit a wide range of data. If so, it has great practical utility, quite apart from the question of its theoretical justification. The same philosophy can be applied to the calculation of the inelastic cross-sections, by retaining the distorted wave formalism and trying to find effective potentials that when inserted into the formalism give cross-sections in agreement with the experimental data.

In the next section the optical potential is specified, and the connection between the potentials for spherical and deformed nuclei is discussed. The optimum potentials in the actinide region are reviewed in section three, and comparisons are made with a range of experimental data.

In practical calculations it is necessary to calculate the cross-sections in other channels as well, in particular those for inelastic scattering. For spherical and near-spherical nuclei this may be done using the distorted wave formalism, but in its standard form this is not sufficiently accurate for strongly deformed nuclei. The effect of the deformation on the elastic scattering may be evaluated by the coupled-channels method, and this provides a useful insight into the connection between the potentials for spherical and deformed nuclei. The appropriate formalism is summarised in section 4 and a modified distorted wave expression is obtained that takes into account the effects of deformation. The implications of this for practical calculations of the direct inelastic scattering are considered in section 5 together with a comparison between the optical model and coupled-channels calculations.

This discussion takes place in the context of the practical need to calculate cross-sections for nuclei and at energies that are difficult to measure directly. While this does exclude the discussion of purely academic points, it must be emphasised that the more fully we understand the physics of the interaction the more reliable our extrapolations to unknown nuclei and energies are likely to be. Furthermore, when many cross-sections are required a fast but somewhat less accurate method of calculation may be preferable to a lengthy but more accurate method, provided that the accuracy obtained is sufficient for the practical purpose for which the calculation is made.

2. The Optical Model

The optical model provides an easy way of calculating the differential cross-sections and polarisations for neutron elastic scattering, together with the associated total and reaction cross-sections. It also provides the distorted wavefunctions that are needed to calculate the cross-sections of inelastic scattering and nuclear transfer reactions.

The optical potential V(r) is defined as the potential that when inserted into the Schrödinger equation

$$\nabla^2\Psi + \frac{2m}{\hbar^2}(E - V(r))\ \Psi=0 \tag{2.1}$$

gives the observed total, reaction and shape elastic cross-sections. In principle V(r) is a highly singular object, certainly non-local and also depending on the neutron energy and the structures of the target, compound and residual nuclei. The remarkable fact, which makes the optical model so useful, is that it is possible to use a very simple form for V(r) and still obtain excellent agreement with the experimental data. Furthermore, the parameters of this potential are very similar for all nuclei, with a rather slight and almost linear energy dependence.

The form most often used for the optical potential is

$$V(r) = Uf_u(r) + iWg(r) + \left(\frac{\hbar}{m_\pi c}\right)^2 U_s \frac{1}{r}\frac{df_s(r)}{dr}\ \mathrm{L.\sigma} \tag{2.2}$$

where U, W and U_s are the real, imaginary and spin-orbit potential depths, and the form factors

$$f_u(r) = \left\{ 1 + \exp\left(\frac{r-R_u}{a_u}\right) \right\}^{-1} \tag{2.3}$$

$$f_s(r) = \left\{ 1 + \exp\left(\frac{r-R_s}{a_s}\right) \right\}^{-1} \tag{2.4}$$

The form factor of the imaginary potential has either the volume form

$$g_v(r) = \left\{ 1 + \exp\left(\frac{r-R_v}{a_v}\right) \right\}^{-1} \tag{2.5}$$

or the surface-peaked radial derivative form

$$g_D(r) = \frac{4a_D \exp\left(\dfrac{r-R_D}{a_D}\right)}{\left\{1 + \exp\left(\dfrac{r-R_D}{a_D}\right)\right\}^2} \tag{2.6}$$

Since the radius parameters are closely proportional to $A^{1/3}$, it is usual to write $R_i = r_i A^{1/3}$. The last term in (2.2) is the spin-orbit term that is necessary to enable the polarisation of the elastically scattered neutrons to be calculated. If polarisations are not required this term may be omitted, thus halving the computing time, with rather small effect on the differential cross-sections.

Many phenomenological analyses of experimental neutron scattering data have been made using this potential, and many useful parameters sets have been found. The details of the calculations and tabulations of the parameters are given in many publications [1]. Several sets are able to fit the data for many nuclei provided the potential is allowed to depend on the incident ehergy and nuclear asymmetry

$$U = U_o + \alpha E + \beta E^2 + \frac{N-Z}{A} U_1 \tag{2.7}$$

In some cases the imaginary potential W and its diffuseness parameter a_W are also allowed to depend linearly or quadratically on the energy, and also on the nuclear asymmetry.

At the same time there have been many theoretical studies of the optical potential with the main aim of deriving it from more fundamental nuclear properties such as the nuclear density distribution and the nucleon-nucleon interaction [2]. These have not yet been developed to the stage where they can given potentials that fit the data better than the phenomenological ones, but they are often useful in providing deeper insight into the physics of the interactions.

A quite early analysis of neutron scattering by Perey and Buck [3] using a non-local potential was very successful in fitting the data for many nuclei from a few MeV to 25 MeV. Subsequently the equivalent local potential was obtained by Wilmore and Hodgson [4] and this was found to be equally successful. Since that time it has been used in many analyses of neutron elastic scattering and reactions.

Detailed comparisons between the predictions of this potential and experiment do however show some discrepancies, notably those due to shell effects at low energies for magic nuclei and to the effects of channel coupling for deformed nuclei. The latter are particularly important for the strongly deformed nuclei in the actinide region. These effects are shown in Fig.1 and it is apparent that the energy dependence of the total neutron cross-sections for spherical nuclei is very well given by the Perey-Buck potential, but that significant discrepancies become apparent for deformed nuclei.

The reason why these discrepancies occur for deformed nuclei is that such nuclei have low-lying collective states strongly coupled to the ground state. This coupling is strong enough to affect the wavefunction in the elastic channel to the extent that the fit to the

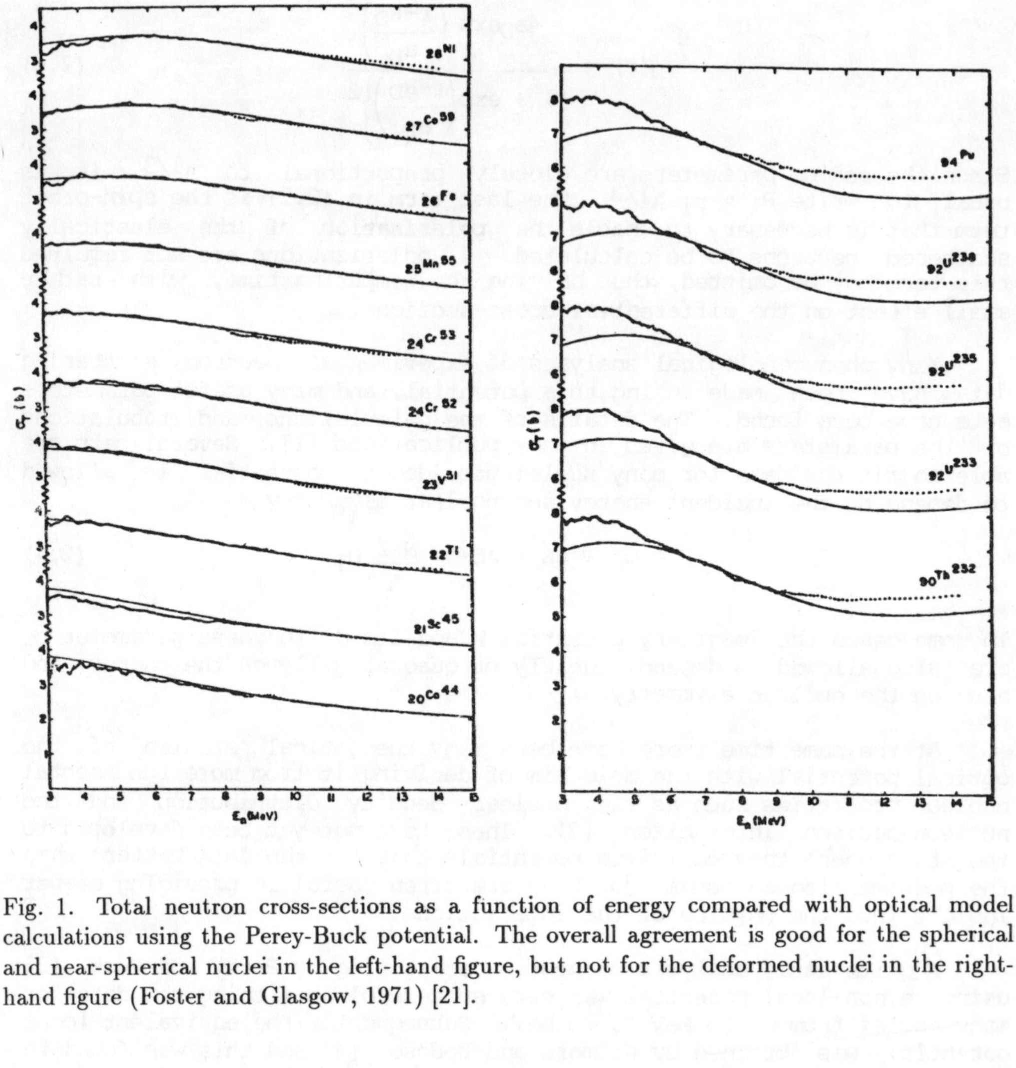

Fig. 1. Total neutron cross-sections as a function of energy compared with optical model calculations using the Perey-Buck potential. The overall agreement is good for the spherical and near-spherical nuclei in the left-hand figure, but not for the deformed nuclei in the right-hand figure (Foster and Glasgow, 1971) [21].

data is significantly worsened. It is however possible to restore the fit by appropriate adjustment of the parameters of the optical model potential, and some analyses of the scattering of neutrons by actinide nuclei made in this way are discussed in the next section.

In subsequent sections the optical potentials defined above are referred to as 'spherical potentials', and the potentials used in the coupled-channels formalism (see §4) are referred to as 'deformed potentials'.

3. Optical Model Analyses of Neutron Elastic Scattering by Actinide Nuclei

Several sets of optical model parameters have been proposed for neutron scattering by various heavy nuclei; these are listed in Table I and compared with the experimental data for the total cross-section for neutrons on ^{241}Am in Fig.2. The comparison is made with the total cross-section because it is the only observable calculable from the optical potential that is directly comparable with the energy-averaged experimental quantity. In all other cases account has to be taken of the fluctuation or compound nucleus component, and this introduces additional uncertainties into the analysis. A similar comparison with the reaction cross-section is also included in Fig.2.

It should be noted that only the potential of Madland and Young (MY) [7] is a spherical potential fitted to actinide data; the other potentials are deformed potentials, apart from that of Wilmore and Hodgson [4], which is a global spherical potential. We thus expect the MY potential to give the best fit; the others are included to show the magnitude of the differences introduced when deformed potentials are introduced into the spherical formalism. The MY spherical potentials were actually obtained as intermediate stages in their search for the optimum global deformed potential, but they have in their own right considerable value for fitting data.

Thus at lower energies the potential of Madland and Young is preferable. At higher energies there is less to choose between them, and that of Wilmore and Hodgson (WH) is preferred because it is known to agree with a wide range of data for spherical and near-spherical nuclei. The effects of the deformation of the actinide nuclei are expected to be most significant at the lower energies and unimportant at higher energies, and the comparison between the data and the WH potential is in accord with this.

The Madland and Young potential is thus a good approximation to the effective optical potential in the actinide region. To test this further, it has been compared with a wide range of data in the actinide region, and the results are shown in Figs.3 - 6.

It should be noted that account was taken of the compound elastic cross-sections in the calculations of Madland and Young. This does contribute to the differential cross-sections, particularly at the lower energies, but is progressively less important as the energy increases. Calculations by Haouat <u>et al</u> [12] show that it contributes 27%, 12% and 0.5% to the 80° differential cross-sections at energies of 0.7, 1.5 and 2.5 MeV respectively. This is in the region of a

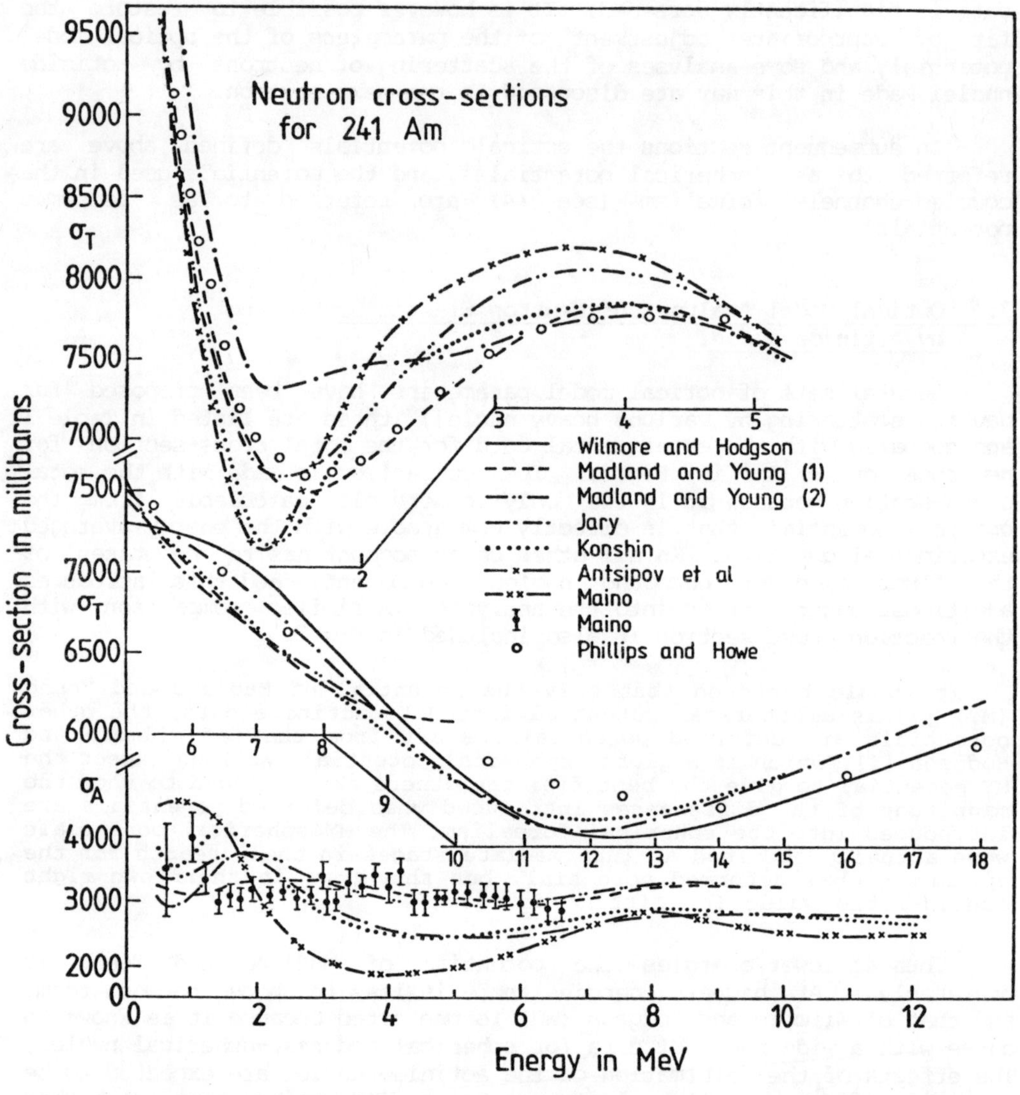

Fig. 2. Total [(a) and (b)] and reaction (c) cross-sections for the interaction of neutrons with 241Am compared with optical model calculations with the potentials of Madland and Young [7], Jary [6], Konshin [8] and Antisipov [5]. The experimental data are taken from the papers of Phillips and Howe [20] and Miano <u>et al</u> [9].

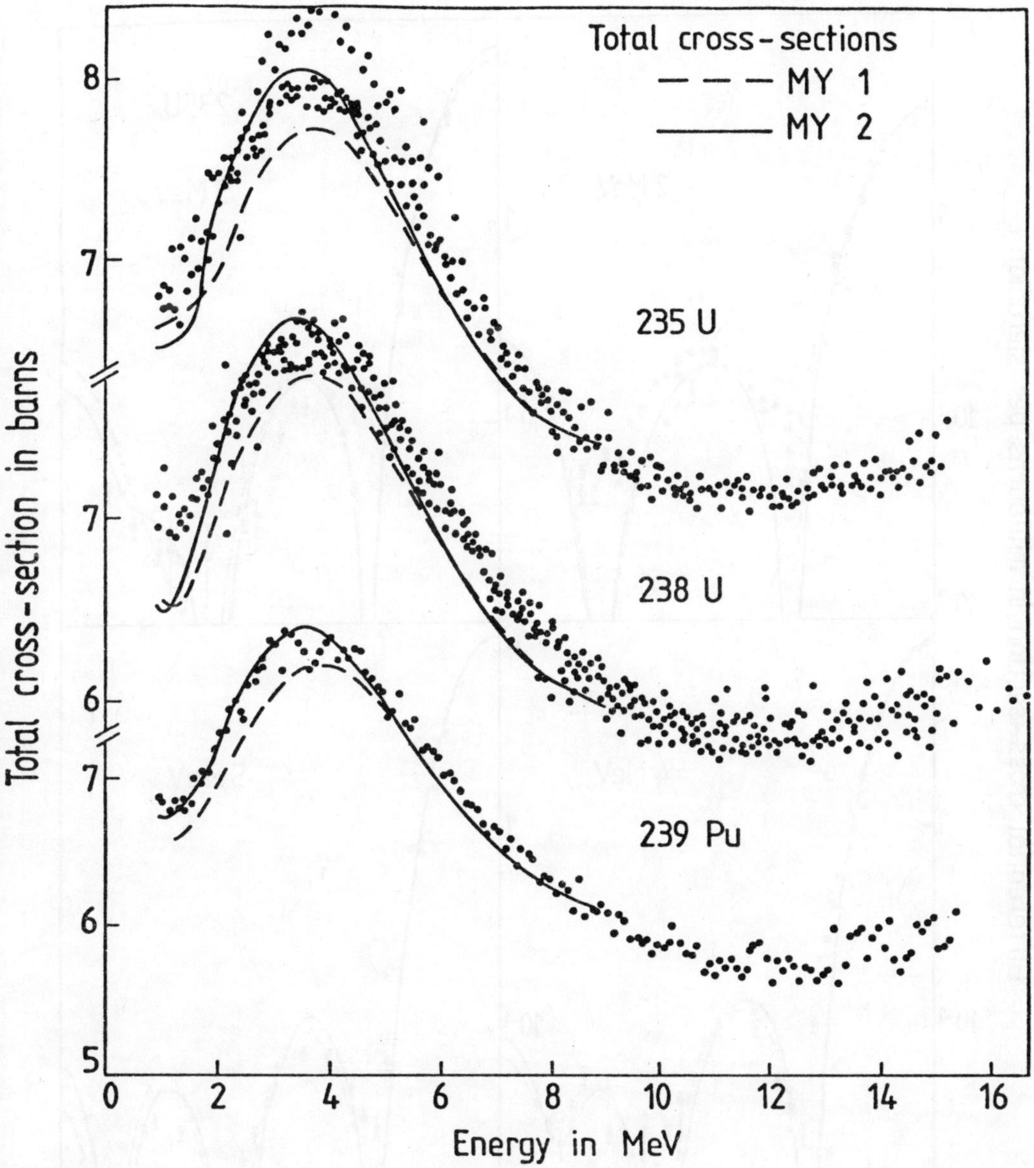

Fig. 3. Total and reaction cross-sections for the interaction of neutrons with U235, U238 ad 239Pu compared with the optical model calculations with the potentials of Madland and Young [7].

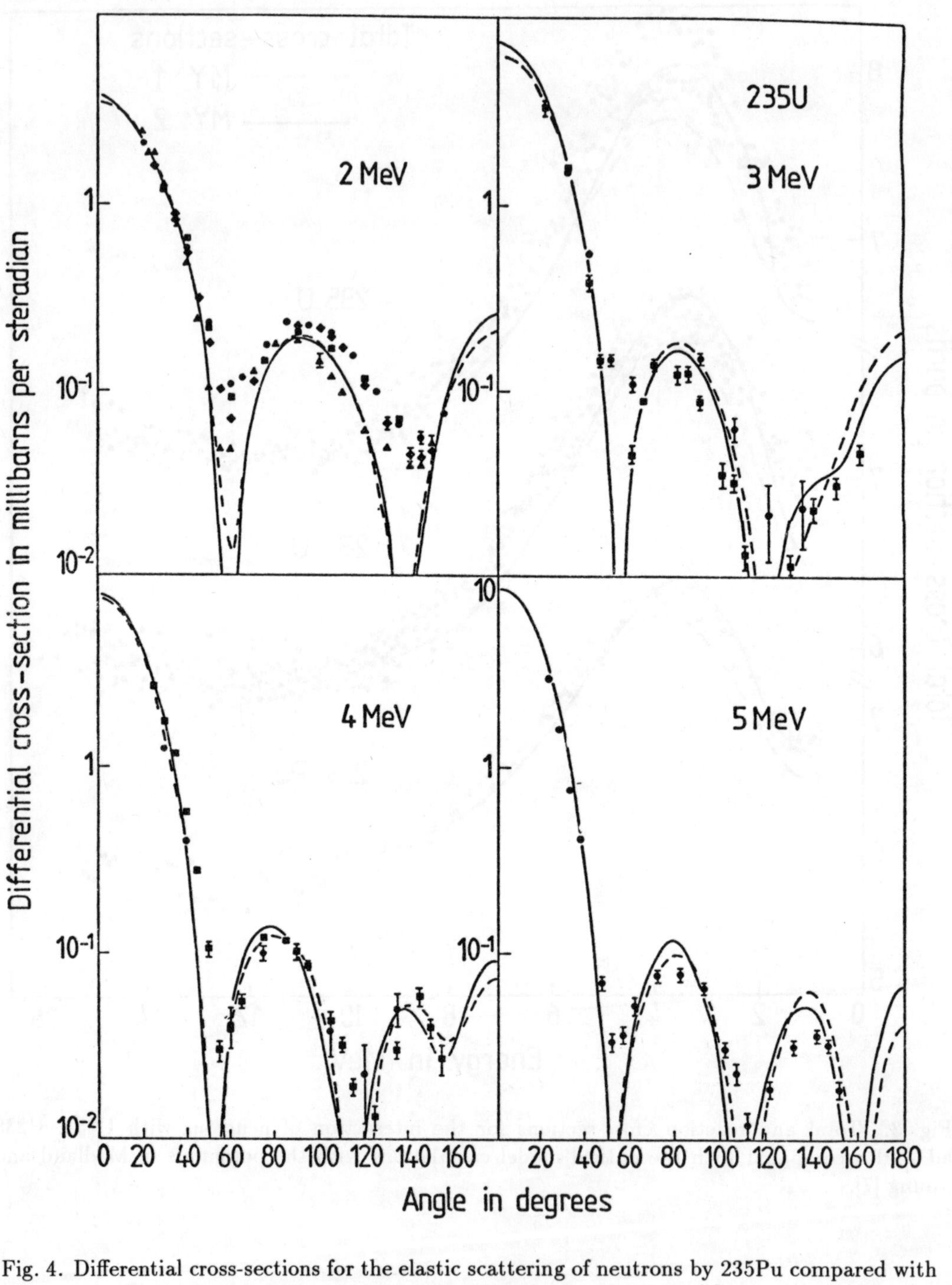

Fig. 4. Differential cross-sections for the elastic scattering of neutrons by 235Pu compared with optical model calculations with the potentials of Madland and Young [7].

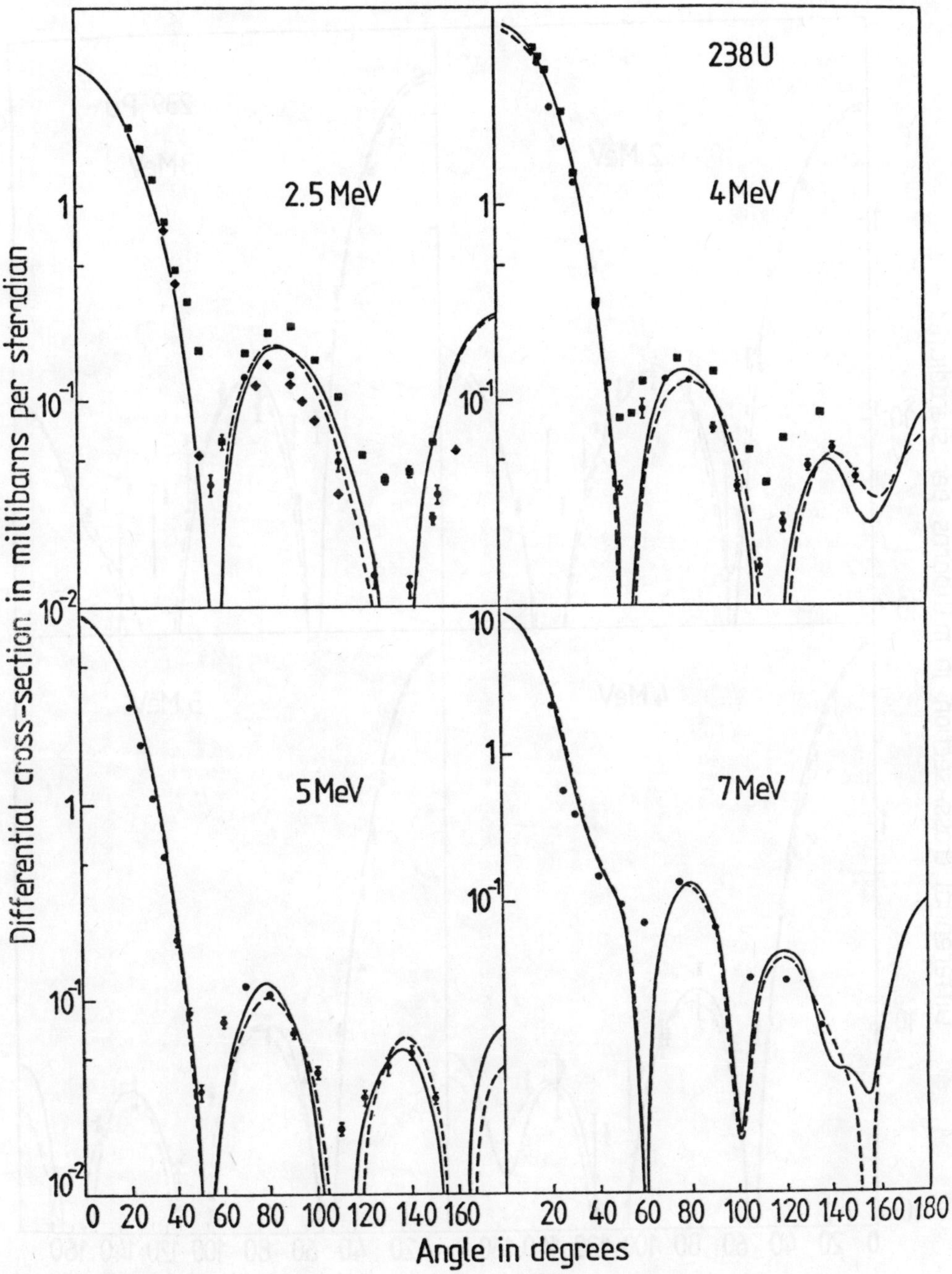

Fig. 5. Differential cross-sections for the elastic scattering of neutrons by 238U compared with optical model calculations with the potentials of Madland and Young [7].

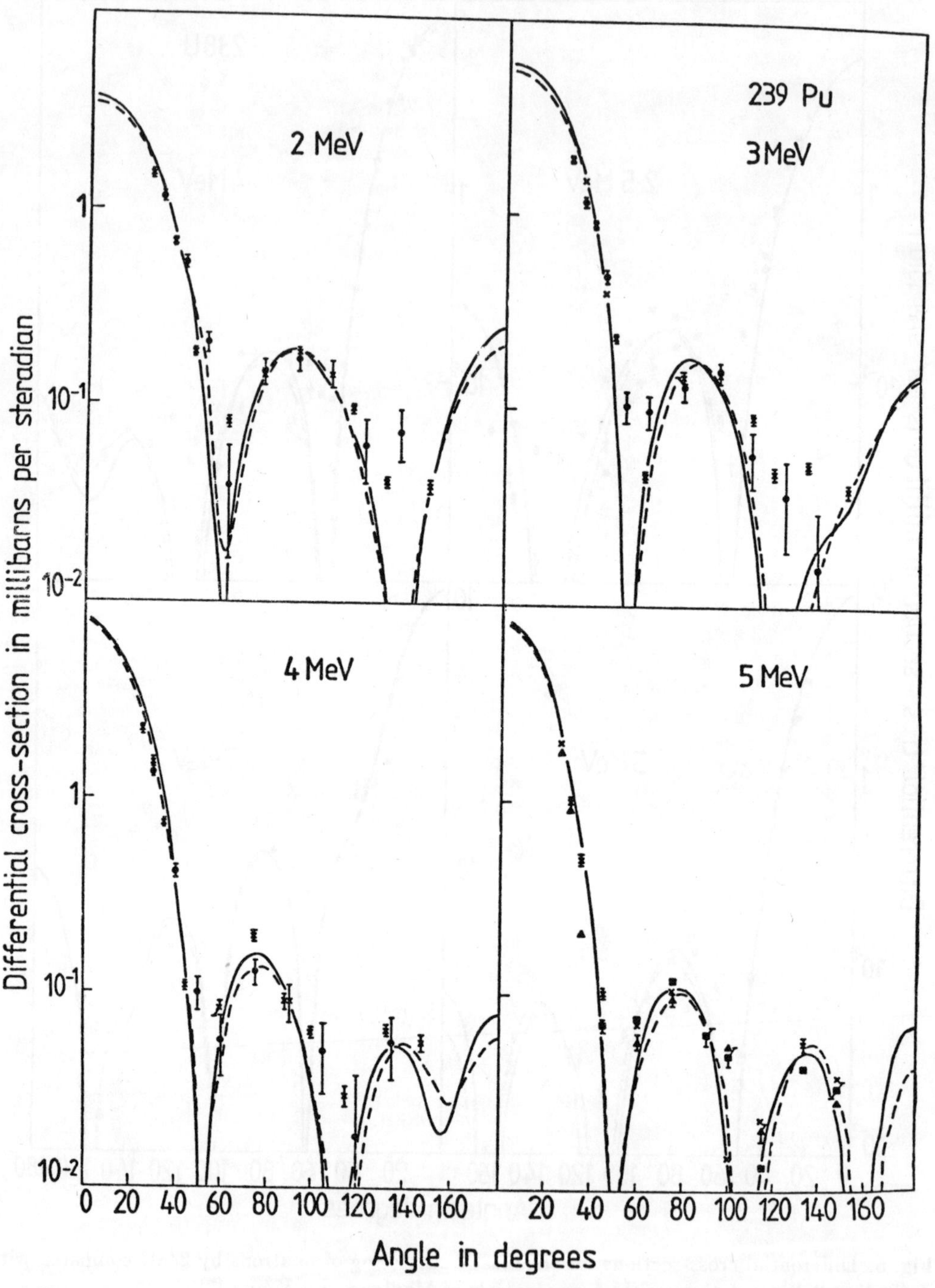

Fig. 6. Differential cross-sections for the elastic scattering of neutrons by 239Pu compared with optical model caluclations with the potentials of Madland and Young [7].

TABLE I

Optical Potentials for Neutron Scattering by Actinide Nuclei

	Antsipov[5]	Jary[6]	MY(1)[7]	MY(2)	Konshin[8]	WH[4]	Maino[9]
U_o	45.87	49.82	53.016	50.378	47.5	47.01	47.01
α	-0.4	-0.3	-0.36	-0.354	-0.3	-0.267	-0.267
β	–	–	–	–	–	-0.00118	-0.00118
U_1	–	-17	-24.62	-27.073	–	–	–
r_u	1.258	1.26	1.203	1.264	1.233	(a)	1.259
a_u	0.626	0.63	0.629	0.612	0.62	0.66	0.66
W	2.95	5.52†; 9.52*	8.983	9.265	2.7	9.52	3.2^x; 2.25$^\odot$
α	0.4	0.4†; 0*	-0.22	-0.232	0.4	-0.053	0^x; 0.42$^\odot$
β	† –	–	0.03252	0.03318	–	–	–
W_1	–	-9	-13.642	-12.666	–	–	–
r_W	1.26	1.26	1.297	1.256	1.26	(b)	1.237
a_W	0.530	0.52	0.555	0.553	0.58	0.48	0.48
α	–	–	0.01428	0.01440	–	–	–
V_{SO}	7.5	6.2	6.2	6.2	7.5	0	7.2
r_s	1.258	1.12	1.1	1.1	1.233	–	1.259
a_s	0.626	0.47	0.75	0.75	0.62	–	0.66

The parameters α and β are the coefficients of the linear and quadratic energy dependence of the parameter immediately above them in the Table.

† $E < 10$ MeV x $E < 2.26$ MeV

* $E > 10$ MeV $\odot$ $E > 2.26$ MeV

(a) $1.322 - 0.00076A + (4\times10^{-6})A^2 - (8\times10^{-9})A^3$

(b) $1.266 - 0.00037A + (2\times10^{-6})A^2 - (4\times10^{-9})A^3$

All radial form factors of imaginary potentials are Saxon-Woods derivative (eqn.(2.6)).

diffraction maximum; between maxima the compound nucleus cross-section, as it varies much less rapidly with angle, makes a proportionately greater contribution and tends to fill in the minima. For precision fitting at lower energies it is therefore necessary to add the compound nucleus cross-section, but otherwise the MY potential can be used directly to calculate total, reaction and differential cross-sections using the simple optical model. It is desirable to consider how this potential might be improved. Three features of the potential may be singled out for discussion, the dependence on energy, asymmetry parameter and deformation. All these effects are included in the potential of Madland and Young and the numerical values of the parameters representing them come from extensive least-squares fitting. As usual in such analyses, there are very considerable ambiguities between the parameter sets, so it is useful to see if these can be reduced by physical considerations. A subsequent parameter search incorporating physically realistic restrictions on the parameters is then likely to be more reliable for extrapolating to other nuclei. Throughout the discussion use will be made of the results of analyses of proton elastic scattering, since there is much evidence of similarity between proton and neutron optical potentials.

Energy Dependence

There have been many studies of the energy dependence of the nucleon optical potentials. As an example, Fig.7 shows the result of a recent global analysis by Bauer et al [10] of the depth of the real potential for protons as a function of energy for both positive (scattering) and negative (bound states) energies. This energy dependence was fitted by linear, quadratic and exponential curves, and of these the quadratic was the best,

$$U(B) = 52.4 - 0.37E + 0.0007E^2$$

This energy dependence is in good agreement with that found by Madland and Young

$$U(MY) = 53.016 - 0.36E$$

Since the form factors of the two potentials are different we should not expect precise agreement between the numerical coefficients in these equations.

It should however be noted that there are significant differences at low energies between the best fit quadratic curve and the experimental data, in particular there is a tendency for the data to show a plateau in the region of the Fermi surface, and also the slope of the energy dependence of the data is greater in the region 5 - 15 MeV. Data in these regions were therefore omitted from the fitting mentioned above. These effects are well understood physically as a result of particle-hole excitations in the region of the Fermi surface, and the figure also shows the results of two calculations based on this mechanism. While however they reproduce the qualitative features of the data it cannot yet be said that they give a quantitative fit. Until this is achieved it would therefore be premature to use these potentials in the actinide region. It is probably sufficiently accurate to use the energy dependence of Madland and Young.

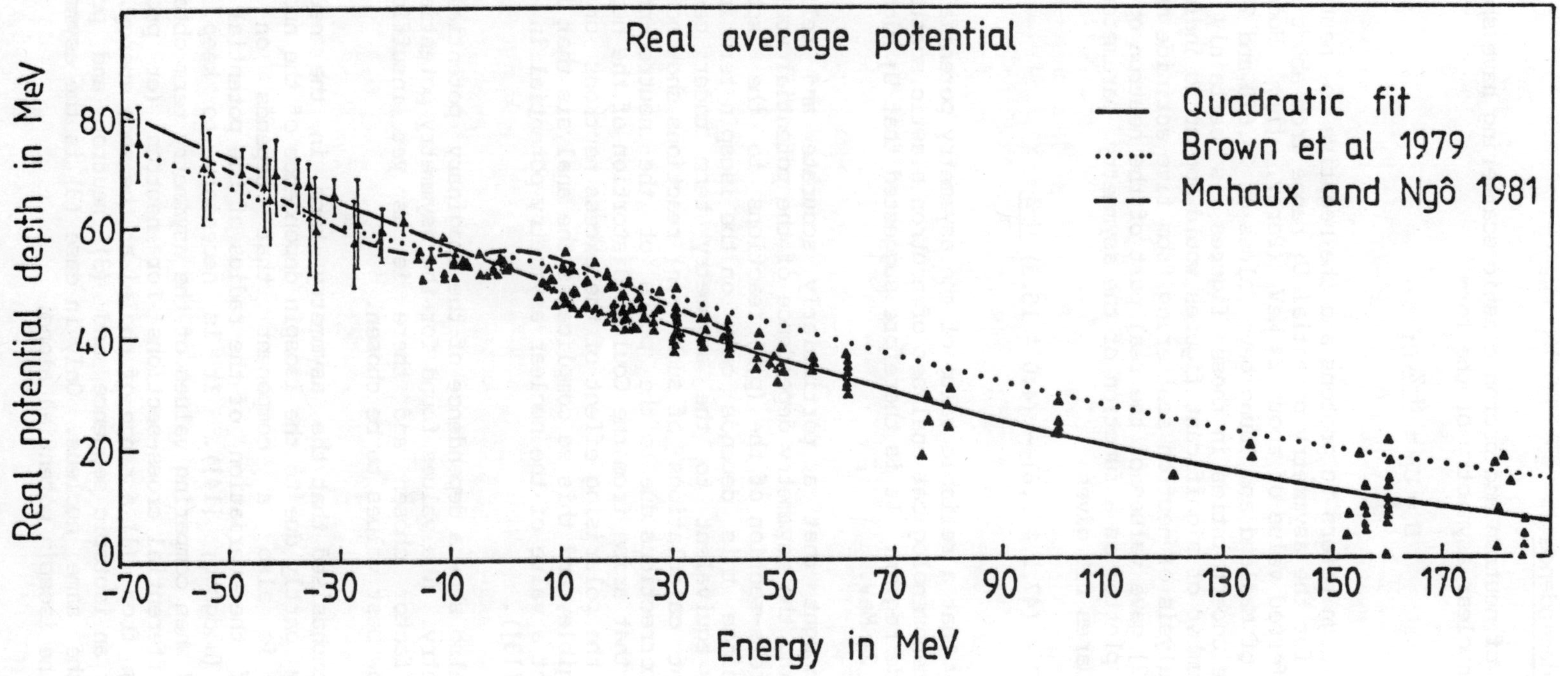

Fig. 7. Depth of the real central optical potential for a range of nuclei with isospin and Coulomb effects removed. The experimental points for positive energies refer to potentials from the compilations of Perey [1] corrected to standard radius and diffuseness, and those for negative energies to potentials fitted to the centroid energies of bound single-particle states. The curves refer to a least-squares quadratic fit (——) and to the theoretical calculations of Brown et al [18] (......) and of Mahaux and Ngô [19] (– – –) (Bauer et al [10]).

Nuclear Asymmetry dependence

Many analyses of neutron and proton elastic scattering have shown a dependence on the nuclear asymmetry of the form

$$U = U_0 \pm \frac{N-Z}{A} U_1$$

where the positive sign refers to protons and the negative to neutrons. The values found for the asymmetry potential U_1 range from about 10 to 30 MeV, with a preferred value of about 24 MeV (Perey, [1]; Hodgson, [11]). The work of Madland and Young gave values of 24.620 and 27.073 MeV for U_1, but the uncertainties in these figures is certainly much greater than the number of significant figures would appear to indicate. A comprehensive analysis of neutron scattering from five actinide nuclei by Haouat et al [12] gave values of the real part of the neutron optical potential that are plotted as a function of the asymmetry parameter in Fig.8. A least-squares fit gives

$$U = (47.2 \pm 3.8) - (4.0 \pm 15.3) \frac{N-Z}{A}$$

It is thus unlikely that a reliable value of the asymmetry potential can be obtained from phenomenological analyses of neutron elastic scattering data in the actinide region. It is therefore suggested that U_1 might be fixed at the value 24 MeV.

It might be thought that a particularly accurate and reliable method of obtaining the asymmetry dependence of the potential would be by analysing the cross-section of the (p,n) reactions to the isobaric analogue state, since this depends only on the isospin term in the potential, which is equivalent to the asymmetry term under certain conditions. Recent calculations of such (p,n) reactions show however that higher-order corrections due to the parts of the neutron-proton density difference that arise from the Coulomb distortion of the nuclear densities and from the polarising effect of the excess neutrons on the core are not negligible, and this so complicates the analysis that it is difficult to extract a value of the nuclear asymmetry potential in this way (Lovas et al, [13]).

Some studies also show a dependence of the imaginary potential on the nuclear asymmetry. The values found for the asymmetry potential W_1 depend on the form factor chosen and there is as yet insufficient information to allow best values to be chosen.

It should be emphasised that the asymmetry term in the neutron potential is only partly due to the isospin dependence of the nucleon potential. There is also a component that depends on the parametrisation of the variation of the radius of the potential from nucleus to nucleus (Hodgson, [14]). It is necessary to keep this distinction in mind when comparing values of the asymmetry term obtained from analyses of differential cross-sections for neutron (or proton) elastic scattering from (1) a range of nuclei at the same energy, (2) the same as (1) for an isotopic sequence and (3) neutron and proton scattering from the same nucleus. Only in case (3) is the asymmetry dependence due to the isospin potential alone.

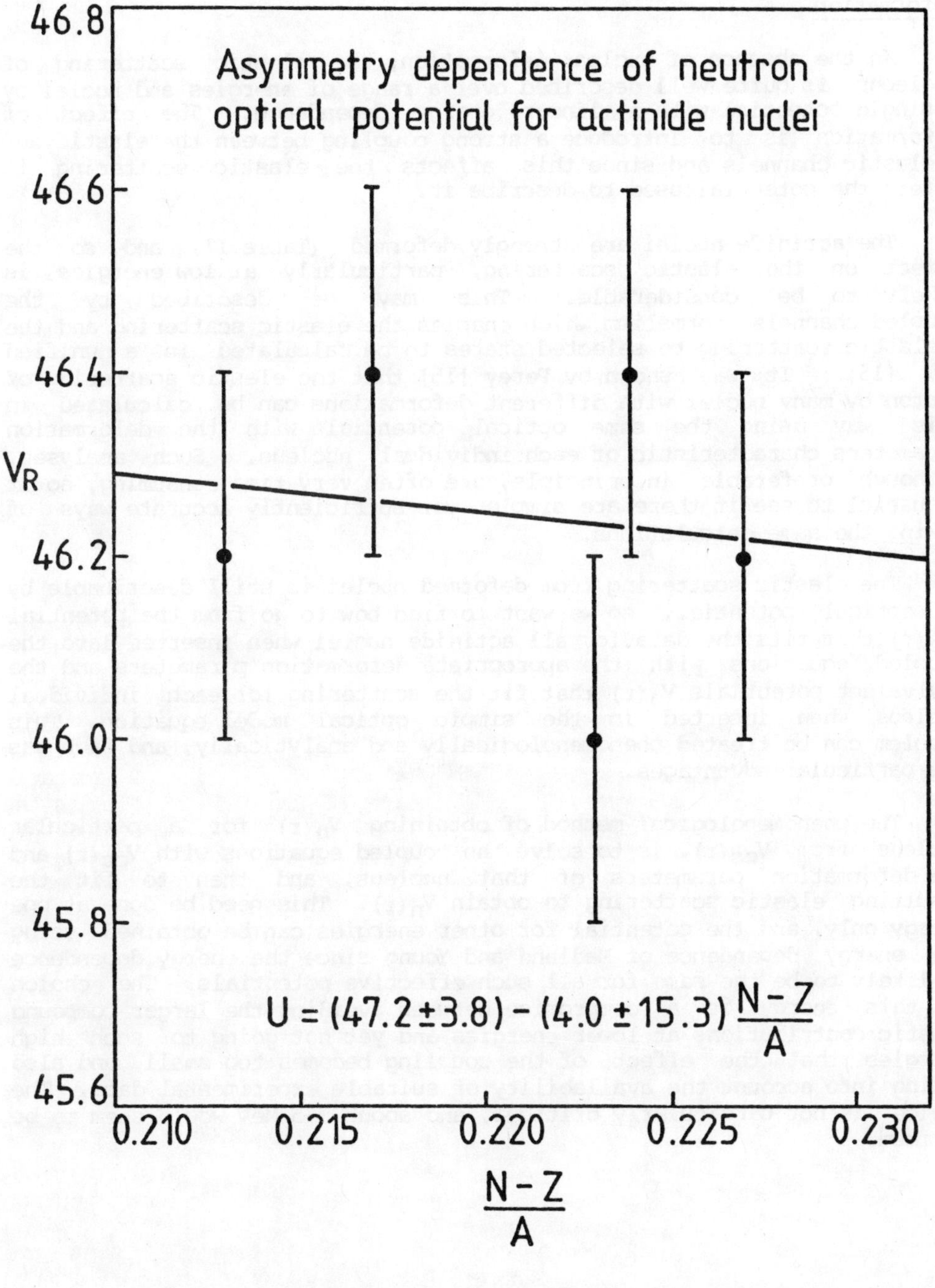

Fig. 8. The asymmetry dependence of the neutron optical potential for actinide nuclei using the parameters of Haouat et al [12].

Deformation

In the absence of nuclear deformation, the elastic scattering of nucleons is quite well described over a range of energies and nuclei by a single potential with a linear energy dependence. The effect of deformation is to introduce a strong coupling between the elastic and inelastic channels and since this affects the elastic scattering it alters the potential used to describe it.

The actinide nuclei are strongly deformed (Table II) and so the effect on the elastic scattering, particularly at low energies, is likely to be considerable. This may be described by the coupled-channels formalism which enables the elastic scattering and the inelastic scattering to selected states to be calculated in a unified way [15]. It was shown by Perey [15] that the elastic scattering of protons by many nuclei with different deformations can be calculated in this way using the same optical potential with the deformation parameters characteristic of each individual nucleus. Such analyses, although preferable in principle, are often very time consuming, so it is useful to see if there are simpler yet sufficiently accurate ways of making the same calculations.

The elastic scattering from deformed nuclei is still describable by an optical potential, so we want to find how to go from the potential $V_{CC}(r)$ that fits the data for all actinide nuclei when inserted into the coupled equations with the appropriate deformation parameters and the equivalent potentials $V_n(r)$ that fit the scattering for each individual nucleus when inserted in the simple optical model equation. This problem can be treated phenomenologically and analytically, and each has its particular advantages.

The phenomenological method of obtaining $V_n(r)$ for a particular nucleus from $V_{CC}(r)$ is to solve the coupled equations with $V_{CC}(r)$ and the deformation parameters of that nucleus, and then to fit the resulting elastic scattering to obtain $V_n(r)$. This need be done at one energy only, and the potential for other energies can be obtained using the energy dependence of Madland and Young since the energy dependence is likely to be the same for all such effective potentials. The choice of this energy is a compromise between avoiding the larger compound elastic contributions at lower energies and yet not going to such high energies that the effect of the coupling becomes too small, and also taking into account the availability of suitable experimental data. The choice is not particularly critical, and about 3-5 MeV would seem to be optimum.

Table II

Deformation parameters of the actinide nuclei (Haouat et al, [12]).

Nucleus	Quadrupole Deformation $\beta 2$	Hexadecapole Deformation $\beta 4$
232 Th	0.190 ± 0.010	0.071 ± 0.007
235 U	0.220 ± 0.011	0.080 ± 0.008
238 U	0.198 ± 0.010	0.057 ± 0.006
239 Pu	0.220 ± 0.011	0.070 ± 0.007
242 Pu	0.204 ± 0.010	0.051 ± 0.005

It would be interesting to test this method on the more familiar actinide nuclei to see if it gives significantly improved fits compared with the potential of Madland and Young. If this is the case, then the method can be applied to other actinide nuclei for which only the deformation is known.

It should however be remarked that the deformation parameters of the actinide nuclei are remarkably similar, so it is likely that the elastic scattering cross-sections are fitted by very similar potentials. This is indeed the likely explanation of the success of the MY potential throughout the actinide region. It is still possible that improved potentials could be obtained by the method described above, but this can only be established by actual calculations. It might be found that the more accurate potentials obtained in this way fall into two groups, for the even A and odd A nuclei respectively, and that an average potential suffices within each group.

In this way the optical model can be used to calculate the cross-sections for the elastic scattering of neutrons by deformed as well as spherical nuclei. Also included in the fitting procedures are the total and reaction cross-sections, so that these are also obtained.

More serious difficulties arise when it is necessary to calculate the inelastic cross-section. The distorted wave Born approximation (DWBA) can be used successfully when the coupling between the elastic and inelastic channels is weak, but it is not sufficiently accurate for strongly deformed nuclei. To overcome this difficulty, the coupled-channels method has been developed, and this is described in the next section.

4. Inelastic Scattering

At low energies, the inelastic cross-section to discrete states of the target nucleus has contributions from both the compound nucleus and the direct interaction processes, as shown in Fig.9. The compound nucleus component can be calculated using the Hauser-Feshbach formalism, and the direct interaction contribution by the distorted wave Born approximation (DWBA) or by the coupled-channels formalism.

If the coupling between the elastic and inelastic channels is weak, the DWBA with wavefunctions calculated from the optical potentials obtained by analyses of elastic scattering gives sufficiently accurate inelastic cross-sections. If however the coupling is strong, as it is for the actinide nuclei, this is no longer the case, and it is necessary to take the coupling into account explicitly by the coupled-channels formalism.

In this section the coupled-channels formalism will be briefly reviewed, and then the question will be considered whether it is possible to modify the DWBA formalism to give equally accurate cross-sections in a simpler way. It should be remarked that the direct contribution, although appreciable, is a rather small fraction of the total cross-section, so even a method of calculating it with somewhat less accuracy than the coupled-channels method would still be of practical value. A more important consideration is the direct contribution of the inelastic scattering to the continuum of unresolved final states of the target nucleus, which may well be quite important at higher energies. This is impossible to calculate by the coupled-channels method, so any method of doing so would be useful.

In the coupled-channels formalism, the total wavefunction is expanded as a sum of the wavefunctions in the elastic and a few selected inelastic channels, each multipled by the wavefunction of the appropriate nuclear state. When this is inserted into the Schrödinger equation it splits into a set of coupled equations for the wavefunctions in the elastic and selected inelastic channels. Making use of a simple model of the collective states, these equations can be solved to give not only the elastic and total cross-sections given by the simple optical model but also the inelastic cross-sections in the selected channels. The formalism and calculational techniques have been described in many papers (D).

The only additional parameters in the coupled-channels calculation are the nuclear deformation parameters for each excited state. With these parameters, the same optical potential can be used for a range of nuclei, even if they are deformed. The main disadvantage of the coupled-channels calculations is that they become extremely time-consuming and ultimately impracticable as the number of excited states that are explicitly included is increased. In practice it is unusual for more that two or three states to be included, since the computing time is approximately proportional to the cube of $\Sigma(2J+1)$, over all the states included.

It is therefore worthwhile enquiring whether the inelastic scattering cross-sections can be calculated to sufficient accuracy for deformed nuclei in a similar way to the elastic scattering; that is by

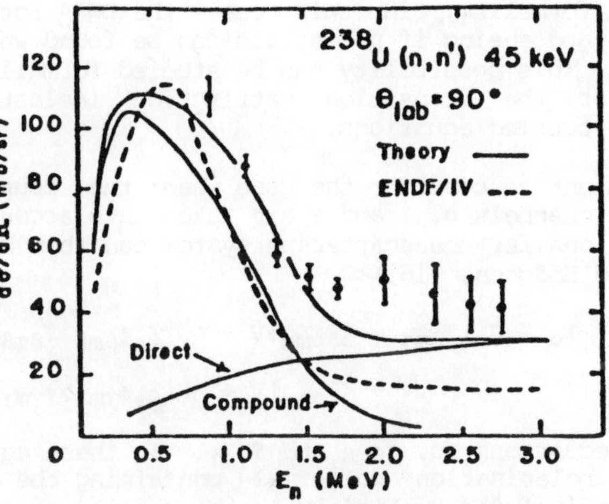

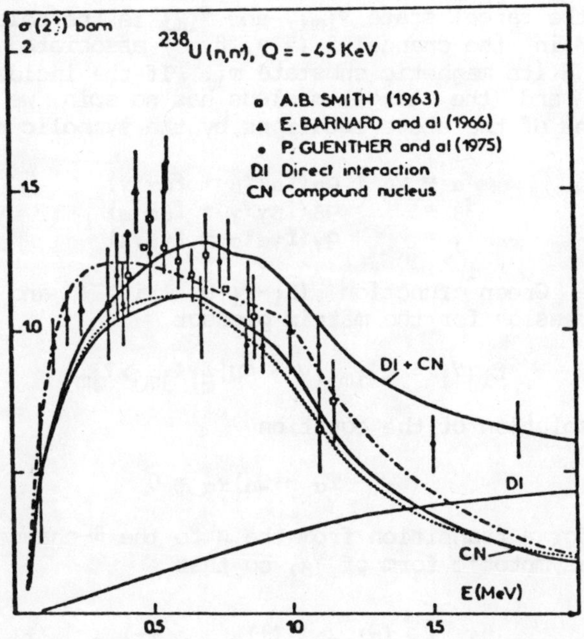

Fig. 9. Excitation functions for the inelastic scattering of neutrons by 238U with excitation of the first 2^+ state at 45 keV compared with calculations of the compound nucleus and direct interaction contributions. The upper figure gives the excitation function at 90° (Egan et al, 1975) [22] and the lower that for the total inelastic cross-section to the 45 keV state (Delaroche et al, 1975) [23].

using the same formalism, in this case the DWBA formalism, as for spherical nuclei, and seeing if potentials can be found which yield the correct results. This possibility can be studied formally by examining the expression for the transition matrix for inelastic scattering obtained from the coupled equations.

It is convenient to consider the case when the couplings between three particular channels α, β and γ are taken into account explicitly. The coupled equations for the scattering system can then be expressed in the form (Kubo and Hodgson, [16])

$$(E_\alpha - T_\alpha - U_\alpha)\Psi_{\alpha m\alpha} = \langle\Phi_{\alpha m\alpha}|V + \Delta U|\sum_{m\beta}\Phi_{\beta m\beta}\rangle\Psi_{\beta m\beta} + \langle\Phi_{\alpha m\alpha}|V + \Delta U|\sum_{m\gamma}\Phi_{\gamma m\gamma}\rangle\Psi_{\gamma m\gamma} \quad (4.1)$$

with two similar equations for $\Psi_{\beta m\beta}$ and $\Psi_{\gamma m\gamma}$. In these equations ΔU is the generalised polarisation potential containing the effects of the other channels not included explicitly in the coupled equations, U_i is the bare optical potential defined by the diagonal matrix element of $(V + \Delta U)$ with the target state Φ_{imi}, and Ψ_{imi} is the relative scattering wave function in the channel i ($\equiv$ α, β, γ) associated with the target spin state i and its magnetic substate m_i. If the incident wave is in the α-channel and the target nucleus has no spin, we can express the formal solutions of the above equations by the symbolic expressions

$$\begin{aligned}\Psi_\alpha &= \chi_\alpha + g_\alpha(f_{\alpha\beta}\Psi_\beta + f_{\alpha\gamma}\Psi_\gamma)\\ \Psi_\beta &= g_\beta(f_{\beta\gamma}\Psi_\gamma + f_{\beta\alpha}\Psi_\alpha)\\ \Psi_\gamma &= g_\gamma(f_{\gamma\alpha}\Psi_\alpha + f_{\gamma\beta}\Psi_\beta)\end{aligned} \quad (4.2)$$

where g_i is the Green function $(E_i - T_i - U_i + i\varepsilon)^{-1}$ and $f_{ij}\Psi_i$ is the simplified expression for the matrix product

$$f_{ij}\Psi_i \equiv \langle\Phi_{imi}|V + \Delta U|\sum_{mi}\Phi_{jmj}\rangle\Psi_{jmj} \quad (4.3)$$

and χ_α is the solution of the equation

$$(E_\alpha - T_\alpha - U_\alpha)\chi_\alpha = 0 \quad (4.4)$$

The amplitude for a transition from the α to the β-channel is then given by taking the asymptotic form of Ψ_β, so that

$$T_{\alpha\beta} = -\frac{\mu_\beta}{2\pi\hbar^2}\{\langle\chi_\beta^{(-)}, f_{\alpha\beta}\Psi_\alpha^{(+)}\rangle + \langle\chi_\beta^{(-)}, f_{\beta\gamma}\Psi_\gamma^{(+)}\rangle\} \quad (4.5)$$

where $\chi_\beta^{(-)}$ is the solution of the equation

$$(E_\beta - T_\beta - U_\beta^{(-)})\,\chi_\beta^{(-)} = 0 \quad (4.6)$$

Now if the coupling between the excited states β and γ is small, the second term in (4.5) may be neglected. We thus obtain the important result that the transition amplitude for inelastic scattering by deformed nuclei is given by the standard distorted wave expression with the exit channel wavefunction obtained from a simple optical model calculation with the deformed potential. The implications of this for practical calculations of neutron inelastic cross-sections are discussed in the following section.

5. Comparison between optical model and coupled-channels calculations

The optical model is perfectly adequate to calculate the shape elastic scattering cross-section. The potential is defined in Section 2 with the requirement that the resulting phase-shifts fit the data, and the analyses summarised in Section 3 show that such a potential can actually be found and expressed numerically in a simple way.

The situation is more complicated for inelastic scattering. If the coupling between the elastic and inelastic channels is weak, then the inelastic scattering can be calculated from the DWBA transition amplitude

$$T_{\alpha\beta}^{(\text{weak})} = \langle \Psi_\beta \, | H' | \, \Psi_\alpha \rangle \tag{5.1}$$

where Ψ_α and Ψ_β are the entrance and exit channel wavefunctions calculated from the optical model potential at the appropriate energy and H' is the perturbing Hamiltonian.

If however the coupling is strong the wavefunction in the elastic channel is perturbed by the coupling and the above expression is no longer valid. It is necessary to solve the coupled equations for the wavefunctions in the elastic and in the inelastic channels of interest. We can however still use the distorted wave formalism, that is much simpler to handle numerically, if we write the transition amplitude for strong coupling

$$T_{\alpha\beta}^{(\text{strong})} = \langle \Psi_\beta' \, | H' | \, \Psi_\alpha' \rangle \tag{5.2}$$

where now the wavefunctions Ψ_β and Ψ_α are calculated from optical model potentials that are <u>defined</u> so that $T_{\alpha\beta}$ reproduces the measured cross-section. The problem is then to find these potentials.

There have been several studies of this problem in connection with the calculation of the inelastic scattering of heavy ions, for which the computing time is far greater than for neutrons. One solution was proposed by Ascuitto <u>et al</u> [17] in connection with their study of the reaction ^{40}Ca(^{16}O,^{16}O')^{40}Ca* (3.90 MeV 2^+, 3.74 MeV 3^-, and 4.49 MeV 5^-) at 60 MeV. They suggested that the transition amplitude is quite well given by a standard DWBA amplitude in which the potential in the coupled-channels equations is used as the exit channel distorting potential, instead of the simple optical model potentials, i.e.

$$T_{\alpha\beta}^{(\text{strong})} = \langle \chi_\beta \, | H' | \, \Psi_\alpha \rangle \tag{5.3}$$

where χ_β is the wavefunction obtained by inserting the coupled-channels potential into the simple optical model equation, and Ψ_α is the optical model wavefunction fitting the elastic scattering, both at the appropriate energy. The formal derivation of this formula is given in Section 4. Ascuitto <u>et al</u> used this prescription to calculate the inelastic scattering cross-section, and obtained a satisfactory fit to the experimental data.

It will be noticed immediately that this does not enable the coupled-channels calculation to be replaced by an equally accurate DWBA calculation, since we have to make a coupled-channels analysis to obtain the potential to calculate χ_β.

However, it could be of great practical utility if this potential is very similar for a large number of transitions. It would then be sufficiently accurate to make a small number of coupled-channels analyses to find this potential, and thereafter simple DWBA calculations would suffice. This could easily be tested by using the MY potential to calculate Ψ_α, and the optimum potential in the coupled-channels analyses of Haouat et al [12] to calculate χ_β. If necessary the potentials could be varied slightly to optimise the fit to a number of transitions for which experimental data are available. If consistent results are obtained, the method could be extended to calculate unknown cross-sections. It should also be remarked that the cross-section of the inelastic scattering to the first few low-lying states is rather small compared with that to the continuum so at higher energies when direct reactions contribute to the continuum it is less important to fit the former well than the latter, so that these modified DWBA calculations could well supercede coupled-channels calculations in practical applications.

It will be noticed that the expression (5.3) for the inelastic cross-section was derived for the case of three coupled-channels. The potential that must be used in the coupled-channels formalism to fit the elastic and inelastic scattering cross-sections depends on the number of channels included in the calculations. Experience shows, however, that this dependence is rather small for channels beyond those corresponding to the first two excited states, so this is unlikely to be a serious practical difficulty. It also remains to be seen how well justified is the assumption that the coupling between the excited states is small. This was used to remove the second term in (5.3) and may not be sufficiently good for rotational nuclei like the actinides. However in this case also we may have to take such deficiencies into account by small adjustments of the potential parameters.

Another possibility, that would render coupled-channels calculations completely superfluous, would be to find a way of calculating the coupled-channels potential analytically, instead of finding it phenomenologically by a coupled-channels analysis. This has recently been done for heavy ion scattering and analytical expressions for the optical potential have been obtained using the following assumptions: (1) The $\chi_\alpha^{(+)}$ and the intermediate states in (4.5) are replaced by plane waves, and (2) only small momentum transfer components are taken into account. The collective vibrational model for both nuclear and Coulomb excitations is used. (Kubo and Hodgson, [16]b). The potentials calculated in this way were found to give cross-sections in excellent agreement with those calculated by solving the coupled equations. Unfortunately, however, the same assumptions are not valid for nucleons, so at present this approach though giving useful insight into the problem, does not provide a practicable method of solution.

References

1. Hodgson, P.E. "The Optical Model of Elastic Scattering", Oxford, 1963.

   Hodgson, P.E. "Nuclear Reactions and Nuclear Structure", Oxford, 1971.

   Perey, C.M. and Perey, F.G. "Compilation of Phenomenological Optical Model Parameters 1954-1975". Atomic Data and Nuclear Tables 17.1-101.1976.

   Perey, F.G. "Optical Model Analysis of Proton Elastic Scattering in the Range of 9 to 22 MeV". Phys.Rev. 131.745-763.1963.

2. Hodgson, P.E. "Summary of the papers presented at the Workshop on Microscopic Optical Potentials, Hamburg 1978. Lecture Notes in Physics No.89. Edited by H.V. von Geramb (Springer-Verlag) 459-481.1979.

3. Perey, F.G. and Buck, B. "A Non-local Potential Model for the Scattering of Neutrons by Nuclei". Nucl.Phys. 32.352-380.1962.

4. Wilmore, D. and Hodgson, P.E. "The Calculation of Neutron Cross-sections from Optical Potentials", Nucl.Phys. 55.673-694.1964.

5. Antsipov, G.V., Zenevich, V.A., Klepatsku, A.B., Konshin, V.A. and Sukhovitsky, E.Sh., "The Influence of Different Level Density Representations on Actinide Neutron Cross-Section Calculations", Proceedings of the IXth International Symposium on the Interaction of Fast Neutrons with Nuclei. Dresden 1979. Edited by D. Seeliger and S. Umholzer, ZfK-410. 141-146.1980.

6. Jary, J., and Lagrange, Ch. "Evaluation par Modeles Nucleaires des Sections Efficaces Neutroniques de 240Pu et 242Pu entre 10 keV et 20 MeV". Ch.2 "Compte Rendu de Travaux sur L'Evaluation de Donnees Nucleaires Relatives aux Actinides". CEA-N-2084; NEANDC(E)203 "L", 6-33. 1979.

7. Madland, D.G. and Young, P.G. "Neutron-Nucleus Optical Potential for the Actinide Region". Proceedings of the International Conference on Neutron Physics and Nuclear Data (OECD Nuclear Energy Agency), 349-354.1978.

8. Konshin, V.A. "Comparison of various theoretical models based on neutron 242Pu cross-section calculations". Proceedings of the IAEA Advisory Group Meeting on Transactinium Nuclear Data INDC(CCP)-135/LN. 43-72.1979.

9. Miano, G., Menapace, E., Motta, M., and Ventura, A. "First and Second Chance Fission Calculations for Actinides and Related Topics". Proceedings of the International Conference on Nuclear Cross-sections for Technology, Knoxville, Tennesee, Edited by J.L. Fowler, C.H. Johnson and C.D. Bowman. National Bureau of Standards. 1980.

10. Bauer, M., Hernandez-Saldana, E., Hodgson, P.E. and Quintanilla, J. "A Unified Potential for Bound and Scattering States". J.Phys.G.(In press) 1982.

11. Hodgson, P.E. "The Symmetry and Coulomb Terms in the Proton Optical Potential". Canad.J.Phys. 45.449-451.1967.

12. Haouat, G., Patin, V., Lachkar, J., Sigaud J., Jary, J. and Lagrange, Ch. "Neutron Scattering Data for Actinide Nuclei". NEANDC(E)211 "L", INDC(FR)41/L 33-40, 1981.

13. Lovas, R.G., Brown, B.A. and Hodgson, P.E. "Coulomb Mixing and Core Polarization in (p,π) Quasi-Elastic Scattering". Nuclear Physics A357.205-227.1981.

14. Hodgson, P.E. "The Isospin Dependence of the Proton Optical Potential". Nucl. Phys. 150.1-10.1970.

15. Hodgson, P.E. "Nuclear Reactions and Nuclear Structure" Oxford 1971, Chapters 3 and 13.
Buck, B., Stamp, A.P. and Hodgson, P.E. "The Excitation of Collective States by Inelastic Scattering: The Extended Optical Model". Phil.Mag. 8.1805-1826. 1963.
Tamura, T., "Analyses of the Scattering of Nuclear Particles by Collective Nuclei in Terms of the Coupled-Channel Calculation", Rev.Mod.Phys. 37.679-708.1965.
Tamura, T., "Coupled-Channel Approach to Nuclear Reactions", Ann.Rev.Nucl.Sci. 99-138.1969.
Buck, B., "Calculation of Elastic and Inelastic Proton Scattering with a Generalised Optical Potential". Phys.Rev. 130.712-726.1963.
Perey, F.G., "Analysis of Elastic Scattering of 17 MeV Protons from Heavy Nuclei", Symposium on Nuclear Spectroscopy with Direct Reactions. Edited by F.E. Throw I. Contributed Papers. ANL-6848.114-117.1964.

16. Kubo, K.-I. and Hodgson, P.E. "Equivalent Bare Optical Potentials in Heavy-Ion Inelastic Scattering". Phys.Lett. 100B.453-456.1981a.
Kubo, K.-I. and Hodgson, P.E. "Exit-Channel Distorting Potentials in Heavy-Ion Inelastic Scattering and the Equivalent Bare Optical Potential". Nucl.Phys. A366.320-338.1981b.

17. Ascuitto, R.J., Petersen, J.F. and Seglie, E.A. "Role of the Exit-Channel Distorting Potential in Heavy-Ion Induced Inelastic Scattering". Phys.Rev.Lett. 41.1159-1163.1978.

18. Brown, G.E., Dehesa, J.S. and Speth, J. "A Dynamical Theory of the Grant Dipole Resonance in Nuclei". Nucl.Phys. A330.290-306.1979.

19. Mahaux, C. and Ngô, H. "Energy Dependence of the Effective Mass in Finite Nuclei". Phys.Lett. 100B.285.1981.

20. Phillips, T.W. and Howe, R.E. "Total Neutron Cross-Section for Americium 241". Nucl.Sci. and Eng. 69.375-377.1979.

21. Glasgow, D.W. and Foster, D.G. "Neutron Total Cross-Sections, 2.5-15 MeV." II "Effects of Nuclear Deformations". Phys.Rev. C3.604-620.1971.

22. Egan, J.J., Kegel, G.H.R., Couchell, G.P., Mittler, A., Barnes, B.K., Schier, W.A., Pullen, D.J., Harihar, P., Marcella, P.V., Sullivan, N.B. and Sheldon, E. "Elastic and Inelastic Differential Neutron Scattering Cross-sections from 238U from 0.9-2.7 MeV". Proceedings of the International Conference on Nuclear Cross-sections and Technology. National Bureau of Standards Special Publication Vol.2.950-952.1975.

23. Delaroche, J.P., Lagrange, Ch. and Salvy, J. "The Optical Model with Particular Consideration of the Coupled-Channel Optical Model". Review Paper 5, IAEA Consultants Meeting on the use of Nuclear Theory in Neutron Nuclear Data Evaluation, International Centre for Theoretical Physics, Trieste, 1975.

THE DISPERSIVE OPTICAL MODEL

P.E. Hodgson

Department of Physics, University of Oxford, U.K.

Abstract:

The advantages of the dispersive optical model for the analysis of neutron interactions are described, and recent analyses reviewed.

1. Introduction

In recent years many studies have shown that the dispersive optical model is able to account for a wide range of data from negative energies (bound states) to positive energies (scattering states) in a unified way (Hodgson, 1989, 1990). It is able to fit the experimental data more accurately than the simple optical model, and thus provides a way of obtaining a more accurate global optical potential. For both these reasons it is potentially of use in nuclear data analyses.

In this review, the advantages of the dispersive optical model are summarised, and the areas where increased predictive power may be expected are discussed. The older work is summarised briefly and recent advances considered in more detail.

2. Historical Review

The simple optical model, with energy-independent form factors and energy-dependent potential depths, has proved able to give differential cross-sections, polarisations and total cross-sections in good accord with the experimental data for many nuclei and energies using parameters that are almost independent of the target nucleus (Rapaport, 1982; Hodgson, 1984). In the case of neutrons several successful global potentials have been proposed, and these have been extensively compared with experimental data (Young, 1986).

The predictions of even the best global potentials show definite divergences from the experimental data, and these can be attributed to nuclear structure effects. It is possible to improve the fits by parameter adjustment for each nucleus, but thereby the predictive advantages of the global potentials are lost. Such parametrisations often require notably different potentials below about 7 MeV, and in particular the radius parameter is greater in this region (Moldauer, 1963; Van Oers *et al*, 1974; Eck and Thompson, 1975; Gyarmati *et al*, 1981; Finlay *et al*, 1984, 1985ab, 1989; Annand *et al*, 1985; Smith, Guenther and Lawson, 1987,1988). More recently, a detailed analysis of the elastic scattering of neutrons and protons by ^{40}Ca and ^{208}Pb by Mahaux and Sartor (1986c, 1990ab) showed that the radial shape of the optical potential changes with energy. All these studies show that the extrapolation of optical potentials from high to low energies assuming that the radial shape does not depend on energy is unreliable. This is confirmed by analysis of the single-particle energies; the gap between the low-lying particle and hole states obtained by a linear extrapolation from higher energies with constant geometry is about twice the experimental value (Brown, Gunn and Gould, 1963; Mahaux and Sartor, 1990).

These difficulties can be overcome in a systematic and physically-realistic way by making use of the dispersion relation that connects the real and imaginary parts of the potential and thus removes the inconsistency in the simple optical model that treats them as independent. Microscopic derivations of the optical potential show that it is a complex non-local function $\mathcal{V}(k, E)$ of the momentum k and the energy E. It is an analytic function of the energy and therefore satisfies the dispersion relation

$$\mathcal{V}(k,E) = \frac{1}{2\pi i}\int_c \frac{\mathcal{V}(k,E')}{E'-E}dE' \tag{2.1}$$

where the contour c encloses the singularity at $E' = E$. The physical basis of the dispersion relation is the requirement of causality. If we impose the condition that the scattered wave cannot start before the incident wave arrives, then the dispersion relation follows as a necessary consequence (Cornwall and Ruderman, 1962; Feshbach, 1958; Hagedorn, 1966; Mahaux *et al*, 1986).

If now we separate the potential into its real and imaginary parts

$$\mathcal{V}(k,E) = V(k,E) + iW(k,E) \tag{2.2}$$

and insert in (2.1), then evaluation of the contour integral and separation into real and imaginary parts gives

$$V(k,E) = \frac{\mathcal{P}}{\pi}\int_{-\infty}^{\infty} \frac{W(k,E')}{E'-E}dE' \tag{2.3}$$

and

$$W(k,E) = -\frac{\mathcal{P}}{\pi}\int_{-\infty}^{\infty} \frac{V(k,E')}{E'-E}dE' \tag{2.4}$$

where $\mathcal{P}$ indicates principal value. To use these relations to analyse data it is assumed that they also hold for the equivalent local potentials.

The real part of the optical potential shows a linear fall with increasing energy, and this is attributed to the non-locality. Superposed on this linear variation is a modulation centred at the Fermi surface which is due to the imaginary part of the potential. The linear part may be identified with the Hartree-Fock potential $V_{HF}(E, r)$ and the modulation obtained from the dispersion relation

$$\begin{aligned} V(E,r) &= V_{HF}(E,r) + \frac{\mathcal{P}}{\pi}\int_{-\infty}^{\infty} \frac{W(E',r)}{E'-E}dE' \\ &= V_{HF}(E,r) + \Delta V(E,r) \end{aligned} \tag{2.5}$$

The great advantage of this relation is that it separates the smoothly varying Hartree-Fock potential from the dispersive term that contains the complicated nuclear structure effects. The Hartree-Fock potential is well described by a

Saxon-Woods form factor with constant radius and diffuseness, and a depth that varies smoothly and monotonically with energy. The dispersive term depends on the imaginary potential that requires a more complicated parametrisation, as described in Section 4.

The imaginary potential has volume and surface components. The volume component adds a small term of the same radial shape to the Hartree-Fock potential. The surface imaginary term generally has a larger radius than the volume term, and so has the effect of adding an energy-dependent surface-peaked term to the edge of the volume potential, thus increasing its radius. If the experimental data are analysed using a potential with a fixed radius the dispersion term produces a modulation of the potential depth as a function of energy that is known as the Fermi surface anomaly. The energy variations of the volume and surface imaginary potentials, and the corresponding dispersive additions to the real potential are shown in fig.1 for neutrons on ^{40}Ca.

Physically, the dispersion term gives the effect on the real part of the potential due to the coupling to the non-elastic channels, principally the inelastic channels associated with the excitation of low-lying collective states. This effect is greatest for incident energies comparable with the energies of these collective states, and thus the dispersion effects are greatest at low energies. The effect of the dispersion term is thus mainly to increase the radius of the real part of the potential at lower energies, in accord with phenomenological studies. It also reduces the total cross-section for neutron scattering, bringing it into closer accord with the experimental data (Zong Di and Hodgson, 1988), and removes the need for an imaginary spin-orbit potential (Honoré *et al*, 1988; Delaroche and Tornow, 1988; Johnson and Mahaux, 1988; Tornow, Chen and Delaroche, 1990; Mahaux and Sartor, 1990b).

These and other deviations from the predictions of the simple optical model could be removed by the introduction of further parameters, or by defining a different potential for low energies (Moldauer, 1963) but this soon becomes unwieldy and unphysical. The dispersive optical model, however, enables the same data to be fitted to a higher accuracy with fewer parameters in a physically-realistic way. It also has improved predictive power for a particular nucleus from one energy to another. The accuracy with which a particular parametrisation of the dispersive term can represent the coupling to excited states has to be studied for each nucleus.

The effects of nuclear structure on the optical potential imply that a perfect global optical potential cannot be obtained. The dispersive optical model does, however, provide the means to define a global optical potential in which the nuclear structure effects are included by a single parameter, namely the strength of the imaginary potential. This possibility is discussed in Section 6.

Early attempts to apply the dispersion relations to the optical potential by Passatore (1968) encountered the difficulty of evaluating the dispersion integral over the whole energy range. A way of doing this was found by Ahmad and Haider (1976), making use of the different energy dependences of the volume and surface imaginary potentials. Subsequently many analyses have been made using the dispersion relation in the following ways:

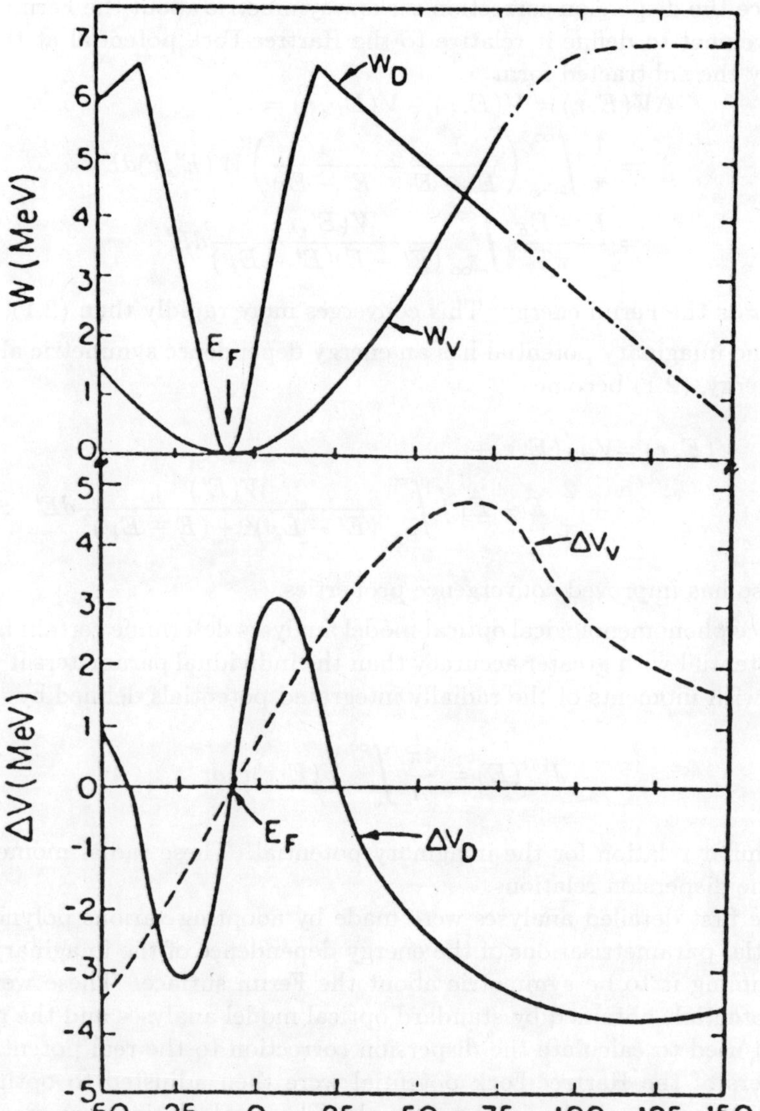

Fig. 1. Energy variations of the volume (W_V) and surface (W_D) imaginary potentials for neutrons on ^{40}Ca, together with the corresponding dispersive additions to the real potential (Delaroche and Tornow, 1988).

(a) Since the dispersion correction is skew-symmetric about the Fermi surface, it is convenient to define it relative to the Hartree-Fock potential at the Fermi energy by the subtracted form

$$\begin{aligned}\Delta V(E,r) &= V(E,r) - V(E_F,r) = \\ &= \frac{1}{\pi}\int_{-\infty}^{\infty}\left(\frac{1}{E'-E} - \frac{1}{E'-E_F}\right)W(E',r)dE' \\ &= \frac{E-E_F}{\pi}\int_{-\infty}^{\infty}\frac{W(E',r)}{(E'-E)(E'-E_F)}dE'\end{aligned} \tag{2.6}$$

where E_F is the Fermi energy. This converges more rapidly than (2.1).

(b) If the imaginary potential has an energy dependence symmetric about the Fermi energy, (2.1) becomes

$$\begin{aligned}V(E,r) =& V_{HF}(E,r)+ \\ &+ \frac{2}{\pi}(E-E_F)\int_{E_F}^{\infty}\frac{W(E')}{(E'-E_F)^2+(E-E_F)^2}dE'\end{aligned} \tag{2.7}$$

which also has improved convergence properties.

(c) Since phenomenological optical model analyses determine certain moments of the potential with greater accuracy than the individual parameters it is useful to work with moments of the radially integrated potentials defined by

$$J_V^{(q)}(E) = \frac{4\pi}{A}\int_0^{\infty} V(E,r)r^q dr \tag{2.8}$$

and a similar relation for the imaginary potential. These radial moments also satisfy the dispersion relations.

The first detailed analyses were made by adopting various polynomial or exponential parametrisations of the energy dependence of the imaginary potential, assuming it to be symmetric about the Fermi surface. These were fitted to the potentials obtained by standard optical model analyses and the resulting potential used to calculate the dispersion correction to the real potential. The parameters of the Hartree-Fock potential were then adjusted to optimise the fit to the phenomenological real potentials. These calculations are more conveniently done using the volume integrals of the potentials. Many of these analyses have been made by Mahaux and Ngô (1981, 1982, 1983, 1984). This established the usefulness of the dispersion relation and gave a detailed understanding of the Fermi surface anomaly.

3. Dispersive Optical Model Analyses

Since the pioneer work of Ahmad and Haider many dispersive optical model analyses have been made. The aim is to find the parametrisation of the optical potential as a function of radial distance and also of energy that will give the best overall fit to the experimental data. The main difficulties are that at negative energies the only data are the bound state energies and sometimes the widths. At positive energies extensive total and differential cross-sections are available, but at the lower energies the interpretation is complicated by the presence of compound elastic scattering, and for protons the Coulomb barrier limits the range of available energies.

There are two ways of carrying out a dispersive optical model analysis: the first parametrises the potential and seeks the optimum fit to selected data, while the second uses as data the phenomenological potentials obtained from standard optical model analyses. These will now be described.

In the first dispersive optical model method (DOMA), the Hartree-Fock potential is parametrised by a Saxon-Woods potential with a fixed radius and diffuseness and a depth that varies linearly or exponentially with energy. The imaginary potential is taken as the sum of volume and surface-peaked terms, usually parametrised by Saxon-Woods and Saxon-Woods derivative radial form factors. The energy variations of these imaginary potentials are separately parametrised; some of the ways of doing this are described in the next section. The values of all the parameters of the potential are either fixed from previous work or adjusted to fit selected experimental data, usually total and differential elastic cross-sections for a range of energies. In these analyses, the dispersive term is calculated from the imaginary potential and added to the Hartree-Fock term to give the real potential as in (2.5). This ensures that, although all the radius parameters are fixed, the effective radius of the real potential varies in the way required by the experimental data. The energy dependence of the volume integral of the real potential then fits the experimental values, as shown in fig.2.

A detailed analysis of the interaction of neutrons with ^{208}Pb from -20 to $+165$ MeV by Johnson, Horen and Mahaux (1987) used straight-line segments (4.4) to represent the energy variation of the volume and surface-peaked imaginary potentials. This showed the usefulness of the dispersive optical model and gave excellent fits to the experimental data. The improved fit to the bound single-particle neutron states in ^{208}Pb due to the addition of the dispersion terms is shown in fig.3. At low energies ($0 < E < 10$ MeV), Jeukenne, Johnson and Mahaux (1988) showed that the fits can be improved by allowing the diffuseness of the surface imaginary potential to decrease with decreasing energies, and its depth to depend on the orbital angular momentum. Providing this is done, the radius of the surface absorption can be kept constant. Almost as good fits can be obtained with surface absorptions that depend on angular momentum but not on energy. The angular momentum dependence has one value of W_s for $L = 1$, 3 and 6 and another for the remaining partial waves. The first group corresponds to those with wave functions that peak in the nuclear surface and so couple to surface excitations, while those in the second group do not. Further evidence of angular momentum dependence has been found by Johnson and Winters (1988).

A similar analysis of the interactions of neutrons with ^{40}Ca from -80 to $+80$ MeV was made by Johnson and Mahaux (1988). The potentials were similar to those found for ^{208}Pb. A small difference between the calculated and experimental total cross-section at low energy could be fitted by allowing the surface diffuseness and radius parameters to depend on energy. The interactions of protons with ^{40}Ca from -60 to $+200$ MeV were analysed by Tornow, Chen and Delaroche (1990), and of neutrons with ^{86}Kr by Johnson, Carlton and Winters (1989). An analysis of the interactions of neutrons and protons with ^{208}Pb by

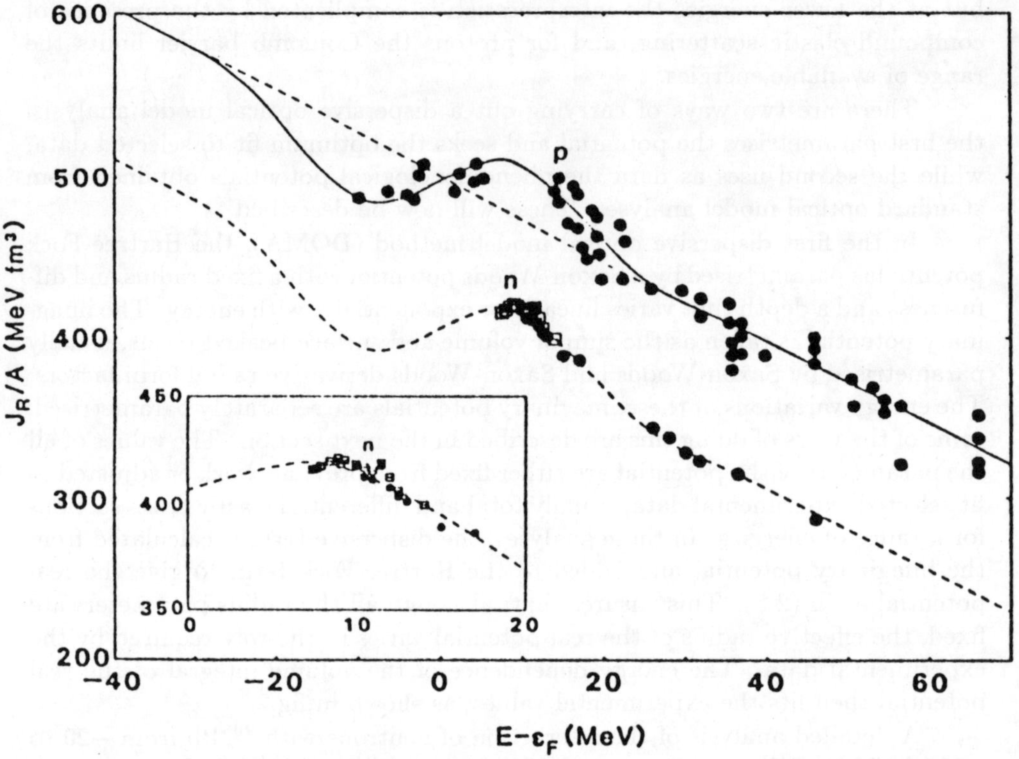

Fig. 2. Energy dependence of the volume integral of the real potential for neutrons and protons on ^{208}Pb showing the Fermi surface anomaly (Finlay *et al*, 1985a).

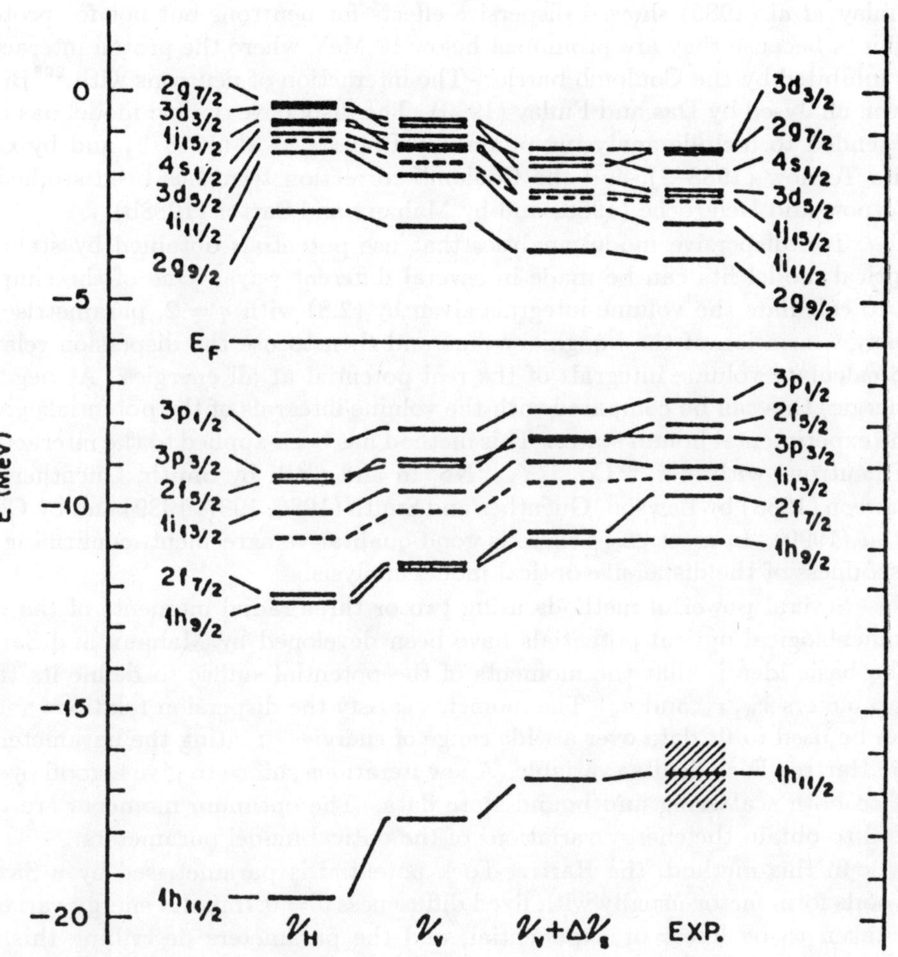

Fig. 3. Energies of bound neutron states in ^{208}Pb compared with Hartree-Fock calculations with and without the dispersion corrections (Johnson, Horen and Mahaux, 1987).

Finlay *et al* (1989) showed dispersive effects for neutrons but not for protons. This is because they are prominent below 10 MeV where the proton interaction is inhibited by the Coulomb barrier. The interaction of neutrons with ^{208}Bi has been analysed by Das and Finlay (1990). The dispersive optical model has been extended to include analysing powers by Roberts *et al* (1988–9) and by Chen and Tornow (1988–9), and the Coulomb correction term has been studied by Tornow and Delaroche (1988) and by Mahaux and Sartor (1988b).

The dispersive model analyses that use potentials obtained by standard optical model fits can be made in several different ways. One of the simplest is to calculate the volume integrals given by (2.8) with $q = 2$, parametrise the energy variation of the imaginary part and then to use the dispersion relation to calculate volume integrals of the real potential at all energies. At negative energies this can be compared with the volume integrals of the potentials giving the experimental bound states. This method has been applied to the interactions of neutrons with ^{51}V, ^{59}Co, ^{89}Y, ^{93}Nb, In and ^{209}Bi by Smith, Guenther and Lawson (1988) by Lawson, Guenther and Smith (1986, 1987, 1989) and by Chiba *et al* (1990). In most cases there is good qualitative agreement, confirming the usefulness of the dispersive optical model analysis.

Several powerful methods using two or three radial moments of the phenomenological optical potentials have been developed by Mahaux and Sartor. The basic idea is that the moments of the potential suffice to define its three parameters V_u, r_u and a_u. The moments satisfy the dispersion relations and so can be used to fit data over a wide range of energies, treating the parameters of the Hartree-Fock field as variable. A few iterations suffice to give a good overall fit to both scattering and bound state data. The optimum moments are then used to obtain the energy variations of the optical model parameters.

In this method, the Hartree-Fock potential is parametrised by a Saxon-Woods form factor usually with fixed diffuseness $a = 0.7$fm. Its energy variation is taken to be linear or exponential, and the parameters describing this, together with the radius parameter, are adjustable. In the first of their studies, Mahaux and Sartor (1986a) calculated the moments (2.8) of the imaginary potential with $q = 0.8$, 2 and 4 for potentials fitted to the elastic scattering of neutrons by ^{208}Pb and then fitted their energy variations with the Brown-Rho parametrisation (4.5a). The dispersion relation was then used to calculate the dispersion correction and this, together with the Hartree-Fock term, gave the moments of the real potential for all energies. These moments were compared with those of the potentials fitted to the bound and scattering state data, and the parameters of the Hartree-Fock potential adjusted to optimise the fit. From the three moments of the optimum real potential the parameters of the equivalent Saxon-Woods potential were obtained, and this confirmed that the main effect of the dispersive term is to modulate the radius of the potential around the Fermi energy.

This method, subsequently called the iterative moment approach (IMA), has been extended and applied to analyse the potentials fitted to data on neutron-^{208}Pb (Mahaux and Sartor, 1987ab, 1989a), proton-^{208}Pb (Mahaux and Sartor 1988a), neutron-^{40}Ca and proton-^{40}Ca (Mahaux and Sartor 1988c), neutron-^{89}Y (Jeukenne *et al*, 1977; Mahaux and Sartor 1987c), neutron-^{59}Co

(Smith, Guenther and Lawson, 1988) and neutron-^{51}V interactions (Lawson, Guenther and Smith, 1989). There are several differences in the ways the analyses were carried out, in particular whether the radius and diffuseness parameters of the Hartree-Fock potential were fixed or allowed to vary, and whether its potential depth was constrained to fit the Fermi energy.

Detailed comparisons between the DOMA and the IMA show that they give almost identical radial moments, but rather different results for the central value of the effective mass. Furthermore, the IMA gives an unphysical energy dependence of the real potential around $E = 0$. These defects may be attributed to inadequacies in the Brown-Rho parametrisation of the radial moments (Mahaux and Sartor, 1989a). In addition, because it does not give the imaginary potentials themselves, the IMA cannot be used to calculate differential cross-sections, spectral functions and occupation probabilities. It was therefore extended so as to include an explicit parametrisation of the imaginary potential, but still using only the moments of the phenomenological potentials (Mahaux and Sartor, 1989a). This allowed the additional observables to be calculated, but other defects remained.

Since the IMA involves only three adjusted parameters compared with seven for the DOMA, it is worthwhile trying to remove its defects by a more sophisticated parametrisation, and this has been done by the variational moment approach (VMA) (Mahaux and Sartor, 1989a).

In the VMA, the total and volume imaginary potentials are parametrised separately, by the BR and JM forms respectively (see (4.5a) and (4.6)). The surface imaginary potential is then given by their difference, which falls to zero at higher energies as required. This is the BR-JM parametrisation (4.9). The parameters are fixed by requiring that their moments fit those of the phenomenological potentials. From the moments the dispersive correction is obtained, and subtraction from the phenomenological moments of the real potential gives those of the Hartree-Fock potential. This potential is taken to have the Saxon-Woods form with surface diffuseness 0.7 fm and depth depending exponentially on energy. The radius parameter is chosen to minimise the difference between the calculated and phenomenological moments. The optical potential is then specified for all energies and can be used to calculate all observable quantities for comparison with experimental data.

In another version of the VMA, the total and volume imaginary potentials are both parametrised by the BR form; this is the BR-BR parametrisation (4.8ab). It is found that while the calculated real potential is not very sensitive to the details of the $W(r, E)$ parametrisation, the total cross-section is better fitted by the BR-BR than by the BR-JM parametrisation (Mahaux and Sartor, 1990, 1991b).

The VMA is simpler than the extended IMA, and also ensures that the dispersion relation is satisfied at each value of the radius, and that the effective mass has physical values. It has been successfully applied to the interactions of neutrons and protons with ^{208}Pb (Mahaux and Sartor, 1989ab). The proton analysis encountered the difficulty that there is no scattering data at low energies due to the Coulomb repulsion, and this was overcome by requiring the extrapolated mean field to reproduce the energy gap between the valence shells.

In both the neutron and proton analyses the overall quality of the fit is very similar to that of the DOMA, and it is achieved with fewer adjustable parameters. It is particularly notable that the VMA fixes the potential from the moments of the scattering data only, and its extrapolation to lower energies fits the bound state data. The method has also been successfully applied to the interactions of neutrons and protons with ^{40}Ca (Mahaux and Sartor, 1991b).

The dispersion relation method has been applied to a well-deformed nucleus by Merchant *et al* (1992). A coupled-channel analysis of the elastic, inelastic and total cross-sections for neutrons on ^{238}U gave a good overall fit to the experimental data, but this was appreciably improved by including the correction to the real potential obtained from the dispersion relation. To do this the imaginary potential was taken to have the form (4.5) and the parameters determined from the best fits to the data.

The energy dependence of both the surface and volume potentials have been taken to have the form (4.7a) by Niizeki *et al* (1989) and the parameters obtained by fitting to the experimental data for neutrons on ^{28}Si at six energies.

4. The Energy Variation of the Imaginary Potential

Evaluation of the dispersion integral requires a knowledge of the imaginary potential over the whole range of energies; for positive energies it can be obtained from optical model analyses of elastic scattering but for negative energies it is only obtainable with difficulty and with less accuracy from the widths of single-particle states. It is therefore usual in dispersive optical model analyses to assume that the imaginary potential is symmetric about the Fermi energy, and this enables the dispersion correction to be written in the form (2.6).

The reliability of this assumption may be studied by examining the energy dependence of the bound-state widths. Figs.4 and 5 show that the widths are given by (Mahaux and Sartor, 1990)

$$\Gamma_{nlj} = \alpha(E - E_F)^2 \tag{4.1}$$

where $\alpha \simeq 0.04$ in the vicinity of the Fermi surface and $\alpha \approx 0.009$ for the deeper states.

The widths of the bound states in a complex potential are related to the imaginary part of the potential by

$$\Gamma = 2\int_0^\infty W(r)|\psi(r)|^2 dr \tag{4.2}$$

If the form factor of the imaginary potential is suitably parametrised, its depth can therefore be obtained using the experimental values of Γ. In the limit of infinite nuclear matter,

$$\Gamma = 2W \tag{4.3}$$

If this relation is applied to finite nuclei, the symmetry assumption is approximately confirmed. Calculations with the more exact expression (4.2) are in progress.

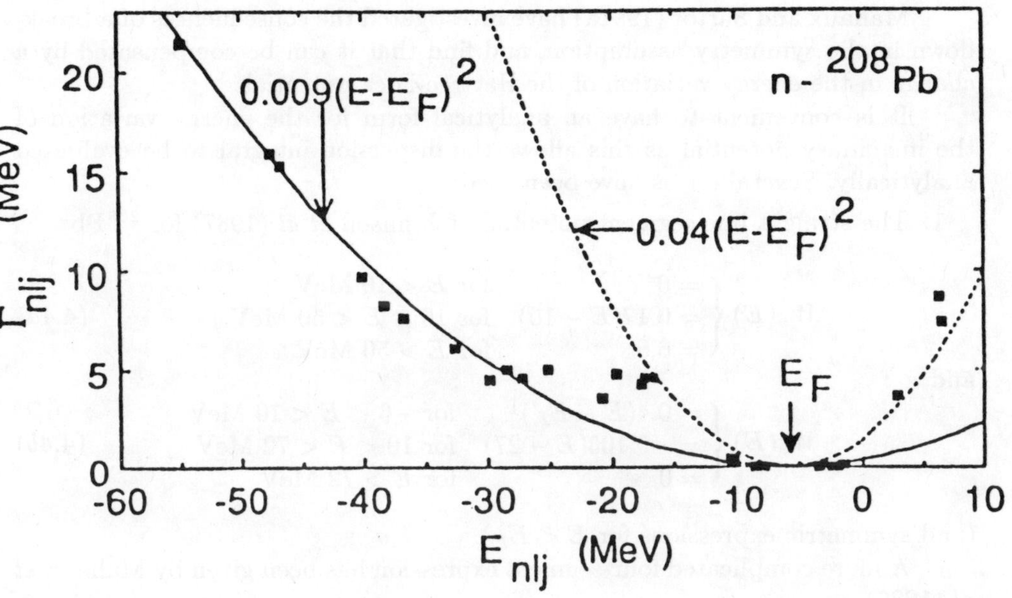

Fig. 4. Calculated quasi-particle bound-state widths as a function of energy for neutrons on ^{208}Pb (Mahaux and Sartor, 1991a).

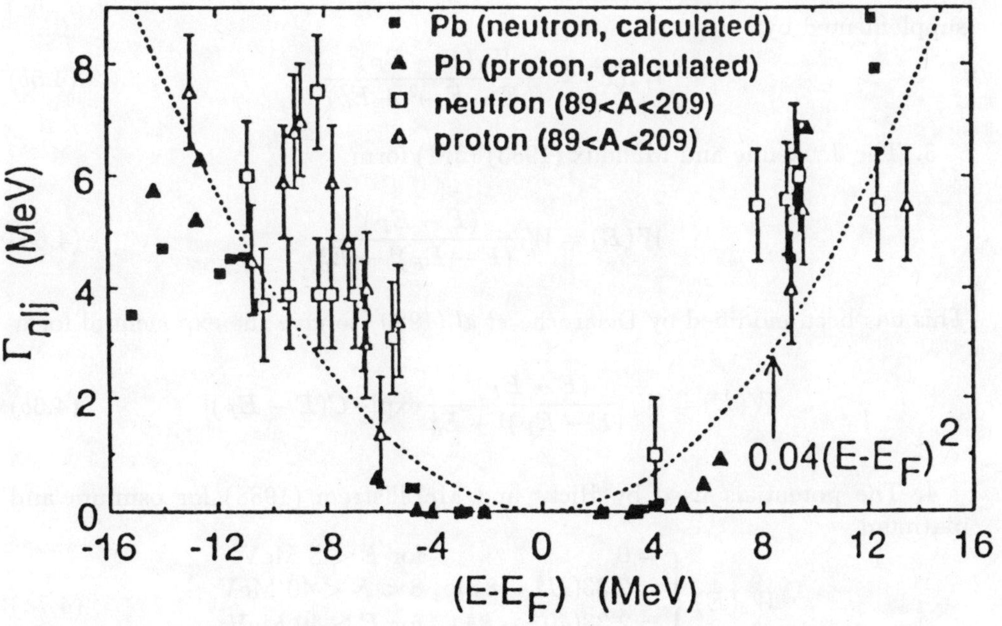

Fig. 5. Empirical and calculated quasi-particle bound-state widths as a function of energy for neutrons and protons on ^{208}Pb (Mahaux and Sartor, 1991a).

Mahaux and Sartor (1991a) have investigated the consequences of a breakdown in the symmetry assumption, and find that it can be compensated by a change in the energy variation of the Hartree-Fock potential.

It is convenient to have an analytical form for the energy variation of the imaginary potential as this allows the dispersion integral to be evaluated analytically. Several forms have been used:

1. The straight-line segment potential of Johnson *et al* (1987) for ^{208}Pb

$$W_v(E)\begin{cases} = 0 & \text{for } E < 10 \text{ MeV} \\ = 0.17(E-10) & \text{for } 10 < E < 50 \text{ MeV} \\ = 6.8 & \text{for } E > 50 \text{ MeV} \end{cases} \tag{4.4a}$$

and

$$W_s(E)\begin{cases} = 0.4(E-E_F) & \text{for } -6 < E < 10 \text{ MeV} \\ = -0.103(E-27) & \text{for } 10 < E < 72 \text{ MeV} \\ = 0 & \text{for } E > 72 \text{ MeV} \end{cases} \tag{4.4b}$$

(and symmetric expressions for $E < E_F$).

A more complicated four-segment expression has been given by Mahaux *et al* (1986).

2. The Brown and Rho (1981) (BR) form

$$W_v(E) = \frac{W_1(E-E_F)^2}{(E-E_F)^2 + E_0^2} \tag{4.5a}$$

supplemented by

$$W_s(E) = \frac{W_1(E-E_F)^2}{((E-E_F)^2 + E_0^2)^2} \tag{4.5b}$$

3. The Jeukenne and Mahaux (1983) (JM) form

$$W(E) = W_1 \frac{(E-E_F)^4}{(E-E_F)^4 + E_0^4} \tag{4.6a}$$

This has been modified by Delaroche *et al* (1989) to give the exponential form

$$W(E) = W_1 \frac{(E-E_F)^4}{(E-E_F)^4 + E_0^4} \exp[-C(E-E_F)] \tag{4.6b}$$

4. The potentials used by Hicks and McEllistrem (1988) for osmium and platinum

$$W_v(E)\begin{cases} = 0 & \text{for } E < 8 \text{ MeV} \\ = 2.33(E^{\frac{1}{2}} - 8^{\frac{1}{2}}) & 8 < E < 40 \text{ MeV} \\ = 2.33(40^{\frac{1}{2}} - 8^{\frac{1}{2}}) & \text{for } E > 40 \text{ MeV} \\ \quad \equiv\approx 8.1 \text{ MeV} & \end{cases} \tag{4.7a}$$

and

$$W_s(E) = \frac{a_0 + a_2(E-E_F)^2}{\{E_0^2 + (E-E_F)^2\}^2} \tag{4.7b}$$

(and symmetric expressions for $E < E_f$).

5. The BR-BR parametrisation of Mahaux and Sartor (1991)

$$W_v(E) = \gamma_v \frac{(E-E_p)^2}{(E-E_p)^2+\mu_v^2} \tag{4.8a}$$

$$W_s(E) = \beta_s \left[\frac{(E-E_p)^2}{(E-E_p)^2+\rho_s^2} - \frac{(E-E_p)^2}{(E-E_p)^2+\mu_v^2} \right] \tag{4.8b}$$

6. The BR-JM parametrisation of Mahaux and Sartor (1989a)

$$W_v(E) = \gamma_v \frac{(E-E_p)^4}{(E-E_p)^4+\mu_v^4} \tag{4.9a}$$

$$W_s(E) = \beta_s \left[\frac{(E-E_p)^2}{(E-E_p)^2+\rho_s^2} - \frac{(E-E_p)^4}{(E-E_p)^4+\mu_v^4} \right] \tag{4.9b}$$

6. Conclusion

The analyses summarised above show that the dispersive optical model is able to give more accurate fits to the neutron data than the simple optical model, and also shows how a better global potential may be constructed. It is therefore desirable that standard optical model computer programs be modified to include the dispersion correction (2.5) to the real potential. This requires incorporation of a parametrised form for the energy dependence of the imaginary potential, and for this one of the analytical forms listed in the previous section can be used. With such a program, the optical model analysis can be carried out in the usual way. Alternatively, the optical model analysis can be made, followed by a variable moment analysis as described in section 3.

An improved global potential could be constructed by performing a dispersive optical model analysis for a large number of nuclei, fixing all the parameters except the depth of the imaginary potential to standard average values. This depth provides the normalisation of the assumed energy variation, and so is just one number for each nucleus that takes account in a global way of the coupling to excited states.

Acknowledgments

I am grateful to Professor C. Mahaux, Professor W. Tornow and Dr A.B. Smith for sending me details of their unpublished work and to Professor R. Keddy and Professor R.H. Lemmer for inviting me to visit the University of the Witwatersrand where much of this paper was written.

References

I. Ahmad and W. Haider, *J. Phys.* **G2**, L157 (1976).
J.R.M. Annand, R.W. Finlay and F.S. Dietrich, *Nucl. Phys.* **A443**, 249 (1985).
G.E. Brown, J.H. Gunn and P. Gould, *Nucl. Phys.* **46**, 598 (1963).
G.E. Brown and M. Rho, *Nucl. Phys.* **A372**, 397 (1981).

Z.P. Chen and W. Tornow, TUNL Report XXVIII 82 (1988–9).

S. Chiba, P.T. Guenther, R.D. Lawson and A.B. Smith, *Phys. Rev.* **C42**, 2487 (1990).

J.M. Cornwall and M.A. Ruderman, *Phys. Rev.* **128**, 1474 (1962).

R.K. Das and R.W. Finlay, *Phys. Rev.* **C42**, 1013 (1990).

J.P. Delaroche and W. Tornow, *Phys. Lett.* **B203**, 4 (1988).

J.P. Delaroche, Y. Wang and J. Rapaport, *Phys. Rev.* **C39**, 391 (1989).

J.S. Eck and W.J. Thompson, *Nucl. Phys.* **A237**, 83 (1975).

H. Feshbach, *Ann. Phys.* (New York) **5**, 357 (1958).

R.W. Finlay, J.R.M. Annand, J.R. Petler and F.S. Dietrich, *Phys. Lett.* **B155**, 313 (1985a); **B157**, 475 (1985b).

R.W. Finlay, J.R.M. Annand, T.S. Cheema, J. Rapaport and F.S. Dietrich, *Phys. Rev.* **C30**, 796 (1984).

R.W. Finlay, J. Wierzbicki, R.K. Das and F.S. Dietrich, *Phys. Rev.* **C39**, 804 (1989).

B.R. Fulton, D.W. Banes, J.S. Lilley, M.A. Nagarajan and I.J. Thompson, *Phys. Lett.* **B162**, 55 (1985).

B. Gyarmati, R.G. Lovas, T. Vertse and P.E. Hodgson, *J. Phys.* **G7**, L209 (1981).

R. Hagedorn, Preludes to Theoretical Physics. Ed. A. de Shalit, H. Feshbach and L. Van Hove, North-Holland (1966), 154.

P.E. Hodgson, *Rep. Prog. Phys.* **47**, 613 (1984); Proceedings of the International Conference on Nuclear Reaction Mechanism, Calcutta. World Scientific (1989) 39. *Cont. Phys.* **31**, 295 (1990).

S.E. Hicks and M.T. McEllistrem, *Phys. Rev.* **C37**, 1787 (1988).

G.M. Honoré, W. Tornow, C.R. Howell, R.S. Pedroni, R.C. Byrd, R.L. Walter and J.P. Delaroche, *Phys. Rev.* **C33**, 1129 (1986).

M. Jaminon, J.-P. Jeukenne and C. Mahaux, *Phys. Rev.* **C34**, 468 (1986).

J.-P. Jeukenne and C. Mahaux, *Nucl. Phys.* **A394**, 445 (1983).

J.-P. Jeukenne, C.H. Johnson and C. Mahaux, *Phys. Rev.* **C38**, 2573 (1988).

J.-P. Jeukenne, A. Lejeune and C. Mahaux, *Phys. Rev.* **C15**, 10 (1977).

C.H. Johnson, R.F. Carlton and R.R. Winters, *Phys. Rev.* **C39**, 415 (1989).

C.H. Johnson, D.J. Horen and C. Mahaux, *Phys. Rev.* **C36**, 2252 (1987).

C.H. Johnson and C. Mahaux, *Phys. Rev.* **C38**, 2589 (1988).

C.H. Johnson and R.R. Winters, *Phys. Rev.* **C37**, 2340 (1988).

R.D. Lawson, P.T. Guenther and A.B. Smith, *Phys. Rev.* **C34**, 1599 (1986); *Phys. Rev.* **C36**, 1298 (1987); *Nucl. Phys.* **A493**, 267 (1989).

C. Mahaux, P.F. Bortignon, R.A. Broglia and C.H. Dasso, *Phys. Rep.* **120**, 1 (1985).

C. Mahaux and H. Ngô, *Phys. Lett.* **B100**, 285 (1981); *Nucl. Phys.* **A378**, 205 (1982); *Nucl. Phys.* **410**, 271 (1983); *Nucl. Phys.* **A431**, 486 (1984).

C. Mahaux, H. Ngô and G.R. Satchler, *Nucl. Phys.* **A449**, 354 (1986).

C. Mahaux and R. Sartor, *Phys. Rev.* **C19**, 229 (1979); *Phys. Rev. Lett.* **57**, 3015 (1986a); *Nucl. Phys.* **A451**, 441 (1986b); *Nucl. Phys.* **A458**, 25 (1986c); *Nucl. Phys.* **A460**, 466 (1986d); *Phys. Rev.* **C34**, 2119 (1986e); *Nucl. Phys.* **A468**, 193 (1987a); *Phys. Rev.* **C36**, 1777 (1987b); *Nucl. Phys.* **A475**, 247 (1987c); *Nucl. Phys.* **A481**, 381 (1988a); *Nucl. Phys.*

A481, 407 (1988b); *Nucl. Phys.* **A484**, 205 (1988c); *Nucl. Phys.* **A493**, 157 (1989a); *Nucl. Phys.* **A503**, 525 (1989b); *Nucl. Phys.* **A516**, 285 (1990); *Advances in Physics*, Ed. J.W. Negele and E. Vogt (1991a) (In press); *Nucl. Phys.* (1991b) (In press).

A.C. Merchant, P.E. Hodgson and H.R. Schelin. *Nucl. Sci. Eng.* **111**, 132 (1992).

P.A. Moldauer, *Nucl. Phys.* **47**, 65 (1963).

M.A. Nagarajan, C. Mahaux and G.R. Satchler, *Phys. Rev. Lett.* **54**, 1136 (1985).

T. Niizeki, H. Orihara, M. Ohura, M. Hosaka, G.C. Jon, K. Ishii, K. Miura and H. Ohnuma, CYRIC Annual Report, Tohoku University 10 (1989).

G. Passatore, *Nucl. Phys.* **A110**, 91 (1968).

J. Rapaport, *Phys. Rep.* **87**, 25 (1982).

M.L. Roberts, Z.M. Chen, P.D. Felsher, D.J. Horen, C.R. Howell, K. Murphy, H.G. Pfützer, W. Tornow and R.L. Walter, TUNL Report XXVIII 70 (1988–9).

A.B. Smith, P.T. Guenther and R.D. Lawson, *Nucl.Phys.* **A455**, 344 (1986); *Nucl. Phys.* **A483**, 50 (1988).

W. Tornow and J.P. Delaroche, *Phys. Lett.* **B210**, 26 (1988).

W. Tornow, Z.P. Chen and J.P. Delaroche, *Phys. Rev.* **C42**, 693 (1990).

W.T.H. van Oers, H. Haw, N.E. Davison, A. Ingemarsson, B. Fagerström and G. Tibell, *Phys. Rev.* **C10**, 307 (1974).

P.G. Young, Proceedings of the Specialists' Meeting on the Use of the Optical Model for the Calculation of Neutron Cross-Sections below 20 MeV., NEANDC–222"U", OECD, Paris (1986), 127.

Su Zong Di and P.E. Hodgson, *J. Phys.* **G14**, 1485 (1988).

RIVISTA DEL NUOVO CIMENTO VOL. 5, N. 1 1982

Nuclear Charge Distributions.

F. MALAGUTI, A. UGUZZONI and E. VERONDINI

Istituto Nazionale di Fisica Nucleare - Sezione di Bologna
Istituto di Fisica dell'Università - Bologna

P. E. HODGSON

Nuclear Physics Laboratory - Oxford

(ricevuto l'11 Settembre 1981)

1. – Introduction.

The precise determination of nuclear charge and matter distributions is one of the central problems of nuclear physics, and over the years continuous efforts have been made to improve the accuracy of our knowledge. The problem can be tackled both experimentally and theoretically, and in recent years consistent results of high precision have been obtained.

The experimental determination relies on the analysis of the differential cross-section of the elastic scattering of various projectiles by nuclei, and useful supplementary information is obtained from muonic atoms. Electrons are particularly useful for determining the charge distribution as they interact only through the well-known electromagnetic field. Protons interact with both the electromagnetic and nuclear fields and so can be used to determine the matter distribution as well, though the accuracy is less because of our poorer knowledge of the nuclear interaction. Alpha-particles are also useful because their scattering depends particularly on the nuclear surface, but the analysis is further complicated by the structure of the alpha-particle itself.

Since we consider nuclei to be composed of protons and neutrons, the charge distribution is almost the same as the proton distribution, though a small correction must be made to include the effect of the charge distribution of the neutrons. The matter distribution is simply the sum of the proton and neutron distributions.

In order to obtain the nuclear charge or matter distribution from the experimental scattering data, the usual procedure is to postulate a particular analytical form for the distribution and to vary its parameters to optimize the fit to the experimental cross-sections. This procedure is inevitably limited by the range of momentum transfer in the particular experimental arrangement used. This can be seen qualitatively from the Born approximation, which expresses the elastic electron scattering cross-section as proportional to the square of a form factor $F(q)$ related to the charge distribution $\varrho(r)$ by

$$\varrho(r) = \frac{Z}{4\pi}\int_0^\infty j_0(qr)\,F(q)\,q^2\,\mathrm{d}q\,, \tag{1.1}$$

where $q = 2K\sin\Theta/2$ is the momentum transfer. Lack of experimental data beyond the maximum experimental momentum transfer thus implies ignorance of the corresponding Fourier component of the charge distribution. The corresponding uncertainty in the charge (or matter) distribution can, however, be reduced by utilizing our theoretical knowledge of their overall characteristics, in particular those derived from the general characteristics of the nucleon-nucleon interaction, which imply that they are continuous, more or less uniform in the nuclear interior and fall exponentially to zero in the surface region.

The early data on electron elastic scattering and muonic atoms were fitted using charge density distribution of the two- or three-parameter Fermi forms [1] and the parameters were adjusted to optimize the fit to the data. This procedure is inadequate, partly because it is not possible to fit the more extensive data now available to sufficient accuracy [2] and partly because the use of such analytic forms introduces undesirable correlations between the densities at different radial distances. These difficulties have now been overcome by the development of the so-called « model independent » methods of extracting the charge distribution from the experimental data, *i.e.* elastic electron scattering and muonic X-rays transition energies. A strictly model-independent determination of the charge density is obviously impossible with a finite number of measurements up to a maximum momentum transfer q_{max}. A « model independent » analysis, therefore, means a method of analysis in which the form of the distribution is expressed in the most physical and general way. As pointed out by FRIAR and NEGELE [3], this can be done in several different ways. The first [4] consists in focusing attention on a theoretically calculated density distribution and asking how such a distribution must be changed to obtain an improved agreement with those features of the charge distribution determined by the experimental data. This implies writing the charge density as

$$\varrho(r) = \varrho_{th}(r) + \sum_{i=1}^{M} C_i \Delta\varrho_i(r) , \tag{1.2}$$

where $\varrho_{th}(r)$ is a theoretical charge distribution and $\Delta\varrho_i(r)$ are M independent functions which are assumed to vanish beyond some distance R and are determined by the available experimental data. An alternative method [5] makes no reference to any specific theoretical calculations, but instead assumes some behaviour of the form factor $F(q)$ beyond q_{max}. An intermediate point of view, in which the density $\varrho(r)$ is expanded in a restricted set of functions, was adopted by SICK [6], who introduced the so-called « sum of Gaussians » (SOG) expansion of the charge density

$$\varrho(r) = \sum_{i=1}^{N} A_i \{\exp[-(r-R_i)^2/\gamma^2] + \exp[-(r+R_i)^2/\gamma^2]\} , \tag{1.3}$$

where the Gaussians centred at $-R_i$ are introduced to ensure the slope to vanish at $r = 0$. In this procedure, A_i and R_i are generated as random numbers and only distributions yielding a normalized χ^2 less than one are retained. The model dependence is contained in the assumptions $R_i < R$, $A_i > 0$ and in the value chosen for the width γ of the Gaussians. Alternative expansions have been discussed by BORYSOWICZ and HETHERINGTON [7]. In the following, we will refer to the charge densities obtained by one of these methods as « model-independent charge densities ».

The theoretical methods of calculating nuclear charge and matter distributions can be divided into two classes. The first approach is based on the independent-particle model of nucleons moving independently in a central field. In its simplest form it approximates the exact nuclear wave function $|\Psi\rangle$ with a unique Slater determinant $|\Phi\rangle$ constructed from an appropriate set of single-particle (s.p.) wave functions. If the nuclear force field is known, the wave functions of all the nucleons and hence their density distributions can readily be calculated. However, the distribution of the nucleons itself determines the central field, and can be used to calculate it, provided we know the appropriate interaction. It is thus possible to proceed iteratively and postulate a distribution, calculate the central field and hence the distribution again, and to continue until self-consistent results are obtained. This procedure was originally devised by HARTREE for atoms and now forms the basis of the Hartree-Fock and other mean-field methods that have been extensively used to calculate nuclear properties. Their application to charge and matter distributions is discussed in sect. **2**.

The great advantage of these mean-field methods is that in principle they depend essentially only on the interaction; once this is fixed, everything else follows. In practice, it is necessary to make certain approximations, such as restricting the number of basis states in order to simplify the calculations, and this may require the use of an effective interaction. It is then found that the latest mean-field theories give results that agree fairly well with a wide range of experimental data. However, if there remains some residual disagreement with the experimental data, there is little that can be done to improve the fit except to alter the parameters of the interaction used in a somewhat *ad hoc* way, and this may produce disagreement with other data. The very generality of the mean-field methods is thus a source of rigidity.

An alternative, and in a sense more appropriate, approach is based on the observation that the knowledge of the full nuclear wave function $|\Psi\rangle$ is not necessary to calculate density distributions. These can be obtained from the more restricted knowledge of the so-called « one-body density matrix » D^1, that carries all physical information on the one-body properties of many-fermion systems.

As the information contained in D^1 is much less than that contained in $|\Psi\rangle$, shifting our attention from $|\Psi\rangle$ to D^1 results in a much less ambitious approach, but one able to give more accurate density distributions, because effects that go beyond the simple independent-particle model, like fractional occupation numbers even in deep orbitals, can naturally be accounted for.

As shown in sect. **3**, the use of the density matrix D^1 results in an expression for the density,

$$\varrho(\boldsymbol{r}) = \sum_{\alpha} \langle \tilde{n}_{\alpha} \rangle |\tilde{\varphi}_{\alpha}(\boldsymbol{r})|^2 , \tag{3.13}$$

that strictly resembles the Hartree-Fock one, but it allows noninteger occupation numbers and uses a very special single-particle basis, the « natural-orbital basis ».

Here we approximate the natural orbitals with the eigenfunctions of a single-particle potential (SPP) consisting of central, spin-orbit and Coulomb terms. The parameters of the potential for each state are adjusted to fit the separation energy and the spin-orbit splitting extracted from the one-nucleon transfer reactions according to the Clement-Baranger definition [8, 9]. From the single-particle wave functions obtained in this way and the knowledge of the experimental occupation numbers it is then possible, as discussed later in sect. **3**, to calculate the nuclear density distributions.

Thus, while the Hartree-Fock method is very general and uses no specific nuclear information, the SPP method is rather particular and uses much information on the nucleus concerned. This is both a strength and a weakness; a strength because it ensures a sound physical basis and allows the calculations to be fitted to the idiosyncrasies of each nucleus; a weakness because this is paid for in loss of generality in the sense that it can be applied directly only to those nuclei for which the requisite experimental data are available.

To overcome this limitation, one can make use of the systematic variation of the single-particle energies across the periodic table. This fact can be used to give physically significant values of those energies which are not directly accessible to experiment. Of course, such a systematics can only describe the general trends of single-particle energies and cannot give details of specific nuclei. Thus it is useful for the SPP density calculations only if the distributions calculated in this way are not strongly sensitive to these details.

The essential feature of the SPP calculations is that in their simplicity they unify the nuclear data from one-nucleon transfer reactions, on the one hand, and from electron scattering, on the other. This increases our knowledge of density distributions particularly in the region of the nuclear surface. In fact, electron scattering and muonic-atom data determine the charge density very well in the inner surface or knee region where it begins to fall from its interior value, but are relatively inaccurate in the interior and far surface region. The densities calculated by the SPP method, however, are very accurate in the far surface region because here they depend on the tails of the s.p. wave functions which are fitted to the experimental separation energies for each orbit. This advantage is important when the method is extended to calculate matter distributions, for many reactions are particularly sensitive to the nuclear surface.

In this paper attention will be paid both to the SPP method (sect. **3**) showing the results obtained for a range of nuclei (sect. **4**) and to the systematics of single-particle states and its use in connection with nuclear density distributions (sect. **5**). Finally, after a specific construction of natural orbitals in a particular case (sect. **6**), a summarizing discussion of the results achieved by the SPP method and of its possible improvements and perspectives will be presented in sect. **7**.

2. – The Hartree-Fock and other mean-field theories.

There have been many calculations of nuclear charge and matter distributions using the Hartree-Fock and other mean-field theories that have been developed to go beyond the Hartree-Fock approximation, while still retaining the one-body description of the system. These calculations have been extensively described in several papers, *e.g.* by NEGELE, SVENNE, QUENTIN and FLOCARD [10]. In this section the accuracy and practicability of these methods will be briefly discussed.

In principle, the mean-field theories are fundamental in the sense that they enable many nuclear properties to be calculated from the nucleon-nucleon interaction. In practice, however, this is not possible because of the presence of a hard core and also because the large basis required for realistic calculations leads to great numerical complexity, beyond the power of the fastest computers. These difficulties may to some extent be overcome, but at the price of introducing some phenomenological elements into the calculations. The set of basis states must be truncated and the parameters optimized for the basis chosen. The results are greatly dependent on the basis, especially for heavy deformed nuclei. Another phenomenological element is that an effective interaction is used with some parameters that are adjusted to fit selected experimental data. In more technical language, the interaction is renormalized by a G-matrix expansion in the Brueckner-Hartree-Fock method, or alternatively some of the higher-order many-body terms are taken into account by introducing a density dependence. The resulting velocity-dependent and density-dependent phenomenological effective interactions have been widely used to analyse nuclear-structure data. Among these the Skyrme interaction is particularly successful. This is expressed in terms of six parameters that are adjusted to fit the volume, surface, symmetry, Coulomb and spin-orbit energies of nuclei. There remains some flexibility relating to the interplay between the velocity dependence, often expressed as an effective mass m^*, and the density dependence. This flexibility can be used to improve the fit in some respects at the expense of that in others, for example to the single-particle energies at the expense of the charge distribution. A rather successful compromise is the S III interaction that can reproduce the charge densities, binding energies, single-particle energies and deformations to good accuracy throughout the periodic table.

The mean-field theories encouter difficulties in fitting both the surface and the deep states with the same effective interaction. The level densities of the states near the Fermi surface require a velocity dependence corresponding to an effective mass m^*/m about unity, while the deep levels require $m^*/m \sim 0.5$. The S III interaction effects a compromise by using $m^*/m \sim 0.75$.

The mean-field theories are most reliable for the hard nuclei with a relatively sharp energy minimum as a function of deformation, such as nuclei with well-closed shells or a permanent static deformation. These nuclei can be well described with rather few basis states. In the case of soft nuclei, on the other hand, with shallow energy minima, the zero-point energy fluctuation implies the superposition of so many configurations that the single Slater determinant configuration of the minimum cannot describe the actual density distribution. This occurs around nickel, the rare-earth region and also for ^{12}C.

The mean-field theories can be applied to deformed nuclei away from closed shells by taking pairing configurations into account following the Hartree-Fock-Bogoliubov scheme. Such an approach introduces further phenomenology into the effective interaction and also greatly increases the numerical complexity of the calculations.

Since the mean-field theories are self-consistent, with each nucleon influencing all the other nucleons, an interaction that gives correct proton distributions is very likely to give correct neutron distributions, and *vice versa*. The differences between the proton and neutron radii are thus expected to be very well given by the mean-field theories and indeed this is found to be so, to an accuracy around $\pm$ 0.02 fm.

Even though the mean-field theories have considerable parameter flexibility, they are not always able to give precise fits to charge distributions. Thus the conditions for the validity of the Hartree-Fock method are best satisfied for ^{208}Pb, and fig. 2.1 shows several calculations of the charge distributions for that nucleus [11]. It is notable that the average shapes and also the peak-to-valley ratios in the interior are quite different. Similar discrepancies are found for other nuclei, as shown in fig. 2.2 [11]. The differences between the charge distributions of isotopes provide a critical test of calculations of nuclear charge densities. Figure 2.3 shows a comparison between the experimental results and Hartree-Fock calculations with the Skyrme III force, which gives the best fit to the data [12]. It is notable that there are marked discrepancies particularly for ^{94}Zr-^{92}Zr, and these may be due to the neglect of pairing effects. In a recent work by Gogny *et al.* [13] these effects have been explicitly introduced in HF + RPA calculations with a noticeable improvement of the fit to the charge densities (see, *e.g.*, fig. 4.10 and paragraph **4**˙2.3).

The nuclear charge and matter distributions are determined by the summation of the probability distributions of the nucleons, and an important test of these distributions is the agreement between the experimental and calculated single-particle energies. On the whole, these are not well given by Hartree-Fock calculations even for ^{208}Pb, as shown in fig. 2.4 [14]. All the calculations give levels that are too widely spaced, by a factor of about two, and in many cases the level ordering is incorrect and sometimes even the gaps between shells are in the wrong places. By contrast, as shown in sect. **5**, a simple energy-inde pendent Saxon-Woods potential gives an excellent overall fit to these energies.

It should be mentioned that the calculated single-particle energies may not be strictly comparable with the experimental centroid energies due principally to difficulties connected with rearrangement energies [15]. The energies

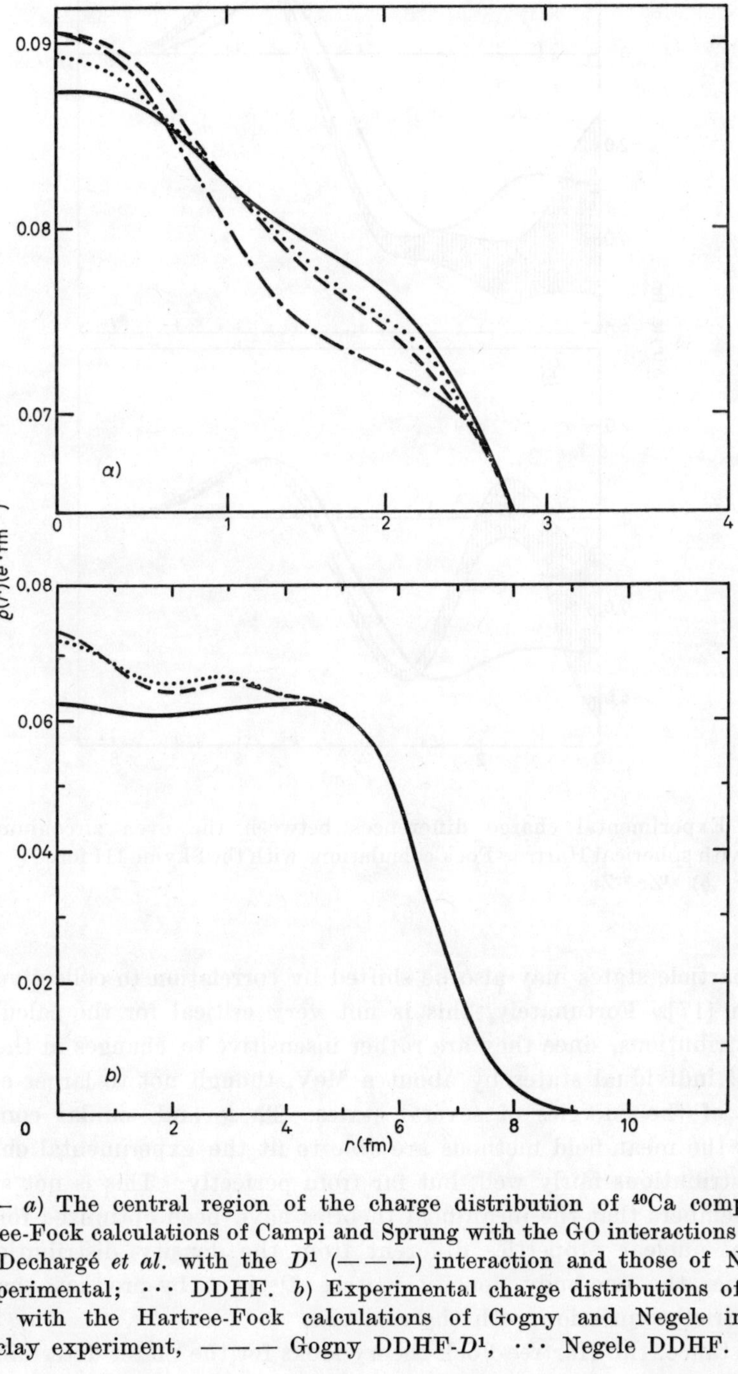

Fig. 2.1. – *a*) The central region of the charge distribution of ^{40}Ca compared with the Hartree-Fock calculations of Campi and Sprung with the GO interactions (—·—·—); those of Dechargé *et al.* with the D^1 (— — —) interaction and those of Negele [11]; —— experimental; ··· DDHF. *b*) Experimental charge distributions of ^{208}Pb [16] compared with the Hartree-Fock calculations of Gogny and Negele in ref. [11]: —— Saclay experiment, — — — Gogny DDHF-D^1, ··· Negele DDHF.

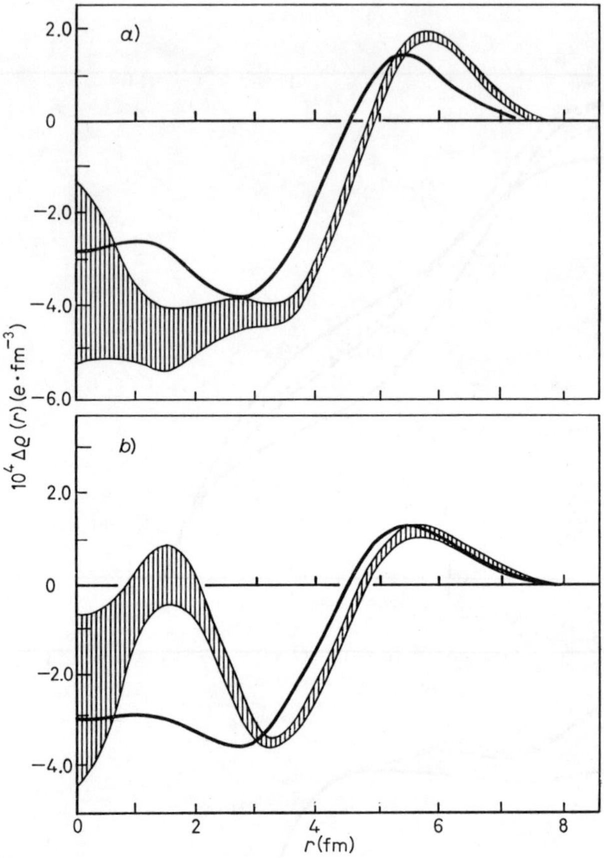

Fig. 2.2. – Experimental charge differences between the even zirconium isotopes compared with spherical Hartree-Fock calculations with the Skyme III force (———) [12]: *a*) ^{92}Zr-^{90}Zr, *b*) ^{94}Zr-^{92}Zr.

of single-particle states may also be shifted by correlation to collective degrees of freedom [17]. Fortunately, this is not very critical for the calculation of charge distributions, since they are rather insensitive to changes in the binding energies of individual states by about a MeV, though not to larger correlated variations of the energies of several states. These and similar comparisons show that the mean-field methods are able to fit the experimental charge and matter distributions fairly well, but far from perfectly. This is not surprising if one remembers that the mean-field theories have been optimized for the calculation of nuclear properties different from the density distributions. For these teories the one-body density matrix D^1 is a by-product that might have no strict connection with the real one.

For instance, the Hartree-Fock theory looks for the Slater determinant that minimizes the energy within a given model space. If the true nuclear wave

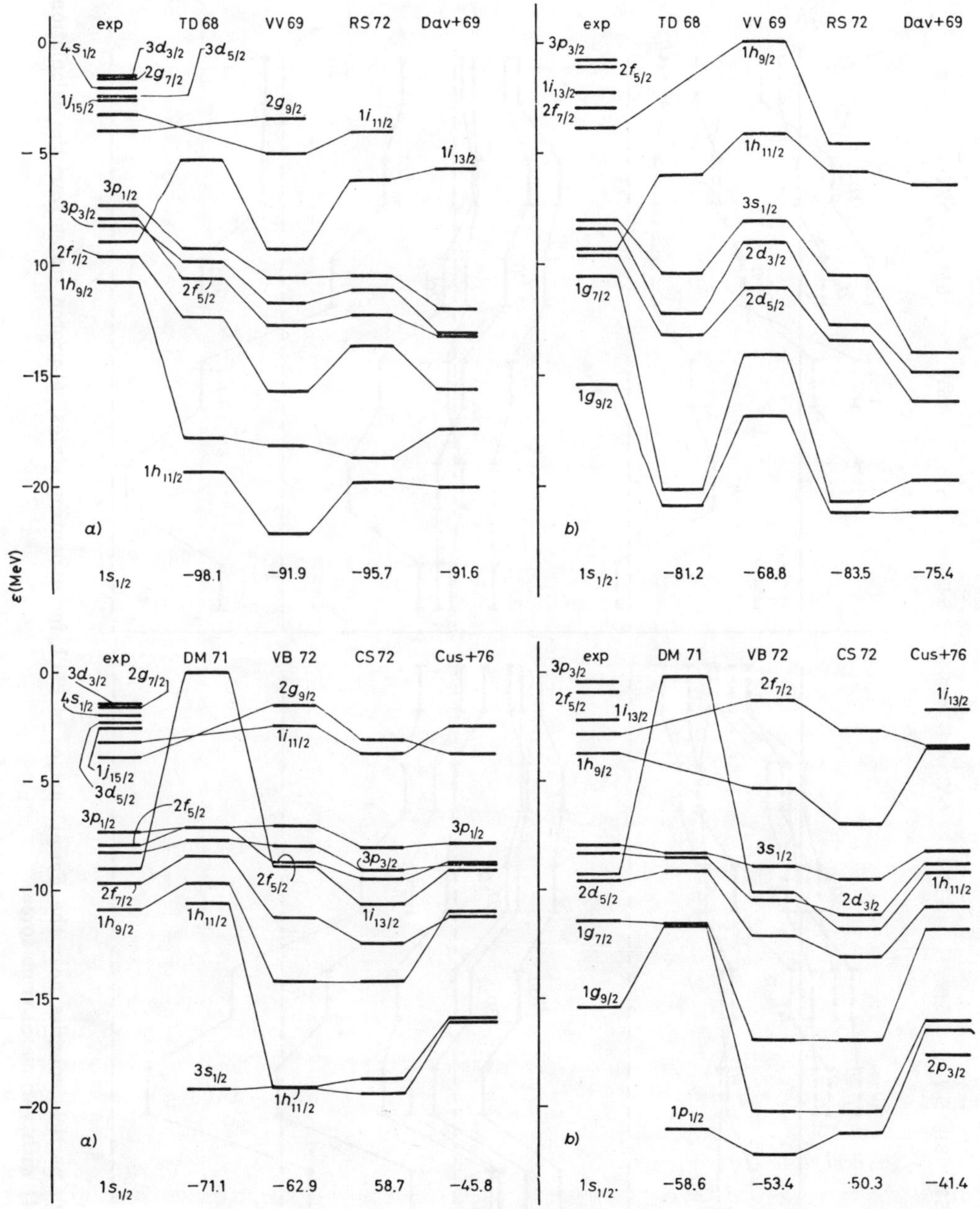

Fig. 2.3. – Single-particle energies near the Fermi surface in ^{208}Pb compared with various Hartree-Fock calculations for *a*) neutron states and *b*) proton states. The two upper graphs show three Hartree-Fock and one Brueckner-Hartree-Fock and the lower show one renormalized Bruckner-Hartree-Fock and three density-dependent Hartree-Fock calculations. The symbols refer to the papers listed by SVENNE [14].

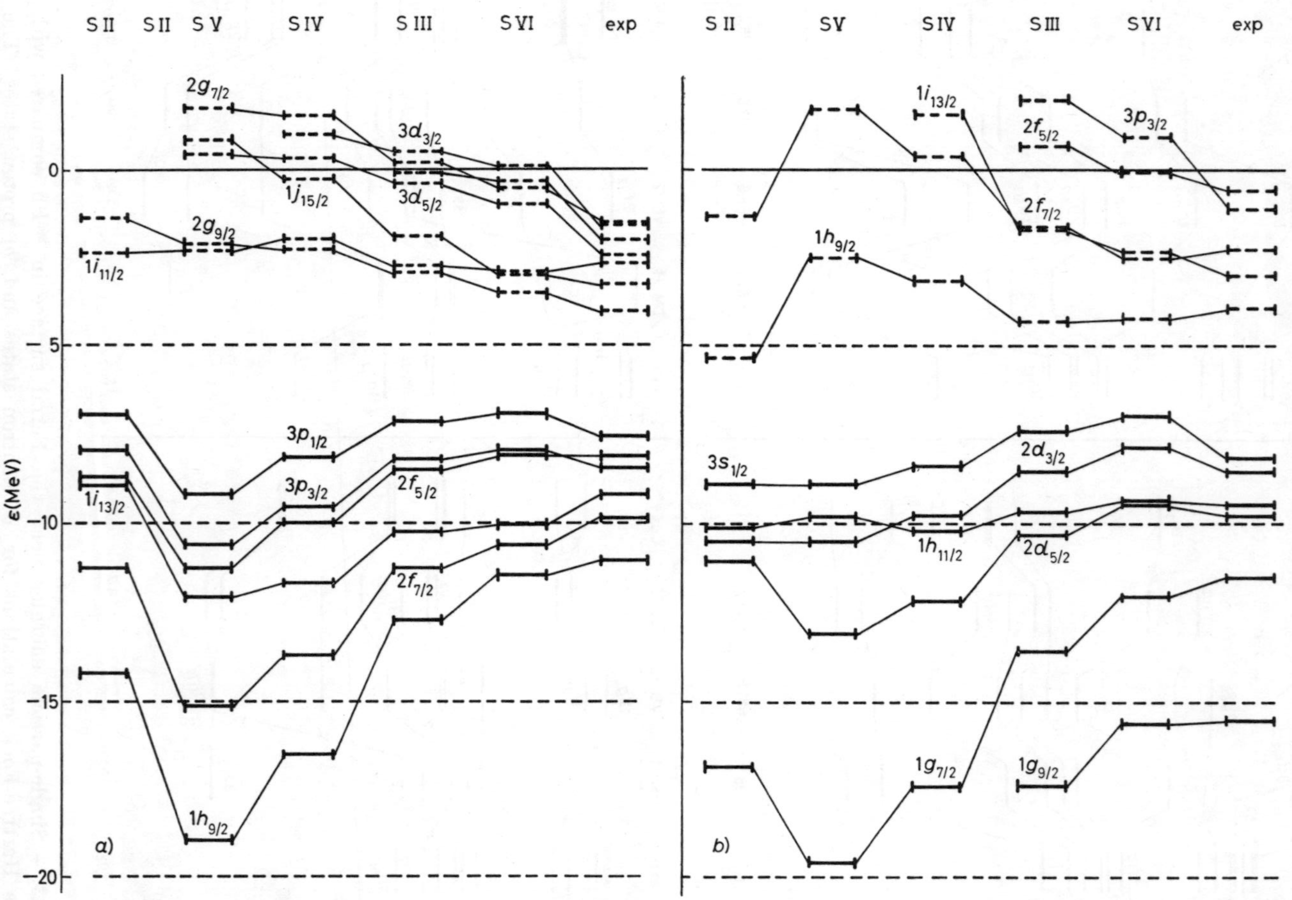

Fig. 2.4. – Single-particle energies near the Fermi surface in ^{208}Pb for *a*) neutron states, *b*) proton states from the calculations by BEINER *et al.* [14] with various Skyrme forces.

function is the HF one plus a small correction, then the total energy is corrected to second order. This means that a good approximation to the energy does not mean necessarily a correspondingly good approximation to the total wave function.

Therefore, a more direct approach to nuclear densities appears useful, that renounces the aim of calculating the full wave function of the nucleus and directly focuses attention to its one-body density matrix.

3. – The single-particle potential method.

3'1. *The one-body density matrix and the natural orbitals.* – We review here the formalism needed to express the nuclear density distributions in terms of the one-body density matrix. This will provide the general framework of the single-particle potential (SPP) method.

The total wave function $|\Psi\rangle$ of a system of A indistinguishable fermions can be expressed in terms of the complete set of Slater determinants, containing all possible particle-hole excitations, constructed from an arbitrary complete set of orthonormal single-particle (s.p.) wave functions $\varphi_\alpha(\boldsymbol{r})$.

For this system we define the Hermitian one-body density matrix D^1 by the equation

$$D^1_{\alpha\beta} = \langle\Psi|a^\dagger_\beta a_\alpha|\Psi\rangle = D^{1*}_{\beta\alpha} . \tag{3.1}$$

Its diagonal elements are the s.p. occupation numbers

$$D^1_{\alpha\alpha} = \langle\Psi|a^\dagger_\alpha a_\alpha|\Psi\rangle = \langle n_\alpha\rangle . \tag{3.2}$$

The $\langle n_\alpha\rangle$ range from zero to unity and satisfy the sum rule

$$\mathrm{Tr}\,(D^1) = \sum_\alpha \langle n_\alpha\rangle = A . \tag{3.3}$$

For this system, any one-body operator $\hat{Q} = \sum_{i=1}^{A} \hat{q}_i$ can be written as

$$\hat{Q} = \sum_{\alpha\beta} q_{\alpha\beta} a^\dagger_\alpha a_\beta \tag{3.4}$$

with

$$q_{\alpha\beta} = \langle\varphi_\alpha|\hat{q}|\varphi_\beta\rangle . \tag{3.5}$$

The expectation value of $\hat{Q}$ on the exact wave function

$$\langle Q\rangle = \langle\Psi|\hat{Q}|\Psi\rangle = \sum_{\alpha\beta} q_{\alpha\beta}\langle\Psi|a^\dagger_\alpha a_\beta|\Psi\rangle = \sum_{\alpha\beta} q_{\alpha\beta} D^1_{\beta\alpha} = \mathrm{Tr}\,(qD^1) . \tag{3.6}$$

Knowledge of D^1 thus enables the expectation value of any one-body operator to be calculated, with no explicit knowledge of the total wave function $|\Psi\rangle$.

Therefore, the one-body density matrix contains all the physical information on the one-body properties of fermion systems.

The D^1-matrix can be diagonalized by a unitary transformation U that defines a new set of single-particle wave functions $\tilde{\varphi}_\alpha(\boldsymbol{r})$

$$\tilde{\varphi}_\alpha(\boldsymbol{r}) = \sum_\beta U_{\alpha\beta}\varphi_\beta(\boldsymbol{r}) \, . \tag{3.7}$$

The transformed D^1-matrix

$$\tilde{D}^1_{\alpha\beta} \equiv (U^\dagger D^1 U)_{\alpha\beta} = \langle \tilde{n}_\alpha \rangle \, \delta_{\alpha\beta} \, . \tag{3.8}$$

The s.p. wave functions $\tilde{\varphi}_\alpha$ obtained from the transformation that diagonalizes D^1 are called « natural orbitals »; their importance in the description of many-fermion systems was stressed already in 1955 by Löwdin [18].

The eigenvalues of the D^1-matrix are the occupation numbers of the natural orbitals; hereafter we will call these « natural occupation numbers ».

In the natural-orbital basis the expectation value (3.6) of any one-body operator has the simple diagonal form

$$\langle Q \rangle = \sum \tilde{q}_{\alpha\alpha} \langle \tilde{n}_\alpha \rangle \, , \tag{3.9}$$

that involves only the occupation numbers and expectation values of $\hat{q}$ in the natural-orbital basis.

The density distribution of a system of A pointlike particles can be easily expressed by using this formalism. The relevant operator is now

$$\hat{\varrho}(\boldsymbol{r}) = \sum_{i=1}^{A} \delta(\boldsymbol{r} - \boldsymbol{r}_i) \tag{3.10}$$

with s.p. matrix elements

$$\varrho_{\alpha\beta}(\boldsymbol{r}) = \langle \varphi_\alpha(\boldsymbol{r}') | \delta(\boldsymbol{r} - \boldsymbol{r}') | \varphi_\beta(\boldsymbol{r}') \rangle = \varphi^*_\alpha(\boldsymbol{r})\, \varphi_\beta(\boldsymbol{r}) \, . \tag{3.11}$$

Therefore, the density distribution from (3.6) and (3.9) becomes

$$\varrho(\boldsymbol{r}) \equiv \langle \Psi | \hat{\varrho}(\boldsymbol{r}) | \Psi \rangle = \sum_{\alpha\beta} \varphi^*_\alpha(\boldsymbol{r}) \varphi_\beta(\boldsymbol{r}) D^1_{\beta\alpha} \quad \text{in an arbitrary s.p. basis,} \tag{3.12}$$

$$\varrho(\boldsymbol{r}) \equiv \langle \Psi | \hat{\varrho}(\boldsymbol{r}) | \Psi \rangle = \sum_{\alpha} \langle \tilde{n}_\alpha \rangle |\tilde{\varphi}_\alpha(\boldsymbol{r})|^2 \quad \text{in the natural-orbital basis.} \tag{3.13}$$

We see that the diagonal expression (3.13) can be used only for the natural orbitals; for an arbitrary s.p. basis $\varrho(\boldsymbol{r})$ also contains off-diagonal contributions.

A simple example of this is a bound spinless particle shared between two states with wave function of the form

$$\psi(\boldsymbol{r}) = c_1\varphi_1(\boldsymbol{r}) + c_2\varphi_2(\boldsymbol{r}) \qquad (c_1^2 + c_2^2 = 1)\,,$$

where c_1, c_2, φ_1 and φ_2 can be taken as real. The occupation numbers are $\langle n_\alpha\rangle = c_\alpha^2$, and the density is

$$\varrho(\boldsymbol{r}) = \psi^2(\boldsymbol{r}) = \sum_{\alpha=1}^{2} \langle n_\alpha\rangle\, \varphi_\alpha^2(\boldsymbol{r}) + 2c_1c_2\varphi_1(\boldsymbol{r})\,\varphi_2(\boldsymbol{r})\,.$$

The natural orbitals here are

$$\tilde{\varphi}_1(\boldsymbol{r}) \equiv \psi(\boldsymbol{r}) = c_1\varphi_1 + c_2\varphi_2 \qquad \text{with } \langle\tilde{n}_1\rangle = 1$$

and

$$\tilde{\varphi}_2(\boldsymbol{r}) = -\,c_2\varphi_1 + c_1\varphi_2 \qquad \text{with } \langle\tilde{n}_2\rangle = 0$$

and for them only the equation

$$\varrho(\boldsymbol{r}) = \sum_{\alpha=1}^{2} \langle\tilde{n}_\alpha\rangle\, \tilde{\varphi}_\alpha^2(\boldsymbol{r})$$

holds.

Equations (3.12) and (3.13) have a particularly simple form for systems having total angular momentum $J = 0$. In this case, any choice of s.p. wave functions having good angular momenta l, j, m and good parity $(-)^l$ leads to the partial diagonalization of D^1:

$$J = 0 \Rightarrow D^1_{\alpha\beta} \propto \delta_{l_\alpha l_\beta}\,\delta_{j_\alpha j_\beta}\,\delta_{m_\alpha m_\beta}\,. \tag{3.14}$$

This follows from definition (3.1) and the angular-momentum and parity orthogonality relations, since now the states $a_\alpha|\Psi\rangle$ and $a_\beta|\Psi\rangle$ have the same angular momenta and parity as φ_α and φ_β.

This shows that the diagonalization leading to the natural orbitals for a $J = 0$ system has to be carried out only within the $\{ljm\}$ subspaces of the total model space $\{\varphi_\alpha\}$. In other words, for a $J = 0$ system, the transformation (3.7) « mixes » only s.p. wave functions having the same l, j, m and different principal quantum numbers n.

Moreover, such a system has spherical symmetry, and the density (3.12) is left unaltered after averaging over all directions of space. Writing explicitly

$$\varphi_\alpha(\boldsymbol{r}) = \sum_{\mu\sigma} \langle l_\alpha s\mu\sigma|j_\alpha m_\alpha\rangle\, \frac{1}{r}\, y_{n_\alpha l_\alpha j_\alpha}(r)\, Y^\mu_{l_\alpha}(\Omega)\, \chi^\sigma_s \tag{3.15}$$

and using (3.14), we find [19]

$$\varrho(r) \equiv \frac{1}{4\pi}\int \mathrm{d}\Omega\, \varrho(\boldsymbol{r}) = \sum_{lj} \varrho_{lj}(r)\,, \tag{3.16}$$

where

$$4\pi r^2 \varrho_{lj}(r) = \sum_{nn'} (2j+1)\, D^1_{n'lj,nlj}\, y^*_{nlj}(r)\, y_{n'lj}(r) \qquad \text{in an arbitrary s.p. basis,} \tag{3.17}$$

$$4\pi r^2 \varrho_{lj}(r) = \sum_{n} (2j+1) \langle \tilde{n}_{nlj} \rangle |\tilde{y}_{nlj}(r)|^2 \qquad \text{in the natural-orbital basis.} \tag{3.18}$$

In this case spherical symmetry ensures that the D^1-matrix does not depend on m; this index has, therefore, been dropped and the summation over m gives the $2j+1$ factor.

To conclude, it is interesting to examine some conditions sufficient to ensure that a particular s.p. basis coincides with a natural-orbital basis. These conditions justify the use of the diagonal density (3.13) in some nuclear models where no explicit reference is made to natural orbitals.

From definition (3.1) of the D^1-matrix it follows:

i) If the total wave function of the system is a unique Slater determinant, the s.p. wave functions are natural orbitals.

Proof. For $\alpha \neq \beta$, $a_\alpha|\Psi\rangle$ and $a_\beta|\Psi\rangle$ are different Slater determinants and their overlap $D^1_{\alpha\beta} = 0$.

This is the case of the Hartree-Fock model.

ii) If, for a $J = 0$ system, all s.p. states except at most one within each $\{ljm\}$ subspace are either completely full or completely empty, then the s.p. wave functions are natural orbitals.

This condition, already discussed in ref. [19], can be shown as follows: the matrix element $D^1_{\alpha\beta} = \langle\Psi|a^\dagger_\beta a_\alpha|\Psi\rangle \neq 0$ for $\alpha \neq \beta$ means that a particle can be shifted from state α to β, and this means $\langle n_\alpha\rangle \neq 0$ and $\langle n_\beta\rangle \neq 1$. But, if $D^1_{\alpha\beta} \neq 0$, then $D^{1*}_{\alpha\beta} = \langle\psi|a^\dagger_\alpha a_\beta|\psi\rangle \neq 0$, and this means $\langle n_\beta\rangle \neq 0$ and $\langle n_\alpha\rangle \neq 1$. Thus both $\langle n_\alpha\rangle, \langle n_\beta\rangle \neq 0$ and $\langle n_\alpha\rangle, \langle n_\beta\rangle \neq 1$. For a $J = 0$ system, all this can be repeated within each $\{ljm\}$ subspace (see eq. (3.14)). Therefore, if there is only one s.p. state in each subspace whose occupation number is neither 0 nor 1, then $D^1_{\alpha\beta} = 0$ for any $\alpha \neq \beta$.

This is the case of the nuclear shell model constructed from a « closed core » and some « valence particles »: $|\Psi\rangle$ is no longer approximated by a single Slater determinant and some « configuration mixing » is introduced. But this mixing is sufficiently weak to involve not more than one s.p. state for each $\{ljm\}$ subspace. An example is provided by the « s-d shell model », where active nucleons occupy the $2s_{\frac{1}{2}}$, $1d_{\frac{5}{2}}$, $1d_{\frac{3}{2}}$ shells outside a closed ^{16}O core.

iii) If all elementary configurations (Slater determinants) entering in the total wave function of the system differ by more than the shift of one particle, then the s.p. wave functions are natural orbitals.

Proof. $D^1_{\alpha\beta} = \langle\Psi|a^\dagger_\beta a_\alpha|\Psi\rangle \neq 0$ for $\alpha \neq \beta$ implies, within $|\Psi\rangle$, at least two configurations that differ by a 1 particle-1 hole excitation.

This is approximately the case for the ground state of $J = 0$ even-even nuclei, where the excitations constructed from the simple « all nucleons coupled to pairs » configuration are more likely to be 2 particles-2 holes, 4p-4h, 6p-6h, .., than 1p-1h, 3p-3h, ...

These conditions will be used in the next subsection.

3'2. *Application to the calculation of charge densities.*

3'2.1. General features of the SPP method. From the properties of the D^1-matrix reviewed in the last subsection, the natural orbitals appear as the best s.p. wave functions for the calculation of nuclear density distributions and in general of the expectation value of any one-body operator.

Unfortunately, the practical use of natural orbitals is limited by the fact that a method to calculate them from the fundamental nucleon-nucleon force has not yet been worked out in detail [20]. We, therefore, try an empirical approach to the problem, observing that the natural orbitals should be similar to the wave functions used in the single-particle shell model, based on a Saxon-Woods potential with little or no configuration mixing.

This assumption is justified by the success of the density distributions calculated around 1970 by Elton and co-workers [21] and by conditions i) to iii) described at the end of the last subsection. Thus we calculate ground-state density distributions of $J = 0$ even-even nuclei using Saxon-Woods wave functions and the diagonal expression (3.18).

We limit the model space $\{\varphi_\alpha\}$ to the bound states for which one-nucleon transfer reactions show a significant occupancy. Therefore, we explicitly neglect contribution to the density coming from continuum single-particle states, whose occupancy may be nonzero as an effect of correlations among particles.

We then vary some parameters of the Saxon-Woods potential to get a best fit to the experimental density distributions.

In this way we find, within the assumed model space, the best Saxon-Woods approximation to natural orbitals.

After that, we can go beyond this approximation and explicitly construct natural orbitals by mixing the SW functions contained in the model space according to eqs. (3.7)-(3.8).

The present work will mainly be concerned with the search of the optimum SW wave functions; an attempt to construct explicit natural orbitals will be described in sect. **6**.

In more detail, we approximate the natural radial functions $\tilde{y}_\alpha(r)$ in eq. (3.18) by the radial eigenfunctions $y_\alpha(r)$ of a local SW potential of standard form and depth V_L,

$$V(r) = -V_L f_1(r) + V_S \left(\frac{\hbar}{m_\pi c}\right)^2 \frac{1}{r}\frac{\mathrm{d}f_2(r)}{\mathrm{d}r}\, \boldsymbol{L}\cdot\boldsymbol{\sigma} + V_C(r)\,, \tag{3.19}$$

where

$$f_{1,2}(r) = \{1 + \exp[(r - R_{1,2})/a_{1,2}]\}^{-1} \tag{3.20}$$

with $R_1 = RA^{\frac{1}{3}}$, $R_2 = 1.1A^{\frac{1}{3}}$ and $a_2 = 0.65$ fm.

R and a_1 are treated as free parameters in the optimizing calculations. V_C (only for protons) is the Coulomb potential of a uniformly charged sphere having the experimentally measured r.m.s. charge radius, while V_L and V_S are chosen so as to reproduce the s.p. energies ε_α of the $j = l \pm \frac{1}{2}$ doublet, measured in one-nucleon transfer reactions. If only the centroid of the doublet is known, the spin-orbit potential V_S is set to 7 MeV [22].

According to Clement and Baranger [8, 9], the experimental binding energy E_α for the s.p. level α is defined by the two-sided centroid

$$E_\alpha \equiv -\varepsilon_\alpha = \Big(\sum_N S_{N\alpha}(E_N - E_0) + \sum_n S_{n\alpha}(E_0 - E_n)\Big)\Big/(2j_\alpha + 1)\,, \tag{3.21}$$

where E_0 is the ground-state energy of the target nucleus A, E_N and E_n are the energies of states in nuclei $A-1$ and $A+1$, respectively, $S_{N\alpha}$ and $S_{n\alpha}$ their s.p. spectroscopic factors, measured by pick-up and stripping reactions from or to the s.p. level α, respectively, and normalized by the sum rule

$$\sum_N S_{N\alpha} + \sum_n S_{n\alpha} = 2j_\alpha + 1\,. \tag{3.22}$$

The use of simple one-sided centroids is probably more popular among experimentalists, but definition (3.21) gives energies that are better fitted by potential (3.19) for nonclosed shells [23].

The natural occupation numbers $\langle\tilde{n}_\alpha\rangle$ to be used in eq. (3.18) are approximated by those extracted from one-nucleon transfer reactions according to the French-MacFarlane sum rules [24]:

$$\langle\tilde{n}_\alpha\rangle \approx \langle n_\alpha\rangle = \Big(\sum_N S_{N\alpha}\Big)\Big/(2j_\alpha + 1) = 1 - \Big(\sum_n S_{n\alpha}\Big)\Big/(2j_\alpha + 1)\,, \tag{3.23}$$

for states near the Fermi level, and are simply taken to be unity (thus neglecting some possible correlation effects) for deep-lying orbits.

The assumption near the Fermi level is partially justified by the work of Clement, who derives (eq. (11) of ref. [25]) an expression for the density similar to our eq. (3.18). If we neglect the centre-of-mass correction, Clement's formula in our notation is

$$4\pi r^2 \varrho_{lj}(r) = \sum_N S_{Nlj} \Phi^2_{Nlj}(r) , \tag{3.24}$$

where $\Phi_N(r)$ is a radial overlap function, *i.e.* the radial part of a s.p. function describing the overlap between the ground state of nucleus A and the N-th state of nucleus $A-1$, excited in a one-nucleon pick-up reaction. In principle, the overlap functions depend on the index N, but, as CLEMENT points out [8], for a group of levels with energies E_N corresponding to removing a nucleon from a s.p. level near the Fermi surface the N-dependence is weak and from (3.23)

$$4\pi r^2 \varrho_{lj}(r) \approx \Phi^2_{lj}(r) \sum_N S_{Nlj} = \Phi^2_{lj}(2j+1)\langle n_{lj}\rangle . \tag{3.25}$$

Therefore, for a group of levels of this kind, Φ_{lj} can be approximately identified with a natural orbital and $\sum_N S_{Nlj}/(2j+1)$ with the corresponding natural occupation number.

Finally, since the physical one-body potential is expected to be nonlocal, the use of the local potential (3.19) appears questionable. As discussed in sect. **5**, a constant nonlocal potential seems to be a better choice. Following a common practice, we, therefore, introduce nonlocality effects into our wave functions $y_{nlj}(r)$ by multiplying them by the Perey factor [26]:

$$y_{nlj}(r) \to y_{nlj}(r)\left[1 + \frac{\mu\beta^2}{2\hbar^2} V_L f_1(r)\right]^{-\frac{1}{2}} , \tag{3.26}$$

where μ is the reduced nucleon mass and $\beta \approx 1$ fm the nonlocality range, that is treated as a free parameter in the fitting calculations.

We have applied the above formalism to the proton states of a series of nuclei, and the resulting charge distributions have been fitted to the model-independent ones obtained from electron scattering. This fitting has been carried out by varying the parameters

R (defining the radius of the central potential as $R_1 = RA^{\frac{1}{3}}$);

a_1, the diffuseness of the central potential;

β, the nonlocality range of the central potential.

In some cases, additional free parameters were one or two occupation numbers or a single-particle energy whose experimental values suffer from large uncertainty.

In general, the calculated charge density is very sensitive to the occupation numbers, but depends only weakly on s.p. energies. This would suggest to use the fit to determine occupation numbers; unfortunately, this cannot be done without ambiguity (see sect. **4** and fig. 3.1).

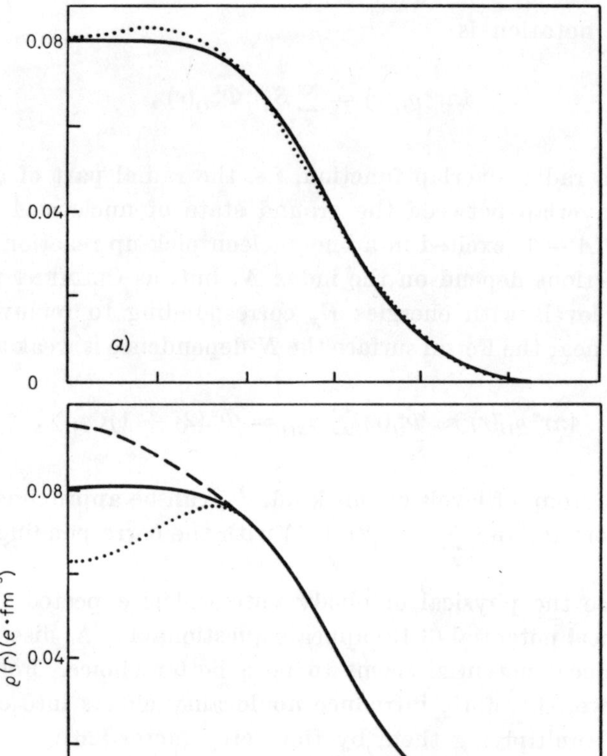

Fig. 3.1. – *a*) Calculated charge density of ^{24}Mg compared with the experimental results of Li *et al.* [27]: ··· experiment; —— calculated: r.m.s. = 2.985 fm, R = 1.3058 fm, a_1 = 0.52 fm; $\langle n_{s_{\frac{1}{2}}}\rangle = 0.60$. *b*) Calculated charge density of ^{24}Mg showing the effect of varying the $2s_{\frac{1}{2}}$ occupation number (r.m.s. = 2.985 fm): — — — $\langle n_{s_{\frac{1}{2}}}\rangle = 1.0$, —— $\langle n_{s_{\frac{1}{2}}}\rangle = 0.6$, ··· $\langle n_{s_{\frac{1}{2}}}\rangle = 0.1$. *c*) Calculated charge density of ^{24}Mg showing the effect of varying the r.m.s. charge radius ($\langle n_{s_{\frac{1}{2}}}\rangle = 0.6$): — — — r.m.s. = 2.88 fm, —— r.m.s. = 2.985 fm, ··· r.m.s. = 3.09 fm.

The calculations were always constrained to reproduce the measured r.m.s. charge radius, and this essentially determines R. Before describing the calculations in detail, we mention now several refinements and corrections that have to be made to the simple formalism just described.

3'2.2. Folding the finite nucleon charge distribution. Nucleons possess an internal structure and, in particular, a charge distribution. Therefore, eq. (3.18) must be folded with a function describing their internal charge density. For the present purposes, it is sufficiently accurate to use a sum of 3-dimensional Gaussians with spherical symmetry

$$h(\boldsymbol{r}) = \sum_k d_k \left(\frac{1}{a_k\sqrt{\pi}}\right)^3 \exp\left[-\frac{r^2}{a_k^2}\right], \tag{3.27}$$

where the parameters d_k and a_k have been found by CHANDRA and SAUER from their analysis of electron-nucleon scattering [28]. For protons they are

$$d_{1,2,3} = 0.506,\ 0.328,\ 0.166\ ; \qquad a_{1,2,3} = 0.657,\ 0.373,\ 1.235\ \text{fm} \tag{3.28}$$

and correspond to a r.m.s. charge radius of 0.88 fm.

Neutrons also, though their total charge is zero, have a charge distribution with a mean square radius of $-0.116\ \text{fm}^2$, and this is sufficiently important to be included in our calculations. It was indeed found by BERTOZZI *et al.* [29] that the apparently anomalous decrease of the charge radius from ^{40}Ca to ^{48}Ca is partly due to the $f_{\frac{7}{2}}$ neutrons. We describe the neutron charge density with eq. (3.27) and [28]

$$d_{1,2} = +1, -1\,, \qquad a_{1,2} = 0.685,\ 0.739\ \text{fm}\,. \tag{3.29}$$

Folding (3.18) with (3.27) and performing the angular integrals yields again (3.18) with a modified form factor [19]

$$\left|\frac{y_\alpha(r)}{r}\right|^2 \to \int_0^\infty \mathrm{d}r'\, g(r, r') \left|\frac{y_\alpha(r')}{r'}\right|^2 \tag{3.30}$$

with

$$g(r, r') = \frac{\pi r'}{r} \sum_k \left\{\pi^{-\frac{3}{2}} d_k a_k^{-1} \left[\exp\left[-\left(\frac{r-r'}{a_k}\right)^2\right] - \exp\left[-\left(\frac{r+r'}{a_k}\right)^2\right]\right]\right\}. \tag{3.31}$$

As this is a small correction, we obtain the neutron point density using occupation numbers given by the extreme s.p. model and s.p. energies given by some average potential described later on (paragraph **3'**2.6). The numerical effect of finite-size corrections is shown for ^{48}Ca by fig. 3.2.

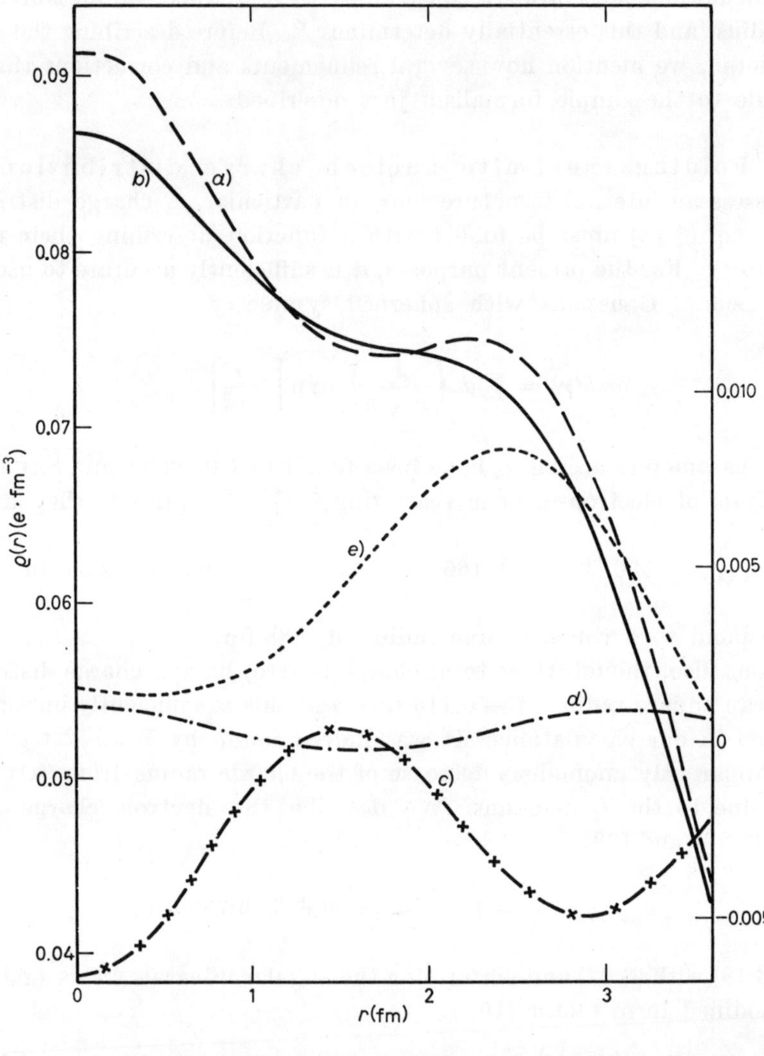

Fig. 3.2. – Corrections to the ^{48}Ca charge density, calculated by using the « standard parameters » described in sect. **4**: Left-hand side scale: *a*) point proton density, *b*) total density, with all corrections. Right-hand side scale: *c*) correction due to the finite proton size+c.m. motion+Darwin-Foldy term, *d*) correction due to the finite neutron size, *e*) 10× spin-orbit correction (neutrons and protons).

3'2.3. Centre-of-mass motion. The densities we want to calculate and compare with experiment are measured in a reference frame in which the centre of mass (c.m.) of the nucleus is fixed. In our calculations, however, like in all shell model or HF calculations, it is the centre of the potential that is fixed, and this introduces a spurious c.m. vibration around the origin, whose effect is a density, $\varrho_{\text{f.p.}}$, having a larger radial extent than that seen in the c.m. frame, $\varrho_{\text{c.m.}}$. It is, therefore, necessary to derive the relation between $\varrho_{\text{c.m.}}$ and $\varrho_{\text{f.p.}}$.

While, in general, this relation is difficult to derive [30], in the particular case of the ground state in a harmonic-oscillator potential it has been found by ELLIOTT and SKYRME [31], *viz.*

$$\varrho_{\text{c.m.}}(r) = \left(b\sqrt{\frac{\pi}{A}}\right)^{-3}\int \mathrm{d}\boldsymbol{r}' \exp\left[A|\boldsymbol{r}-\boldsymbol{r}'|^2/b^2\right]\varrho_{\text{f.p.}}(\boldsymbol{r}')\,, \tag{3.32}$$

where $b=(\hbar/m\omega)^{\frac{1}{2}}$, m is the nucleon mass, ω the oscillator frequency and A the nuclear mass number.

After this correction, $\varrho_{\text{c.m.}}$ has to be folded again with the internal density of the nucleon. It is here that form (3.27) for the nucleon density appears very useful, because a straightforward calculation shows that unfolding (3.32) plus finite-size correction is equivalent to a single finite-size folding, but with a reduced range parameter

$$a_k^2 \to a_k^2 - b^2/A\,. \tag{3.33}$$

We have used this correction, that holds for a harmonic-oscillator potential, for our Saxon-Woods proton densities; the approximation introduced here gives reasonable results for electron scattering up to $q_{\max} \approx 3\ \text{fm}^{-1}$ [32].

For each nucleus, we determined the parameter b by the condition that a simple harmonic-oscillator model of the nucleus (without configuration mixing) reproduces the measured r.m.s. charge radius.

3·2.4. Relativistic corrections. The high precision of the experimental charge densities makes it necessary to consider also the small ($\approx 1\%$) relativistic corrections. Among these, the oldest and best known are the ones to the electron-nucleus interaction which arise from reducing the nucleon's 4-component s.p. current to a 2-component (nonrelativistic) form. These have been described by DE FOREST and WALECKA [33] and worked out in detail for a $J=0$ nucleus by BERTOZZI *et al.* [29] and by CHANDRA and SAUER [28]. The corrections are of two types, called the « Darwin-Foldy term » and the « spin-orbit contribution ». Details of derivation are given in ref. [28, 29], we here only give the results. The Darwin-Foldy term can be combined with the Gaussian folding of the proton size and the c.m. motion by using eq. (3.27) with parameters (3.28) corrected as (see also eq. (3.33))

$$a_k^2 \to a_k^2 - b^2/A + 2.21\cdot 10^{-2}\ \text{fm}^2\,. \tag{3.34}$$

The spin-orbit correction results in an « effective » contribution to the charge density that can be written as in eq. (3.3) of ref. [34]:

$$\varrho_{\text{s.o.}}(r) = -\frac{1}{2}\left(\frac{\hbar}{mc}\right)^2 \frac{1}{r^2}\frac{\mathrm{d}}{\mathrm{d}r}\left[r\sum_{lj\tau}\langle n_{lj\tau}\rangle\mu_\tau\langle\boldsymbol{\sigma}\cdot\boldsymbol{l}\rangle_{lj}\varrho_{lj\tau}\right]. \tag{3.35}$$

The numerical importance of these corrections is shown for ^{48}Ca in fig. 3.2.

3˙2.5. Wave function orthogonalization. The formalism described in subsect. **3˙1** is correct only if the s.p. wave functions form a complete orthonormal set. Since we fit the SW potential to each s.p. energy in turn, we actually use different potentials and orthogonality is not ensured. For states with different l or j or m, orthogonality is ensured by the angular part of the wave functions, but within the same $\{ljm\}$ subspace it is not guaranteed.

Of course, it is possible to modify the formalism of subsect. **3˙1** so as to work with nonorthogonal wave functions, but in practice it is easier to perform an algebraic transformation on the wave functions of each $\{ljm\}$ subspace that makes them orthogonal. This change of basis is arbitrary, provided one transforms the occupation numbers accordingly. We choose to use the Gram-Schmidt method starting from the most bound level and following the order of decreasing binding energy. The details are given in subsect. **2˙2** of ref. [19] and the transformation of the occupation numbers is given by eq. (23) of that paper. This shows that, under the general conditions met by these calculations, all transformed occupation numbers are unity except possibly that of the least bound state in each subspace. In nearly all cases of practical interest, the $\{ljm\}$ subspaces contain two levels (the first exception in the periodic table is the 1, 2, $3s_{\frac{1}{2}}$ levels in ^{208}Pb), thus the transformation may be written as

$$\langle n_1\rangle = 1\,,\qquad \langle n_2\rangle = 1 - \frac{1-\langle\bar{n}_2\rangle}{|u_{22}|^2}\,, \tag{3.36}$$

where the bar in $\bar{n}_2$ means « before orthogonalization ».

By means of eq. (15) of ref. [19], $\langle n_2\rangle$ becomes

$$\langle n_2\rangle = \frac{\langle\bar{n}_2\rangle - |o_{12}|^2}{1-|o_{12}|^2} \approx \langle\bar{n}_2\rangle + |o_{12}|^2(\langle\bar{n}_2\rangle - 1) \tag{3.37}$$

to the 2nd order in o_{12}, the overlap between the two wave functions before orthogonalization. This overlap can be large (up to $\approx 20\,\%$) when the wave functions are first generated in potential (3.19), but it is then strongly reduced by the nonlocality correction (3.26). The reason is that the Perey factor corresponds to a unique nonlocal potential (hence orthogonal wave functions) and its local equivalent depends linearly on energy, that is approximately our case.

Typical values for o_{12} in our calculations after the Perey correction (3.26) range from 1 % to 6 % and, since $\langle \overline{n}_2 \rangle - 1$ is of order unity, from (3.37)

$$\langle n_2 \rangle \approx \langle \overline{n}_2 \rangle \tag{3.38}$$

to a very good approximation.

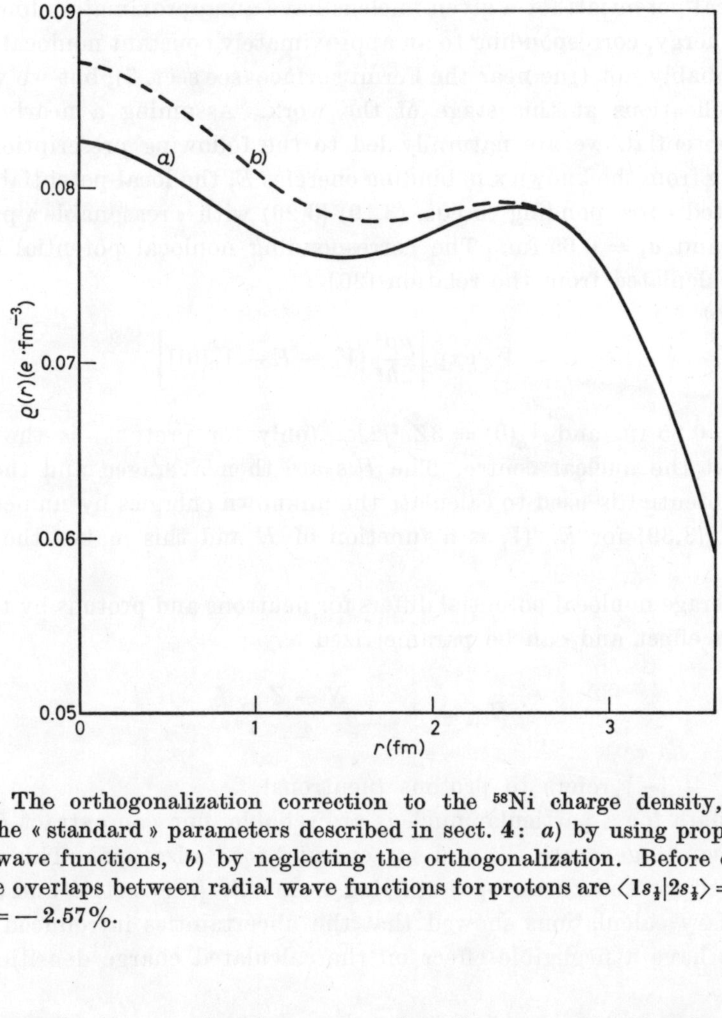

Fig. 3.3. – The orthogonalization correction to the ^{58}Ni charge density, calculated by using the « standard » parameters described in sect. 4: *a*) by using properly orthogonalized wave functions, *b*) by neglecting the orthogonalization. Before orthogonalization, the overlaps between radial wave functions for protons are $\langle 1s_{\frac{1}{2}}|2s_{\frac{1}{2}}\rangle = -7.05\%$, $\langle 1p_{\frac{3}{2}}|2p_{\frac{3}{2}}\rangle = -2.57\%$.

To summarize, our nonlocal s.p. wave functions are only slightly nonorthogonal with maximum overlap $\approx 6\%$; we orthogonalize them by the Gram-Schmidt method, that changes them linearly in the overlap; in principle, we should also change the occupation numbers defined by eq. (3.23), but the correction is quadratic in the overlap and can be neglected. The numerical effect of orthogonalization is shown for ^{58}Ni in fig. 3.3.

3.2.6. The energies of the deep states. The spectroscopic data needed for the application of eq. (3.21) exist only for states near the Fermi surface. For medium-weight nuclei (up to ^{58}Ni) one can make some use of (p, 2p) knock-out reactions, but thereafter (and for neutron states) experimental energies of the deep levels are unknown.

To evaluate the deep energies, we make use of their systematic behaviour, as discussed in sect. **5**.

The local potentials for a given nucleus have an approximately linear dependence on energy, corresponding to an approximately constant nonlocal potential. This is probably not true near the Fermi surface (see sect. **5**), but we will ignore such complications at this stage of the work. Assuming a nearly constant nonlocal potential, we are naturally led to the following prescription.

Starting from the known s.p. binding energies E, the local potential depths V_L are extracted corresponding to eqs. (3.19)-(3.20) with « reasonable » parameters $R = 1.25$ and $a_1 = 0.65$ fm. The corresponding nonlocal potential depths U are then calculated from the relation [26]

$$U = V_L \exp\left[\frac{\mu\beta^2}{2\hbar^2}\left[V_L - E - V_C(0)\right]\right], \tag{3.39}$$

where $\beta = 0.85$ fm and $V_C(0) = 3Ze^2/2R_C$ (only for protons) is the Coulomb potential at the nuclear centre. The U's are then averaged and the average nonlocal potential is used to calculate the unknown energies by numerical solution of eq. (3.39) for E. (V_L is a function of E and this makes the equation implicit.)

The average nonlocal potential differs for neutrons and protons by the known asymmetry effect and can be parametrized as

$$U_{p,n} = U_0 \pm \frac{N-Z}{A} U_1 , \tag{3.40}$$

where the $+$ ($-$) refers to protons (neutrons).

If the data for a particular nucleus are reliable, one can extract U_0 and U_1 and use them to generate U's and s.p. energies for its isotopes. This procedure was followed for the Zr isotopes starting from ^{90}Zr (see sect. **4** and ref. [35]).

Subsidiary calculations showed that the uncertainties introduced by these procedures have a negligible effect on the calculated charge densities.

4. – Charge distribution calculations.

4.1. *General remarks and exploratory calculations.* – The single-particle potential method outlined in the previous section has been applied to the calculations of the charge density distributions of ^{58}Ni [19], ^{39}K, ^{40,48}Ca [36] and ^{90,92,94}Zr [35] isotopes.

The relevant spectroscopic information about single-particle states of these nuclei is given in the tables of appendix A. Generally one can observe that the proton s.p. energies of the nickel and calcium isotopes are quite well known, while for the deeper states of the zirconium isotopes this information is lacking. Therefore, while in most cases the single-particle potential is fitted to the s.p. energies, for the latter states the procedure described in paragraph **3**˙2.6 has been adopted.

There are some uncertainties concerning the occupation number of valence states extracted from particle transfer reactions. Their relevance to charge density calculations will be discussed in the following. In all cases examined here the neutron contribution to charge densities is calculated once for all according to the method of paragraph **3**˙2.2.

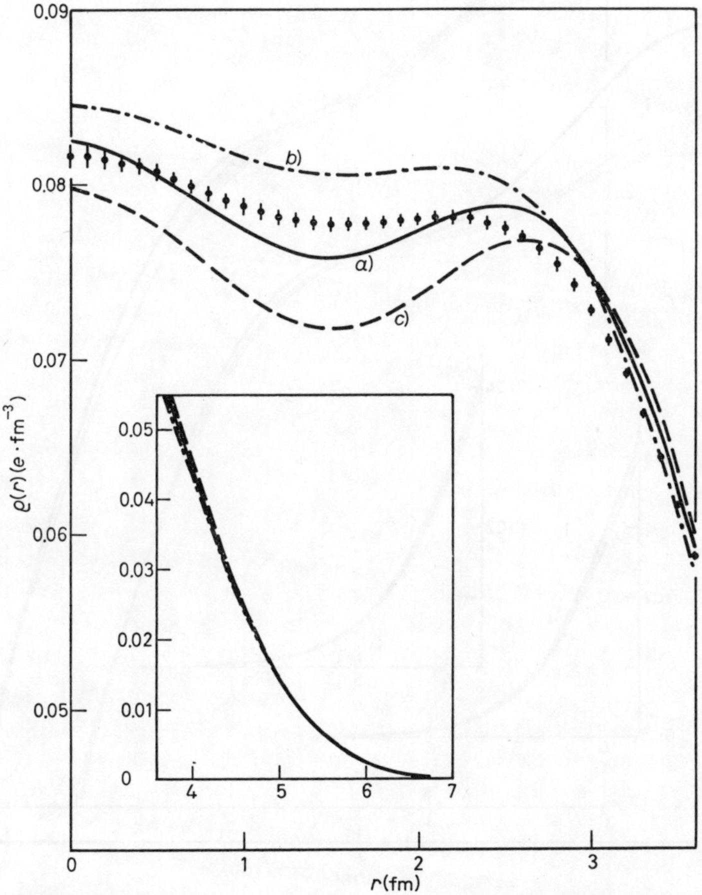

Fig. 4.1. – Charge density distribution of ^{58}Ni. The points show the model-independent charge density and the curves *a*) and *b*) correspond to standard calculations with $a_1 = 0.65$ fm and $n_p = 0.25$ and $n_p = 1.12$, respectively. The curve *c*) corresponds to $a_1 = 0.52$ fm and $n_p = 0.25$.

The potentialities of the present method of constructing charge densities are shown by fig. 4.1, 4.2 and 4.3, where a comparison is displayed between the model-independent charge densities extracted from electron scattering [37-39] and the results of the calculations corresponding to the following « standard » choices of the parameters of the SPP method.

The occupation numbers are fixed to the « experimental » values reported in the tables of appendix A. The nonlocality parameter and the central potential diffuseness are taken, respectively, as $\beta = 0.85$ fm and $a_1 = 0.65$ fm. The

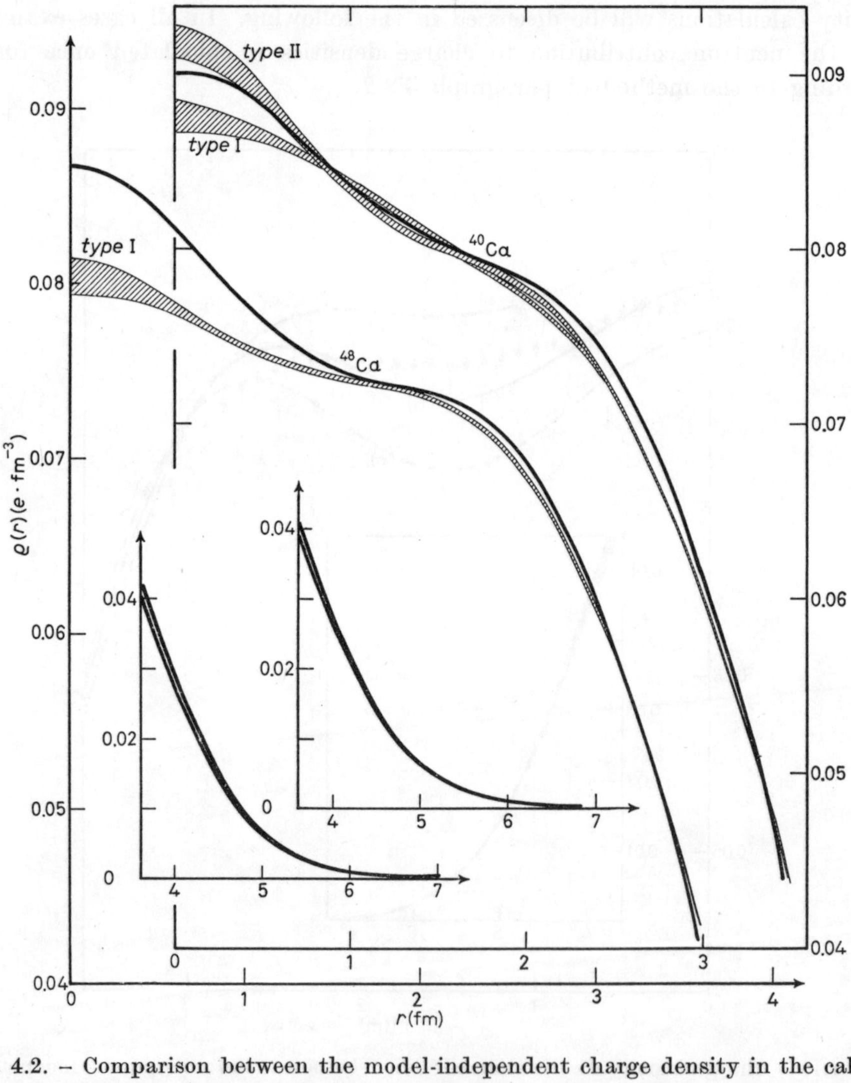

Fig. 4.2. – Comparison between the model-independent charge density in the calcium isotopes and the results of « standard » calculations as described in the text. For ^{40}Ca both the type-I and the type-II densities of ref. [37] are shown and for ^{48}Ca only the type-I density is shown.

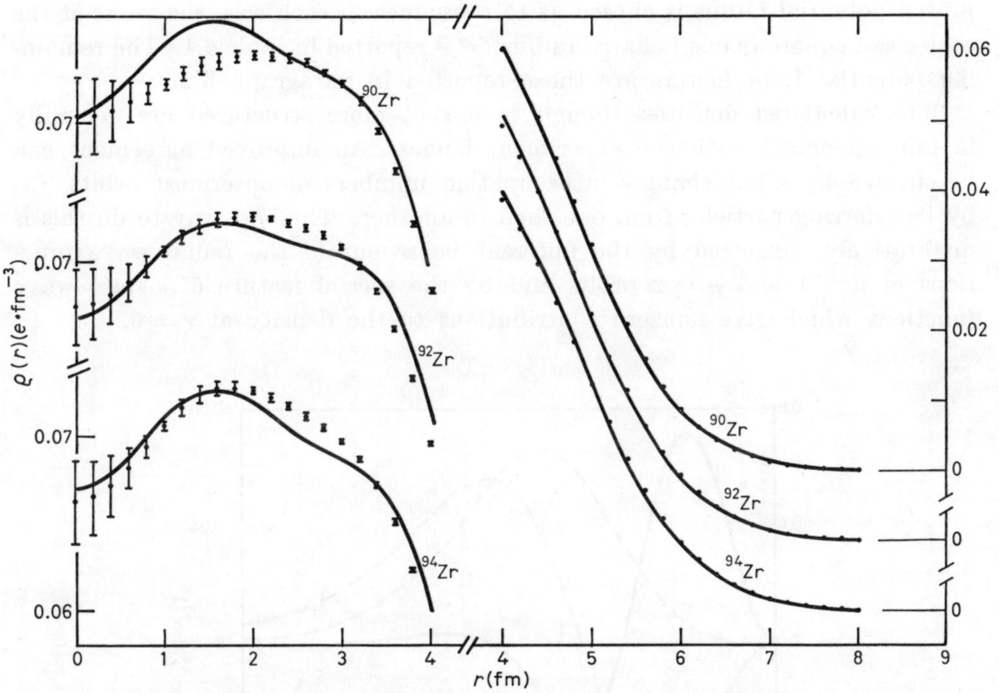

Fig. 4.3. – Comparison between the model-independent charge densities of the zirconium isotopes and the results of « standard » calculations (solid lines).

TABLE 4.1. – *Root mean square radius* $\langle r^2\rangle^{\frac{1}{2}}$ *of the model-independent nuclear charge density distributions.*

Nucleus	$\langle r^2\rangle^{\frac{1}{2}}$ (fm)
^{39}K	3.436 ([a])
^{40}Ca	3.475 ([b])
	3.472 ([a])
^{48}Ca	3.465 ([b])
^{58}Ni	3.775 ([c])
^{90}Zr	4.264 ([d])
^{92}Zr	4.296 ([d])
^{94}Zr	4.319 ([d])
^{208}Pb	5.495 ([e])

(*a*) From ref. [37] and I. SICK: private communication, type-II density.
(*b*) From ref. [37, 40] and I. SICK: private communication, type-I density.
(*c*) From ref. [38] and I. SICK: private communication.
(*d*) From ref. [39] and H. ROTHHAAS: private communication.
(*e*) From ref. [17] and I. SICK: private communication.

proton potential radius is chosen as to reproduce in each case the value of the root mean square (r.m.s.) charge radius $\langle r^2\rangle^{\frac{1}{2}}$ reported in table 4.1. The remaining potential form factors are those reported in paragraph **3**˙2.1.

The calculated densities though somewhat more structured are generally in fair agreement with the experimental ones. An improved agreement can be effected by slight changes in occupation numbers of outermost orbits, *i.e.* by transferring particles from one shell to another. The best way to do this is qualitatively suggested by the different behaviour of the radial wave functions of $n = 1$ and $n = 2$ orbits and by the special feature of s-state wave functions which give nonzero contributions to the density at $r = 0$.

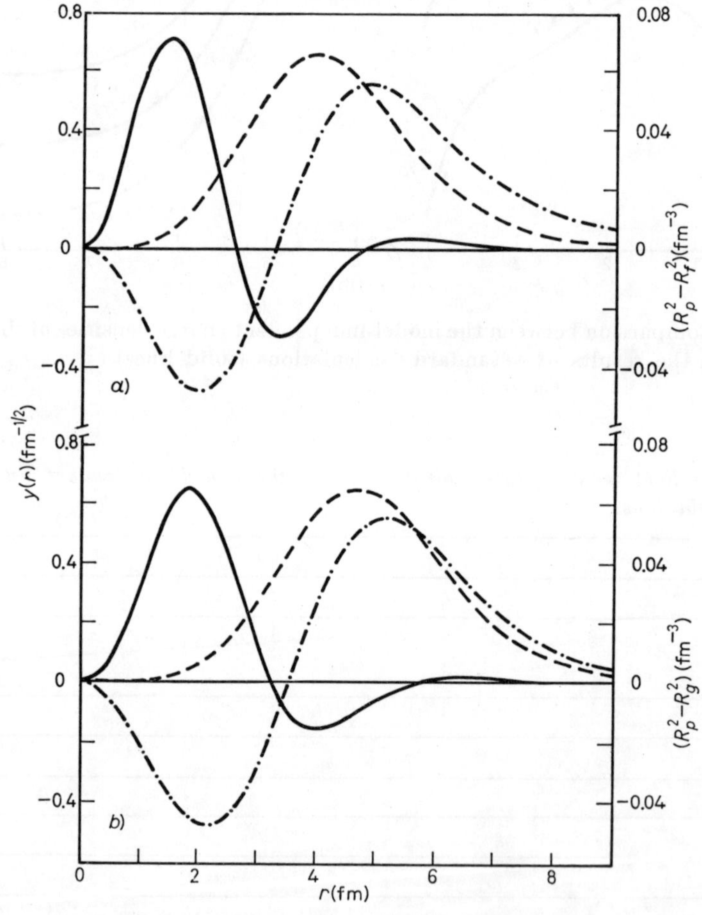

Fig. 4.4. – Comparison between the reduced radial wave functions $y(r)$ for *a*) $2p_{\frac{3}{2}}$ (—·—·—) and $1f_{\frac{7}{2}}$ (— — —) states in ^{58}Ni, *b*) $2p_{\frac{1}{2}}$ (—·—·—) and $1g_{\frac{9}{2}}$ (— — —) states in ^{90}Zr. The solid lines (right-hand scale) show the differences between the square of the radial wave function $R(r) = y(r)/r$ for *a*) the $2p_{\frac{3}{2}}$ and $1f_{\frac{7}{2}}$ states in ^{58}Ni, *b*) the $2p_{\frac{1}{2}}$ and $1g_{\frac{9}{2}}$ states in ^{90}Zr.

Figure 4.4 shows the difference between the squared moduli of *a*) $2p$-$1f$ wave functions in ^{58}Ni, *b*) $2p$-$1g$ wave functions in ^{90}Zr. This shows that a particle transfer from the $2p_{\frac{1}{2}}$ to the $1g_{\frac{9}{2}}$ orbit in ^{90}Zr causes a depression in the $(0 \div 2)$ fm region which smooths the prominent maximum at 1.5 fm, so that a better agreement may be found with the experimental density. In a similar way the large peak at $r = 0$ in ^{48}Ca can be substantially reduced by a reasonable reduction in the $2s_{\frac{1}{2}}$ occupation number.

The occupation numbers are not the only parameters which significantly affect the calculated densities. The results are generally sensitive to the diffuseness parameter of the central potential. In calculations constrained to give a fixed r.m.s. radius, a decrease of the diffuseness parameter causes a depression of the density in the interior region (compensated by a slight increase in the surface). This effect is shown for ^{58}Ni in fig. 4.1, where the curve *c*) corresponds to the choice $a_1 = 0.52$ fm, which was used in our first work [19], and to the value $n_p = 0.25$ obtained from the pick-up reaction for the $2p_{\frac{3}{2}}$ occupation number. By inspection of fig. 4.1 and fig. 4.4*a*), one can see the interplay between the choice of n_p and a_1 in the search for a better agreement between calculated and experimental densities in the case of ^{58}Ni. Thus, while with the standard value $a_1 = 0.65$ fm the pick-up value for n_p gives better results, the opposite conclusion would be reached for the lower value 0.52 fm for a_1 [19]. A similar situation was found in the case of ^{40}Ca [36] for the roles of a_1 and $n_{2p_{\frac{3}{2}}}$, while it does not occur in the Zr isotopes [35].

This point was carefully examined [36] and it was shown to be the main reason why the first shell model calculation by ELTON and WEBB [21] introduced a much larger configuration mixing than is suggested by experiments.

The above discussion indicates that, in order to find a soundly based way of improving the results of the SPP model and also to allow for experimental and theoretical uncertainties in the single-particle energies E_{nlj}, it is necessary to perform in each case a detailed study of the sensitivity of the charge densities to reasonable variations of these quantities and of the parameters of the model. These studies have shown that generally the results are almost insensitive to changes of E_{nlj} within the experimental uncertainties for the outermost states and within a larger range for the deeper states. The results are also very slightly affected by variations of the potential form factors R_2, a_2, which were then fixed to the standard values.

As expected, a stronger dependence is found on the occupation numbers of some valence states (different from nucleus to nucleus) and on the diffuseness parameter of the central potential. A fair, but weaker sensitivity to the variation of the nonlocality parameter β is also present, because the Perey factor reduces the nonlocal wave functions with respect to the local ones for small r and increases them in the tail region.

In the next subsections the various cases will be discussed in detail.

4.2. *Best-fit results.*

4.2.1. Fitting procedure. The sensitivity studies suggested for each case the proper choice of the parameters to be allowed to vary when fitting the model-independent charge densities. Subsequently, these parameter variations were carried out by the Minuits routine [41] which minimizes a χ^2-type function referring to the fit from $r = 0$ to some cut-off radius R_0, constrained by a penalty function [42] to reproduce the r.m.s. radius $\langle r^2 \rangle^{\frac{1}{2}}$ of the experimental density. This approaoch was chosen, since the $\langle r^2 \rangle^{\frac{1}{2}}$ constraint is by itself alone highly effective in determining the density behaviour in the tail region, while, owing to larger percentage errors, the interior region $0 \leqslant r \leqslant R_0$, where quantum fluctuations are present, has comparatively much less importance. The actual choice of R_0 was made so that the results are only slightly affected by a variation of about 1 fm.

4.2.2. The ^{58}Ni charge density. The nucleus ^{58}Ni is a good instance of the application of the SPP method as the mean separation energies of most of the states are experimentally determined and a model-independent charge density is available [38]. This nucleus is also challenging as many standard Hartree-Fock calculations [43] with different kinds of effective forces failed to reproduce the structure in the charge density between 0 and 2.5 fm.

In our first calculations [19] the a_1-dependence discussed in subsect. 4.1 was not analysed. Then, by taking $a_1 = 0.52$ fm, as suggested by DOVER and

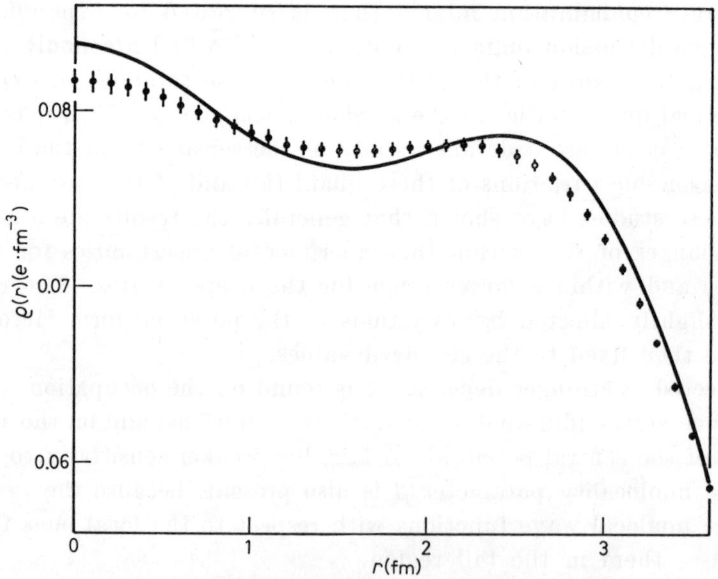

Fig. 4.5. – Comparison between the best-calculated charge density in ^{58}Ni and the model-independent experimental density from ref. [38].

Van Giai [44], a best-fit value $n_p = 1.33$ was found for the occupation number of the $2p_{\frac{3}{2}}$ state (the other fitting parameters were β, R and the centroid E_d of the d-states which has a large experimental uncertainty around 21 MeV). R_0 was taken to be 3 fm. On the other hand, if the uncertainties in the determination of n_p from stripping and pick-up reactions are properly taken into account, the pick-up value $n_p = 0.25$ must be considered more reliable.

Since a full parameter search varying both n_p and a_1 as well as those already listed is ambiguous, a simplified anlysis was then performed [36] where only a_1, β (and R) were treated as fitting parameters, while E_d and n_p were fixed at $E_d = 21.0$ MeV and $n_p = 0.25$, respectively. The best-fit density is shown in fig. 4.5. The corresponding best-fit values of the parameters are reported in table 4.2.

Table 4.2. – *Best-fit values of the* SPP *parameters for the* ^{58}Ni *charge distribution.*

a_1 (fm)	R (fm)	β (fm)
0.67	1.239	0.847

4.2.3. The charge densities of the calcium isotopes. Because of their importance for nuclear theory, the calcium isotopes have been studied extensively by electron elastic scattering. A model-independent combined analysis by Sick [40] of electron scattering cross-sections and muonic X-ray data has removed a previous ambiguity between two different charge density distributions (referred as type I and type II, respectively [37]) in favour of type-I density, whereas some years ago type-II density was preferred.

On the basis of the exploratory calculations described in subsect. 4.1 (compare fig. 4.2) and of sensitivity studies, we choose as fitting parameters the occupation numbers n_p, n_s of the $2p_{\frac{3}{2}}$ and $2s_{\frac{1}{2}}$ states, and the potential parameters a_1 and R. Figure 4.6*a*) shows the best results obtained with the parameter values reported in the first row of table 4.3 and $R_0 = 2.3$ fm [36].

A similar procedure was also followed for the isotone ^{39}K and for ^{48}Ca. The results are shown in fig. 4.6*b*), *c*) and in the second and third row of table 4.3, respectively [36].

Table 4.3. – *Best-fit values of the* SPP *parameters for the charge distributions of* ^{40}Ca, ^{39}K *and* ^{48}Ca.

Nucleus	a_1 (fm)	R (fm)	$n_{2p_{\frac{3}{2}}}$	$n_{2s_{\frac{1}{2}}}$
^{40}Ca	0.622	1.299	0.150	1.755
^{39}K	0.654	1.302	0.099	1.695
^{48}Ca	0.619	1.281	0.150	1.599

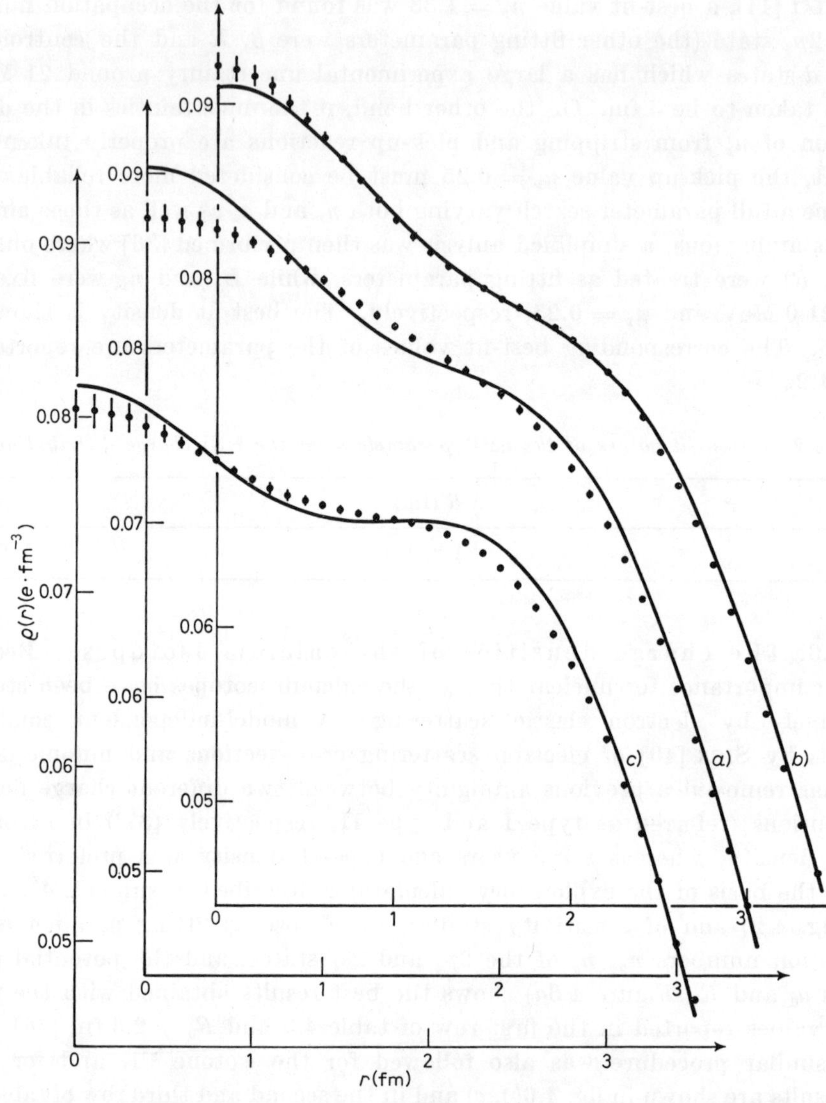

Fig. 4.6. – Comparison between the model-independent charge densities of ref. [37] and the results of best-fit calculations for *a*) ^{40}Ca, *b*) ^{39}K, *c*) ^{48}Ca.

It can be noticed that for ^{40}Ca there is a good agreement between the configuration mixing resulting from our best-fit calculations and that deduced from transfer reactions, in marked contrast with the result of an earlier analysis by Elton and Webb [21]. As already pointed out in subsect. 4·1, this discrepancy is mainly related to the interplay between n_p and a_1, which was fixed to 0.45 fm in ref. [21].

For ^{39}K and ^{48}Ca in our best-fit calculation we found about 0.1 and 0.15 particles, respectively, in the $2p_{\frac{3}{2}}$ orbit instead of zero. In particular, ^{39}K appears to be nearly pure $1d_{\frac{3}{2}}$ proton hole coupled to ^{40}Ca.

The problem of accounting for the charge density difference between the two isotopes ^{40}Ca and ^{48}Ca is a crucial test for any theoretical calculation. For instance, the most recent self-consistent calculations of Gogny and Grammaticos [45], while improving the fit to individual densities, are not able to reproduce the experimental results, even if RPA correlations are included. However, we find, as shown in fig. 4.7, that the difference between our fits to charge densities of ^{40}Ca and ^{48}Ca is in very good agreement with the experimental one, without any further parameter variation.

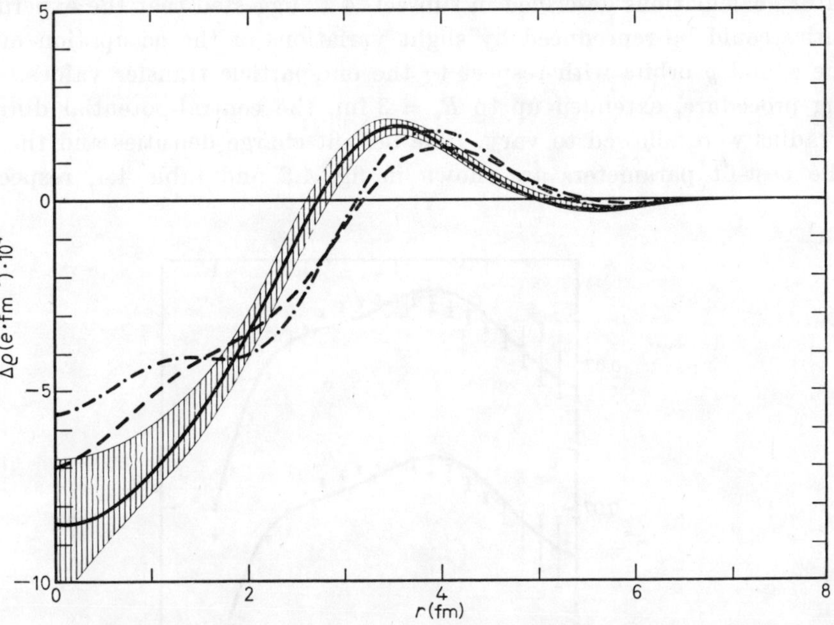

Fig. 4.7. – Comparison among the experimental charge density differences $\Delta\varrho$ between ^{48}Ca and ^{40}Ca [37, 40] and the results of different types of calculations as discussed in the text: ▥ experiment, ――― HF, —·—·— HF+RPA, ―― present work.

In these calculations the neutron contributions have been evaluated following the average nonlocal prescription of paragraph 3˙2.6 and assuming for the equivalent local potential radial parameters the values $R = 1.319$ fm and $R = 1.306$ fm for ^{40}Ca and ^{48}Ca, respectively (according to the suggestion of Dover and Van Giai [44]).

The calculations have been extended to ^{42}Ca and ^{44}Ca and the results are reported in a previous work [36]. A comparison was also made between the calculated charge density differences and those derived from electron scattering by FROSCH *et al* [46]. As this experiment was limited in momentum transfer range up to $q_{max} = 2.3\ fm^{-1}$ and the analysis was not model-independent, the fair quality of the agreement we have found is not very significant.

4˙2.4. **The charge densities of the zirconium isotopes.** A recent Mainz electron scattering experiment gave the charge densities of the even isotopes ^{90}Zr, ^{92}Zr and ^{94}Zr in a model-independent way [39]. Also the changes in the density due to successive addition of neutron pairs have been independently determined from the electron scattering cross-section ratios for the isotopic pairs (^{92}Zr, ^{90}Zr) and (^{94}Zr, ^{92}Zr) [39, 47].

The SPP method was, therefore, applied also to these cases [35]. The exploratory calculations described in subsect. 4˙1 suggested that the experimental densities could be reproduced by slight variations of the occupation numbers of the *p* and *g* orbits with respect to the one-particle transfer values. In the fitting procedure, extended up to $R_0 = 3$ fm, the central-potential diffuseness and radius were allowed to vary. The best-fit charge densities and the values of the best-fit parameters are shown in fig. 4.8 and table 4.4, respectively.

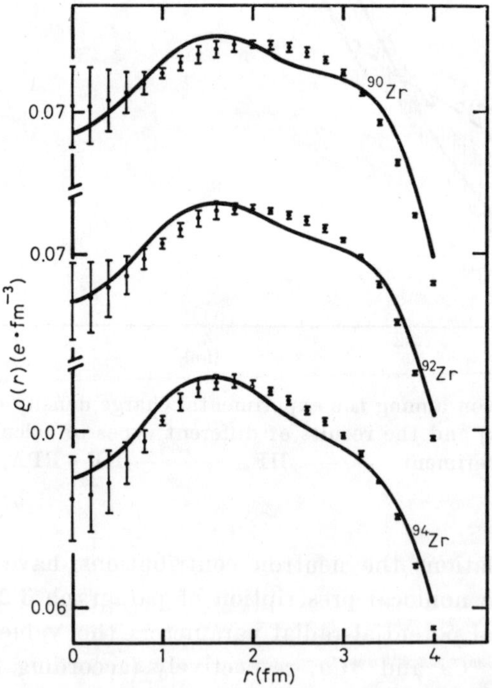

Fig. 4.8. – Comparison between the experimental charge density fo zirconium isotopes and the best-fit calculations. The tail regions are not shown as the fits there are essentially the same as in fig. 4.3.

TABLE 4.4. – *Best-fit values of the* SPP *parameters for the charge distribution of the zirconium isotopes.*

Nucleus	a_1 (fm)	R (fm)	$n_{2p_{\frac{1}{2}}}$
^{90}Zr	0.615	1.261	0.97
^{92}Zr	0.653	1.270	1.12
^{94}Zr	0.665	1.277	1.42

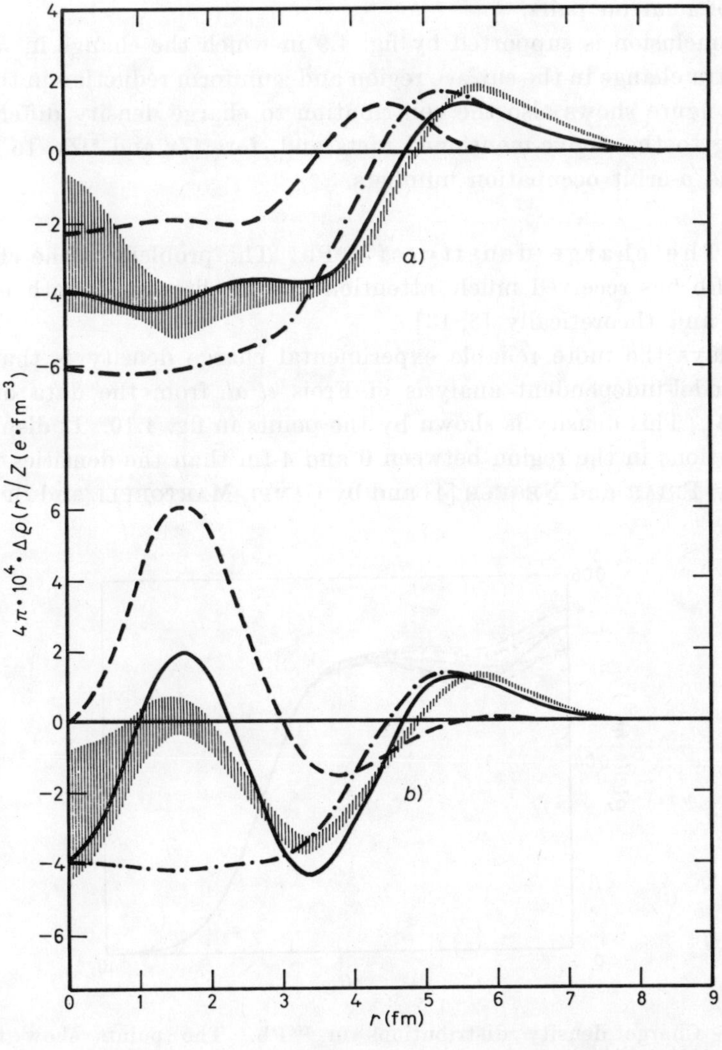

Fig. 4.9. – The contributions to the charge density differences in zirconium isotopes. The dash-dotted curves show the effect of the radial expansion. The dashed curves show the effect of the difference in best-fit diffuseness for ^{90}Zr-^{92}Zr (a)) and the effect of a 0.3 particle transfer from the 1g to the 2p orbit for ^{94}Zr-^{92}Zr (b)). The cumulative effects are shown by the solid lines.

The occupation number of the p-states turned out to be $\simeq 4.6$ both for ^{90}Zr and ^{92}Zr and to increase by about 0.3 from ^{92}Zr to ^{94}Zr, while the stripping reactions give 4.8, 4.4, 4.8, respectively, for ^{90}Zr, ^{92}Zr, ^{94}Zr.

The values for the radius and diffuseness of the central potential were found to increase along the isotopic sequence. This fact may be connected with the macroscopic effects pointed out by Rothhaas [47], the change in diffuseness corresponding to the change in the intrinsic deformation and the increase of the potential radius to the macroscopic core polarization effects following the addition of neutron pairs.

This conclusion is supported by fig. 4.9 in which the change in R is shown to give extra change in the surface region and a uniform reduction in the interior. The same figure shows also the contribution to charge density differences corresponding to the above-mentioned facts and, for ^{94}Zr and ^{92}Zr, to the difference in the p-orbit occupation numbers.

4.2.5. The charge density of ^{208}Pb. The problem of the charge density of ^{208}Pb has received much attention in the literature both experimentally [16] and theoretically [3, 13].

Nowadays the more reliable experimental charge density is that deduced by the model-independent analysis of Frois *et al.* from the data of different groups [16]. This density is shown by the points in fig. 4.10. It displays much less oscillations in the region between 0 and 4 fm than the densities previously derived by Friar and Negele [4] and by Campi, Martorell and Sprung [48].

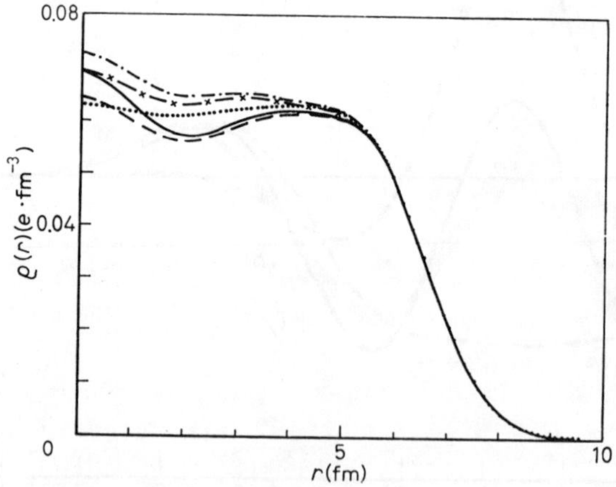

Fig. 4.10. – Charge density distributions in ^{208}Pb. The points show the model-independent density of ref. [16]. The RPA (HF) curves correspond to the Hartree-Fock calculations of ref. [45] with (without) RPA corrections. The SPP curves show the density calculated by the present method with *a*) $n_{3s_{\frac{1}{2}}} = 2$, *b*) $n_{3s_{\frac{1}{2}}} = 1.75$, $n_{1h_{\frac{9}{2}}} = 0.25$ (see the text). ··· experiment, —×—×— RPA, —·—·— HF, —— SPP (*a*)), – – – SPP (*b*)).

This distinctive feature is quite difficult to account for in any single-particle calculation. In fact, as shown in fig. 4.10, HF calculations are not able to reproduce the density, even if some correlations are included in a RPA approach [13, 45]. The possibility of applying the present method to this problem was investigated, but the results are not so good as in the cases previously reviewed.

In this nucleus the knowledge of the s.p. state energies is limited to the surface states down to ~ 15 MeV binding energy, so that for the deeper states the procedure described in paragraph 3˙2.6 was adopted in the calculations. The experimental data do not exclude the possibility that the $3s_{\frac{1}{2}}$ state may be partly unoccupied [49]. This allows us to reduce the height of the $r = 0$ peak in the calculated density which is due to s-states.

In fig. 4.10, curve *a*) represents the results of « standard » calculations with $n_{3s_{\frac{1}{2}}} = 2$, while curve *b*) corresponds to a transfer of 0.25 particles from the $3s_{\frac{1}{2}}$ to the $1h_{\frac{9}{2}}$ state. If one compares the quality of the results of these exploratory calculations with the corresponding ones in the other cases examined in this paper (see fig. 4.1-4.3) one can hardly expect to achieve a good fit to the experimental density of ^{208}Pb by subsequent parameter variations. For this reason no best-fit search was attempted.

5. – The systematics of nuclear single-particle states.

5˙1. *Energy dependence of the single-particle potentials.* – In the case of the heavier nuclei, the energies of the deeper s.p. states cannot be determined experimentally, so it is necessary to have a way of calculating them. We, therefore, seek a potential that gives a good overall fit to the known s.p. energies, so that we can use it with some confidence to calculate the energies that are not known.

Many detailed analyses [22, 23, 50, 51] have been made of the systematic behaviour of the s.p. energies near the Fermi surface, whereas the deep states have only been studied in global analyses [52] including all bound states.

The local potentials corresponding to the deep states are strongly energy dependent and this feature has often been put in the form of a nonlocality. Several nonlocal potentials that give approximately the experimental energies of the bound states have been found.

In order to find the best overall potential, the experimental s.p. energies of the nuclei ^{12}C, ^{16}O, ^{40}Ca, ^{48}Ca, ^{58}Ni, ^{90}Zr and ^{208}Pb were analysed. The first analysis was made in terms of a nonlocal potential with depth depending on the asymmetry parameter as in eq. (3.40):

$$U^{p,n}(\text{nonlocal}) = U_0 \pm \frac{N-Z}{A} U_1 . \tag{5.1}$$

This asymmetry term is a convenient way of combining proton and neutron potentials and, provided the neutron and proton form factors are the same, it may be identified as an isospin potential.

The nonlocal potential depth is connected with the equivalent local depth by relation (3.39), *viz.*

$$U = V_L \exp\left[\frac{\mu\beta^2}{2\hbar^2}\,[V_L - E - V_C(0)]\right], \tag{5.2}$$

where U, $V_L > 0$ and $E > 0$ for bound states. The form factor parameters were fixed to the values $r_0 = 1.236$ and $a = 0.62$ fm, and the spin-orbit potential set to an average value of 7 MeV. A subsidiary study of the deep states of ^{40}Ca, ^{48}Ca and ^{58}Ni showed that the best overall fit to their energy variation is given by $\beta \simeq 0.9$ fm.

By means of relation (5.2), the local potentials fitted to the experimental centroid energies were used to obtain U_0 and U_1. The values obtained varied markedly from one nucleus to another, and the U_1's were unexpectedly high. Inspection of the actual values of the local and nonlocal potentials (table 5.1) showed the reason: the proton local potentials are energy dependent (apart from ^{208}Pb), whereas the neutron local potentials show no significant energy dependence. More precisely, the local potentials corresponding to states near the Fermi surface are independent of energy, while those corresponding to the deep states (which are known only for protons) depend on energy.

TABLE 5.1. – *Local and equivalent nonlocal potentials for states in ^{58}Ni and ^{90}Zr.*

Nucleon	^{58}Ni			^{90}Zr		
	State	Local	Nonlocal	State	Local	Nonlocal
proton	$1f_{\frac{7}{2}}$	56.2	85.9	$1d_{\frac{3}{2}}$	67.3	88.5
	$2s_{\frac{1}{2}}$	54.0	74.4	$1d_{\frac{5}{2}}$	63.4	80.3
	$1d_{\frac{3}{2}}$	57.7	80.1	$1p_{\frac{1}{2}}$	74.6	90.1
	$1d_{\frac{5}{2}}$	63.1	84.5	$1p_{\frac{3}{2}}$	72.9	86.6
	$1p_{\frac{1}{2}}$	72.1	91.3	$1s_{\frac{1}{2}}$	77.7	86.7
	$1p_{\frac{3}{2}}$	74.0	91.6			
	$1s_{\frac{1}{2}}$	80.8	92.6			
neutron	$2p_{\frac{1}{2}}$	50.2	69.6	$1g_{\frac{7}{2}}$	50.0	69.1
	$2p_{\frac{3}{2}}$	50.3	68.6	$2d_{\frac{3}{2}}$	51.2	71.3
	$1f_{\frac{5}{2}}$	52.6	73.6	$3s_{\frac{1}{2}}$	51.3	71.0
	$1f_{\frac{7}{2}}$	51.6	67.4	$2d_{\frac{5}{2}}$	51.1	69.5
				$1g_{\frac{9}{2}}$	50.4	65.0
				$2p_{\frac{1}{2}}$	48.1	60.2

This effect, of course, has been known since a long time and is usually described by an increase of effective nucleon mass across the Fermi surface [53]. It is thus necessary to distinguish in the analysis between the « surface » states for which the local potential is nearly independent of energy and the « deep » states for which the local potential has a strong, and approximately linear, energy dependence. It is clear that the distinction between these two types of states is not a sharp one, but nothing is known about the transition region, so a gradual change from one to the other would be devoid of theoretical basis and would unnecessarily increase the number of parameters. It was, therefore, decided to make a sharp though unphysical separation between the two classes of states. The energy at which this transition takes place is not critical or well defined by the data. Certainly it is between 10 MeV and 20 MeV, thus we fixed the boundary at 15 MeV. Subsequently we found that this is just the energy obtained in a theoretical calculation by BERTSCH and KUO of the energy variation of effective mass due to core polarization [54]. More recently, this effect has been studied by LEJEUNE [55] and by GIANNINI *et al.* [56]. We thus decided to abandon, at least temporarily, the search for an optimum nonlocal potential and instead to seek the best local-potential depths of the forms for surface states

$$V_L = V_0^{p,n} \qquad \text{for } 0 < E < 15 \text{ MeV} \tag{5.3}$$

and for deep states

$$V_L = V_0^{p,n} + \alpha(E - 15) \qquad \text{for } E > 15 \text{ MeV}, \tag{5.4}$$

where the superscripts refer to proton and neutron states, respectively, and $E > 0$ for bound states.

As before, we expressed the values of V_0 for neutrons and protons in terms of a nuclear asymmetry potential V_1

$$V_0^{p,n} = V_0 \pm \frac{N-Z}{A} V_1 . \tag{5.5}$$

The usual $0.4\, ZA^{-\frac{1}{3}}$ term [57] was not included in the expression for V_0^p, because V_0 does not depend on energy for the surface states and there is in any case some theoretical uncertainty concerning this term [58]. The energy dependence of the deep states would give rise to a $ZA^{-\frac{1}{3}}$ term, but this was not included, because the energies of these states are not known with sufficient accuracy to justify this. Correspondingly, the Coulomb energy E_C for protons was also neglected at this stage of the work (ideally, one should use $E + E_C$ instead of E in eq. (5.4)).

To find the parameters of the best overall potential, the local potentials corresponding to the s.p. states of the nuclei mentioned were fitted to eqs. (5.3) to (5.5), and the values for $V_0^{p,n}$, V_0, V_1 and α are given in table 5.2. The parameters V_0, V_1 and α are essentially the same for all nuclei within the statistical uncertainties, and we adopt the average values

$$V_0 = 55.3 \text{ MeV}, \quad V_1 = 39.3 \text{ MeV}, \quad \alpha = 0.51 . \tag{5.6}$$

The fitted lines are compared with the potentials in fig. 5.1 showing the overall validity of the potential for the deep states [59]. The results for ^{12}C and ^{16}O are not included as they have a large scatter.

TABLE 5.2. – *Local-potential depths and their energy dependence.*

Nucleus	V_0^p	V_0^n	V_0	V_1	α
^{12}C	56.3	57.0	56.7	—	—
^{16}O	56.4	55.5	56.0	—	—
^{40}Ca	56.1	55.8	56.0	—	0.61
^{48}Ca	61.2	48.5	54.9	38.1	0.32
^{58}Ni	56.0	53.2	54.6	40.0	0.64
^{90}Zr	59.1	49.6	54.4	42.8	0.47
^{208}Pb	62.1	46.8	54.5	36.2	—
average values			55.3	39.3	0.51

For the purpose of calculating nuclear charge densities, the experimental energies of the surface states are used wherever available and, if not, they are taken from the expressions describing their systematic behaviour [22, 23, 50, 51]. The energies of the deep states, if not known experimentally, may be obtained either from the energy-dependent local potential described here or from the energy-independent nonlocal potential described in paragraph **3**'2.6.

From a practical point of view, the difference between the energies calculated in these two ways does not significantly affect the charge densities.

5'2. *The nonlocality correction.* – The energy dependence of the local s.p. potential just described raises some problems about the consistency of the wave functions used in this work.

It is known from very general considerations [60] that the effective nucleon-nucleus potential is partly local and partly nonlocal, so that in the s.p. Schrödinger equation one has to make the substitution

$$V(\boldsymbol{r})\varphi(\boldsymbol{r}) \to \int d\boldsymbol{r}'\, V(\boldsymbol{r}, \boldsymbol{r}')\varphi(\boldsymbol{r}') , \tag{5.7}$$

where the nonlocal potential $V(\boldsymbol{r}, \boldsymbol{r}')$ probably contains a local part proportional to $\delta(\boldsymbol{r} - \boldsymbol{r}')$.

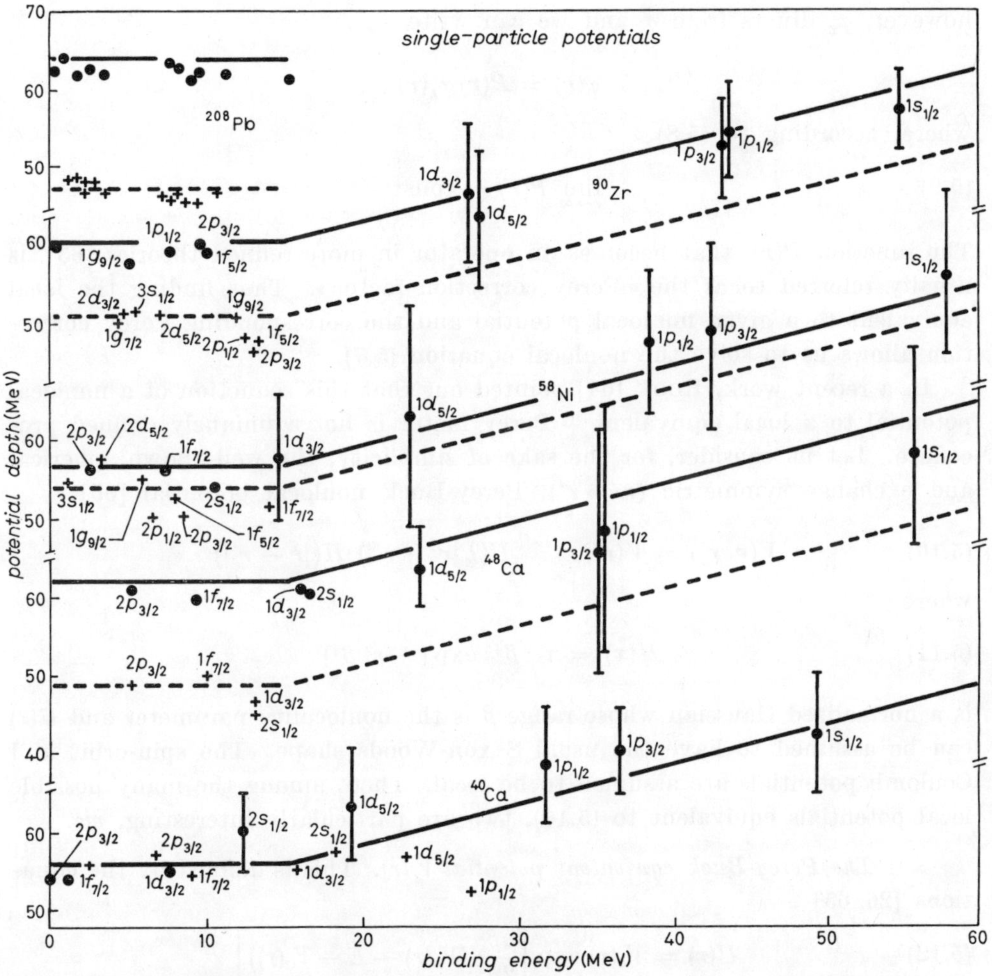

Fig. 5.1. – Nuclear-potential depths for single-particle states of ^{40}Ca, ^{48}Ca, ^{58}Ni, ^{90}Zr and ^{208}Pb compared with the best-fit lines, eqs. (5.3), (5.4), (5.6). The values for the 1*s*, 1*p* and 1*d* states of ^{90}Zr are from Vautherin and Brink [61] obtained by interpolation from James *et al.* [62]. • ——— proton, + — — — neutron.

Instead of including a potential of the form (5.7), the Schrödinger equation is often written in terms of a « local equivalent potential », *i.e.* a potential that, if inserted in a local Schrödinger equation, gives a wave function $\varphi_L(\boldsymbol{r})$ having the same asymptotic behaviour as $\varphi(\boldsymbol{r})$

(5.8) $$\varphi_L(\boldsymbol{r}) \sim \varphi(\boldsymbol{r}) , \qquad \text{as } r \to \infty .$$

This ensures that the local equation gives the same binding energies for bound states and the same phase shifts for scattering states. For finite values of r, however, φ_L differs from φ and we can write

$$\varphi(\boldsymbol{r}) = P(\boldsymbol{r})\,\varphi_L(\boldsymbol{r})\,,$$

where, according to (5.8),

$$\lim_{r\to\infty} P(\boldsymbol{r}) = \text{const}\,. \tag{5.9}$$

The function $P(\boldsymbol{r})$, that becomes an operator in more refined theories [63], is usually referred to as the « Perey correction factor ». Thus finding the local equivalent to a given nonlocal potential and the corresponding Perey correction allows us to solve the nonlocal equation (5.7).

In a recent work, ROOK [64] pointed out that this reduction of a nonlocal potential to a local equivalent + Perey factor is not a uniquely defined procedure. Let us consider, for the sake of simplicity, the well-known spherical and exchange symmetric ($\boldsymbol{r}\leftrightarrow\boldsymbol{r}'$), Perey-Buck nonlocal potential [65]

$$V(\boldsymbol{r},\boldsymbol{r}') = V(\boldsymbol{r}',\boldsymbol{r}) = U\big(\tfrac{1}{2}\,|\boldsymbol{r}+\boldsymbol{r}'|\big)\cdot H\big(|\boldsymbol{r}-\boldsymbol{r}'|\big)\,, \tag{5.10}$$

where

$$H(x) = \pi^{-\frac{3}{2}}\beta^{-3}\exp\left[-x^2/\beta^2\right] \tag{5.11}$$

is a normalized Gaussian whose range β is the nonlocality parameter and $U(r)$ can be assumed to have the usual Saxon-Woods shape. The spin-orbit and Coulomb potentials are assumed to be local. Then, among the many possible local potentials equivalent to (5.10), two are particularly interesting, *viz.*

i) *The Perey-Buck equivalent potential* $V_L(r)$. This is defined by the equations [26, 65]

$$U(r) = V_L(r)\exp\left[\frac{\mu\beta^2}{2\hbar^2}\left(V_L(r)-E-V_C(r)\right)\right] \tag{5.12}$$

(see also eqs. (5.2), (3.19) and (3.39)) and

$$P_L(r) = C\left[1+\frac{\mu\beta^2}{2\hbar^2}\,V_L f_1(r)\right]^{-\frac{1}{2}} \tag{5.13}$$

(see also eq. (3.26)), where C is a normalization constant. If U does not depend explicitly on energy, then numerical solution of eq. (5.12) gives a local potential depth $V_L(0)$, that is, to a good approximation, a linear function of E with slope α depending on the nonlocality range β. Useful corresponding values are

$$\begin{cases} \beta = 0 \quad\ \text{fm} \Leftrightarrow \alpha = 0\,, \\ \beta = 0.85\ \text{fm} \Leftrightarrow \alpha \approx 0.38\,, \\ \beta \approx 1.05\ \text{fm} \Leftrightarrow \alpha = 0.51\,. \end{cases} \tag{5.14}$$

One should remember that the Perey-Buck equivalent potential and correction factor are only approximations, though accurate ones [66], to the original nonlocal equation.

ii) *The trivially equivalent potential* $V_T(r)$. This is defined by the equations

$$V_T(\boldsymbol{r})\,\varphi(\boldsymbol{r}) = \int d\boldsymbol{r}'\, V(\boldsymbol{r}, \boldsymbol{r}')\,\varphi(\boldsymbol{r}') \tag{5.15}$$

and

$$P_T(\boldsymbol{r}) = 1 \qquad \text{(no Perey correction).} \tag{5.16}$$

Clearly, V_T diverges at the zeroes of φ, but this is not a serious difficulty [64]. Moreover, the structure of V_T is complicated by an explicit dependence on the orbital angular momentum l introduced by its definition. In a phenomenological calculation where one fits the local equivalent potential to the experimental data, it is not clear what kind of potential has been chosen, *e.g.* if V_L or V_T, and what is the corresponding Perey factor. Expression (5.13) is justified only for the Perey-Buck potential (5.12), that is a linear function of energy, in contradiction with the results of subsect. **5**˙1. Therefore, the use of (5.13) seems hardly justified.

Another point to be considered here concerns the orthogonality of the wave functions. If they were properly constructed in a unique, energy-independent nonlocal potential, they would be orthogonal and the Gram-Schmidt correction (paragraph **3**˙2.5) unnecessary. Therefore, small—or possibly vanishing—overlaps are a necessary condition for the nonlocal wave functions to be constructed in a consistent way.

All this leads us to three possible extreme prescriptions for the Perey correction, none of which is probably realistic:

a) Use no Perey correction, as though the empirical potential were the trivially equivalent one (5.15).

b) Use the Perey factor (5.13) with the same value for β for all states, as though the empirical potential were the Perey-Buck one. The fit to the charge densities of many nuclei (sect. **4**) suggests $\beta = 0.85$ fm.

c) Use eq. (5.13) with $\beta = 1.05$ fm for deep states ($E > 15$ MeV) and local wave functions for surface states, as suggested by subsect. **5**˙1 and eq. (5.14).

If one calculates the overlaps between our radial wave functions having the same l, j and different n-values, one finds that prescription *a*) gives large overlaps ($(5\div10)\,\%$), while *b*) is much better ($(1\div3)\,\%$) and *c*) is the best ($\approx 1\,\%$).

Prescriptions *a*), *b*) and *c*) have also been applied to the calculation of the ^{40}Ca and ^{208}Pb charge densities, and the results compared with experiment in fig. 5.2. From the figure one can conclude that the numerical results appear

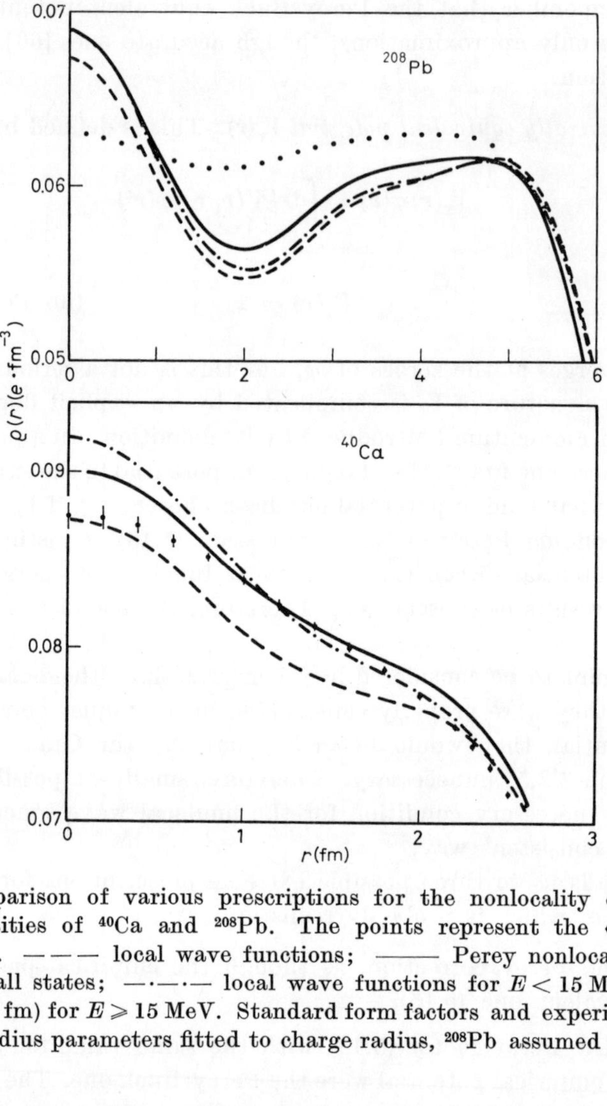

Fig. 5.2. – Comparison of various prescriptions for the nonlocality correction with the charge densities of ^{40}Ca and ^{208}Pb. The points represent the « experimental » densities [16, 40]. ——— local wave functions; ——— Perey nonlocality correction, $\beta = 0.85$ fm for all states; —·—·— local wave functions for $E < 15$ MeV, Perey correction ($\beta = 1.05$ fm) for $E \geqslant 15$ MeV. Standard form factors and experimental occupation numbers, radius parameters fitted to charge radius, ^{208}Pb assumed to be « closed ».

quite similar and none of them substantially improves the fit to experiment. Prescription *a*) appears again the worst, particularly for ^{40}Ca. Though the subject certainly deserves further study, the results for the overlaps between wave functions and for the ^{40}Ca and ^{208}Pb densities show that prescription *b*), in spite of its simplicity and the ambiguities connected with the use of eq. (5.13), seems to be sufficiently accurate for the present purposes.

As discussed by Rook [64], this might be because our phenomenological potential, being *l*-independent and giving rise to small overlaps if using prescription *b*),is probably nearer to the Perey-Buck equivalent potential than to the trivially equivalent one.

6. – Natural orbitals in ^{40}Ca.

As discussed in paragraph **3**·2.1, the wave functions used in this work are the best Saxon-Woods approximations to natural orbitals in a model space restricted to the bound states for which one-nucleon transfer reactions show a significant occupancy.

It is now interesting to go beyond this approximation and, while still keeping for simplicity this restricted model space, explicitly construct natural orbitals by mixing SW functions according to eq. (3.7).

For a $J = 0$ nucleus eq. (3.14) shows that the natural orbitals are simply admixtures of s.p. wave functions having the same l, j, m and different principal quantum numbers n. For nearly all cases of practical interest, there are at most two bound s.p. wave functions with the same l, j, m (the first exception in the periodic table are the 1, 2, $3s_{\frac{1}{2}}$ states in ^{208}Pb), so the unitary transformation leading from the best SW radial functions $y(r)$ to the natural radial functions $\tilde{y}(r)$ can be seen as a rotation in the (1, 2) plane, *viz.*

$$\begin{cases} \tilde{y}_1(r) = \cos\chi y_1(r) + \sin\chi y_2(r)\,, \\ \tilde{y}_2(r) = -\sin\chi y_1(r) + \cos\chi y_2(r)\,. \end{cases} \tag{6.1}$$

Here χ is the appropriate mixing parameter and the indices l, j, m will be dropped for simplicity whenever possible. The use of bound states ensures that (6.1) involves only real numbers. Therefore, according to eq. (3.18), the lj contribution to the density becomes

$$\begin{aligned} \varrho_{lj}(r) &= \frac{2j+1}{4\pi r^2}\,[\langle\tilde{n}_1\rangle\,\tilde{y}_1^2(r) + \langle\tilde{n}_2\rangle\,\tilde{y}_2^2(r)] = \\ &= \frac{2j+1}{4\pi r^2}\,[\langle n_1\rangle\,y_1^2(r) + \langle n_2\rangle\,y_2^2(r) + \varepsilon y_1(r)\,y_2(r)]\,, \end{aligned} \tag{6.2}$$

where

$$\begin{cases} \langle n_1\rangle = \langle\tilde{n}_1\rangle\cos^2\chi + \langle\tilde{n}_2\rangle\sin^2\chi\,, \\ \langle n_2\rangle = \langle\tilde{n}_1\rangle\sin^2\chi + \langle\tilde{n}_2\rangle\cos^2\chi\,, \\ \varepsilon = [\langle\tilde{n}_1\rangle - \langle\tilde{n}_2\rangle]\sin 2\chi\,. \end{cases} \tag{6.3}$$

If χ is small, to first order $\langle n_1\rangle \approx \langle\tilde{n}_1\rangle$, $\langle n_2\rangle \approx \langle\tilde{n}_2\rangle$, and the unique correction to the equations used up to now is the « cross-term » $\varepsilon y_1 y_2$.

Similar equations, but involving three harmonic-oscillator wave functions, have been proposed by Boffi and Pacati [67].

We have applied the above expressions to ^{40}Ca, using the « standard » model functions $y(r)$ described in subsect. **4**·1. The corresponding « diagonal » (with no cross-terms) density (dashed line) is compared with experiment in fig. 6.1. If we insist on considering only bound single-particle states, the sole $\{lj\}$ subspaces containing 2 n-values are the $\{s_{\frac{1}{2}}\}$ and the $\{p_{\frac{3}{2}}\}$ ones (forcing $2p_{\frac{3}{2}}$, slightly unbound—see the appendix—to be bound). Thus the only two possible cross-terms are those proportional to $y_{1s_{\frac{1}{2}}} y_{2s_{\frac{1}{2}}}$ and to $y_{1p_{\frac{3}{2}}} y_{2p_{\frac{3}{2}}}$. These are also displayed in fig. 6.1; none of the corrections described in paragraphs 3˙2.2 to

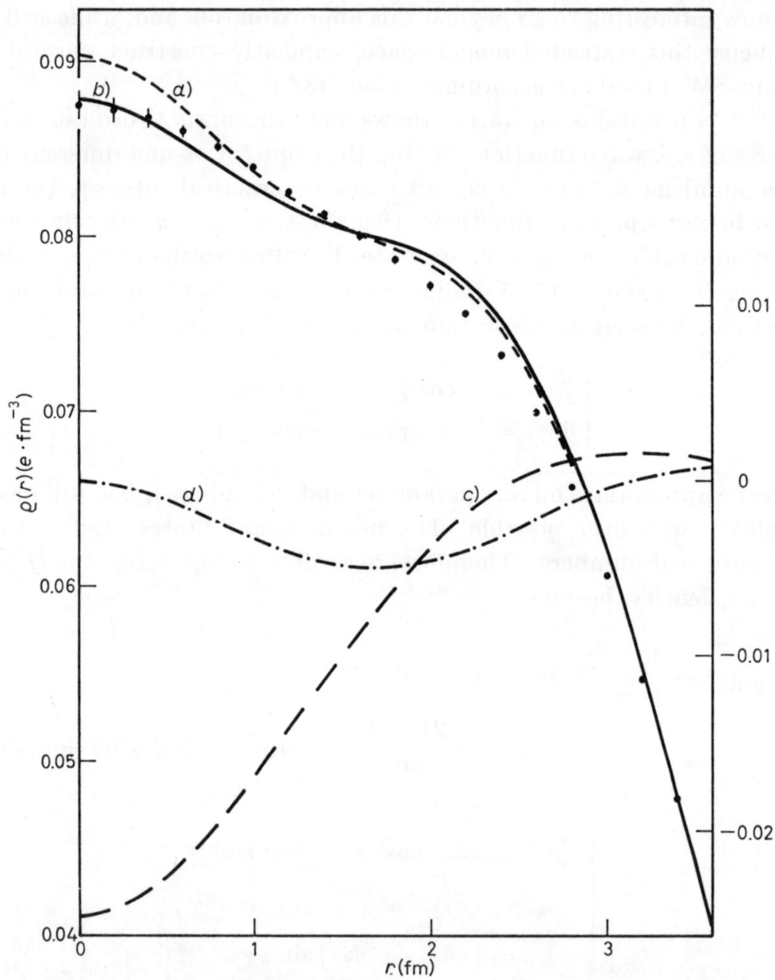

Fig. 6.1. – Construction of natural orbitals in ^{40}Ca. The points with error bars represent the « experimental » charge density [40]. Curve a) is the « standard » calculation: experimental occupation numbers from the appendix, $\beta = 0.85$ fm, $a_1 = 0.65$ fm, radius parameter fitted to charge radius, no « cross-terms ». Curve c) is the $y_{1s_{\frac{1}{2}}} y_{2s_{\frac{1}{2}}}/4\pi r^2$ cross-term. Curve d) is the $y_{1p_{\frac{3}{2}}} y_{2p_{\frac{3}{2}}}/4\pi r^2$ cross-term. Curve b) is the standard calculation $+\,0.1\cdot c) - 0.2\cdot d)$, that reproduces experiment at $r = 0$ and $r \approx 1.6$ fm. The left-hand side scale refers to curves a) and b), the right-hand side scale to curves c) and d).

3'2.5 was applied to them, since they are 2nd-order corrections to the total density if the cross-term is small.

Inspection of fig. 6.1 shows that taking $(2j+1)\varepsilon = 0.1$ for the $s_{\frac{1}{2}}$ cross-term forces the calculated density to coincide with experiment at the origin $r = 0$. The $p_{\frac{3}{2}}$ cross-term reaches its maximum value for $r \approx 1.6$ fm and vanishes at $r = 0$: if we assume for it $(2j+1)\varepsilon = -0.2$, we force the calculated density to experiment at $r \approx 1.6$ fm.

The full line in fig. 6.1 shows the density corrected with cross-terms $(2j+1)\,\varepsilon(s_{\frac{1}{2}}) = 0.1$, $(2j+1)\,\varepsilon(p_{\frac{3}{2}}) = -0.2$. It fits experiment much better than the « standard » calculation in the region $(0\div 2)$ fm. The fit is slightly worse for $r > 2$ fm, but the inclusion of more cross-terms (involving unbound orbits) would certainly cure this.

In this way, fitting eq. (6.2) to experiment we construct the 1, $2s_{\frac{1}{2}}$ and 1, $2p_{\frac{3}{2}}$ natural orbitals in ^{40}Ca starting from Saxon-Woods model functions. Solving eqs. (6.3) for χ, we find

i) for the $s_{\frac{1}{2}}$ states $(2j+1=2)$

$2\langle\tilde{n}_1\rangle = 2$, $2\langle\tilde{n}_2\rangle = 1.7$ from the appendix, $2\varepsilon = 0.1$ as above,

$$\chi = 0.17\,, \qquad \sin\chi = 0.17\,, \qquad \cos\chi = 0.99\,; \tag{6.4}$$

ii) for the $p_{\frac{3}{2}}$ states $(2j+1=4)$

$4\langle\tilde{n}_1\rangle = 4$, $4\langle\tilde{n}_2\rangle = 0.15$ from the appendix, $4\varepsilon = -0.2$ as above,

$$\chi = -2.6\cdot 10^{-2}\,, \qquad \sin\chi = -2.6\cdot 10^{-2}\,, \qquad \cos\chi = 1\,. \tag{6.5}$$

As expected, the natural orbitals are only slightly different from the starting model wave functions, and the similarity between $\langle\tilde{n}\rangle$ and $\langle n\rangle$ justifies *a posteriori* the use of $\langle\tilde{n}\rangle$ instead of $\langle n\rangle$ in the « standard » calculation.

7. – Discussion.

7'1. *Summary of the main results.* – The previous sections provide some basis for the usefulness of the SPP method and the progress made towards the objectives mentioned in the introduction.

The first objective was to find a physically based method of calculating nuclear charge distributions that makes use of nuclear properties determined by one-nucleon transfer reactions. It was hoped that as more accurate data for each particular nucleus are incorporated in the calculations, the accuracy of the resulting distributions would improve. Attention was focused on the one-body density matrix D^1 and its eigenvectors, the natural orbitals $\tilde{\varphi}_\alpha(\boldsymbol{r})$. In this basis, the (spherically symmetric) density distribution of a $J = 0$ nucleus was shown to have the intuitive diagonal form

$$\varrho(r) = \frac{1}{4\pi r^2} \sum_{lj} (2j+1) \sum_n \langle \tilde{n}_{nlj} \rangle |\tilde{y}_{nlj}(r)|^2 , \tag{3.16-18}$$

where $\langle \tilde{n}_{nlj} \rangle$ is a natural occupation number and $\tilde{y}_{nlj}$ the corresponding radial wave function. In an arbitrary s.p. basis there are also off-diagonal contributions to the density, eq. (3.17), that are sometimes forgotten in this kind of calculation.

In the present work, the natural orbitals were approximated by the eigenfunctions of a local Saxon-Woods potential of standard form, corrected for nonlocality by the usual Perey factor, eq. (3.26), written in terms of a nonlocality range β. The potential depth was adjusted for each state to the single-particle energy determined, according to the Clement-Baranger definition [8, 9], by a combined analysis of one-nucleon stripping and pick-up reactions. The natural occupation numbers were identified, for states near the Fermi surface, with the appropriate sums of one-nucleon spectroscopic factors [24] and were taken to be unity for deep states.

After applying some refinements, like relativistic contributions, centre-of-mass and nucleon finite-size corrections, the resulting charge densities were fitted to the « experimental » ones by varying the following parameters:

R_1, the radius of the central SW potential;

a_1, the diffuseness of the central SW potential;

β, the nonlocality range.

The results of many calculations (sect. 4) show that the best-fit parameters are very near to the following average values:

$$\begin{cases} R_1 \text{ varied to reproduce the measured r.m.s. charge radius,} \\ a_1 = 0.64 \text{ fm}, \\ \beta = 0.85 \text{ fm}. \end{cases} \tag{7.1}$$

The difference between the optimum fits and those obtained with parameters (7.1) is generally negligible.

The quality of the fits is very satisfactory for light and medium-weight nuclei apart from the knee region where the density begins to fall from its interior value to zero. For heavy nuclei, the example of ^{208}Pb shows that the accuracy of the present method seems to be poor. Therefore, as far as the first objective is concerned, we can conclude that the parameters (7.1) and the prescription for the s.p. energies give a good Saxon-Woods approximation to the proton natural orbitals for light and medium-weight nuclei.

The second objective was to find a simple prescription for calculating reliable densities for nuclei for which no measurements are available. As the occupation numbers are highly effective in determining the results (the s.p. energies are much less important), here the essential question is the degree of accuracy that can be obtained with what degree of complexity of calculation. At one extreme there are very simple calculations that give only approximate distributions starting from s.p. energies generated in an average potential (see subsect. **5**˙1) and from occupation numbers given by the extreme s.p. model. Some of these calculations are displayed in fig. 2 of the work of Bear and Hodgson [59]. At the other extreme there are highly complicated calculations that make use of s.p. energies and occupation numbers obtained in refined shell model or Hartree-Fock approaches (*) and that should, in principle, give accurate distributions.

7˙2. *Possible refinements.* – There is still much work that can be done to improve the present calculations in their ability to fit known charge distributions or to predict unknown ones.

a) *As for the fit to known charge densities, i.e.* the construction of natural orbitals starting from a one-body potential, there still is the problem of the best form for the nonlocality correction, discussed in subsect. **5**˙2. Since the nonlocal properties of the nucleon-nucleus potential are not exactly known, we used for simplicity the Perey correction to local wave functions, though it is not consistent with the known energy dependence of the local potentials. This prescription appears accurate enough for the present practical purposes, but the subject certainly needs further insight starting from a microscopic approach.

Another step towards the construction of better s.p. wave functions has been taken by Brown, Massen and Hodgson [68], who have proposed a method that retains the flexibility of the SPP calculations thereby introducing some self-consistency in the spirit of the HF method. They write the proton and neutron potentials as

$$V_p(r) = \left(V_0 + V_1 \frac{\varrho_1(r)}{\varrho_0(r)}\right) f(r) + V_C(r)\,, \tag{7.2}$$

$$V_n(r) = \left(V_0 - V_1 \frac{\varrho(r)}{\varrho_0(r)}\right) f(r)\,, \tag{7.3}$$

(*) The s.p. wave functions used in HF and shell model do not coincide with the true natural orbitals, therefore the corresponding occupation numbers are not the natural occupation numbers. In practice, however, the difference is negligible, because the model wave functions are near to the natural orbitals and the change in the occupation numbers is a second-order correction if the wave functions are changed to first order. The reason for that lies in the transformation properties of the D^1-matrix (whose diagonal elements are the occupation numbers) under a change of basis. See paragraph **3**˙2.5 and sect. **6** for examples.

where

$$\varrho_{0,1}(r) = \varrho_n(r) \pm \varrho_p(r) \,. \tag{7.4}$$

The calculation begins in the way already described for the SPP method, and the resulting values of $\varrho_n(r)$ and $\varrho_p(r)$ are inserted in (7.2) and (7.3) and the calculation is iterated until self-consistent densities are obtained. The nuclear-symmetry term in (7.2) and (7.3) is similar to the more familiar $(N-Z)/A$ term, but is more refined. For $N = Z$ nuclei it is not zero, thus it enables the small differences between the neutron and proton potentials in such nuclei to be taken into account. The calculation thus enables the neutron and proton distributions, and hence charge and matter distributions, to be calculated together in a consistent way.

The method may be extended to give the distributions for a sequence of isotopes, without further parameter variation. To do this, the potentials for nuclei with valence particles are written in the form

$$V_p(r) = \{V_0\varrho(r) + V_1\varrho_1(r)\}\,F(r) + V_C(r) \,, \tag{7.5}$$

$$V_n(r) = \{V_0\varrho(r) - V_1\varrho_1(r)\}\,F(r) \,, \tag{7.6}$$

where $F(r) = f(r)/\varrho_0^c(r)$, where $\varrho_0^c(r)$ refers to the core nucleus without valence nucleons. These potentials are identical with (7.2) and (7.3) for the core nuclei and the other isotopes have extra terms due to the valence nucleons. The iterative calculations are made as before for each isotope and take into account the interaction between the core and valence nucleons in a fully self-consistent way. As in the standard SPP calculations, the values of the occupation numbers may be taken from nucleon transfer experiments and, if these are not available, the results of shell model calculations may be used instead.

Preliminary results obtained in this way are encouraging.

b) To improve the ability of the SPP *method of predicting* unknown charge densities, two essential steps remain to be done. The first is to find a simple prescription to calculate occupation numbers if they are not known from experiment. Therefore, some understanding of their systematics is necessary. A preliminary study was made by plotting the experimental occupation numbers for the states near the Fermi surface for the nuclei we have studied. The scatter of the points is, however, so great that any simple prescription for all the nuclei would be almost as likely to worsen the fit as to improve it. Another possible approach is to use occupation numbers given by a full shell model calculation, but this is a complicated procedure that would destroy the simplicity of the SPP method.

The second step is to find a prescription for the radius R_1 of the central SW potential, that is highly effective in determining the charge density, but does not show a simple systematical behaviour along the periodic table (see eq. (7.1)). Of course, the saturation property of nuclear matter tells us that, if we write

$$R_1 = RA^{\frac{1}{3}}\,, \tag{7.7}$$

then R is approximately a constant. If we take the average of our best-fit results (sect. **4**), we find

$$\langle R\rangle = 1.28\ \text{fm}\,, \tag{7.8}$$

but the dispersion is sufficiently great to make the value of (7.8) useless.

Particularly important is the isotopic effect: along a chain of isotopes R_1 does usually not increase as $A^{\frac{1}{3}}$ and R is not constant. The most well-known example is probably the tin isotopes [69], but similar trends can be found in the calcium and zirconium isotopes (sect. **4**, [23], [35] and references therein). Semi-quantitative explanations can be given in terms of additional binding (repulsion) for protons (neutrons) due to the isovector potential, or of non-$A^{\frac{1}{3}}$ behaviour of the matter r.m.s. radius at the beginning or at the end of a shell [70], but the subject is far from being settled.

We, therefore, conclude that further progress towards the prediction of charge distributions must await a better knowledge of the systematics of occupation numbers and of potential radii.

7'3. *Prospects for the future.* – This work can be extended in several directions. Among these, the most interesting seem to be the following:

a) *Nuclear matter distributions.* The SPP method can easily be used to calculate neutron densities that, together with proton densities, generate matter distributions. Unfortunately, the comparison with experimental data cannot be so precise, due to the difficulty of interpreting the elastic-scattering experiments of high-energy strongly interacting projectiles, like protons or pions. For a recent review, see the work of Thomas [71].

It is, however, possible to calculate nuclear matter distributions from the single-particle potentials obtained from analysis of the charge distributions, by using occupation numbers obtained from the appropriate one-nucleon transfer reactions. These matter distributions can be used together with a suitable nuclear-reaction formalism, such as that of Glauber or of Kerman, McManus and Thaler [72], to calculate the differential cross-sections for proton elastic scattering which can be compared with the experimental data. This has been done by Ray and Hodgson [73] and on the whole the agreement is satisfactory, thus confirming the overall correctness of the method.

Some flexibility can be introduced into the calculation by allowing the radius of the neutron potential to vary to optimize the fit to the proton elastic-scattering data, and this allows the extraction of the optimum value of the difference between the neutron and proton r.m.s. radii. It is not yet possible to obtain information on the finer details of the nuclear matter distributions, although it may be noted that the neutron distributions for several nuclei calculated from the best-fit parameters are in excellent accord with those obtained by the density matrix expansion method of Negele and Vautherin [74]. A comparison between SPP calculations and model-independent matter distributions has been made by STREETS *et al.* [75].

b) Calculation of optical potentials. The nuclear densities calculated by the SPP method are likely to be accurate especially in the nuclear-surface region, since the wave functions are generated by potentials adjusted to give the correct separation energies. This has important applications to their use for the calculation of heavy-ion optical potentials by the folding method, since heavy-ion reactions are particularly sensitive to the far surface region. Our method of calculation has already been used to calculate alpha-particle and heavy-ion potentials with encouraging results. In particular, the departures of the potential radius R_1 from the simple $A^{\frac{1}{3}}$ dependence have been quantitatively reproduced in the calculations by LOZANO, MADURGA and HODGSON [70]. These calculations show that the r.m.s. radius of the matter distribution increases more rapidly than $A^{\frac{1}{3}}$ when a major shell is just beginning to be filled and less rapidly when it is nearly full. Thus the SPP method links the structure of the nucleus with the scattering potential in a direct and quantitative way.

c) Determination of the nuclear one-body density matrix. As discussed in subsect. **3·1**, the knowledge of its one-body density matrix D^1 permits the calculation of all single-particle properties of a fermion system. For a $J = 0$ nucleus, they are essentially limited to the monopole neutron and proton distributions, but if $J \neq 0$ they extend to the magnetic and electric moments. Comparison of D^1-matrices of neighbouring nuclei may permit the calculation of single-particle properties appearing in nuclear reactions, like transition densities for electron scattering, spectroscopic factors for one-nucleon transfer reactions, and so on. Also the total energy of the nucleus, though not a single-particle property, is related to the D^1-matrix by the Koltun sum rule [76].

The D^1 matrix, therefore, provides a unified framework for connecting a large body of experimental data. Of course, this has been known for a long time, but it has sometimes been obscured by the great success of the HF and shell models, which try to calculate the whole nuclear wave function.

As pointed out by CLEMENT and others [25, 77], perhaps now the time is coming for a direct extraction of D^1 from experiment, without reference to a particular nuclear model.

Our SPP calculations may provide a starting point for this goal. Since our SW wave functions are a good approximation to the natural orbitals, they can be used to construct a model space in which D^1 is nearly diagonal. Then, by comparing the best-fit densities obtained by the SPP method with the experimental distributions, one can find the (small) off-diagonal elements of D^1. If D^1 is nearly diagonal, we hope that this procedure can be accomplished unambiguously as for ^{40}Ca in sect. **6**. That calculation is in a very preliminary and early stage, being limited to the $\{s_{\frac{1}{2}}\}$ and $\{p_{\frac{3}{2}}\}$ proton subspaces only, but appears promising. Extension to the neutron states needs knowledge of the neutron density, and extension to the other subspaces requires a way of removing the oscillations in the density introduced, as $r \to \infty$, by the use of unbound SW orbitals.

From a more fundamental point of view, one can, however, expect that the D^1-matrix might in the future reveal itself able to connect one-body properties only in an approximate way, thus showing the need to go beyond the neutron-proton picture of the nucleus, as suggested by high-energy physics.

* * *

We are indebted to many friends and colleagues for their help. Among these, we thank I. SICK, H. ROTHHAAS and B. GRAMMATICOS for having permitted the use of their results before pubblication, M. DI TORO and K. BEAR, who have contributed to sect. **2** and **5**, respectively, B. A. BROWN and F. CANNATA for many suggestions and R. ROOK for valuable discussions concerning the nonlocality of the potentials. Special thanks are due to S. BOFFI, who stressed to us the importance of the one-body density matrix. Most of the figures have been accurately drawn by D. BULGARELLI.

APPENDIX

Single-particle data.

The experimental information on single-particle energies and occupation numbers we use in this work comes mainly from one-nucleon transfer and, to a lesser extent, from (p, 2p) and (e, e′p) reactions.

For neutron transfer on a target having pure isospin T, the pick-up reaction can excite final states having either isospin $T + \frac{1}{2}$ ($T_>$) or $T - \frac{1}{2}$ ($T_<$), while the stripping reaction can excite only $T + \frac{1}{2}$ ($T_>$) states (we assume $t_3 = +\frac{1}{2}$ for the neutron and $t_3 = -\frac{1}{2}$ for the proton).

Similarly, proton pick-up can go only to $T_>$ states and proton stripping both to $T_<$ and $T_>$ ones. It is convenient to classify the final states according to their isospin ($T_<$ or $T_>$) and to the quantum numbers n, l, j of the transferred particle, since sums of s.p. spectroscopic factors relative to states of the same isospin are related to the average numbers of neutrons and protons, $\langle\nu\rangle$ and $\langle\pi\rangle$, in the n, l, j, shell by the French-MacFarlane sum rules [24], *viz.*

TABLE A.1. – *Single-particle states in* ^{58}Ni.

Shell	Proton states [78, 79]						E	$\langle\pi\rangle$
	Hole $T_>$		Particle $T_<$		Particle $T_>$			
	BE_1	S_1	BE_2	S_2	BE_3	S_3		
$2p_{\frac{3}{2}}$	9.84	0.25	2.56	2.42	−0.48	0.46	2.69	0.25 (^{a}) — 1.12 (^{b})
$1f_{\frac{7}{2}}$	8.72	7.37	1.79	0.66	−2.94 (^{c})	0.42 (^{c})	7.60	6.92 — 7.37
$2s_{\frac{1}{2}}$	11.16	1.31		0		0	11.16	≈2
$1d_{\frac{3}{2}}$	14 ± 2			0		0	14 ± 2 (^{d})	
$1d_{\frac{5}{2}}$	23 ± 5			0		0	23 ± 5	
$1p_{\frac{1}{2}}$	38 ± 5			0		0	38 ± 5	
$1p_{\frac{3}{2}}$	42 ± 7			0		0	42 ± 7	
$1s_{\frac{1}{2}}$	57 ± 7			0		0	57 ± 7	

Shell	Neutron states [80]						E	$\langle\nu\rangle$
	Hole $T_<$		Hole $T_>$		Particle $T_>$			
	BE_1	S_1	BE_2	S_2	BE_3	S_3		
$2p_{\frac{1}{2}}$	13.31	0.15	—	0	6.84	2.96	7.33 (^{e})	≈0.15
$2p_{\frac{3}{2}}$	12.20	0.79	—	≈0.08 (^{f})	8.91	3.56	9.51	≈0.87
$1f_{\frac{5}{2}}$	12.97	0.55	—	—	8.43	4.74	8.90	≈0.55
$1f_{\frac{7}{2}}$	15.07	3.94	17.78	2.79	6.39	1.21	14.70	6.76
$2s_{\frac{1}{2}}$	17.78	1.13	20.44	1.04	—	—	19.06	2

(*a*) This figure comes from the pick-up reaction [78].
(*b*) This figure comes from the stripping reaction [79].
(*c*) The entries for the $1f_{\frac{7}{2}}\,T_>$ states were evaluated [81] from the energy and spectroscopic factor of the analogue states [82] and the sum rules, eqs. (A.1)-(A.6).
(*d*) The errors for the energies of the $1d_{\frac{3}{2}}$ and the deeper proton shells come from ref. [83]. The centroid of the $1d$ states, referred to in sect. **4**, is $E_d = 19.4 \pm 3.1$.
(*e*) Assuming $S_2 = 0$ and $S_3 = 1.85$ to satisfy sum rules if $\langle\pi\rangle\,(2p_{\frac{1}{2}}) \approx 0$.
f) From sum rules and $\langle\pi\rangle\,(2p_{\frac{3}{2}}) \approx 0.25$.

for neutron states

(A.1) $$S_1 = \sum_N S_N(T_<) = \langle\nu\rangle - \frac{\langle\pi\rangle}{N-Z+1}\ (\text{pick-up},\ T_<)\,,$$

(A.2) $$S_2 = \sum_N S_N(T_>) = \frac{\langle\pi\rangle}{N-Z+1}\ (\text{pick-up},\ T_>)\,,$$

(A.3) $$S_3 = \sum_n S_n(T_>) = 2j+1-\langle\nu\rangle\ (\text{stripping})\,,$$

TABLE A.2. – *Single-particle states in* ^{40}Ca.

Shell	Proton states [84]						E	$\langle\pi\rangle$
	Hole				Particle $T_>$ (a)			
	Pick-up		p, 2p	e, e′p	BE_3	S_3		
	BE_1	S_1	BE_1	BE_1				
$2p_{\frac{3}{2}}$	13.19	0.16			−0.70	3.76	−0.13	0.15
$1f_{\frac{7}{2}}$	2.82	0.58			1.08	7.36	1.21	0.56
$1d_{\frac{3}{2}}$	8.33	3.70	$8.5^{+2}_{-0.2}$	10.9±0.7	−1.02	0.32	7.59	3.59
$2s_{\frac{1}{2}}$	10.94	1.75	12 ±1	14.4±0.3			12 ±1	1.70
$1d_{\frac{5}{2}}$	16^{+3}_{-1}	6?	16^{+3}_{-2}	19.0±1.1			19 ±2 (b)	
$1p_{\frac{1}{2}}$			31.5±3.5	35 ±1			31.5±3.5	
$1p_{\frac{3}{2}}$			36 ±3				36 ±3	
$1s_{\frac{1}{2}}$			48.5±5	59 ±3			48.5±5 (c)	

Shell	Neutron states [85]				E	$\langle\nu\rangle$
	Hole $T_>$ (a)		Particle $T_>$			
	BE_2	S_2	BE_3	S_3		
$2p_{\frac{3}{2}}$	21.20	0.1	6.29	4.88	6.67 (d)	0.1
$1f_{\frac{7}{2}}$	18.44	0.41	8.36	6.87	8.93	0.4
$1d_{\frac{3}{2}}$	15.63	4.3			15.63	3.5
$2s_{\frac{1}{2}}$	18.19	1.54			18.19	

(a) The $T_<$ fragment does not exist, since $T = 0$ for the target.
(b) 19 is to be preferred to 16, since the pick-up reaction is probably missing some fragments.
(c) 48.5 is to be preferred to 59, since it fits better to the systematics of $1s_{\frac{1}{2}}$ energies.
(d) Using $S_3 = 3.9$ instead of 4.88 to satisfy the sum rule $\sum_{i=1}^{3} S_i = 2j + 1$.

TABLE A.3. – *Single-particle states in* ^{39}K.

Shell	Proton states [86]						E	$\langle\pi\rangle$(d)
	Hole $T_>$		Particle $T_<$		Particle $T_>$			
	BE_1	S_1	BE_2	S_2	BE_3	S_3		
$2p_{\frac{3}{2}}$	—	0	< 1.57 (b)	> 0.71	−1.37 (a)	2	≈0.2	0
$1f_{\frac{7}{2}}$	11.94	0.51	3.33	4.02	0.21	4.20	2.33	0.48
$1d_{\frac{3}{2}}$	8.28	3.20	8.33	0.78	—	0 (c)	8.29	3.04
$2s_{\frac{1}{2}}$	11.29	1.56						1.48

(a) From the energy of the analogue state and $Q(p, n)$ following ref. [81].
(b) There are missing fragments.
(c) From simple shell model and sum rules, eqs. (A.1)-(A.6).
(d) The values for $\langle\pi\rangle$ are normalized to 5 particles.

TABLE A.4. – *Single-particle states in* ^{48}Ca.

Shell	Proton states [87]							E	$\langle\pi\rangle$
	Hole $T_>$			Particle $T_<$		Particle $T_>$			
	Pickup		p, 2p	BE_2	S_2	BE_3	S_3		
	BE_1	S_1	BE_1						
$2p_{\frac{3}{2}}$				6.08	3.96	−1.83	0.44 ([a])	5.29	0
$1f_{\frac{7}{2}}$				9.62	8	—	0	9.62	0.3 ([b])
$1d_{\frac{3}{2}}$	16.17	3.88						16.17	3.9
$2s_{\frac{1}{2}}$	16.45	1.79						16.45	1.8
$1d_{\frac{5}{2}}$	21.5^{+2}_{-0}	4.34						23.5 ± 2	
$\langle 1p\rangle$			35 ± 10					35 ± 10	
$1s_{\frac{1}{2}}$			55 ± 10					55 ± 10	

Shell	Neutron states [88]						E	$\langle\nu\rangle$
	Hole $T_<$		Hole $T_>$		Particle $T_>$			
	BE_1	S_1	BE_2	S_2	BE_3	S_3		
$2p_{\frac{3}{2}}$	11.97	0.05			5.14	4.12	5.22	0.05
$1f_{\frac{7}{2}}$	10.02	7.03	28.06	0.035			10.11	7.95

(a) From extreme s.p. model and sum rules, eqs. (A.1)-(A.6).
(b) From neutron pick-up $T_>$ and sum rules, eqs. (A.1)-(A.6).

for proton states

$$S_1 = \sum_N S_N(T_>) = \langle\pi\rangle \text{ (pick-up)}\,, \tag{A.4}$$

$$S_2 = \sum_n S_n(T_<) = 2j+1-\langle\pi\rangle - \frac{2j+1-\langle\nu\rangle}{N-Z+1} \text{ (stripping, } T_<)\,, \tag{A.5}$$

$$S_3 = \sum_n S_n(T_>) = \frac{2j+1-\langle\nu\rangle}{N-Z+1} \text{ (stripping, } T_>)\,. \tag{A.6}$$

Here N and Z refer to the target nucleus, the indices n, l, j have been dropped for simplicity and $\langle\nu\rangle$, $\langle\pi\rangle$ are to be identified with $2j+1$ times the occupation numbers defined by eq. (3.23).

TABLE A.5. – *Single-particle states in* ^{90}Zr.

Shell	Proton states [89]						E	$\langle\pi\rangle$ (^{a})
	Hole $T_>$		Particle $T_<$		Particle $T_>$			
	BE_1	S_1	BE_2	S_2	BE_3	S_3 (^{c})		
$2d_{5/2}$			<1.97	2.45 ± 0.22				≈ 0
$1g_{9/2}$	9.291	1.10	5.27	$8.7\ \pm0.4$		≈ 0	5.72	1.45
$2p_{1/2}$	8.381	1.91 (^{b})	5.17	0.84 ± 0.06		≈ 0	7.03 (^{b})	1.17
$2p_{3/2}$	9.891	4.25 (^{b})	3.68	0.40 ± 0.05		≈ 0	9.27 (^{b})	3.61
$1f_{5/2}$	10.131	7.80 (^{b})	3.43	0.23 ± 0.05		≈ 0	9.87 (^{b})	5.77

Shell	Neutron states [90]						E	$\langle\nu\rangle$ (^{d})
	Hole $T_<$		Hole $T_>$		Particle $T_>$			
	BE_1	S_1	BE_2	S_2	BE_3	S_3		
$1g_{7/2}$					4.362	8.57	4.36	≈ 0
$2d_{3/2}$					4.779	4.63	4.78	0.07
$3s_{1/2}$					5.529	1.89	5.53	0.04
$2d_{5/2}$					7.098	5.99	7.10	0.11
$1g_{9/2}$	12.10	9.91	21.023	0.077			12.17	≈ 10
$2p_{1/2}$	12.573	2.1	20.093	0.13	≈ 2.5	0.10	12.86 (^{e})	5.91
$2p_{3/2}$	13.291	3.51	21.603	0.26			13.67 (^{e})	
$1f_{5/2}$	13.838	4.8	21.843	0.43	2.68	0.14	14.19	5.87
$1f_{7/2}$	16.783	7.7	25.083	0.9			17.65	≈ 8

(*a*) From the stripping $T_<$ spectroscopic factors, renormalized to 10 holes in the $1g_{9/2}$, $2p$ and $1f_{5/2}$ shells.
(*b*) In calculating E, S_1 has been taken as $2j+1-S_2$, since the experimental value is clearly over-estimated.
(*c*) From the sum rules, eqs. (A.1)-(A.6), and near closure of the corresponding neutron shells.
(*d*) From the stripping spectroscopic factors, renormalized to 20 holes in the $1g_{7/2}$, $3s$, $2d$, $2p$ and $1f_{5/2}$ shells. The 0.22 particles missing from the $2p$ and $1f$ shells have been arbitrarily distributed among the $3s$ and $2d$ shells in a statistical way, *i.e.* proportional to the number of available states.
(*e*) Neglecting the spin splitting of the small stripping fragment.

TABLE A.6. – *Single-particle states in* ^{92}Zr.

Shell	Proton states [89]						E	$\langle\pi\rangle$ (^{c})
	Hole $T_>$		Particle $T_<$		Particle $T_>$			
	BE_1	S_1	BE_2	S_2	BE_3	S_3 (^{b})		
$2d_{\frac{5}{2}}$			$<$ 4.33	1.34$\pm$0.08				≈ 0 (^{a})
$1g_{\frac{9}{2}}$	9.949	1.09	5.979	8.3 $\pm$0.4		$\approx$0	6.44	1.81
$2p_{\frac{1}{2}}$	9.398	1.33	6.002	1.06$\pm$0.06		$\approx$0	7.89	0.96
$2p_{\frac{3}{2}}$	10.805	3.33	5.006	0.58$\pm$0.08		$\approx$0	9.94	3.43
$1f_{\frac{5}{2}}$	10.914	7.99(?)		$\approx$0.20 (^{d})		$\approx$0	10.91	5.80

Shell	Neutron states [91]						E	$\langle\nu\rangle$
	Hole $T_<$		Hole $T_>$ (^{e})		Particle $T_>$			
	BE_1	S_1	BE_2	S_2	BE_3	S_3		
$1g_{\frac{7}{2}}$	10.815	0.20		$\approx$0	4.388	7.54	4.55	0.21 (^{g})
$2d_{\frac{3}{2}}$	10.675	0.07		$\approx$0	$<$4.669	3.362	$<$ 4.79	0.08 (^{g})
$3s_{\frac{1}{2}}$	9.839	0.06		$\approx$0	5.549	1.736	5.69	0.06 (^{g})
$2d_{\frac{5}{2}}$	8.897	2$\pm$0.33		$\approx$0	6.222	4.046	7.11	2.15 (^{g})
$1g_{\frac{9}{2}}$	11.505	9.13	$\approx$21.45	$\approx$0.14			11.66	9.99 (^{g})
$2p_{\frac{1}{2}}$	$>$12.52	4.84	$\approx$20.9	$\approx$0.074	3.164	0.12	$>$12.91	5.88 (^{f})
$2p_{\frac{3}{2}}$			$\approx$22.3	$\approx$0.264				
$1f_{\frac{5}{2}}$					3.337	0.37		5.63 (^{f})

(*a*) From extreme s.p. model.
(*b*) From sum rules, eqs. (A.1)-(A.6), and near closure of corresponding neutron shells.
(*c*) From the stripping $T_<$ spectroscopic factors, renormalized to 10 holes in the $1g_{\frac{9}{2}}$, $2p$ and $1f_{\frac{5}{2}}$ shells.
(*d*) Assumed, in analogy with ^{90}Zr and ^{94}Zr.
(*e*) From energy and occupation number of the analogue (proton) state, sum rules (A.1)-(A.6) and $Q(p, n)$ [81].
(*f*) From stripping spectroscopic factors, renormalized to 14 holes in the $1g_{\frac{7}{2}}$, $3s$, $2d_{\frac{5}{2}}$, $2p$ and $1f_{\frac{5}{2}}$ shells.
(*g*) From pick-up spectroscopic factors, renormalized to 12.49 particles in the $1g$, $2d$ and $3s$ shells (0.49 particles come from $2p$ and $1f$, see (^{f})).

The s.p. binding energy E, defined by eq. (3.21), can now be rewritten as

$$E = \sum_{i=1}^{3} S_i BE_i \Big/ \sum_{i=1}^{3} S_i\,, \tag{A.7}$$

where BE_i is the mean removal energy corresponding to a group of final levels having the same isospin and S_i is the relative total spectroscopic factor, eqs. (A.1) to (A.6). Overall normalization errors in the experimental S_i are partly removed by definition (A.7). The zero for energies is set at the target ground state and all energies will be given in MeV. See ref. [23] for more details.

TABLE A.7. – *Single-particle states in* ^{94}Zr.

Shell	Proton states [89]						E	$\langle\pi\rangle$ (^{c})
	Hole $T_>$		Particle $T_<$		Particle $T_>$			
	BE_1	S_1	BE_2	S_2	BE_3	S_3 (^{a})		
$1g_{\frac{9}{2}}$	11.086	0.81	6.806	8.6 $\pm$0.9		≈ 0	7.17	1.4
$2p_{\frac{1}{2}}$	10.306	1.58	6.546	0.64$\pm$0.06		≈ 0	9.22	1.36
$2p_{\frac{3}{2}}$	11.900	3.57	5.967	0.58$\pm$0.04		≈ 0	11.07	3.42
$1f_{\frac{5}{2}}$	$>$ 11.47	5.70	5.816	0.18$\pm$0.05		≈ 0	$>$ 11.30 (^{b})	5.82

Shell	Neutron states [92]						E	$\langle\nu\rangle$
	Hole $T_<$		Hole $T_>$		Particle $T_>$			
	BE_1	S_1	BE_2	S_2 (^{a})	BE_3	S_3		
$1g_{\frac{7}{2}}$	10.150	0.63		≈ 0	3.643	2.392		0.65 (^{e})
$2d_{\frac{3}{2}}$	10.315	0.43		≈ 0	4.331	3.876	4.93	0.44 (^{e})
$3s_{\frac{1}{2}}$	9.275	0.376		≈ 0	5.057	2.164	5.68	0.39 (^{e})
$2d_{\frac{5}{2}}$	8.191	3.41		≈ 0	6.475	1.80	7.60	3.49 (^{e})
$1g_{\frac{9}{2}}$								—
$2p_{\frac{1}{2}}$								5.47 (^{d})
$2p_{\frac{3}{2}}$					4.153	0.592		
$1f_{\frac{5}{2}}$					2.872	0.496		5.56 (^{d})

(a) From sum rules (A.1)-(A.6) and occupation number of the corresponding neutron (proton) shells.
(b) The $f_{\frac{5}{2}}$ energy systematics in the remaining Zr isotopes suggests $E \approx 11.7$.
(c) From the stripping $T_<$ spectroscopic factors, normalized to 10 holes.
(d) From stripping spectroscopic factors, renormalized to 8 holes in the $2d$, $3s$, $2p$ and $1f_{\frac{5}{2}}$ shells.
(e) From pick-up spectroscopic factors, renormalized to 4.97 particles in the $1g_{\frac{7}{2}}$, $2d$ and $3s$ shells (0.97 particles come from $2p$ and $1f$, see (^{d})).

In tables A.1 to A.8, we collect the experimental information we have used to generate E's and $\langle\pi\rangle$'s to be used as input in the calculation of charge densities. In the tables, « hole » means pick-up (= hole state), « part » means stripping (= particle state). Small discrepancies among different experimental determinations for $\langle\pi\rangle$ have been solved by the personal judgement of the compiler.

TABLE A.8. – *Single-particle states in* ^{208}Pb.

Protons states [49, 93]					
Particle $T_<$			Hole $T_>$		
Shell	$E \approx BE_2$ (c)	$\langle\pi\rangle$ (a)	Shell	$E \approx BE_1$ (b)	$\langle\pi\rangle$ (a)
$3p_{\frac{1}{2}}$	−0.20	≈0	$3s_{\frac{1}{2}}$	8.11	1.5 ÷ 2?
$3p_{\frac{3}{2}}$	0.68	≈0	$2d_{\frac{3}{2}}$	8.46	≈ 4
$2f_{\frac{5}{2}}$	0.98	≈0	$1h_{\frac{11}{2}}$	9.45	≈12
$1i_{\frac{13}{2}}$	2.13	≈0	$2d_{\frac{5}{2}}$	> 9.8 (d)	≈ 6
$2f_{\frac{7}{2}}$	2.33	≈0?	$1g_{\frac{7}{2}}$	> 11.6 (d)	≈ 8
$1h_{\frac{9}{2}}$	3.80	≈0?			

Neutron states [93, 94]					
Particle $T_>$			Hole $T_<$		
Shell	$E \approx BE_3$ (c)	$\langle\nu\rangle$ (a)	Shell	$E \approx BE_1$ (b)	$\langle\nu\rangle$ (a)
$3d_{\frac{3}{2}}$	1.42	≈0	$3p_{\frac{1}{2}}$	7.38	≈ 2
$2g_{\frac{7}{2}}$	1.45	≈0	$2f_{\frac{5}{2}}$	7.95	≈ 6
$4s_{\frac{1}{2}}$	1.91	≈0	$3p_{\frac{3}{2}}$	8.27	≈ 4
$3d_{\frac{5}{2}}$	2.36	≈0	$1i_{\frac{13}{2}}$	9.01	≈14
$1j_{\frac{15}{2}}$	2.53	≈0	$2f_{\frac{7}{2}}$	9.71	≈ 8
$1i_{\frac{11}{2}}$	3.15	≈0	$1h_{\frac{9}{2}}$	10.81	≈10
$2g_{\frac{9}{2}}$	BE_1=10.11	S_1= ?	BE_3=3.94	S_3= ?	$\langle\nu\rangle$= ?

(*a*) Spectroscopic factors for transfer reactions on ^{208}Pb are only poorly known; nevertheless, the s.p. states are generally believed to be completely full or completely empty, apart from the $2g_{\frac{9}{2}}$ neutron state, that is excited both by stripping, and pick-up. Some doubts exist also for the $3s_{\frac{1}{2}}$ proton state, that is not excited by stripping, while the pick-up reaction has spectroscopic factor S_1 ranging from 3.1 to 2.1 in different experiments, so one cannot exclude that it is slightly depleted.
(*b*) The s.p. energy E is identified with the pick-up centroid under the hypothesis that the s.p. state is completely occupied. Moreover, for the neutron states the $T_>$ centroid does not exist according to eq. (A.2), since the corresponding proton orbits are unoccupied.
(*c*) The s.p. energy is identified with the stripping centroid under the hypothesis that the state ie completely empty. Moreover, for protons the $T_>$ centroid does not exist according to eq. (A.6), sincs the corresponding neutron orbits are full.
(*d*) There are certainly missing fragments at higher binding energy.

Since the neutron contribution to charge densities is very small, we always took the $\langle\nu\rangle$'s from the simple shell model instead of from experiment. The proton occupation numbers $\langle\pi\rangle$ for deep orbits were assumed to be $2j+1$ when experimentally unknown. Some s.p. energies for deep proton orbits and many for neutrons have been calculated starting from experimentally known ones using the « constant nonlocal potential recipe » described in paragraph **3**·2.6 or some other reasonable approximation. They are collected in tables A.9 to A.11.

TABLE A.9. – *Single-particle energies not determined by transfer reactions for* ^{39}K, ^{40}Ca, ^{48}Ca *and* ^{58}Ni.

Shell	Proton states	Neutron states			
	^{39}K (^{a})	^{40}Ca (^{b})	^{39}K (^{a})	^{48}Ca (^{c})	^{58}Ni (^{d})
$1f_{\frac{7}{2}}$			8.0		
$2s_{\frac{1}{2}}$	12.5		17.2	20.7	
$1d_{\frac{3}{2}}$			14.7	19.6	20.2
$1d_{\frac{5}{2}}$	19.6	25.4	24.5	26.5	26.3
$1p_{\frac{1}{2}}$	32.2	35.6	34.7	36.0	36.3
$1p_{\frac{3}{2}}$	36.8	40.1	39.4	40.0	39.6
$1s_{\frac{1}{2}}$	49.9	54.4	53.5	53.1	52.5

(*a*) All energies are scaled from ^{40}Ca by using « standard from factors » and an asymmetry potential of 28 $((N-Z)/A)$ MeV.
(*b*) From a constant nonlocal potential with depth $V_N = 75$ MeV and $R_1 = 1.319$ fm.
(*c*) From a constant nonlocal potential with depth $V_N = 71$ MeV and $R_1 = 1.306$ fm. V_N was determined from the ^{40}Ca value assuming $V_N = V_0 + (N-Z)/A\ V_1$, with $V_1 = 25$ MeV.
(*d*) From a constant nonlocal potential with depth $V_N = 70$ MeV and $R_1 = 1.236$ fm.

TABLE A.10. – *Single-particle energies not determined by transfer reactions: the zirconium isotopes.*

Shell	Proton states			Neutron states		
	^{90}Zr	^{92}Zr	^{94}Zr	^{90}Zr	^{92}Zr	^{94}Zr
$1g_{\frac{7}{2}}$						4
$1g_{\frac{9}{2}}$						12.18
$2p_{\frac{1}{2}}$					14.58	14.63
$2p_{\frac{3}{2}}$					16.63	16.66
$1f_{\frac{5}{2}}$					16.02	16.10
$1f_{\frac{7}{2}}$	14.00 (^{b})	14.96 (^{b})	15.85 (^{b})		23.22	23.20
$2s_{\frac{1}{2}}$	22.52	23.45	24.37	28.90	28.81	28.71
$1d_{\frac{3}{2}}$	23.70	24.70	25.67	29.40	29.34	29.25
$1d_{\frac{5}{2}}$	28.78	29.72	30.65	34.24	34.10	33.94
$1p_{\frac{1}{2}}$	36.88	37.84	38.77	42.34	42.09	41.82
$1p_{\frac{3}{2}}$	39.44	40.36	41.25	44.74	44.45	44.14
$1s_{\frac{1}{2}}$	49.00	49.89	50.74	54.40	53.95	53.49
U (^{a})	74.22	74.93	75.59	66.23	65.57	64.87

(*a*) The energies are calculated according to the prescription of paragraph 3.2.6 with the nonlocal potential parameters determined in ref. [35].
(*b*) From the $1f_{\frac{7}{2}}$ energy systematics [23].

TABLE A.11. – *Single-particle energies not determined by transfer reactions for* ^{208}Pb.

Shell	Proton states (^{a})	Neutron states (^{b})	Shell	Proton states (^{a})	Neutron states (^{b})
$3s_{\frac{1}{2}}$		17.7	$1f_{\frac{7}{2}}$	29.0	35.7
$2d_{\frac{3}{2}}$		17.9	$1f_{\frac{5}{2}}$	24.6	31.7
$2d_{\frac{5}{2}}$		20.3	$2s_{\frac{1}{2}}$	32.1	40.1
$1i_{\frac{9}{2}}$		18.0	$1d_{\frac{3}{2}}$	35.0	41.6
$1g_{\frac{9}{2}}$		27.0	$1d_{\frac{5}{2}}$	37.6	43.9
$1g_{\frac{7}{2}}$		21.1	$1p_{\frac{1}{2}}$	44.1	50.5
$2p_{\frac{1}{2}}$	20.1	28.8	$1p_{\frac{3}{2}}$	46.6	51.5
$2p_{\frac{3}{2}}$	21.6	30.2	$1s_{\frac{1}{2}}$	51.6	58.2

(a) From a constant nonlocal potential with depth $V_N = 80.8$ MeV and $R_1 = 1.215$ fm.
(b) From a constant nonlocal potential with depth $V_N = 65.6$ MeV and $R_1 = 1.215$ fm.

REFERENCES

[1] R. HERMAN and R. HOFSTADTER: in *High Energy Electron Scattering Tables* (Stanford, Cal., 1960).

[2] J. B. BELLICARD, P. BOUNIN, R. F. FROSCH, R. HOFSTADTER, J. S. MCCARTHY, F. J. URHANE, M. R. YEARIAN, B. C. CLARK, R. HERMAN and D. G. RAVENHALL: *Phys. Rev. Lett.*, **19**, 527 (1967).

[3] J. L. FRIAR and J. W. NEGELE: *Adv. Nucl. Phys.*, **8**, 219 (1975).

[4] J. L. FRIAR and J. W. NEGELE: *Nucl. Phys. A*, **212**, 93 (1973); **240**, 301 (1975).

[5] J. BORYSOWICZ and J. H. HETHERINGTON: *Nucl. Phys. A*, **219**, 221 (1974).

[6] I. SICK: *Nucl. Phys. A*, **218**, 509 (1974).

[7] J. BORYSOWICZ and J. H. HETHERINGTON: *Phys. Rev. C*, **7**, 2293 (1973).

[8] C. F. CLEMENT: *Phys. Lett. B*, **28**, 395, 398 (1969).

[9] M. BARANGER: *Nucl. Phys. A*, **149**, 225 (1970).

[10] J. W. NEGELE: *Phys. Rev. C*, **1**, 1260 (1970); *Proceedings of the Meeting on Diffusion d'Electrons à Energie Intermediare*, edited by J.B. BELLICARD (Saclay, 1975); *Proceedings of the Conference on Radial Shapes of Nuclei*, edited by A. BUDZANOWSKI (Cracow, 1976), p. 79; *Proceedings of the VIII International Conference on High Energy Physics and Nuclear Structure*, edited by M. P. LOCHER (Zurich, 1977); *Proceedings of the Conference on Modern Trends in Elastic Electron Scattering* (Amsterdam, 1978), p. 73; J. P. SVENNE: *Adv. Nucl. Phys.*, **11**, 179 (1979); P. QUENTIN and H. FLOCARD: *Annu. Rev. Nucl. Sci.*, **28**, 523 (1978).

[11] B. FROIS: *Proceedings of the Conference on Modern Trends in Elastic Electron Scattering* (Amsterdam, 1978), p. 1; J. DECHARGÈ, M. GIROD and D. GOGNY: *Proceedings of the Conference on Modern Trends in Elastic Electron Scattering* (Amsterdam, 1978); X. CAMPI and D. W. L. SPRUNG: *Nucl. Phys. A*, **194**, 401 (1972); J. W. NEGELE: *Phys. Rev. C*, **1**, 1260 (1970).

[12] P. G. REINHARD and D. DRECHSEL: *Proceedings of the Conference on Modern Trends in Elastic Electron Scattering* (Amsterdam, 1978), p. 135.

[13] D. GOGNY: in *Nuclear Physics with Electromagnetic Interactions, Lecture Notes*

in Physics, Vol. **108**, edited by H. ARENHOVEL and D. DRECHSEL (Berlin, 1979), p. 88, and references therein.
[14] M. BEINER, H. FLOCARD, NGUYEN VAN GIAI and P. QUENTIN: *Nucl. Phys. A*, **238**, 29 (1975); J. P. SVENNE: *Adv. Nucl. Phys.*, **11**, 179 (1979).
[15] P. E. HODGSON: *Rep. Prog. Phys.*, **38**, 847 (1975).
[16] B. FROIS, J. B. BELLICARD, J. M. CAVEDON, M. HUET, P. LECONTE, P. LUDEAU, A. NAKADA, PHAM XUAN HO and I. SICK: *Phys. Rev. Lett.*, **38**, 152 (1977), and references therein.
[17] I. HAMAMOTO and P. J. SIEMENS: *Nucl. Phys. A*, **269**, 199 (1976).
[18] P. O. LÖWDIN: *Phys. Rev.*, **97**, 1474 (1955).
[19] F. MALAGUTI, A. UGUZZONI, E. VERONDINI and P. E. HODGSON: *Nucl. Phys. A*, **297**, 287 (1978).
[20] See, for example, L. SCHÄFER and H. A. WEIDENMÜLLER: *Nucl. Phys. A*, **174**, 1 (1971).
[21] L. R. B. ELTON and A. SWIFT: *Nucl. Phys. A*, **94**, 57 (1967); L. R. B. ELTON and S. J. WEBB: *Phys. Rev. Lett.*, **24**, 145 (1970).
[22] D. J. MILLENER and P. E. HODGSON: *Nucl. Phys. A*, **209**, 59 (1973); F. MALAGUTI and P. E. HODGSON: *Nucl. Phys. A*, **215**, 243 (1973).
[23] F. MALAGUTI: *Nucl. Phys. A*, **308**, 125 (1978).
[24] J. B. FRENCH and M. H. MACFARLANE: *Nucl. Phys.*, **26**, 168 (1961).
[25] C. F. CLEMENT: *Nucl. Phys. A*, **213**, 493 (1973); see also pages 469 and 510.
[26] F. G. PEREY: in *Direct Interactions and Nuclear Reaction Mechanism* (New York, N. Y., 1963), p. 125; *Phys. Rev.*, **131**, 745 (1963).
[27] G. C. LI, M. R. YEARIAN and I. SICK: *Phys. Rev. C*, **9**, 1861 (1974).
[28] H. CHANDRA and G. SAUER: *Phys. Rev. C*, **13**, 245 (1976).
[29] W. BERTOZZI, J. FRIAR, J. HEISENBERG and J. W. NEGELE: *Phys. Lett. B*, **41**, 408 (1972).
[30] H. J. LIPKIN: *Phys. Rev.*, **110**, 1395 (1958).
[31] J. P. ELLIOT and T. H. R. SKYRME: *Proc. R. Soc. London Ser. A*, **232**, 561 (1953).
[32] S. GAMBA, G. RICCO and G. ROTTIGNI: *Nucl. Phys. A*, **213**, 383 (1973).
[33] T. DE FOREST and J. D. WALECKA: *Adv. Phys.*, **15**, 1 (1966).
[34] X. CAMPI, D. W. L. SPRUNG and J. MARTORELL: *Nucl. Phys. A*, **223**, 541 (1974).
[35] F. MALAGUTI, A. UGUZZONI, E. VERONDINI and P. E. HODGSON: *Nuovo Cimento A*, **53**, 1 (1979).
[36] F. MALAGUTI, A. UGUZZONI, E. VERONDINI and P. E. HODGSON: *Nuovo Cimento A*, **49**, 412 (1979).
[37] I. SICK: *Phys. Lett. B*, **53**, 15 (1974).
[38] I. SICK, J. B. BELLICARD, M. BERNHEIM, B. FROIS, M. HUET, PH. LECONTE, J. MOUGEY, PHAM XUAN HO, D. ROYER and S. TURCK: *Phys. Rev. Lett.*, **35**, 910 (1975).
[39] H. ROTHHAAS: Dissertation (Institut für Kernphysik Universität Mainz, 1976).
[40] I. SICK, J. B. BELLICARD, J. M. CAVEDON, B. FROIS, M. HUET, P. LECONTE, PHAM XUAN HO and S. PLATCHKOV: presented at the *Conference on Nuclear Physics with Electromagnetic Interactions* (1979); I. SICK: private communication.
[41] F. JAMES and M. ROOS: CERN Computer Centre Programme Library D506 (1976).
[42] A. V. FIACCO and G. P. MCCORMICK: *Nonlinear Programming: Sequential Unconstrained Minimization Techniques* (New York, N. Y., 1968).
[43] R. M. TARBUTTON and K. T. R. DAVIES: *Nucl. Phys. A*, **120**, 1 (1968); D. S. ONLEY: *Nucl. Phys. A*, **118**, 436 (1968); M. CHEMTOB, F. J. MONIZ and M. RHO: *Phys. Rev. C*, **10**, 344 (1974).
[44] C. B. DOVER and N. VAN GIAI: *Nucl. Phys. A*, **190**, 373 (1972).

[45] B. Grammaticos: private communication; B. Frois: in *Nuclear Physics with Electromagnetic Interactions, Lecture Notes in Physics*, Vol. **108**, edited by H. Arenhovel and D. Drechsel (Berlin, 1979), p. 52, and references therein.
[46] R. F. Frosch, R. Hofstadter, J. S. McCarthy, G. K. Nöldeke, K. J. Van Oostrum, M. R. Yearian, B. C. Clark, R. Herman and D. G. Ravenhall: *Phys. Rev.*, **174**, 1380 (1968).
[47] H. Rothhaas: *Proceedings of the Conference on Modern Trends in Elastic Electron Scattering* (Amsterdam, 1978), p. 135.
[48] D. W.L. Sprung, J. Martorell and X. Campi: *Nucl. Phys. A*, **268**, 301 (1976).
[49] M. R. Schmorak: *Nucl. Data Sheets for A*=207, **22**, 487 (1977); M. J. Martin: *Nucl. Data Sheets for A*=209, **22**, 545 (1977).
[50] T. Ishimatsu, H. Saito, M. Yamabe, T. Awaya and H. Ohmura: *Nucl. Phys. A*, **246**, 21 (1975).
[51] A. Calboreanu and S. Mancas: *Nucl. Phys. A*, **266**, 72 (1976).
[52] J. Janiszewski and I. E. McCarthy: *Nucl. Phys. A*, **192**, 85 (1972); M. Bauer and J. J. E. Herrera: *J. Phys. G*, **4**, 67 (1978).
[53] G. E. Brown, J. H. Gunn and P. Gould: *Nucl. Phys.*, **46**, 598 (1963); C. J. Batty and G. W. Greenlees: *Nucl. Phys. A*, **133**, 673 (1969).
[54] G. F. Bertsch and T. T. S. Kuo: *Nucl. Phys. A*, **112**, 204 (1968).
[55] A. Lejeune: *Nucl. Phys. A*, **339**, 317 (1980).
[56] M. M. Giannini, G. Ricco and A. Zucchiatti: *Ann. Phys.* (*N. Y.*), **124**, 208 (1980); M. M. Giannini and G. Ricco: *Ann. Phys.* (*N. Y.*), **102**, 458 (1976).
[57] A recent discussion is given by R. C. Barrett and D. F. Jackson: in *Nuclear Sizes and Structure* (Oxford, 1977), p. 18.
[58] J. P. Jeukenne, A. Lejeune and C. Mahaux: *Phys. Rev. C*, **15**, 10 (1977).
[59] K. Bear and P. E. Hodgson: *J. Phys. G*, **4**, L287 (1978).
[60] W. E. Frahn and R. H. Lemmer: *Nuovo Cimento*, **5**, 523, 1564 (1957), and references therein.
[61] D. Vautherin and D. M. Brink: *Phys. Rev. C*, **5**, 626 (1972).
[62] A. N. James, P. T. Andrews, P. Kirkby and B. G. Lowe: *Nucl. Phys. A*, **138**, 145 (1969).
[63] T. de Forest jr.: *Nucl. Phys. A*, **163**, 237 (1971).
[64] J. R. Rook: *Nucl. Phys.*, in press (1981).
[65] F. G. Perey and B. Buck: *Nucl. Phys.*, **32**, 353 (1962).
[66] H. Schultz, J. Wiebicke and R. Reif: *Nucl. Phys. A*, **101**, 577 (1967).
[67] S. Boffi and F. D. Pacati: *Nucl. Phys. A*, **204**, 485 (1973).
[68] B. A. Brown, S. E. Massen and P. E. Hodgson: *J. Phys. G*, **5**, 1655 (1979); *Phys. Lett. B*, **85**, 167 (1979).
[69] R. Chapman, M. Hyland, J. L. Durell, J. N. Mo, M. MacPhail, H. Sharma and N. H. Merrill: *Nucl. Phys. A*, **316**, 40 (1979); S. Y. Van der Werf, M. N. Harakeh, L. W. Put, O. Scholten and R. H. Siemssen: *Nucl. Phys. A*, **289**, 141 (1977); F. D. Becchetti, W. S. Gray, J. Jänecke, E. R. Sugarbaker and R. S. Tickle: *Nucl. Phys. A*, **271**, 77 (1976).
[70] M. Lozano, G. Madurga and P. E. Hodgson: *Phys. Lett. B*, **82**, 170 (1979).
[71] A. W. Thomas: *Proceedings of the International Conference on Nuclear Physics, Berkeley, Cal., August 1980*, edited by R. M. Diamond and J. O. Rasmussen, *Nucl. Phys. A*, **354**, 51c (1981).
[72] A. Kerman, H. McManus and R. Thaler: *Ann. Phys.* (*N. Y.*), **8**, 551 (1959).
[73] L. Ray and P. E. Hodgson: *Phys. Rev. C*, **20**, 2403 (1979).
[74] J. W. Negele and D. Vautherin: *Phys. Rev. C*, **5**, 1472 (1972).
[75] J. Streets, B. A. Brown and P. E. Hodgson: *J. Phys G.* **8**, 839 (1982).

[76] D. S. KOLTUN: *Phys. Rev. Lett.*, **28**, 182 (1972).
[77] M. C. BIRSE and C. F. CLEMENT: *Nucl. Phys. A*, **351**, 112 (1981); C. F. CLEMENT and S. M. PEREZ: *Nucl. Phys. A*, **362**, 86 (1981); C. V. CHARZEWSKI, R. M. DREIZLER and H. T. FORTUNE: *Nucl. Phys. A*, **213**, 429 (1973).
[78] J. RAPAPORT: *Nucl. Data Sheets for A*=57, *Nucl. Data Sect. B*, **3**, 103 (1969); A. G. BLAIR and D. D. ARMSTRONG: *Phys. Rev.*, **151**, 930 (1966).
[79] H. J. KIM: *Nucl. Data Sheets*, **17**, 485 (1976); J. BOMMER, H. FUCHS, K. GRABISCH, H. KLUGE, W. RIBBE and G. RÖSCHERT: *Nucl. Phys. A*, **199**, 115 (1973); D. J. PULLEN and B. ROSNER: *Phys. Rev.*, **170**, 1034 (1968).
[80] P. ROUSSEL, G. BRUGE, A. BUSSIÈRE, H. FARAGGI and J. E. TESTONI: *Nucl. Phys. A*, **155**, 306 (1970); M. S. CHOWDURI and H. M. SEN GUPTA: *Lett. Nuovo Cimento*, **6**, 587 (1973); *Nucl. Phys. A*, **205**, 454 (1973); F. M. EDWARDS, J. J. KRAUSHAAR and B. W. RIDLEY: *Nucl. Phys. A*, **199**, 463 (1973).
[81] B. F. BAYMAN: *Proc. S.I.F.*, Course XL (New York, N. Y., and London, 1969), p. 380; A. H. WAPSTRA and N. B. GOVE: *Nucl. Data Tables*, **9**, 303 (1971).
[82] F. M. EDWARDS, J. J. KRAUSHAAR and B. W. RIDLEY: *Nucl. Phys. A*, **199**, 463 (1973).
[83] R. L. BECKER: *Phys. Lett. B*, **32**, 263 (1970).
[84] R. BOCK, H. H. DUHM and R. STOCK: *Phys. Lett.*, **18**, 61 (1965); J. C. HIEBERT, E. NEWMAN and R. H. BASSEL: *Phys. Rev.*, **154**, 898 (1967); R. SANTO, R. STOCK, J. H. BJERREGAARD, O. HANSEN, O. NATHAN, R. CHAPMAN and S. HINDS: *Nucl. Phys. A*, **118**, 409 (1968); A. N. JAMES, P. T. ANDREWS, P. KIRKBY and B. G. LOWE: *Nucl. Phys. A*, **138**, 145 (1969); D. H. YOUNGBLOOD, R. L. KOZUB, R. A. KENEFICK and J. C. HIEBERT: *Phys. Rev. C*, **2**, 477 (1970); R. L. BECKER: *Phys. Lett. B*, **32**, 263 (1970); P. DOLL, G. J. WAGNER, K. T. KNÖPFLE and G. MAIRLE: *Nucl. Phys. A*, **263**, 210 (1976); K. NAKAMURA, S. HIRAMATSU, T. KAMAE, H. MURAMATSU, N. IZUTSU and Y. WATASE: *Nucl. Phys. A*, **271**, 221 (1976); P. DOLL and G. J. WAGNER: *Proceedings of the International Conference on the Physics of Medium-Light Nuclei* (Florence, 1977), p. 88.
[85] T. A. BELOTE, A. SPERDUTO and W. W. BÜCHNER: *Phys. Rev. Sect. B*, **139**, 80 (1965); T. A. BELOTE, W. E. DORENBUSCH and J. RAPAPORT: *Nucl. Phys. A*, **120**, 401 (1968); J. RAPAPORT, W. E. DORENBUSCH and T. A. BELOTE: *Nucl. Phys. A*, **177**, 307 (1971); G. HANSER, R. LÖHKEN, G. NOWICKI, H. REBEL, G. SCHATZ, G. SCHWEIMER and J. SPECHT: *Nucl. Phys. A*, **182**, 1 (1972); P. MARTIN, M. BUENERD, Y. DUPONT and M. CHABRE: *Nucl. Phys. A*, **185**, 465 (1972); J. KÄLLNE: Uppsala University preprint GW1-PH 2/74 (1974); P. DOLL, G. J. WAGNER, K. T. KNÖPFLE and G. MAIRLE: *Nucl. Phys. A*, **263**, 210 (1976).
[86] J. R. ERSKINE: *Phys. Rev.*, **149**, 854 (1966); W. S. GRAY, P. J. ELLIS, T. WEI, P. M. POLICHAR and J. JÄNECKE: *Nucl. Phys. A*, **140**, 494 (1970); M. E. CAGE, R. R. JOHNSON, P. D. KUNZ and D. A. LIND: *Nucl. Phys. A*, **162**, 657 (1971); P. M. ENDT and C. VAN DER LEUN: *Nucl. Phys. A*, **214**, 400 (1973).
[87] J. R. ERSKINE, A. MARINOV and J. P. SCHIFFER: *Phys. Rev.*, **142**, 633 (1966); G. SARTORIS and L. ZAMICK: *Phys. Rev.*, **167**, 1035 (1968); A. N. JAMES, P. T. ANDREWS, P. KIRKBY and B. G. LOWE: *Nucl. Phys. A*, **138**, 145 (1969); P. DOLL, G. J. WAGNER, K. T. KNÖPFLE and G. MAIRLE: *Nucl. Phys. A*, **263**, 210 (1976); P. DOLL and G. J. WAGNER: *Proceedings of the International Conference on the Physics of Medium-Light Nuclei* (Florence, 1977), p. 88.
[88] E. KASHY, A. SPERDUTO, H. A. ENGE and W. W. BUECHNER: *Phys. Rev. Sect. B*, **135**, 865 (1964); J. L. YNTEMA: *Phys. Rev.*, **186**, 1144 (1969); J. RAPAPORT, W. E. DORENBUSCH and T. A. BELOTE: *Nucl. Phys. A*, **177**, 307 (1971); P. MARTIN, M. BUENERD, Y. DUPONT and M. CHABRE: *Nucl. Phys. A*, **185**, 465 (1972).

[89] B. M. PREEDOM and E. NEWMAN: *Phys. Rev.*, **166**, 1156 (1968); M. R. CATES, J. B. BALL and E. NEWMAN: *Phys. Rev.*, **187**, 1682 (1969).
[90] J. KÄLLNE: preprint GWI-PH 2/74, Uppsala University; H. TAKETANI, M. ADACHI, M. OGAWA and K. ASHIBE: *Nucl. Phys. A*, **204**, 385 (1973); A. GRAUE, L. H. HERLAND, K. J. LERVIK and J. T. NESSE: *Nucl. Phys. A*, **187**, 141 (1972).
[91] J. B. BALL and C. B. FULMER: *Phys. Rev.*, **172**, 1199 (1968); D. C. KOCHER: *Nucl. Data Sheets Sect. B*, **8**, 527 (1972).
[92] B. L. COHEN and O. V. CHUBINSKY: *Phys. Rev.*, **131**, 2184 (1963); L. R. MEDSKER and D. J. HOREN: *Nucl. Data Sheets Sect. B*, **8**, 29 (1972).
[93] B. H. WILDENTHAL, B. M. PREEDOM, E. NEWMAN and M. R. CATES: *Phys. Rev. Lett.*, **19**, 960 (1967); T. TAMURA and K. S. LOW: *Phys. Rev. Lett.*, **31**, 1356 (1973); W. C. PARKINSON, D. L. HENDRIE, H. H. DUHM, J. MAHONEY, J. SAUNDINOS and G. R. SATCHLER: *Phys. Rev.*, **178**, 1976 (1963).
[94] M. DOST, W. R. HERING and W. R. SMITH: *Nucl. Phys. A*, **93**, 357 (1967); D. ROYER, M. DOST and H. DOUBRE: *Z. Phys.*, **265**, 371 (1973).

AUTHOR INDEX

SUBJECT INDEX